MW01202073

Birkhäuser

Trends in Mathematics

Trends in Mathematics is a series devoted to the publication of volumes arising from conferences and lecture series focusing on a particular topic from any area of mathematics. Its aim is to make current developments available to the community as rapidly as possible without compromise to quality and to archive these for reference.

Proposals for volumes can be submitted using the Online Book Project Submission Form at our website www.birkhauser-science.com.

Material submitted for publication must be screened and prepared as follows:

All contributions should undergo a reviewing process similar to that carried out by journals and be checked for correct use of language which, as a rule, is English. Articles without proofs, or which do not contain any significantly new results, should be rejected. High quality survey papers, however, are welcome.

We expect the organizers to deliver manuscripts in a form that is essentially ready for direct reproduction. Any version of TEX is acceptable, but the entire collection of files must be in one particular dialect of TEX and unified according to simple instructions available from Birkhäuser.

Furthermore, in order to guarantee the timely appearance of the proceedings it is essential that the final version of the entire material be submitted no later than one year after the conference.

For further volumes:
http://www.springer.com/series/4961

Quaternion and Clifford Fourier Transforms and Wavelets

Eckhard Hitzer
Stephen J. Sangwine
Editors

Birkhäuser

Editors
Eckhard Hitzer
Department of Material Science
International Christian University
Tokyo, Japan

Stephen J. Sangwine
School of Computer Science
and Electronic Engineering
University of Essex
Colchester, United Kingdom

ISBN 978-3-0348-0777-7 ISBN 978-3-0348-0603-9 (eBook)
DOI 10.1007/978-3-0348-0603-9
Springer Basel Heidelberg New York Dordrecht London

Mathematics Subject Classification (2010): 11R52, 15A66, 42A38, 65T60, 42C40, 68U10, 94A08, 94A12

Printed on acid-free paper

Springer Basel is part of Springer Science+Business Media (www.springer.com)

Contents

Quaternion and Clifford–Fourier Transforms and Wavelets
Trends in Mathematics, vii–x

Preface

One hundred and seventy years ago (in 1843) W.R. Hamilton formally introduced the four-dimensional quaternions, perceiving them as one of the major discoveries of his life. One year later, in 1844, H. Grassmann published the first version of his Ausdehnungslehre, now known as Grassmann algebra, without any dimensional limitations. Circa thirty years later (in 1876) W.K. Clifford supplemented the Grassmann product of vectors with an inner product, which fundamentally unified the preceding works of Hamilton and Grassmann in the form of Clifford's geometric algebras or Clifford algebras. A Clifford algebra is a complete algebra of a vector space and all its subspaces, including the measurement of volumes and dihedral angles between any pair of subspaces.

To work in higher dimensions with quaternion and Clifford algebras allows us to systematically generalize known concepts of symmetry, phase, analytic signal and holomorphic function to higher dimensions. And as demonstrated in the current proceedings, it successfully generalizes Fourier and wavelet transformations to higher dimensions. This is interesting both for the development of analysis in higher dimensions, as well as for a broad range of applications in multi-dimensional signal, image and color image processing. Therefore a wide variety of readers from pure mathematicians, keen to learn about the latest developments in quaternion and Clifford analysis, to physicists and engineers in search of dimensionally appropriate and efficient tools in concrete applications, will find many interesting contributions in this book.

The contributions in this volume originated as papers in a session on *Quaternion and Clifford–Fourier transforms and wavelets* of the *9th International Conference on Clifford Algebras and their Applications* (*ICCA9*), which took place from 15th to 20th July 2011 at the Bauhaus-University in Weimar, Germany. The session was organized by the editors of this volume.

After the conference we asked the contributors to prepare expanded versions of their works for this volume, and many of them agreed to participate. The expanded submissions were subjected to a further round of reviews (in addition to the original reviews for the ICCA9 itself) in order to ensure that each contribution was clearly presented and worthy of publication. We are very grateful to all those reviewers whose efforts contributed significantly to the quality of the final chapters by asking the authors to revise, clarify or to expand on points in their drafts.

The contributions have been edited to achieve as much uniformity in presentation and notation as can reasonably be achieved across the somewhat different traditions that have arisen in the quaternion and Clifford communities. We hope that this volume will contribute to a growing unification of ideas across the expanding field of hypercomplex Fourier transforms and wavelets.

The book is divided into two parts: Chapters 1 to 6 deal exclusively with quaternions $\mathbb{H}$, while Chapters 7 to 16 mainly deal with Clifford algebras $C\ell_{p,q}$, but sometimes include high-dimensional complex as well as quaternionic results in several subsections. This is natural, since complex numbers ($\mathbb{C} \cong C\ell_{0,1}$) and quaternions ($\mathbb{H} \cong C\ell_{0,2}$) are low-dimensional Clifford algebras, and often appear as subalgebras, *e.g.*, $\mathbb{C} \cong Cl^+_{2,0}$, $\mathbb{H} \cong Cl^+_{3,0}$, etc. The first chapter was written especially for this volume to provide some background on the history of the subject, and to show how the contributions that follow relate to each other and to prior work. We especially thank Fred Brackx (Ghent/Belgium) for agreeing to contribute to this chapter at a late stage in the preparation of the book.

The quaternionic part begins with an exploration by Ell (Chapter 1) of the evolution of quaternion Fourier transform (QFT) definitions as a framework for problems in vector-image and vector-signal processing, ranging from NMR problems to applications in colour image processing. Next, follows an investigation by Hitzer and Sangwine (Chapter 2) into a steerable quaternion algebra split, which leads to: a local phase rotation interpretation of the classical two-sided QFT, efficient fast numerical implementations and the design of new steerable QFTs.

Then Le Bihan and Sangwine (Chapter 3) perform a quaternionic spectral analysis of non-stationary improper complex signals with possible correlation of real and imaginary signal parts. With a one-dimensional QFT they introduce a hyperanalytic signal closely linked to the geometric features of improper complex signals. In the field of low level image processing Moya-Sánchez and Bayro-Corrochano (Chapter 4) employ quaternionic atomic functions to enhance geometric image features and to analytically express image processing operations like low-pass, steerable and multiscale filtering, derivatives, and local phase computation.

In the next two chapters on quaternion analysis Georgiev and Morais (Chapter 5) characterize a class of quaternion Bochner functions generated via a quaternion Fourier–Stieltjes transform and generalize Bochner's theorem to quaternion functions. In Chapter 6 Georgiev, Morais, Kou and Spröβig study the asymptotic behavior of the QFT, apply the QFT to probability measures, including positive definite measures, and extend the classical Bochner–Minlos theorem to the framework of quaternion analysis.

The Clifford algebra part begins with Chapter 7 by Hitzer, Helmstetter and Abłamowicz, who establish a detailed algebraic characterization of the continuous manifolds of (multivector) square roots of -1 in all real Clifford algebras $C\ell_{p,q}$, including as examples detailed computer generated tables of representative square roots of -1 in dimensions $n = p + q = 5, 7$ with signature $s = p - q = 3 (\mathrm{mod}\ 4)$.

Their work is fundamental for any form of Clifford–Fourier transform (CFT) using multivector square roots of -1 instead of the complex imaginary unit. Based on this Bujack, Scheuermann and Hitzer (Chapter 8) introduce a general (Clifford) geometric Fourier transform covering most CFTs in the literature. They prove a range of standard properties and specify the necessary conditions in the transform design.

A series of four chapters on image processing begins with Batard and Berthier's (Chapter 9) on spinorial representation of images focusing on edge- and texture detection based on a special CFT for spinor fields, that takes into account the Riemannian geometry of the image surface. Then Girard, Pujol, Clarysse, Marion, Goutte and Delachartre (Chapter 10) investigate analytic signals in Clifford algebras of n-dimensional quadratic spaces, and especially for three-dimensional video ($2D + T$) signals in (complex) biquaternions ($\cong C\ell_{3,0}$). Generalizing from the right-sided QFT to a rotor CFT in $C\ell_{3,0}$, which allows a complex fast Fourier transform (FFT) decomposition, they investigate the corresponding analytic video signal including its generalized six biquaternionic phases. Next, Bernstein, Bouchot, Reinhardt and Heise (Chapter 11) undertake a mathematical overview of generalizations of analytic signals to higher-dimensional complex and Clifford analysis together with applications (and comparisons) for artificial and real-world image samples.

Soulard and Carré (Chapter 12) define a novel colour monogenic wavelet transform, leading to a non-marginal multiresolution colour geometric analysis of images. They show a first application through the definition of a full colour image denoising scheme based on statistical modeling of coefficients.

Motivated by applications in optical coherence tomography, Bernstein (Chapter 13) studies inverse scattering for Dirac operators with scalar, vector and quaternionic potentials, by writing Maxwell's equations as Dirac equations in Clifford algebra (*i.e.*, complex biquaternions). For that she considers factorizations of the Helmholtz equation and related fundamental solutions; standard- and Faddeev's Green functions.

In Chapter 14 Bahri introduces a windowed CFT for signal functions $f : \mathbb{R}^n \to C\ell_{0,n}$, and investigates some of its properties. For a different type of windowed CFT for signal functions $f : \mathbb{R}^n \to C\ell_{n,0}, n = 2, 3 (\mathrm{mod}\ 4)$, Fu, Kähler and Cerejeiras establish in Chapter 15 a Balian–Low theorem, a strong form of Heisenberg's classical uncertainty principle. They make essential use of Clifford frames and the Clifford–Zak transform.

Finally, Li and Qian (Chapter 16) employ a compressed sensing technique in order to introduce a new kind of sparse representation of signals in a Hardy space dictionary (of elementary wave forms) over a unit disk, together with examples illustrating the new algorithm.

We thank all the authors for their enthusiastic participation in the project and their enormous patience with the review and editing process. We further thank the organizer of the ICCA9 conference K. Guerlebeck and his dedicated team for their strong support in organizing the ICCA9 session on Quaternion and Clifford–Fourier Transforms and Wavelets. We finally thank T. Hempfling and B. Hellriegel of Birkhäuser Springer Basel AG for venturing to accept and skillfully accompany this proceedings with a still rather unconventional theme, thus going one more step in fulfilling the 170 year old visions of Hamilton and Grassmann.

Eckhard Hitzer
Tokyo, Japan

Stephen Sangwine
Colchester, United Kingdom

October 2012

Quaternion and Clifford–Fourier Transforms and Wavelets
Trends in Mathematics, xi–xxvii

History of Quaternion and Clifford–Fourier Transforms and Wavelets

Fred Brackx, Eckhard Hitzer and Stephen J. Sangwine

Abstract. We survey the historical development of quaternion and Clifford–Fourier transforms and wavelets.

Mathematics Subject Classification (2010). Primary 42B10; secondary 15A66, 16H05, 42C40, 16-03.

Keywords. Quaternions, Clifford algebra, Fourier transforms, wavelet transforms.

The development of hypercomplex Fourier transforms and wavelets has taken place in several different threads, reflected in the overview of the subject presented in this chapter. We present in Section 1 an overview of the development of quaternion Fourier transforms, then in Section 2 the development of Clifford–Fourier transforms. Finally, since wavelets are a more recent development, and the distinction between their quaternion and Clifford algebra approach has been much less pronounced than in the case of Fourier transforms, Section 3 reviews the history of both quaternion and Clifford wavelets.

We recognise that the history we present here may be incomplete, and that work by some authors may have been overlooked, for which we can only offer our humble apologies.

1. Quaternion Fourier Transforms (QFT)

1.1. Major Developments in the History of the Quaternion Fourier Transform

Quaternions [51] were first applied to Fourier transforms by Ernst [49, § 6.4.2] and Delsuc [41, Eqn. 20] in the late 1980s, seemingly without knowledge of the earlier work of Sommen [90, 91] on Clifford–Fourier and Laplace transforms further explained in Section 2.2. Ernst and Delsuc's quaternion transforms were two-dimensional (that is they had two independent variables) and proposed for application to nuclear magnetic resonance (NMR) imaging. Written in terms of two independent time variables[1] t_1 and t_2, the forward transforms were of the following form[2]:

$$\mathcal{F}(\omega_1, \omega_2) = \int_{-\infty}^{\infty} \int_{-\infty}^{\infty} f(t_1, t_2) e^{\boldsymbol{i}\omega_1 t_1} e^{\boldsymbol{j}\omega_2 t_2} \mathrm{d}t_1 \mathrm{d}t_2 \,. \tag{1.1}$$

Notice the use of different quaternion basis units $\boldsymbol{i}$ and $\boldsymbol{j}$ in each of the two exponentials, a feature that was essential to maintain the separation between the two dimensions (the prime motivation for using a quaternion Fourier transformation was to avoid the mixing of information that occurred when using a complex Fourier transform – something that now seems obvious, but must have been less so in the 1980s). The signal waveforms/samples measured in NMR are complex, so the quaternion aspect of this transform was essential only for maintaining the separation between the two dimensions. As we will see below, there was some unused potential here.

The fact that exponentials in the above formulation do not commute (with each other, or with the 'signal' function f), means that other formulations are possible[3], and indeed Ell in 1992 [45, 46] formulated a transform with the two exponentials positioned either side of the signal function:

$$\mathcal{F}(\omega_1, \omega_2) = \int_{-\infty}^{\infty} \int_{-\infty}^{\infty} e^{\boldsymbol{i}\omega_1 t_1} f(t_1, t_2)\, e^{\boldsymbol{j}\omega_2 t_2} \,\mathrm{d}t_1 \mathrm{d}t_2 \,. \tag{1.2}$$

Ell's transform was a theoretical development, but it was soon applied to the practical problem of computing a holistic Fourier transform of a colour image [84] in which the signal samples (discrete image pixels) had three-dimensional values (represented as quaternions with zero scalar parts). This was a major change from the previously intended application in nuclear magnetic resonance, because now the two-dimensional nature of the transform mirrored the two-dimensional nature of the image, and the four-dimensional nature of the algebra used followed naturally from the three-dimensional nature of the image pixels.

Other researchers in signal and image processing have followed Ell's formulation (with trivial changes of basis units in the exponentials) [27, 24, 25], but as with the NMR transforms, the quaternion nature of the transforms was applied essentially to separation of the two independent dimensions of an image (Bülow's work [24, 25] was based on greyscale images, that is with one-dimensional pixel values). Two new ideas emerged in 1998 in a paper by Sangwine and Ell [86]. These were, firstly, the choice of a general root μ of -1 (a unit quaternion with zero scalar part) rather than a basis unit ($\boldsymbol{i}$, $\boldsymbol{j}$ or $\boldsymbol{k}$) of the quaternion algebra,

[1]The two independent time variables arise naturally from the formulation of two-dimensional NMR spectroscopy.

[2]Note, that Georgiev *et al.* use this form of the quaternion Fourier transform (QFT) in Chapter 6 to extend the Bochner–Minlos theorem to quaternion analysis. Moreover, the same form of QFT is extended by Georgiev and Morais in Chapter 5 to a quaternion Fourier–Stieltjes transform.

[3]See Chapter 1 by Ell in this volume with a systematic review of possible forms of quaternion Fourier transformations.

and secondly, the choice of a single exponential rather than two (giving a choice of ordering relative to the quaternionic signal function):

$$\mathcal{F}(\omega_1, \omega_2) = \int_{-\infty}^{\infty}\int_{-\infty}^{\infty} e^{\mu(\omega_1 t_1 + \omega_2 t_2)} f(t_1, t_2) \mathrm{d}t_1 \mathrm{d}t_2 \,. \tag{1.3}$$

This made possible a quaternion Fourier transform of a one-dimensional signal:

$$\mathcal{F}(\omega) = \int_{-\infty}^{\infty} e^{\mu\omega t} f(t) \mathrm{d}t \,. \tag{1.4}$$

Such a transform makes sense only if the signal function has quaternion values, suggesting applications where the signal has three or four independent components. (An example is vibrations in a solid, such as rock, detected by a sensor with three mutually orthogonal transducers, such as a vector geophone.)

Very little has appeared in print about the interpretation of the Fourier coefficients resulting from a quaternion Fourier transform. One interpretation is components of different symmetry, as explained by Ell in Chapter 1. Sangwine and Ell in 2007 published a paper about quaternion Fourier transforms applied to colour images, with a detailed explanation of the Fourier coefficients in terms of elliptical paths in colour space (the n-dimensional space of the values of the image pixels in a colour image) [48].

1.2. Splitting Quaternions and the QFT

Following the earlier works of Ernst, Ell, Sangwine (see Section 1.1), and Bülow [24, 25], Hitzer thoroughly studied the quaternion Fourier transform (QFT) applied to quaternion-valued functions in [54]. As part of this work a quaternion split

$$q_{\pm} = \frac{1}{2}(q \pm \boldsymbol{i} q \boldsymbol{j}), \qquad q \in \mathbb{H}, \tag{1.5}$$

was devised and applied, which led to a better understanding of $GL(\mathbb{R}^2)$ transformation properties of the QFT spectrum of two-dimensional images, including colour images, and opened the way to a generalization of the QFT concept to a full spacetime Fourier transformation (SFT) for spacetime algebra $C\ell_{3,1}$-valued signals.

This was followed up by the establishment of a fully *directional* (opposed to componentwise) uncertainty principle for the QFT and the SFT [58]. Independently Mawardi *et al.* [77] established a componentwise uncertainty principle for the QFT.

The QFT with a Gabor window was treated by Bülow [24], a study which has been continued by Mawardi *et al.* in [1].

Hitzer reports in [59] initial results (obtained in co-operation with Sangwine) about a further generalization of the QFT to a general form of orthogonal 2D planes split (OPS-) QFT, where the split (1.5) with respect to two orthogonal pure quaternion units $\boldsymbol{i}, \boldsymbol{j}$ is generalized to a steerable split with respect to any two pure unit quaternions $f, g \in \mathbb{H}, f^2 = g^2 = -1$. This approach is fully elaborated upon in a contribution to the current volume (see Chapter 2). Note that the Cayley–Dickson form [87] of quaternions and the related simplex/perplex split [47] are obtained for $f = g = \boldsymbol{i}$ (or more general $f = g = \mu$), which is employed in Chapter 3 for a novel spectral analysis of non-stationary improper complex signals.

2. Clifford–Fourier Transformations in Clifford's Geometric Algebra

W.K. Clifford introduced (Clifford) geometric algebras in 1876 [28]. An introduction to the vector and multivector calculus, with functions taking values in Clifford algebras, used in the field of Clifford–Fourier transforms (CFT) can be found in [53, 52]. A tutorial introduction to CFTs and Clifford wavelet transforms can be found in [55]. The Clifford algebra application survey [65] contains an up to date section on applications of Clifford algebra integral transforms, including CFTs, QFTs and wavelet transforms[4].

2.1. How Clifford Algebra Square Roots of −1 Lead to Clifford–Fourier Transformations

In 1990 Jancewicz defined a trivector Fourier transformation

$$\mathcal{F}_3\{g\}(\boldsymbol{\omega}) = \int_{\mathbb{R}^3} g(\mathbf{x}) e^{-i_3 \mathbf{x}\cdot\boldsymbol{\omega}} d^3\mathbf{x}, \quad i_3 = \boldsymbol{e}_1\boldsymbol{e}_2\boldsymbol{e}_3, \quad g : \mathbb{R}^3 \to C\ell_{3,0}, \tag{2.1}$$

for the electromagnetic field[5] replacing the imaginary unit $i \in \mathbb{C}$ by the central trivector i_3, $i_3^2 = -1$, of the geometric algebra $C\ell_{3,0}$ of three-dimensional Euclidean space $\mathbb{R}^3 = \mathbb{R}^{3,0}$ with orthonormal vector basis $\{\boldsymbol{e}_1, \boldsymbol{e}_2, \boldsymbol{e}_3\}$.

In [50] Felsberg makes use of signal embeddings in low-dimensional Clifford algebras $\mathbb{R}_{2,0}$ and $\mathbb{R}_{3,0}$ to define his Clifford–Fourier transform (CFT) for one-dimensional signals as

$$\mathcal{F}_1^{fe}[f](\underline{u}) = \int_{\mathbb{R}} \exp\left(-2\pi i_2 \underline{u}\,\underline{x}\right) f(\underline{x})\, d\underline{x}, \quad i_2 = \boldsymbol{e}_1\boldsymbol{e}_2, \quad f : \mathbb{R} \to \mathbb{R}, \tag{2.2}$$

where he uses the pseudoscalar $i_2 \in C\ell_{2,0}$, $i_2^2 = -1$. For two-dimensional signals[6] he defines the CFT as

$$\mathcal{F}_2^{fe}[f](\underline{u}) = \int_{\mathbb{R}^2} \exp\left(-2\pi i_3 < \underline{u}, \underline{x} >\right) f(\underline{x})\, d\underline{x}, \quad f : \mathbb{R}^2 \to \mathbb{R}^2, \tag{2.3}$$

---

[4] Fourier and wavelet transforms provide alternative signal and image representations. See Chapter 9 for a spinorial representation and Chapter 16 by Li and Qian for a sparse representation of signals in a Hardy space dictionary (of elementary wave forms) over a unit disk.

[5] Note also Chapter 11 in this volume, in which Bernstein considers optical coherence tomography, formulating the Maxwell equations with the Dirac operator and Clifford algebra.

[6] Note in this context the spinor representation of images by Batard and Berthier in Chapter 9 of this volume. The authors apply a CFT in $C\ell_{3,0}$ to the spinor represenation, which uses in the exponential kernel an adapted choice of bivector, that belongs to the orthonormal frame of the tangent bundle of an oriented two-dimensional Riemannian manifold, isometrically immersed in $\mathbb{R}^3$.

where he uses the pseudoscalar $i_3 \in C\ell_{3,0}$. It is used amongst others to introduce a concept of two-dimensional analytic signal. Together with Bülow and Sommer, Felsberg applied these CFTs to image stucture processing (key-notion: structure multivector) [50, 24].

Ebling and Scheuermann [44, 43] consequently applied to vector signal processing in two- and three dimensions, respectively, the following two-dimensional CFT

$$\mathcal{F}_2\{f\}(\boldsymbol{\omega}) = \int_{\mathbb{R}^2} f(\mathbf{x}) e^{-i_2 \mathbf{x}\cdot\boldsymbol{\omega}} d^2\mathbf{x}, \quad f : \mathbb{R}^2 \to \mathbb{R}^2, \tag{2.4}$$

with Clifford–Fourier kernel

$$\exp\left(-\boldsymbol{e}_1\boldsymbol{e}_2(\omega_1 x_1 + \omega_2 x_2)\right), \tag{2.5}$$

and the three-dimensional CFT (2.1) of Jancewicz with Clifford–Fourier kernel

$$\exp\left(-\boldsymbol{e}_1\boldsymbol{e}_2\boldsymbol{e}_3(\omega_1 x_1 + \omega_2 x_2 + \omega_3 x_3)\right). \tag{2.6}$$

An important integral operation defined and applied in this context by Ebling and Scheuermann was the Clifford convolution. These Clifford–Fourier transforms and the corresponding convolution theorems allow Ebling and Scheuermann for amongst others the analysis of vector-valued patterns in the frequency domain.

Note that the latter Fourier kernel (2.6) has also been used by Mawardi and Hitzer in [78, 63, 78] to define their Clifford–Fourier transform of three-dimensional multivector signals: that means, they researched the properties of $\mathcal{F}_3\{g\}(\boldsymbol{\omega})$ of (2.1) in detail when applied to full multivector signals $g : \mathbb{R}^3 \to C\ell_{3,0}$. This included an investigation of the uncertainty inequality for this type of CFT. They subsequently generalized $\mathcal{F}_3\{g\}(\boldsymbol{\omega})$ to dimensions $n = 3 (\mathrm{mod}\ 4)$, *i.e.*, $n = 3, 7, 11, \ldots$,

$$\mathcal{F}_n\{g\}(\boldsymbol{\omega}) = \int_{\mathbb{R}^n} g(\mathbf{x}) e^{-i_n \mathbf{x}\cdot\boldsymbol{\omega}} d^n\mathbf{x}, \quad g : \mathbb{R}^n \to C\ell_{n,0}, \tag{2.7}$$

which is straightforward, since for these dimensions the pseudoscalar $i_n = \boldsymbol{e}_1 \ldots \boldsymbol{e}_n$ squares to -1 and is central [64], *i.e.*, it commutes with every other multivector belonging to $C\ell_{n,0}$. A little less trivial is the generalization of $\mathcal{F}_2\{f\}(\boldsymbol{\omega})$ of (2.4) to

$$\mathcal{F}_n\{f\}(\boldsymbol{\omega}) = \int_{\mathbb{R}^n} f(\mathbf{x}) e^{-i_n \mathbf{x}\cdot\boldsymbol{\omega}} d^n\mathbf{x}, \quad f : \mathbb{R}^n \to C\ell_{n,0}, \tag{2.8}$$

with $n = 2 (\mathrm{mod}\ 4)$, *i.e.*, $n = 2, 6, 10 \ldots$, because in these dimensions the pseudoscalar $i_n = \boldsymbol{e}_1 \ldots \boldsymbol{e}_n$ squares to -1, but it ceases to be central. So the relative order of the factors in $\mathcal{F}_n\{f\}(\boldsymbol{\omega})$ becomes important, see [66] for a systematic investigation and comparison.

In the context of generalizing quaternion Fourier transforms (QFT) *via algebra isomorphisms* to higher-dimensional Clifford algebras, Hitzer [54] constructed a spacetime Fourier transform (SFT) in the full algebra of spacetime $C\ell_{3,1}$, which includes the CFT (2.1) as a partial transform of space. Implemented analogously (isomorphicaly) to the orthogonal 2D planes split of quaternions, the SFT permits a natural spacetime split, which algebraically splits the SFT into right and left propagating multivector wave packets. This analysis allows to compute the effect of Lorentz transformations on the spectra of these wavepackets, as well as a 4D directional spacetime uncertainty formula [58] for spacetime signals.

Mawardi *et al.* extended the CFT $\mathcal{F}_2\{f\}(\boldsymbol{\omega})$ of (2.4) to a windowed CFT in [76]. Fu *et al.* establish in Chapter 15 a strong version of Heisenberg's uncertainty principle for Gabor-windowed CFTs.

In Chapter 8 in this volume, Bujack, Scheuermann, and Hitzer, expand the notion of Clifford–Fourier transform to include multiple left and right exponential kernel factors, in which commuting (or anticommuting) blades, that square to -1, replace the complex unit $i \in \mathbb{C}$, thus managing to include most practically used CFTs in a single comprehensive framework. Based on this they have also constructed a general CFT convolution theorem [23].

Spurred by the systematic investigation of (complex quaternion) biquaternion square roots of -1 in $C\ell_{3,0}$ by Sangwine [85], Hitzer and Ablamowicz [62] systematically investigated the explicit equations and solutions for square roots of -1 in all real Clifford algebras $C\ell_{p,q}, p+q \leq 4$. This investigation is continued in the present volume in Chapter 7 by Hitzer, Helmstetter and Ablamowicz for all square roots of -1 in all real Clifford algebras $C\ell_{p,q}$ without restricting the value of $n = p+q$. One important motivation for this is the relevance of the Clifford algebra square roots of -1 for the general construction of CFTs, where the imaginary unit $i \in \mathbb{C}$ is replaced by a $\sqrt{-1} \in C\ell_{p,q}$, without restriction to pseudoscalars or blades.

Based on the knowledge of square roots of -1 in real Clifford algebras $C\ell_{p,q}$, [60] develops a general CFT in $C\ell_{p,q}$, wherein the complex unit $i \in \mathbb{C}$ is replaced by any square root of -1 chosen from any component and (or) conjugation class of the submanifold of square roots of -1 in $C\ell_{p,q}$, and details its properties, including a convolution theorem. A similar general approach is taken in [61] for the construction of two-sided CFTs in real Clifford algebras $C\ell_{p,q}$, freely choosing two square roots from any one or two components and (or) conjugation classes of the submanifold of square roots of -1 in $C\ell_{p,q}$. These transformations are therefore generically steerable.

This algebraically motivated approach may in the future be favorably combined with group theoretic, operator theoretic and spinorial approaches, to be discussed in the following.

2.2. The Clifford–Fourier Transform in the Light of Clifford Analysis

Two robust tools used in image processing and computer vision for the analysis of scalar fields are convolution and Fourier transformation. Several attempts have been made to extend these methods to two- and three-dimensional vector fields and even multi-vector fields. Let us give an overview of those generalized Fourier transforms.

In [25] Bülow and Sommer define a so-called quaternionic Fourier transform of two-dimensional signals $f(x_1, x_2)$ taking their values in the algebra $\mathbb{H}$ of real quaternions. Note that the quaternion algebra $\mathbb{H}$ is nothing else but (isomorphic to) the Clifford algebra $C\ell_{0,2}$ where, traditionally, the basis vectors are denoted by $\boldsymbol{i}$ and $\boldsymbol{j}$, with $\boldsymbol{i}^2 = \boldsymbol{j}^2 = -1$, and the bivector by $\boldsymbol{k} = \boldsymbol{ij}$. In terms of these basis vectors this quaternionic Fourier transform takes the form

$$\mathcal{F}^q[f](u_1, u_2) = \int_{\mathbb{R}^2} \exp\left(-2\pi \boldsymbol{i} u_1 x_1\right) f(x_1, x_2) \exp\left(-2\pi \boldsymbol{j} u_2 x_2\right) d\underline{x}. \tag{2.9}$$

Due to the non-commutativity of the multiplication in $\mathbb{H}$, the convolution theorem for this quaternionic Fourier transform is rather complicated, see also [23].

This is also the case for its higher-dimensional analogue, the so-called Clifford–Fourier transform[7] in $C\ell_{0,m}$ given by

$$\mathcal{F}^{cl}[f](\underline{u}) = \int_{\mathbb{R}^m} f(\underline{x}) \exp\left(-2\pi \boldsymbol{e}_1 u_1 x_1\right) \ldots \exp\left(-2\pi \boldsymbol{e}_m u_m x_m\right) d\underline{x}. \tag{2.10}$$

Note that for $m = 1$ and interpreting the Clifford basis vector $\boldsymbol{e}_1$ as the imaginary unit i, the Clifford–Fourier transform (2.10) reduces to the standard Fourier transform on the real line, while for $m = 2$ the quaternionic Fourier transform (2.9) is recovered when restricting to real signals.

Finally Bülow and Sommer also introduce a so-called commutative hypercomplex Fourier transform given by

$$\mathcal{F}^h[f](\underline{u}) = \int_{\mathbb{R}^m} f(\underline{x}) \exp\left(-2\pi \sum_{j=1}^m \tilde{\boldsymbol{e}}_j u_j x_j\right) d\underline{x} \tag{2.11}$$

where the basis vectors $(\tilde{\boldsymbol{e}}_1, \ldots, \tilde{\boldsymbol{e}}_m)$ obey the commutative multiplication rules $\tilde{\boldsymbol{e}}_j \tilde{\boldsymbol{e}}_k = \tilde{\boldsymbol{e}}_k \tilde{\boldsymbol{e}}_j$, $j, k = 1, \ldots, m$, while still retaining $\tilde{\boldsymbol{e}}_j^2 = -1$, $j = 1, \ldots, m$. This commutative hypercomplex Fourier transform offers the advantage of a simple convolution theorem.

The hypercomplex Fourier transforms $\mathcal{F}^q$, $\mathcal{F}^{cl}$ and $\mathcal{F}^h$ enable Bülow and Sommer to establish a theory of multi-dimensional signal analysis and in particular to introduce the notions of multi-dimensional analytic signal[8], Gabor filter, instantaneous and local amplitude and phase, *etc.*

In this context the Clifford–Fourier transformations by Felsberg [50] for one- and two-dimensional signals, by Ebling and Scheuermann for two- and three-dimensional vector signal processing [44, 43], and by Mawardi and Hitzer for general multivector signals in $C\ell_{3,0}$ [78, 63, 78], and their respective kernels, as already reviewed in Section 2.1, should also be considered.

The above-mentioned Clifford–Fourier kernel of Bülow and Sommer

$$\exp\left(-2\pi \boldsymbol{e}_1 u_1 x_1\right) \cdots \exp\left(-2\pi \boldsymbol{e}_m u_m x_m\right) \tag{2.12}$$

was in fact already introduced in [19] and [89] as a theoretical concept in the framework of Clifford analysis. This generalized Fourier transform was further elaborated by Sommen in [90, 91] in connection with similar generalizations of the Cauchy, Hilbert and Laplace transforms. In this context also the work of Li, McIntosh and Qian should be mentioned; in [72] they generalize the standard

[7] Note that in this volume Mawardi establishes in Chapter 14 a windowed version of the CFT (2.10).

[8] See also Chapter 10 by Girard *et al.* and Chapter 11 by Bernstein *et al.* in this volume.

multi-dimensional Fourier transform of a function in $\mathbb{R}^m$, by extending the Fourier kernel $\exp\left(i\left\langle \underline{\xi}, \underline{x}\right\rangle\right)$ to a function which is holomorphic in $\mathbb{C}^m$ and monogenic[9] in $\mathbb{R}^{m+1}$.

In [15, 16, 18] Brackx, De Schepper and Sommen follow another philosophy in their construction of a Clifford–Fourier transform. One of the most fundamental features of Clifford analysis is the factorization of the Laplace operator. Indeed, whereas in general the square root of the Laplace operator is only a pseudo-differential operator, by embedding Euclidean space into a Clifford algebra, one can realize $\sqrt{-\Delta_m}$ as the Dirac operator $\partial_{\underline{x}}$. In this way Clifford analysis spontaneously refines harmonic analysis. In the same order of ideas, Brackx *et al.* decided to not replace nor to improve the classical Fourier transform by a Clifford analysis alternative, since a refinement of it automatically appears within the language of Clifford analysis. The key step to making this refinement apparent is to interpret the standard Fourier transform as an operator exponential:

$$\mathcal{F} = \exp\left(-i\frac{\pi}{2}\mathcal{H}\right) = \sum_{k=0}^{\infty} \frac{1}{k!}\left(-i\frac{\pi}{2}\right)^k \mathcal{H}^k\,, \tag{2.13}$$

where $\mathcal{H}$ is the scalar operator

$$\mathcal{H} = \frac{1}{2}\left(-\Delta_m + r^2 - m\right)\,. \tag{2.14}$$

This expression links the Fourier transform with the Lie algebra $\mathfrak{sl}_2$ generated by Δ_m and $r^2 = |\underline{x}|^2$ and with the theory of the quantum harmonic oscillator determined by the Hamiltonian $-\frac{1}{2}\left(\Delta_m - r^2\right)$. Splitting the operator $\mathcal{H}$ into a sum of Clifford algebra-valued second-order operators containing the angular Dirac operator Γ, one is led, in a natural way, to a *pair* of transforms $\mathcal{F}_{\mathcal{H}^\pm}$, the harmonic average of which is precisely the standard Fourier transform:

$$\mathcal{F}_{\mathcal{H}^\pm} = \exp\left(\frac{i\pi m}{4}\right)\exp\left(\mp\frac{i\pi\Gamma}{2}\right)\exp\left(\frac{i\pi}{4}\left(\Delta_m - r^2\right)\right). \tag{2.15}$$

For the special case of dimension two, Brackx *et al.* obtain a closed form for the kernel of the integral representation of this Clifford–Fourier transform leading to its internal representation

$$\mathcal{F}_{\mathcal{H}^\pm}[f](\underline{\xi}) = \mathcal{F}_{\mathcal{H}^\pm}[f](\xi_1,\xi_2) = \frac{1}{2\pi}\int_{\mathbb{R}^2} \exp\left(\pm \boldsymbol{e}_{12}(\xi_1 x_2 - \xi_2 x_1)\right) f(\underline{x})\, d\underline{x}\,, \tag{2.16}$$

which enables the generalization of the calculation rules for the standard Fourier transform both in the L_1 and in the L_2 context. Moreover, the Clifford–Fourier transform of Ebling and Scheuermann

$$\mathcal{F}^e[f](\underline{\xi}) = \int_{\mathbb{R}^2} \exp\left(-\boldsymbol{e}_{12}(x_1\xi_1 + x_2\xi_2)\right) f(\underline{x})\, d\underline{x}\,, \tag{2.17}$$

[9] See also in this volume Chapter 4 by Moya-Sánchez and Bayro-Corrochano on the application of atomic function based monogenic signals.

can be expressed in terms of the Clifford–Fourier transform:

$$\mathcal{F}^e[f](\underline{\xi}) = 2\pi \mathcal{F}_{\mathcal{H}^\pm}[f](\mp\xi_2, \pm\xi_1) = 2\pi \mathcal{F}_{\mathcal{H}^\pm}[f](\pm \boldsymbol{e}_{12}\underline{\xi}), \tag{2.18}$$

taking into account that, under the isomorphism between the Clifford algebras $C\ell_{2,0}$ and $C\ell_{0,2}$, both pseudoscalars are isomorphic images of each other.

The question whether $\mathcal{F}_{\mathcal{H}^\pm}$ can be written as an integral transform is answered positively in the case of even dimension by De Bie and Xu in [39]. The integral kernel of this transform is not easy to obtain and looks quite complicated. In the case of odd dimension the problem is still open.

Recently, in [35], De Bie and De Schepper have studied the fractional Clifford–Fourier transform as a generalization of both the standard fractional Fourier transform and the Clifford–Fourier transform. It is given as an operator exponential by

$$\mathcal{F}_{\alpha,\beta} = \exp\left(\frac{i\alpha m}{2}\right)\exp\left(i\beta\Gamma\right)\exp\left(\frac{i\alpha}{2}\left(\Delta_m - r^2\right)\right). \tag{2.19}$$

For the corresponding integral kernel a series expansion is obtained, and, in the case of dimension two, an explicit expression in terms of Bessel functions.

The above, more or less chronological, overview of generalized Fourier transforms in the framework of quaternionic and Clifford analysis, gives the impression of a medley of ad hoc constructions. However there is a structure behind some of these generalizations, which becomes apparent when, as already slightly touched upon above, the Fourier transform is linked to group representation theory, in particular the Lie algebras $\mathfrak{sl}_2$ and $\mathfrak{osp}(1|2)$. This unifying character is beautifully demonstrated by De Bie in the overview paper [34], where, next to an extensive bibliography, also new results on some of the transformations mentioned below can be found. It is shown that using realizations of the Lie algebra $\mathfrak{sl}_2$ one is lead to scalar generalizations of the Fourier transform, such as:

(i) the fractional Fourier transform, which is, as the standard Fourier transform, invariant under the orthogonal group; this transform has been reinvented several times as well in mathematics as in physics, and is attributed to Namias [81], Condon [30], Bargmann [2], Collins [29], Moshinsky and Quesne [80]; for a detailed overview of the theory and recent applications of the fractional Fourier transform we refer the reader to [82];

(ii) the Dunkl transform, see, *e.g.*, [42], where the symmetry is reduced to that of a finite reflection group;

(iii) the radially deformed Fourier transform, see, *e.g.*, [71], which encompasses both the fractional Fourier and the Dunkl transform;

(iv) the super Fourier transform, see, *e.g.*, [33, 31], which is defined in the context of superspaces and is invariant under the product of the orthogonal with the symplectic group.

Realizations of the Lie algebra $\mathfrak{osp}(1|2)$, on the contrary, need the framework of Clifford analysis, and lead to:

(v) the Clifford–Fourier transform and the fractional Clifford–Fourier transform, both already mentioned above; meanwhile an entire class of Clifford–Fourier transforms has been thoroughly studied in [36];

(vi) the radially deformed hypercomplex Fourier transform, which appears as a special case in the theory of radial deformations of the Lie algebra $\mathfrak{osp}(1|2)$, see [38, 37], and is a topic of current research, see [32].

3. Quaternion and Clifford Wavelets

3.1. Clifford Wavelets in Clifford Analysis

The interest of the Ghent Clifford Research Group for generalizations of the Fourier transform in the framework of Clifford analysis, grew out from the study of the multidimensional Continuous Wavelet Transform in this particular setting. Clifford-wavelet theory, however restricted to the continuous wavelet transform, was initiated by Brackx and Sommen in [20] and further developed by N. De Schepper in her PhD thesis [40]. The Clifford-wavelets originate from a mother wavelet not only by translation and dilation, but also by rotation, making the Clifford-wavelets appropriate for detecting directional phenomena. Rotations are implemented as specific actions on the variable by a spin element, since, indeed, the special orthogonal group $\mathrm{SO}(m)$ is doubly covered by the spin group $\mathrm{Spin}(m)$ of the real Clifford algebra $C\ell_{0,m}$. The mother wavelets themselves are derived from intentionally devised orthogonal polynomials in Euclidean space. It should be noted that these orthogonal polynomials are not tensor products of one-dimensional ones, but genuine multidimensional ones satisfying the usual properties such as a Rodrigues formula, recurrence relations, and differential equations. In this way multidimensional Clifford wavelets were constructed grafted on the Hermite polynomials [21], Laguerre polynomials [14], Gegenbauer polynomials [13], Jacobi polynomials [17], and Bessel functions [22].

Taking the dimension m to be even, say $m = 2n$, introducing a complex structure, *i.e.*, an $\mathrm{SO}(2n)$-element squaring up to -1, and considering functions with values in the complex Clifford algebra $\mathbb{C}_{2n}$, so-called Hermitian Clifford analysis originates as a refinement of standard or Euclidean Clifford analysis. It should be noticed that the traditional holomorphic functions of several complex variables appear as a special case of Hermitian Clifford analysis, when the function values are restricted to a specific homogeneous part of spinor space. In this Hermitian setting the standard Dirac operator, which is invariant under the orthogonal group $\mathrm{O}(m)$, is split into two Hermitian Dirac operators, which are now invariant under the unitary group $\mathrm{U}(n)$. Also in this Hermitian Clifford analysis framework, multidimensional wavelets have been introduced by Brackx, H. De Schepper and Sommen [11, 12], as kernels for a Hermitian Continuous Wavelet Transform, and (generalized) Hermitian Clifford–Hermite polynomials have been devised to generate the corresponding Hermitian wavelets [9, 10].

3.2. Further Developments in Quaternion and Clifford Wavelet Theory

Clifford algebra multiresolution analysis (MRA) has been pioneered by M. Mitrea [79]. Important are also the electromagnetic signal application oriented developments of Clifford algebra wavelets by G. Kaiser [70, 67, 68, 69].

Quaternion MRA Wavelets with applications to image analysis have been developed in [92] by Traversoni. Clifford algebra multiresolution analysis has been applied by Bayro-Corrochano [5, 3, 4] to: Clifford wavelet neural networks (information processing), also considering quaternionic MRA, a quaternionic wavelet phase concept, as well as applications to (*e.g.*, robotic) motion estimation and image processing.

Beyond this Zhao and Peng [94] established a theory of quaternion-valued admissible wavelets. Zhao [93] studied Clifford algebra-valued admissible (continuous) wavelets using the complex Fourier transform for the spectral representation. Mawardi and Hitzer [74, 75] extended this to continuous Clifford and Clifford–Gabor wavelets in $C\ell_{3,0}$ using the CFT of (2.1) for the spectral representation. They also studied a corresponding Clifford wavelet transform uncertainty principle. Hitzer [56, 57] generalized this approach to continous admissible Clifford algebra wavelets in real Clifford algebras $C\ell_{n,0}$ of dimensions $n = 2, 3 (\mathrm{mod}\ 4)$, *i.e.*, $n = 2, 3, 6, 7, 10, 11, \ldots$. Restricted to $C\ell_{n,0}$ of dimensions $n = 2 (\mathrm{mod}\ 4)$ this approach has also been taken up in [73].

Kähler *et al.* [26] treated monogenic (Clifford) wavelets over the unit ball. Bernstein studied Clifford continuous wavelet transforms in $L_{0,2}$ and $L_{0,3}$ [6], as well as monogenic kernels and wavelets on the three-dimensional sphere [7]. Bernstein *et al.* [8] further studied Clifford diffusion wavelets on conformally flat cylinders and tori. In the current volume Soulard and Carré extend in Chapter 12 the theory and application of monogenic wavelets to colour image denoising.

References

[1] M. Bahri, E. Hitzer, R. Ashino, and R. Vaillancourt. Windowed Fourier transform of two-dimensional quaternionic signals. *Applied Mathematics and Computation*, 216(8):2366–2379, June 2010.

[2] V. Bargmann. On a Hilbert space of analytic functions and an associated integral transform. *Communications on Pure and Applied Mathematics*, 14(3):187–214, Aug. 1961.

[3] E. Bayro-Corrochano. Multi-resolution image analysis using the quaternion wavelet transform. *Numerical Algorithms*, 39(1–3):35–55, 2005.

[4] E. Bayro-Corrochano. The theory and use of the quaternion wavelet transform. *Journal of Mathematical Imaging and Vision*, 24:19–35, 2006.

[5] E. Bayro-Corrochano and M.A. de la Torre Gomora. Image processing using the quaternion wavelet transform. In *Proceedings of the Iberoamerican Congress on Pattern Recognition, CIARP'* 2004, pages 612–620, Puebla, Mexico, October 2004.

[6] S. Bernstein. Clifford continuous wavelet transforms in $l_{0,2}$ and $l_{0,3}$. In T.E. Simos, G. Psihoyios, and C. Tsitouras, editors, *NUMERICAL ANALYSIS AND APPLIED*

MATHEMATICS: International Conference on Numerical Analysis and Applied Mathematics, volume 1048 of *AIP Conference Proceedings*, pages 634–637, Psalidi, Kos, Greece, 16–20 September 2008.

[7] S. Bernstein. Spherical singular integrals, monogenic kernels and wavelets on the 3D sphere. *Advances in Applied Clifford Algebras*, 19(2):173–189, 2009.

[8] S. Bernstein, S. Ebert, and R.S. Kraußhar. Diffusion wavelets on conformally flat cylinders and tori. In Simos et al. [88], pages 773–776.

[9] F. Brackx, H. De Schepper, N. De Schepper, and F. Sommen. The generalized Hermitean Clifford–Hermite continuous wavelet transform. In T.E. Simos, G. Psihoyios, and C. Tsitouras, editors, *NUMERICAL ANALYSIS AND APPLIED MATHEMATICS: International Conference of Numerical Analysis and Applied Mathematics*, volume 936 of *AIP Conference Proceedings*, pages 721–725, Corfu, Greece, 16–20 September 2007.

[10] F. Brackx, H. De Schepper, N. De Schepper, and F. Sommen. Generalized Hermitean Clifford–Hermite polynomials and the associated wavelet transform. *Mathematical Methods in the Applied Sciences*, 32(5):606–630, 2009.

[11] F. Brackx, H. De Schepper, and F. Sommen. A Hermitian setting for wavelet analysis: the basics. In *Proceedings of the 4th International Conference on Wavelet Analysis and Its Applications*, 2005.

[12] F. Brackx, H. De Schepper, and F. Sommen. A theoretical framework for wavelet analysis in a Hermitean Clifford setting. *Communications on Pure and Applied Analysis*, 6(3):549–567, 2007.

[13] F. Brackx, N. De Schepper, and F. Sommen. The Clifford–Gegenbauer polynomials and the associated continuous wavelet transform. *Integral Transforms and Special Functions*, 15(5):387–404, 2004.

[14] F. Brackx, N. De Schepper, and F. Sommen. The Clifford–Laguerre continuous wavelet transform. *Bulletin of the Belgian Mathematical Society – Simon Stevin*, 11(2):201–215, 2004.

[15] F. Brackx, N. De Schepper, and F. Sommen. The Clifford–Fourier transform. *Journal of Fourier Analysis and Applications*, 11(6):669–681, 2005.

[16] F. Brackx, N. De Schepper, and F. Sommen. The two-dimensional Clifford–Fourier transform. *Journal of Mathematical Imaging and Vision*, 26(1):5–18, 2006.

[17] F. Brackx, N. De Schepper, and F. Sommen. Clifford–Jacobi polynomials and the associated continuous wavelet transform in Euclidean space. In Qian et al. [83], pages 185–198.

[18] F. Brackx, N. De Schepper, and F. Sommen. The Fourier transform in Clifford analysis. *Advances in Imaging and Electron Physics*, 156:55–201, 2009.

[19] F. Brackx, R. Delanghe, and F. Sommen. *Clifford Analysis*, volume 76. Pitman, Boston, 1982.

[20] F. Brackx and F. Sommen. The continuous wavelet transform in Clifford analysis. In F. Brackx, J.S.R. Chisholm, and V. Souček, editors, *Clifford Analysis and its Applications*, pages 9–26. Kluwer, 2001.

[21] F. Brackx and F. Sommen. The generalized Clifford–Hermite continuous wavelet transform. *Advances in Applied Clifford Algebras*, 11(S1):219–231, Feb. 2001.

[22] F. Brackx and F. Sommen. Clifford–Bessel wavelets in Euclidean space. *Mathematical Methods in the Applied Sciences*, 25(16–18):1479–1491, Nov./Dec. 2002.

[23] R. Bujack, G. Scheuermann, and E. Hitzer. A general geometric Fourier transform convolution theorem. *Advances in Applied Clifford Algebras*, 23(1):15–38, 2013.

[24] T. Bülow. *Hypercomplex Spectral Signal Representations for the Processing and Analysis of Images.* PhD thesis, University of Kiel, Germany, Institut für Informatik und Praktische Mathematik, Aug. 1999.

[25] T. Bülow and G. Sommer. Hypercomplex signals – a novel extension of the analytic signal to the multidimensional case. *IEEE Transactions on Signal Processing*, 49(11):2844–2852, Nov. 2001.

[26] P. Cerejeiras, M. Ferreira, and U. Kähler. Monogenic wavelets over the unit ball. *Zeitschrift für Analysis und ihre Anwendungen*, 24(4):841–852, 2005.

[27] V.M. Chernov. Discrete orthogonal transforms with data representation in composition algebras. In *Proceedings Scandinavian Conference on Image Analysis*, pages 357–364, Uppsala, Sweden, 1995.

[28] W.K. Clifford. On the classification of geometric algebras. In R. Tucker, editor, *Mathematical Papers* (1882), pages 397–401. Macmillian, London, 1876. Found amongst Clifford's notes, of which the forewords are the abstract that he communicated to the London Mathematical Society on 10th March 1876 (see A. Diek, R. Kantowski, `www.nhn.ou.edu/~ski/papers/Clifford/history.ps`).

[29] S.A. Collins. Lens-system diffraction integral written in terms of matrix optics. *Journal of the Optical Society of America*, 60:1168–1177, 1970.

[30] E.U. Condon. Immersion of the Fourier transform in a continuous group of functional transformations. *Proceedings of the National Academy of Sciences USA*, 23:158–164, 1937.

[31] K. Coulembier and H. De Bie. Hilbert space for quantum mechanics on superspace. *Journal of Mathematical Physics*, 52, 2011. 063504, 30 pages.

[32] H. De Bie. The kernel of the radially deformed Fourier transform. Preprint arXiv.org/abs/1303.2979, Mar. 2013.

[33] H. De Bie. Fourier transform and related integral transforms in superspace. *Journal of Mathematical Analysis and Applications*, 345:147–164, 2008.

[34] H. De Bie. Clifford algebras, Fourier transforms and quantum mechanics. *Mathematical Methods in the Applied Sciences*, 35(18):2198–2225, 2012.

[35] H. De Bie and N. De Schepper. The fractional Clifford–Fourier transform. *Complex Analysis and Operator Theory*, 6(5):1047–1067, 2012.

[36] H. De Bie, N. De Schepper, and F. Sommen. The class of Clifford–Fourier transforms. *Journal of Fourier Analysis and Applications*, 17:1198–1231, 2011.

[37] H. De Bie, B. Ørsted, P. Somberg, and V. Souček. The Clifford deformation of the Hermite semigroup. Symmetry, integrability and geometry-methods and applications, vol. 9, 010, 2013. doi: 10.3482/SIGMA.2013.010

[38] H. De Bie, B. Ørsted, P. Somberg, and V. Souček. Dunkl operators and a family of realizations of $\mathfrak{osp}(1|2)$. *Transactions of the American Mathematical Society*, 364:3875–3902, 2012.

[39] H. De Bie and Y. Xu. On the Clifford–Fourier transform. *International Mathematics Research Notices*, 2011(22):5123–5163, 2011.

[40] N. De Schepper. *Multidimensional Continuous Wavelet Transforms and Generalized Fourier Transforms in Clifford Analysis.* PhD thesis, Ghent University, Belgium, 2006.

[41] M.A. Delsuc. Spectral representation of 2D NMR spectra by hypercomplex numbers. *Journal of magnetic resonance*, 77(1):119–124, Mar. 1988.

[42] C.F. Dunkl. Differential-difference operators associated to reflection groups. *Transactions of the American Mathematical Society*, 311:167–183, 1989.

[43] J. Ebling and G. Scheuermann. Clifford Fourier transform on vector fields. *IEEE Transactions on Visualization and Computer Graphics*, 11(4):469–479, July 2005.

[44] J. Ebling and J. Scheuermann. Clifford convolution and pattern matching on vector fields. In *Proceedings IEEE Visualization*, volume 3, pages 193–200, Los Alamitos, CA, 2003. IEEE Computer Society.

[45] T.A. Ell. *Hypercomplex Spectral Transformations.* PhD thesis, University of Minnesota, June 1992.

[46] T.A. Ell. Quaternion-Fourier transforms for analysis of 2-dimensional linear time-invariant partial-differential systems. In *Proceedings of the 32nd Conference on Decision and Control*, pages 1830–1841, San Antonio, Texas, USA, 15–17 December 1993. IEEE Control Systems Society.

[47] T.A. Ell and S.J. Sangwine. Decomposition of 2D hypercomplex Fourier transforms into pairs of complex Fourier transforms. In M. Gabbouj and P. Kuosmanen, editors, *Proceedings of EUSIPCO 2000, Tenth European Signal Processing Conference*, volume II, pages 1061–1064, Tampere, Finland, 5–8 Sept. 2000. European Association for Signal Processing.

[48] T.A. Ell and S.J. Sangwine. Hypercomplex Fourier transforms of color images. *IEEE Transactions on Image Processing*, 16(1):22–35, Jan. 2007.

[49] R.R. Ernst, G. Bodenhausen, and A. Wokaun. *Principles of Nuclear Magnetic Resonance in One and Two Dimensions.* International Series of Monographs on Chemistry. Oxford University Press, 1987.

[50] M. Felsberg. *Low-Level Image Processing with the Structure Multivector.* PhD thesis, Christian-Albrechts-Universität, Institut für Informatik und Praktische Mathematik, Kiel, 2002.

[51] W.R. Hamilton. On a new species of imaginary quantities connected with a theory of quaternions. *Proceedings of the Royal Irish Academy*, 2:424–434, 1843. Online: `http://www.maths.tcd.ie/pub/HistMath/People/Hamilton/Quatern1/Quatern1.html`.

[52] E. Hitzer. Multivector differential calculus. *Advances in Applied Clifford Algebras*, 12(2):135–182, 2002.

[53] E. Hitzer. Vector differential calculus. *Memoirs of the Faculty of Engineering, Fukui University*, 50(1):109–125, 2002.

[54] E. Hitzer. Quaternion Fourier transform on quaternion fields and generalizations. *Advances in Applied Clifford Algebras*, 17(3):497–517, May 2007.

[55] E. Hitzer. Tutorial on Fourier transformations and wavelet transformations in Clifford geometric algebra. In K. Tachibana, editor, *Lecture Notes of the International*

Workshop for 'Computational Science with Geometric Algebra' (*FCSGA*2007), pages 65–87, Nagoya University, Japan, 14–21 February 2007.

[56] E. Hitzer. Clifford (geometric) algebra wavelet transform. In V. Skala and D. Hildenbrand, editors, *Proceedings of GraVisMa* 2009, pages 94–101, Plzen, Czech Republic, 2–4 September 2009. Online: `http://gravisma.zcu.cz/GraVisMa-2009/Papers_2009/!_2009_GraVisMa_proceedings-FINAL.pdf`.

[57] E. Hitzer. Real Clifford algebra $cl(n,0), n = 2,3 (\text{mod } 4)$ wavelet transform. In Simos et al. [88], pages 781–784.

[58] E. Hitzer. Directional uncertainty principle for quaternion Fourier transforms. *Advances in Applied Clifford Algebras*, 20(2):271–284, 2010.

[59] E. Hitzer. OPS-QFTs: A new type of quaternion Fourier transforms based on the orthogonal planes split with one or two general pure quaternions. In T.E. Simos, G. Psihoyios, C. Tsitouras, and Z. Anastassi, editors, *NUMERICAL ANALYSIS AND APPLIED MATHEMATICS ICNAAM* 2011*: International Conference on Numerical Analysis and Applied Mathematics*, volume 1389 of *AIP Conference Proceedings*, pages 280–283, Halkidiki, Greece, 19–25 September 2011.

[60] E. Hitzer. The Clifford Fourier transform in real Clifford algebras. In K. Guerlebeck, T. Lahmer, and F. Werner, editors, *Proceedings* 19*th International Conference on the Application of Computer Science and Mathematics in Architecture and Civil Engineering*, Weimar, Germany, 4–6 July 2012.

[61] E. Hitzer. Two-sided Clifford Fourier transform with two square roots of -1 in $C\ell_{p,q}$. In *Proceedings of the* 5*th Conference on Applied Geometric Algebras in Computer Science and Engineering* (*AGACSE* 2012), La Rochelle, France, 2–4 July 2012.

[62] E. Hitzer and R. Abłamowicz. Geometric roots of -1 in Clifford algebras $C\ell_{p,q}$ with $p+q \leq 4$. *Advances in Applied Clifford Algebras*, 21(1):121–144, 2010. Published online 13 July 2010.

[63] E. Hitzer and B. Mawardi. Uncertainty principle for the Clifford geometric algebra $cl(3,0)$ based on Clifford Fourier transform. In T.E. Simos, G. Sihoyios, and C. Tsitouras, editors, *International Conference on Numerical Analysis and Applied Mathematics* 2005, pages 922–925, Weinheim, 2005. Wiley-VCH.

[64] E. Hitzer and B. Mawardi. Uncertainty principle for Clifford geometric algebras $C\ell_{n,0}, n = 3 (\text{mod}\, 4)$ based on Clifford Fourier transform. In Qian et al. [83], pages 47–56.

[65] E. Hitzer, T. Nitta, and Y. Kuroe. Applications of Clifford's geometric algebra. *Advances in Applied Clifford Algebras*, 2013. accepted.

[66] E.M.S. Hitzer and B. Mawardi. Clifford Fourier transform on multivector fields and uncertainty principles for dimensions $n = 2 (\text{mod}\, 4)$ and $n = 3 (\text{mod}\, 4)$. *Advances in Applied Clifford Algebras*, 18(3-4):715–736, 2008.

[67] G. Kaiser. Communications via holomorphic Green functions. In *Clifford Analysis and its Applications*, Kluwer NATO Science Series, 2001.

[68] G. Kaiser. Complex-distance potential theory, wave equations, and physical wavelets. *Mathematical Methods in the Applied Sciences*, 25:1577–1588, 2002. Invited paper, Special Issue on Clifford Analysis in Applications.

[69] G. Kaiser. Huygens' principle in classical electrodynamics: a distributional approach. *Advances in Applied Clifford Algebras*, 22(3):703–720, 2012.

[70] G. Kaiser, E. Heyman, and V. Lomakin. Physical source realization of complex-source pulsed beams. *Journal of the Acoustical Society of America*, 107:1880–1891, 2000.

[71] T. Kobayashi and G. Mano. Integral formulas for the minimal representation of $o(p, 2)$. *Acta Applicandae Mathematicae*, 86:103–113, 2005.

[72] X. Li. On the inverse problem for the Dirac operator. *Inverse Problems*, 23:919–932, 2007.

[73] B. Mawardi, S. Adji, and J. Zhao. Clifford algebra-valued wavelet transform on multivector fields. *Advances in Applied Clifford Algebras*, 21(1):13–30, 2011.

[74] B. Mawardi and E. Hitzer. Clifford algebra $Cl(3, 0)$-valued wavelets and uncertainty inequality for Clifford Gabor wavelet transformation. Preprints of Meeting of the Japan Society for Industrial and Applied Mathematics, 16–18 September 2006.

[75] B. Mawardi and E. Hitzer. Clifford algebra $Cl(3, 0)$-valued wavelet transformation, Clifford wavelet uncertainty inequality and Clifford Gabor wavelets. *International Journal of Wavelets, Multiresolution and Information Processing*, 5(6):997–1019, 2007.

[76] B. Mawardi, E. Hitzer, and S. Adji. Two-dimensional Clifford windowed Fourier transform. In E. Bayro-Corrochano and G. Scheuermann, editors, *Applied Geometric Algebras in Computer Science and Engineering*, pages 93–106. Springer, London, 2010.

[77] B. Mawardi, E. Hitzer, A. Hayashi, and R. Ashino. An uncertainty principle for quaternion Fourier transform. *Computers and Mathematics with Applications*, 56(9):2411–2417, 2008.

[78] B. Mawardi and E.M.S. Hitzer. Clifford Fourier transformation and uncertainty principle for the Clifford algebra $C\ell_{3,0}$. *Advances in Applied Clifford Algebras*, 16(1):41–61, 2006.

[79] M. Mitrea. *Clifford Wavelets, Singular Integrals, and Hardy Spaces*, volume 1575 of *Lecture notes in mathematics*. Springer, Berlin, 1994.

[80] M. Moshinsky and C. Quesne. Linear canonical transformations and their unitary representations. *Journal of Mathematical Physics*, 12:1772–1780, 1971.

[81] V. Namias. The fractional order Fourier transform and its application to quantum mechanics. *IMA Journal of Applied Mathematics*, 25(3):241–265, 1980.

[82] H. Ozaktas, Z. Zalevsky, and M. Kutay. *The Fractional Fourier Transform*. Wiley, Chichester, 2001.

[83] T. Qian, M.I. Vai, and Y. Xu, editors. *Wavelet Analysis and Applications*, Applied and Numerical Harmonic Analysis. Birkhäuser Basel, 2007.

[84] S.J. Sangwine. Fourier transforms of colour images using quaternion, or hypercomplex, numbers. *Electronics Letters*, 32(21):1979–1980, 10 Oct. 1996.

[85] S.J. Sangwine. Biquaternion (complexified quaternion) roots of -1. *Advances in Applied Clifford Algebras*, 16(1):63–68, June 2006.

[86] S.J. Sangwine and T.A. Ell. The discrete Fourier transform of a colour image. In J.M. Blackledge and M.J. Turner, editors, *Image Processing II Mathematical Methods, Algorithms and Applications*, pages 430–441, Chichester, 2000. Horwood Publishing for Institute of Mathematics and its Applications. Proceedings Second IMA Conference on Image Processing, De Montfort University, Leicester, UK, September 1998.

[87] S.J. Sangwine and N. Le Bihan. Quaternion polar representation with a complex modulus and complex argument inspired by the Cayley–Dickson form. *Advances in Applied Clifford Algebras*, 20(1):111–120, Mar. 2010. Published online 22 August 2008.

[88] T.E. Simos, G. Psihoyios, and C. Tsitouras, editors. *NUMERICAL ANALYSIS AND APPLIED MATHEMATICS: International Conference on Numerical Analysis and Applied Mathematics* 2009, volume 1168 of *AIP Conference Proceedings*, Rethymno, Crete (Greece), 18–22 September 2009.

[89] F. Sommen. A product and an exponential function in hypercomplex function theory. *Applicable Analysis*, 12:13–26, 1981.

[90] F. Sommen. Hypercomplex Fourier and Laplace transforms I. *Illinois Journal of Mathematics*, 26(2):332–352, 1982.

[91] F. Sommen. Hypercomplex Fourier and Laplace transforms II. *Complex Variables*, 1(2–3):209–238, 1983.

[92] L. Traversoni. Quaternion wavelet problems. In *Proceedings of* 8th *International Symposium on Approximation Theory*, Texas A& M University, Jan. 1995.

[93] J. Zhao. Clifford algebra-valued admissible wavelets associated with admissible group. *Acta Scientarium Naturalium Universitatis Pekinensis*, 41(5):667–670, 2005.

[94] J. Zhao and L. Peng. Quaternion-valued admissible wavelets associated with the 2D Euclidean group with dilations. *Journal of Natural Geometry*, 20(1/2):21–32, 2001.

Fred Brackx
Department of Mathematical Analysis
Faculty of Engineering
Ghent University
Galglaan 2
B-9000 Gent, Belgium
e-mail: `Freddy.Brackx@ugent.be`

Eckhard Hitzer
College of Liberal Arts
Department of Material Science
International Christian University
181-8585 Tokyo, Japan
e-mail: `hitzer@icu.ac.jp`

Stephen J. Sangwine
School of Computer Science
and Electronic Engineering
University of Essex
Wivenhoe Park
Colchester, CO4 3SQ, UK
e-mail: `sjs@essex.ac.uk`

Part I

Quaternions

Quaternion and Clifford–Fourier Transforms and Wavelets
Trends in Mathematics, 3–14

1 Quaternion Fourier Transform: Re-tooling Image and Signal Processing Analysis

Todd Anthony Ell

'Did you ask a good question today?' – *Janet Teig*

Abstract. Quaternion Fourier transforms (QFT's) provide expressive power and elegance in the analysis of higher-dimensional linear invariant systems. But, this power comes at a cost – an overwhelming number of choices in the QFT definition, each with consequences. This chapter explores the evolution of QFT definitions as a framework from which to solve specific problems in vector-image and vector-signal processing.

Mathematics Subject Classification (2010). Primary 11R52; secondary 42B10.

Keywords. Quaternion, Fourier transform.

1. Introduction

In recent years there has been an increasing recognition on the part of engineers and investigators in image and signal processing of holistic vector approaches to spectral analysis. Generally speaking, this type of spectral analysis treats the vector components of a system not in an iterated, channel-wise fashion but instead in a holistic, gestalt fashion. The Quaternion Fourier transform (QFT) is one such analysis tool.

One of the earliest documented attempts (1987) at describing this type of spectral analysis was in the area of two-dimensional nuclear magnet resonance. Ernst, *et al.* [6, pp. 307–308] briefly discusses using a hypercomplex Fourier transform as a method to independently adjust phase angles with respect to two frequency variables in two-dimensional spectroscopy. After introducing the concept they immediately fall back to an iterated approach leaving the idea unexplored. For similar reasons, Ell [2] in 1992 independently explored the use of QFTs as a tool in the analysis of linear time-invariant systems of partial differential equations (PDEs). Ell specifically 'designed' a quaternion Fourier transform whose spectral operators allowed him to disambiguate partial derivatives with respect to two different independent variables. Ell's original QFT was given by

$$H\left[\boldsymbol{j}\omega,\boldsymbol{k}\nu\right]=\int_{\mathbb{R}^2}e^{-\boldsymbol{j}\omega t}h\left(t,\tau\right)e^{-\boldsymbol{k}\nu\tau}dtd\tau\,, \tag{1.1}$$

where $H\left[\boldsymbol{j}\omega,\boldsymbol{k}\nu\right]\in\mathbb{H}$ (the set of quaternions), $\boldsymbol{j}$ and $\boldsymbol{k}$ are Hamilton's hypercomplex operators, and $h\left(t,\tau\right):\mathbb{R}\times\mathbb{R}\to\mathbb{R}$ (the set of reals). The partial-differential equivalent spectral operators for this transform are given by

$$\frac{\partial}{\partial t}h\left(t,\tau\right)\Leftrightarrow\boldsymbol{j}\omega\,H\left[\boldsymbol{j}\omega,\boldsymbol{k}\nu\right],\quad\frac{\partial}{\partial\tau}h\left(t,\tau\right)\Leftrightarrow H\left[\boldsymbol{j}\omega,\boldsymbol{k}\nu\right]\boldsymbol{k}\nu. \tag{1.2}$$

These two differentials have clearly different spectral signatures in contrast to the two-dimensional iterated complex Fourier transform where

$$\frac{\partial}{\partial t}h\left(t,\tau\right)\Leftrightarrow\boldsymbol{j}\omega\,H\left[\boldsymbol{j}\omega,\boldsymbol{j}\nu\right],\quad\frac{\partial}{\partial\tau}h\left(t,\tau\right)\Leftrightarrow\boldsymbol{j}\nu H\left[\boldsymbol{j}\omega,\boldsymbol{j}\nu\right], \tag{1.3}$$

especially when $\omega=\nu$, at which point the complex spectral domain responses are indistinguishable. This was the first step towards stability analysis in designing controllers for systems described by PDEs.

The slow adoption of QFTs at the present time by the engineering community is due in part to their lack of practical understanding of its properties. This slow adoption is further exacerbated by the variety of transform definitions available. But, in the middle of difficulty lies opportunity. Instead of attempting to find *the* single best QFT (which cannot meet every design engineer's needs) we provide instead the means to allow the designer to select the definition most appropriate to his specific problem. That means, allow him to *re-tool* for the analysis problem at hand.

For example, when QFTs were later applied to colour-image processing [4], where each colour pixel in an image is treated as a 3-vector with basis $\{\boldsymbol{i},\boldsymbol{j},\boldsymbol{k}\}\in\mathbb{H}$, it became apparent that there was no preferential association of colour-space axes with either the basis or the QFT's exponential-kernel axis. This lead to the next generalization of the QFT defined as

$$\mathcal{F}^{+}\left[\omega,\nu\right]=\int_{\mathbb{R}^2}e^{-\boldsymbol{\mu}(\omega t+\nu\tau)}f\left(t,\tau\right)dtd\tau\,, \tag{1.4}$$

where the transform kernel axis $\boldsymbol{\mu}$ is *any* pure unit quaternion, *i.e.*,

$$\boldsymbol{\mu}\in\left\{\boldsymbol{i}x+\boldsymbol{j}y+\boldsymbol{k}z\in\mathbb{H}\mid x^2+y^2+z^2=1\right\}$$

so that $\boldsymbol{\mu}^2=-1$. Still later it was realized [9] that since there is no preferred direction of indexing the image's pixels then the sign on the transform kernel is also arbitrary, so that a *forward* QFT could also be defined as

$$\mathcal{F}^{-}\left[\omega,\nu\right]=\int_{\mathbb{R}^2}e^{+\boldsymbol{\mu}(\omega t+\nu\tau)}f\left(t,\tau\right)dtd\tau\,, \tag{1.5}$$

and the two definitions could be intermixed without concern of creating a non-causal set of image processing filters. This led to several simplifications of a spectral form of the vector correlation operation on two images [8].

Bearing in mind such diverse application of various QFTs, the focus of this work is to detail as broad a set of QFT definitions as possible, and where known, some of the issues associated with applying them to problems in signal and image processing. It also includes a review of approaches taken to define the interrelations between the various QFT definitions.

2. Preliminaries

To provide a basis for discussion this section gives nomenclature, basic facts on quaternions, and some useful subsets and algebraic equations.

2.1. Just the Facts

The quaternion algebra over the reals $\mathbb{R}$, denoted by

$$\mathbb{H} = \{q = r_0 + \boldsymbol{i}r_1 + \boldsymbol{j}r_2 + \boldsymbol{k}r_3 \mid r_0, r_1, r_2, r_3 \in \mathbb{R}\}, \tag{2.1}$$

is an associative non-commutative four-dimensional algebra, which obeys Hamilton's multiplication rules

$$\boldsymbol{ij} = \boldsymbol{k} = -\boldsymbol{jk},\ \boldsymbol{jk} = \boldsymbol{i} = -\boldsymbol{kj},\ \boldsymbol{ki} = \boldsymbol{j} = -\boldsymbol{ik}, \tag{2.2}$$

$$\boldsymbol{i}^2 = \boldsymbol{j}^2 = \boldsymbol{k}^2 = \boldsymbol{ijk} = -1. \tag{2.3}$$

The quaternion conjugate is defined by

$$\bar{q} = r_0 - \boldsymbol{i}r_1 - \boldsymbol{j}r_2 - \boldsymbol{k}r_3\,, \tag{2.4}$$

which is an anti-involution, *i.e.*, $\bar{\bar{q}} = q$, $\overline{p+q} = \bar{p} + \bar{q}$, and $\overline{qp} = \bar{p}\,\bar{q}$. The norm of a quaternion is defined as

$$|q| = \sqrt{q\bar{q}} = \sqrt{r_0^2 + r_1^2 + r_2^2 + r_3^2}. \tag{2.5}$$

Using the conjugate and norm of q, one can define the inverse of $q \in \mathbb{H} \setminus \{0\}$ as

$$q^{-1} = \bar{q}\big/|q|^2. \tag{2.6}$$

Two classical operators on quaternions are the vector- and scalar-part, $V[.]$ and $S[.]$, respectively; these are defined as

$$V[q] = \boldsymbol{i}r_1 + \boldsymbol{j}r_2 + \boldsymbol{k}r_3, \qquad S[q] = r_0. \tag{2.7}$$

2.2. Useful Subsets

Various subsets of the quaternions are of interest and used repeatedly throughout this work. The 3-vector subset of $\mathbb{H}$ is the set of pure quaternions defined as

$$V[\mathbb{H}] = \{q = \boldsymbol{i}r_1 + \boldsymbol{j}r_2 + \boldsymbol{k}r_3 \in \mathbb{H}\}. \tag{2.8}$$

The set of pure, unit length quaternions is denoted $\mathbb{S}^3_{\mathbb{H}}$, *i.e.*,

$$\mathbb{S}^3_{\mathbb{H}} = \left\{\boldsymbol{\mu} = \boldsymbol{i}r_1 + \boldsymbol{j}r_2 + \boldsymbol{k}r_3 \in \mathbb{H} \mid r_1^2 + r_2^2 + r_3^2 = 1\right\}. \tag{2.9}$$

Each element of $\mathbb{S}^3_{\mathbb{H}}$ creates a distinct copy of the complex numbers because $\boldsymbol{\mu}^2 = -1$, that is, each creates an injective ring homomorphism from $\mathbb{C}$ to $\mathbb{H}$. So, for each $\boldsymbol{\mu} \in \mathbb{S}^3_{\mathbb{H}}$, we associate a complex sub-field of $\mathbb{H}$ denoted

$$\mathbb{C}_{\boldsymbol{\mu}} = \left\{\alpha + \beta\boldsymbol{\mu}; \mid \alpha, \beta \in \mathbb{R}, \boldsymbol{\mu} \in \mathbb{S}^3_{\mathbb{H}}\right\}. \tag{2.10}$$

2.3. Useful Algebraic Equations

In various quaternion equations the non-commutativity of the multiplication causes difficulty, however, there are algebraic forms which assist in making simplifications. The following three defined forms appear to be the most useful.

Definition 2.1 (Even-Odd Form). Every $f : \mathbb{R}^2 \to \mathbb{H}$ can be split into even and odd parts along the x- and y-axis as

$$f(x,y) = f_{\mathrm{ee}}(x,y) + f_{\mathrm{eo}}(x,y) + f_{\mathrm{oe}}(x,y) + f_{\mathrm{oo}}(x,y) \tag{2.11}$$

where f_{eo} denotes the part of f that is even with respect to x and odd with respect to y, *etc.*, given as

$$\begin{aligned}
f_{\mathrm{ee}}(x,y) &= \tfrac{1}{4}\left(f(x,y) + f(-x,y) + f(x,-y) + f(-x,-y)\right),\\
f_{\mathrm{eo}}(x,y) &= \tfrac{1}{4}\left(f(x,y) + f(-x,y) - f(x,-y) - f(-x,-y)\right),\\
f_{\mathrm{oe}}(x,y) &= \tfrac{1}{4}\left(f(x,y) - f(-x,y) + f(x,-y) - f(-x,-y)\right),\\
f_{\mathrm{oo}}(x,y) &= \tfrac{1}{4}\left(f(x,y) - f(-x,y) - f(x,-y) + f(-x,-y)\right).
\end{aligned} \tag{2.12}$$

Definition 2.2 (Symplectic Form [3]). Every $q = r_0 + \boldsymbol{i}r_1 + \boldsymbol{j}r_2 + \boldsymbol{k}r_3 \in \mathbb{H}$ can be rewritten in terms of a new basis of operators $\{\boldsymbol{\mu}_1, \boldsymbol{\mu}_2, \boldsymbol{\mu}_3\}$ as

$$q = r'_0 + \boldsymbol{\mu}_1 r'_1 + \boldsymbol{\mu}_2 r'_2 + \boldsymbol{\mu}_3 r'_3 = \left(r'_0 + \boldsymbol{\mu}_1 r'_1\right) + \left(r'_2 + \boldsymbol{\mu}_1 r'_3\right)\boldsymbol{\mu}_2, \tag{2.13}$$

where $\boldsymbol{\mu}_1\boldsymbol{\mu}_2 = \boldsymbol{\mu}_3$, $\boldsymbol{\mu}_{1,2,3} \in \mathbb{S}^3_{\mathbb{H}}$, hence they form an orthogonal triad.

Remark 2.3. The mapping $\{r_1, r_2, r_3\} \to \{r'_1, r'_2, r'_3\}$ is a change in basis from $\{\boldsymbol{i}, \boldsymbol{j}, \boldsymbol{k}\}$ to $\{\boldsymbol{\mu}_1, \boldsymbol{\mu}_2, \boldsymbol{\mu}_3\}$ *via*

$$r'_0 = r_0, \quad r'_n = -\tfrac{1}{2}\left(V[q]\,\boldsymbol{\mu}_n + \boldsymbol{\mu}_n V[q]\right), \quad n = \{1,2,3\}. \tag{2.14}$$

Remark 2.4. The symplectic form essentially decomposes a quaternion with respect to a specific complex sub-field. That is

$$q = \left(r'_0 + \boldsymbol{\mu}_1 r'_1\right) + \left(r'_2 + \boldsymbol{\mu}_1 r'_3\right)\boldsymbol{\mu}_2 = c_1 + c_2\boldsymbol{\mu}_2, \tag{2.15}$$

where $c_{1,2} \in \mathbb{C}_{\boldsymbol{\mu}_1}$. The author coined the terms *simplex* and *perplex* parts of q, for c_1 and c_2, respectively.

Remark 2.5. The symplectic form works for any permutation of the basis $\{\boldsymbol{\mu}_1, \boldsymbol{\mu}_2, \boldsymbol{\mu}_3\}$ so that the simplex and complex parts can be taken from any complex sub-field $\mathbb{C}_{\boldsymbol{\mu}_n}$ Further, the *swap rule* applies to the last term, *i.e.*, $c_2\boldsymbol{\mu}_2 = \boldsymbol{\mu}_2\overline{c_2}$, where the over bar denotes both quaternion and complex sub-field conjugation.

Definition 2.6 (Split Form [7]). Every $q \in \mathbb{H}$ can be split as

$$q = q_+ + q_-, \quad q_\pm = \tfrac{1}{2}\left(q \pm \boldsymbol{\mu}_1 q \boldsymbol{\mu}_2\right), \tag{2.16}$$

where $\boldsymbol{\mu}_1\boldsymbol{\mu}_2 = \boldsymbol{\mu}_3$, and $\boldsymbol{\mu}_{1,2,3} \in \mathbb{S}^3_{\mathbb{H}}$.

Remark 2.7. The split form allows for the explicit ordering of factors with respect to the operators. So, for example, $q = r_0 + \boldsymbol{\mu}_1 r_1 + \boldsymbol{\mu}_2 r_2 + \boldsymbol{\mu}_3 r_3$ becomes

$$\begin{aligned} q_\pm &= \{(r_0 \pm r_3) + \boldsymbol{\mu}_1 (r_1 \pm r_2)\} \frac{1 \pm \boldsymbol{\mu}_3}{2} \\ &= \frac{1 \pm \boldsymbol{\mu}_3}{2} \{(r_0 \pm r_3) - \boldsymbol{\mu}_2 (r_1 \pm r_2)\} . \end{aligned} \tag{2.17}$$

Euler's formula holds for quaternions, so any unit length quaternion can be written as $\cos a + \boldsymbol{\mu} \sin a = e^{\boldsymbol{\mu} a}$, for $a \in \mathbb{R}$ and $\boldsymbol{\mu} \in \mathbb{S}^3_{\mathbb{H}}$. Here $\boldsymbol{\mu}$ is referred to as the (Eigen-) axis and a as the (Eigen-) phase angle. Although in general $e^q\, e^p \neq e^{q+p}$ for $p, q \in \mathbb{H}$, their exponential product is a linear combination of exponentials of the sum and difference of their phase angles. This can be written in two ways as shown in the following two propositions.

Proposition 2.8 (Exponential Split). *Let* $\boldsymbol{\mu}_{1,2} \in \mathbb{S}^3_{\mathbb{H}}$ *and* $a, b \in \mathbb{R}$, *then*

$$e^{\boldsymbol{\mu}_1 a} e^{\boldsymbol{\mu}_2 b} = e^{\boldsymbol{\mu}_1 (a-b)} \frac{1 + \boldsymbol{\mu}_3}{2} + e^{\boldsymbol{\mu}_1 (a+b)} \frac{1 - \boldsymbol{\mu}_3}{2} \tag{2.18}$$

and

$$e^{\boldsymbol{\mu}_1 a} e^{\boldsymbol{\mu}_2 b} = \frac{1 + \boldsymbol{\mu}_3}{2} e^{\boldsymbol{\mu}_2 (b-a)} + \frac{1 - \boldsymbol{\mu}_3}{2} e^{\boldsymbol{\mu}_2 (b+a)}, \tag{2.19}$$

where $\boldsymbol{\mu}_1\boldsymbol{\mu}_2 = \boldsymbol{\mu}_3$ *and* $\boldsymbol{\mu}_3 \in \mathbb{S}^3_{\mathbb{H}}$.

Proof. Application of split form to the exponential product. □

Proposition 2.9 (Exponential Modulation). *Let* $\boldsymbol{\mu}_{1,2} \in \mathbb{S}^3_{\mathbb{H}}$ *and* $a, b \in \mathbb{R}$, *then*

$$e^{\boldsymbol{\mu}_1 a} e^{\boldsymbol{\mu}_2 b} = \tfrac{1}{2}\left(e^{\boldsymbol{\mu}_1 (a+b)} + e^{\boldsymbol{\mu}_1 (a-b)}\right) - \tfrac{1}{2}\boldsymbol{\mu}_1 \left(e^{\boldsymbol{\mu}_1 (a+b)} - e^{\boldsymbol{\mu}_1 (a-b)}\right) \boldsymbol{\mu}_2 \tag{2.20}$$

and

$$e^{\boldsymbol{\mu}_1 a} e^{\boldsymbol{\mu}_2 b} = \tfrac{1}{2}\left(e^{\boldsymbol{\mu}_2 (b+a)} + e^{\boldsymbol{\mu}_2 (b-a)}\right) - \tfrac{1}{2}\boldsymbol{\mu}_1 \left(e^{\boldsymbol{\mu}_2 (b+a)} - e^{\boldsymbol{\mu}_2 (b-a)}\right) \boldsymbol{\mu}_2 . \tag{2.21}$$

Proof. Direct application of Euler's formula and trigonometric identities. □

Remark 2.10. The sandwich terms (*i.e.*, $\boldsymbol{\mu}_1(.)\boldsymbol{\mu}_2$) in the exponential equations introduce 4-space rotations into the interpretation of the product [1]. For if $p = \boldsymbol{\mu}_1 q \boldsymbol{\mu}_2$, then p is a rotated version of q about the $(\boldsymbol{\mu}_1, \boldsymbol{\mu}_2)$-plane by $\frac{\pi}{2}$.

3. Quaternion Fourier Transforms

The purpose of this section is to enumerate a list of *possible* definitions for a quaternion Fourier transform. This is followed by a discussion regarding various operator properties used in the engineering fields that require simple Fourier transform pairs between the non-transformed operation and the equivalent Fourier domain operation, *i.e.*, the so-called *operator pairs* as seen in most engineering textbooks on Fourier analysis. Finally, a discussion on how the inter-relationship between QFT definitions are explored, not so as to reduce them to a single canonical form, but to provide the investigator a tool to cross between definitions when necessary so as to gain insight into operator properties.

3.1. Transform Definitions

Although there has been much use of the QFT forms currently in circulation, there are however more available. Not all 'degrees-of-freedom' have been exploited. The non-commutativity of the quaternion multiplication gave rise to the left- and right-handed QFT kernels. The infinite number of square-roots of -1 (the cardinality of $\mathbb{S}_{\mathbb{H}}$) gave rise to the two-sided, or sandwiched kernel. One concept left unexplored is the implication of the exponential product of two quaternions, *i.e.*, $e^p e^q \neq e^{p+q}$. When this is also taken into account, the list expands to *eight* distinct QFTs as enumerated in Table 1.

TABLE 1. QFT kernel definitions for $f : \mathbb{R}^2 \to \mathbb{H}$.

	Left	Right	Sandwich
Single-axis	$e^{-\mu_1(\omega x+\nu y)}\, f(.)$	$f(.)\, e^{-\mu_1(\omega x+\nu y)}$	$e^{-\mu_1\omega x}\, f(.)\, e^{-\mu_1\nu y}$
Dual-axis	$e^{-(\mu_1\omega x+\mu_2\nu y)}\, f(.)$	$f(.)\, e^{-(\mu_1\omega x+\mu_2\nu y)}$	–
Factored	$e^{-\mu_1\omega x}e^{-\mu_2\nu y}\, f(.)$	$f(.)\, e^{-\mu_1\omega x}e^{-\mu_2\nu y}$	$e^{-\mu_1\omega x}\, f(.)\, e^{-\mu_2\nu y}$

Depending on the value space of $f(x,y)$, the available number of distinct QFT forms changes. Table 1 shows the options when $f : \mathbb{R}^2 \to \mathbb{H}$. However, if $f : \mathbb{R}^2 \to \mathbb{R}$, then all chirality options (left, right, and sandwiched) collapse to the same form, leaving three distinct choices: single-axis and factored and un-factored dual-axis forms.

If neither x nor y are time-like, so that causality of the solution is not a factor, then the number of given QFTs doubles. Variations created by conjugating the quaternion-exponential kernel of both the forward and inverse transform are usually a matter of convention – the signs must be opposites. For non-causal systems, however, the sign on the kernel can be taken both ways; each defining its own *forward* transform from the spatial to spatial-frequency domain. To distinguish the two versions of the forward transform, one is called *forward* the other *reverse*. Of course, the *inverse* transform is still obtained by conjugating the corresponding forward (or reverse) kernel. Hence, one may define the single-axis, left- and right-sided, forward and reverse transforms as follows[1].

Definition 3.1 (Single-axis, Left-sided QFT). The single-axis, left-sided, forward ($\mathcal{F}^{+L}$) and reverse ($\mathcal{F}^{-L}$) QFTs are defined as

$$\mathcal{F}^{\pm L}\left[f(x,y)\right] = \iint\limits_{\mathbb{R}^2} e^{\mp\boldsymbol{\mu}_1(\omega x+\nu y)} f(x,y)\, dxdy = F^{\pm L}\left[\omega,\nu\right]. \tag{3.1}$$

Definition 3.2 (Single-axis, Right-sided QFT). The single-axis, right-sided, forward ($\mathcal{F}^{+R}$) and reverse ($\mathcal{F}^{-R}$) QFTs are defined as

$$\mathcal{F}^{\pm R}\left[f(x,y)\right] = \iint\limits_{\mathbb{R}^2} f(x,y)\, e^{\mp\boldsymbol{\mu}_1(\omega x+\nu y)} dxdy = F^{\pm R}\left[\omega,\nu\right]. \tag{3.2}$$

All of the entries in Table 1 exploit the fact that unit length complex numbers act as rotation operators within the complex plane. There is, however, another rotation operator – the 3-space rotation operator for which quaternions are famous. Table 2 lists additional definitions under the provision that f takes on values restricted to $V[\mathbb{H}]$. Note the factor of $\frac{1}{2}$ in the kernel exponent, this is included so that the frequency scales between the various definitions align.

TABLE 2. QFT kernel definitions exclusively for $f : \mathbb{R}^2 \to V[\mathbb{H}]$.

	3-Space Rotator
Single-axis	$e^{-\boldsymbol{\mu}_1\omega x/2}\ f(.)\ e^{+\boldsymbol{\mu}_1\nu y/2}$
Dual-axis	$e^{-(\boldsymbol{\mu}_1\omega x+\boldsymbol{\mu}_2\nu y)/2}\ f(.)\ e^{+(\boldsymbol{\mu}_1\omega x+\boldsymbol{\mu}_2\nu y)/2}$
Dual-axis, factored	$e^{-\boldsymbol{\mu}_1\omega x/2}e^{-\boldsymbol{\mu}_2\nu y/2}\ f(.)\ e^{+\boldsymbol{\mu}_2\nu y/2}e^{+\boldsymbol{\mu}_1\omega x/2}$

Taking all these permutations in mind, one arrives at 22 unique QFT definitions.

3.2. Functional Relationships

There are several properties used in *complex* Fourier transform (CFT) analysis that one hopes will carry over to the QFT in some fashion. These are listed in Table 3 from which we will discuss the challenges which arise in QFT analysis. In what follows let $f(x,y) \Leftrightarrow F[\omega,\nu]$ denote transform pairs, *i.e.*, $\mathcal{F}[f(x,y)] = F[\omega,\nu]$ is the forward (or reverse) transform and $\mathcal{F}^{-1}[F(\omega,\nu)] = f(x,y)$ is its inversion.

Inversion. Every transform should be invertible. Although this seems obvious, there are instances where a given transform is not. For example, if the restriction of $f : \mathbb{R}^2 \to V[\mathbb{H}]$ were not imposed on the inputs of Table 2, then

[1]Note, that in (3.1) and (3.2) the arguments x and y of f in $\mathcal{F}^{\pm L,R}[f(x,y)]$ are shown for clarity, but are actually dummy arguments, which are integrated out. A more mathematical notation would be $\mathcal{F}^{\pm L,R}\{f\}(\omega,\nu) = F^{\pm L,R}(\omega,\nu)$.

TABLE 3. Fourier transform $\mathcal{F}$ properties. $[\alpha, \beta, \gamma, \delta \in \mathbb{R}]$

Property	Definition
Inversion	$\mathcal{F}^{-1}\left[\mathcal{F}\left[f\left(x,y\right)\right]\right] = f\left(x,y\right)$
Linearity	$\alpha f(x,y) + \beta g(x,y)$
Complex Degenerate	$(\boldsymbol{\mu}_1 = \boldsymbol{i}$ and $f : \mathbb{R}^2 \to \mathbb{C}_{\boldsymbol{i}}) \to (\text{QFT}\cong\text{CFT})$
Convolution	$f \circ g(x,y) = (?)$
Correlation	$f \star g(x,y) = (?)$
Modulation	$e^{\boldsymbol{\mu}_1 \omega_0 x} f(.),\ e^{\boldsymbol{\mu}_2 \omega_0 x} f(.),\ f(.)e^{\boldsymbol{\mu}_1 \nu_0 y},\ f(.)e^{\boldsymbol{\mu}_2 \nu_0 y}$, *etc.*
Scaling	$f(x/\alpha, y/\beta)$
Translation	$f(x - x_0, y - y_0)$
Rotation	$f(x\cos\alpha - y\sin\alpha, x\sin\alpha + y\cos\alpha)$
Axis-reversal	$f(-x,y), f(x,-y), f(-x,-y)$
Re-coordinate	$f(\alpha x + \beta y, \gamma x + \delta y)$
Conjugation	$\overline{f(x,y)}$
Differentials	$\frac{\partial}{\partial x}, \frac{\partial}{\partial y}, \frac{\partial^2}{\partial x \partial y}$, *etc.*

every transform of that table would cease to be invertible. This is because any real-valued function, or the scalar part of full quaternion valued functions, commute with the kernel factors which then vanish from under the integral.

Linearity. $\alpha f(x,y) + \beta g(x,y) \Leftrightarrow \alpha F[\omega,\nu] + \beta G[\omega,\nu]$ where $\alpha, \beta \in \mathbb{R}$. A quick check verifies that this property holds for all proposed QFT definitions.

Complex Degenerate. For the single-axis transforms, if $\boldsymbol{\mu}_1 = \boldsymbol{i}$ and $f : \mathbb{R}^2 \to \mathbb{C}_{\boldsymbol{i}}$, then the QFT should ideally degenerate to the twice iterated complex Fourier transform. This degenerate property cannot apply to the dual axis, factored forms of Tables 1 and 2 since if $\boldsymbol{\mu}_1 = \boldsymbol{\mu}_2 = \boldsymbol{i}$, these forms reduce to their single-axis versions.

Convolution (Faltung) theorem. Rarely does the standard, complex transform type pair $f \circ g\,(x,y) \Leftrightarrow F\,[\omega,\nu]\,G\,[\omega,\nu]$ exist in such a simple form for the QFT. Even the very *definition* of convolution needs an update since $f \circ g \neq g \circ f$ when f and g are $\mathbb{H}$-valued. The definition is altered again based on which of the two functions is translated, *i.e.*, is the integrand $f(x - x', y - y')g(x,y)$ or $f(x,y)g(x - x', y - y')$. This gives rise to at least four distinct convolution definitions. This will also alter the spectral operator pair.

Further, if f is an input function, then g is typically related to the *impulse response* of a system. But, if $g : \mathbb{R}^2 \to V[\mathbb{H}]$, is a *single* impulse response sufficient to describe such a system? Or does it take at least two orthogonally oriented impulses, say $\boldsymbol{\mu}_1\delta(x,y)$ and $\boldsymbol{\mu}_2\delta(x,y)$, where $\boldsymbol{\mu}_1 \perp \boldsymbol{\mu}_2$ and $\delta(x,y)$ is the Dirac delta function?

Correlation. Consider the correlation definition of two $\mathbb{R}$-valued functions f and g (let $f, g : \mathbb{R} \to \mathbb{R}$ for simplicity of discussion)

$$f \star g\,(t) = \int_{\mathbb{R}} f\,(\tau)\, g\,(\tau - t)\, d\tau = \int_{\mathbb{R}} f\,(\tau + t)\, g\,(\tau)\, d\tau, \tag{3.3}$$

where substituting $\tau - t = \tau'$ yields the second form. For the correlation of real-valued functions this is entirely sufficient.

However, for $\mathbb{C}$-valued functions a conjugation operation is required to ensure the relation of the autocorrelation functions $(f \star f)$ to the power spectrum as required by the Wiener-Khintchine theorem. This effectively ensures that the power spectrum of a complex auto-correlation is $\mathbb{R}$-valued. The complex extension to the cross-correlation function can then be given as

$$f \star g\,(t) = \int_{\mathbb{R}} f\,(\tau)\, \overline{g\,(\tau - t)}\, d\tau, \tag{3.4}$$

or alternatively as

$$f \star g\,(t) = \int_{\mathbb{R}} \overline{f\,(\tau)}\, g\,(\tau + t)\, d\tau. \tag{3.5}$$

In general, the literature does not give significance to the direction of the shifted signal $(\tau \pm t)$. However, in the case of vector correlation matching problems, such as colour image registration, direction is fundamental.

Taking this into consideration, for $\mathbb{H}$-valued functions the equivalent correlation could be either

$$f \star g\,(x, y) = \int_{\mathbb{R}^2} f\,(x', y')\, \overline{g\,(x' - x, y' - y)} dx' dy', \tag{3.6}$$

or

$$f \star g\,(x, y) = \int_{\mathbb{R}^2} f\,(x' + x, y' + y)\, \overline{g\,(x', y')} dx' dy', \tag{3.7}$$

depending on which shift direction is required. For more details see [9, 4]. Note that the correlation result is not necessarily $\mathbb{R}$-valued.

Modulation. There are multiple types of frequency modulation that need to be addressed. The modulating exponential can be applied from the left or right, can be driven as a function of either input parameter (*i.e.*, x or y), and be pointing along one of the kernel axes (*i.e.*, $\boldsymbol{\mu}_1$ or $\boldsymbol{\mu}_2$). Some options are detailed in the Karnaugh map of Table 4.

In summary, when addressing the QFT operator properties one often needs to regress back to the basic operator definitions and their underlying assumptions, then verify they are still valid for generalization to quaternion forms. Either the operator definition itself needs to be modified (as in the case of correlation) or the number of permutations on the definition increases (as in convolution and modulation).

TABLE 4. Frequency Modulations

	Left	Right	
x	$e^{\boldsymbol{\mu}_2\omega_0 x} f(.)$	$f(.)e^{\boldsymbol{\mu}_2\omega_0 x}$	$\boldsymbol{\mu}_2$
	$e^{\boldsymbol{\mu}_1\omega_0 x} f(.)$	$f(.)e^{\boldsymbol{\mu}_1\omega_0 x}$	$\boldsymbol{\mu}_1$
y	$e^{\boldsymbol{\mu}_1\boldsymbol{\nu}_0 y} f(.)$	$f(.)e^{\boldsymbol{\mu}_1\boldsymbol{\nu}_0 y}$	
	$e^{\boldsymbol{\mu}_2\boldsymbol{\nu}_0 y} f(.)$	$f(.)e^{\boldsymbol{\mu}_2\boldsymbol{\nu}_0 y}$	$\boldsymbol{\mu}_2$

3.3. Relationships between Transforms

At the heart of all methods for determining inter-relationships between various QFTs is a decomposing process, of either the input function $f(.)$ or the exponential-kernel, so that their parts can be commuted into an alternate QFT form. Ell and Sangwine [5] used the *symplectic form* to link the single-axis, left and right, forward and reverse forms of the QFT *via* simplex and perplex complex sub-fields. Yeh [10] reworked these relationships and made further connections to the dual-axis, factored form QFT, but instead used *even-odd decomposition* of the input function. This approach essentially split each QFT into cosine and sine QFTs. Hitzer's [7] approach was to use the *split form* to factor the input function and kernel into factors with respect to the hypercomplex operators, so as to manipulate the result to an alternate QFT.

The inter-relationships between the various transform definitions not only give insight into the subsequent Fourier analysis, they are also used to simplify operator pairs. For example, the inter-relationships between the single-axis forms as given in Definitions 3.2 and 3.1 were used with the symplectic form (Def. 2.2) by Ell and Sangwine [5] to arrive at operator pairs for the convolution operator. Let the single-sided convolutions be defined as follows.

Definition 3.3 (Convolution [5]). The left- and right-sided convolution are defined, respectively, as

$$\begin{aligned} h_L \circ f(x,y) &= \iint_{\mathbb{R}^2} h_L(x',y')\, f(x-x',y-y')\, dx'dy', \\ f \circ h_R(x,y) &= \iint_{\mathbb{R}^2} f(x-x',y-y')\, h_R(x',y')\, dx'dy'. \end{aligned} \tag{3.8}$$

Now, let the QFT of the input function f be symplectically decomposed with respect to $\boldsymbol{\mu}_1$ as

$$\mathcal{F}^{\pm L}[f(x,y)] = F_1^{\pm L}[\omega,\nu] + F_2^{\pm L}[\omega,\nu]\,\boldsymbol{\mu}_2$$

and

$$\mathcal{F}^{\pm(L,R)}[h_R(x,y)] = H_R^{\pm(L,R)}[\omega,\nu],$$

then the right-convolution operator can be written as

$$\mathcal{F}^{\pm L}\left[f \circ h_R\right](\omega,\nu) = F_1^{\pm L}\left[\omega,\nu\right] H_R^{\pm L}\left[\omega,\nu\right] + F_2^{\pm L}\left[\omega,\nu\right] \boldsymbol{\mu}_2 H_R^{\mp L}\left[\omega,\nu\right].$$

Note the use of both forward and reverse QFT transforms. Such a compact operator formula would not be possible without the intermixing of QFT definitions.

4. Conclusions

The three currently defined quaternion Fourier transforms have been shown to be incomplete. By careful consideration of the underlying reasons for those three forms, this list has been extended to no less than twenty-two unique definitions. Future work may show that some of these definitions hold little of practical value or, without loss of generality, they may be reduced to but a few. The shift from iterated, channel-wise vector analysis to gestalt vector-image and vector-signal analysis shows promise. This promise raises several challenges:

1. Are there other, more suitable quaternion Fourier transform definitions?
2. Can these transforms be reduced to a salient few?
3. Are there additional decomposition methods, like the even-odd, split, and symplectic discussed herein, which can be used?
4. All the decomposition methods used to simplify the operator formulas are at odds with the very gestalt, holistic approach espoused, can this be done otherwise?

These questions will be the focus of future efforts.

Acknowledgment

Many thanks to *Iesus propheta a Nazareth Galilaeae.*

References

[1] H.S.M. Coxeter. Quaternions and reflections. *The American Mathematical Monthly*, 53(3):136–146, Mar. 1946.

[2] T.A. Ell. *Hypercomplex Spectral Transformations.* PhD thesis, University of Minnesota, June 1992.

[3] T.A. Ell and S.J. Sangwine. Decomposition of 2D hypercomplex Fourier transforms into pairs of complex Fourier transforms. In M. Gabbouj and P. Kuosmanen, editors, *Proceedings of EUSIPCO* 2000, *Tenth European Signal Processing Conference*, volume II, pages 1061–1064, Tampere, Finland, 5–8 Sept. 2000. European Association for Signal Processing.

[4] T.A. Ell and S.J. Sangwine. Hypercomplex Wiener-Khintchine theorem with application to color image correlation. In *IEEE International Conference on Image Processing* (*ICIP* 2000), volume II, pages 792–795, Vancouver, Canada, 11–14 Sept. 2000. IEEE.

[5] T.A. Ell and S.J. Sangwine. Hypercomplex Fourier transforms of color images. *IEEE Transactions on Image Processing*, 16(1):22–35, Jan. 2007.

[6] R.R. Ernst, G. Bodenhausen, and A. Wokaun. *Principles of Nuclear Magnetic Resonance in One and Two Dimensions.* International Series of Monographs on Chemistry. Oxford University Press, 1987.

[7] E. Hitzer. Quaternion Fourier transform on quaternion fields and generalizations. *Advances in Applied Clifford Algebras*, 17(3):497–517, May 2007.

[8] C.E. Moxey, S.J. Sangwine, and T.A. Ell. Vector phase correlation. *Electronics Letters*, 37(25):1513–1515, Dec. 2001.

[9] C.E. Moxey, S.J. Sangwine, and T.A. Ell. Hypercomplex correlation techniques for vector images. *IEEE Transactions on Signal Processing*, 51(7):1941–1953, July 2003.

[10] M.-H. Yeh. Relationships among various 2-D quaternion Fourier transforms. *IEEE Signal Processing Letters*, 15:669–672, Nov. 2008.

Todd Anthony Ell
Engineering Fellow
Goodrich, Sensors and Integrated Systems
14300 Judicial Road
Burnsville, MN 55306, USA
e-mail: `t.ell@ieee.org`

Quaternion and Clifford–Fourier Transforms and Wavelets
Trends in Mathematics, 15–39

2 The Orthogonal 2D Planes Split of Quaternions and Steerable Quaternion Fourier Transformations

Eckhard Hitzer and Stephen J. Sangwine

Abstract. The two-sided quaternionic Fourier transformation (QFT) was introduced in [2] for the analysis of 2D linear time-invariant partial-differential systems. In further theoretical investigations [4, 5] a special split of quaternions was introduced, then called ±split. In the current chapter we analyze this split further, interpret it geometrically as an *orthogonal* 2D *planes split* (OPS), and generalize it to a freely steerable split of $\mathbb{H}$ into two orthogonal 2D analysis planes. The new general form of the OPS split allows us to find new geometric interpretations for the action of the QFT on the signal. The second major result of this work is a variety of *new steerable forms* of the QFT, their geometric interpretation, and for each form, OPS split theorems, which allow fast and efficient numerical implementation with standard FFT software.

Mathematics Subject Classification (2010). Primary 16H05; secondary 42B10, 94A12, 94A08, 65R10.

Keywords. Quaternion signals, orthogonal 2D planes split, quaternion Fourier transformations, steerable transforms, geometric interpretation, fast implementations.

1. Introduction

The two-sided quaternionic Fourier transformation (QFT) was introduced in [2] for the analysis of 2D linear time-invariant partial-differential systems. Subsequently it has been applied in many fields, including colour image processing [8]. This led to further theoretical investigations [4, 5], where a special split of quaternions was introduced, then called the ±split. An interesting physical consequence was that this split resulted in a left and right travelling multivector wave packet analysis, when generalizing the QFT to a full spacetime Fourier transform (SFT). In the current chapter we investigate this split further, interpret it geometrically and generalize it to a *freely steerable*[1] split of $\mathbb{H}$ into two orthogonal 2D analysis planes. For reasons to become obvious we prefer to call it from now on the *orthogonal* 2D *planes split* (OPS).

The general form of the OPS split allows us to find new geometric interpretations for the action of the QFT on the signal. The second major result of this work is a variety of new forms of the QFT, their detailed geometric interpretation, and for each form, an OPS split theorem, which allows fast and efficient numerical implementation with standard FFT software. A preliminary formal investigation of these new OPS-QFTs can be found in [6].

The chapter is organized as follows. We first introduce in Section 2 several properties of quaternions together with a brief review of the $\pm$-*split* of [4, 5]. In Section 3 we generalize this split to a *freely steerable orthogonal* 2D *planes split* (OPS) of quaternions $\mathbb{H}$. In Section 4 we use the general OPS of Section 3 to generalize the two-sided QFT to a *new two-sided QFT* with freely *steerable analysis planes*, complete with a detailed *local geometric transformation interpretation.* The geometric interpretation of the OPS in Section 3 further allows the construction of a new type of *steerable QFT with a direct phase angle interpretation.* In Section 5 we finally investigate *new steerable QFTs involving quaternion conjugation.* Their local geometric interpretation crucially relies on the notion of 4D *rotary reflections.*

2. Orthogonal Planes Split of Quaternions with Two Orthonormal Pure Unit Quaternions

Gauss, Rodrigues and Hamilton's four-dimensional (4D) quaternion algebra $\mathbb{H}$ is defined over $\mathbb{R}$ with three imaginary units:

$$\begin{gathered} \boldsymbol{ij} = -\boldsymbol{ji} = \boldsymbol{k}, \qquad \boldsymbol{jk} = -\boldsymbol{kj} = \boldsymbol{i}, \qquad \boldsymbol{ki} = -\boldsymbol{ik} = \boldsymbol{j}, \\ \boldsymbol{i}^2 = \boldsymbol{j}^2 = \boldsymbol{k}^2 = \boldsymbol{ijk} = -1. \end{gathered} \tag{2.1}$$

Every quaternion can be written explicitly as

$$q = q_r + q_i\boldsymbol{i} + q_j\boldsymbol{j} + q_k\boldsymbol{k} \in \mathbb{H}, \quad q_r, q_i, q_j, q_k \in \mathbb{R}, \tag{2.2}$$

and has a *quaternion conjugate* (equivalent[2] to Clifford conjugation in $C\ell_{3,0}^+$ and $C\ell_{0,2}$)

$$\overline{q} = q_r - q_i\boldsymbol{i} - q_j\boldsymbol{j} - q_k\boldsymbol{k}, \quad \overline{pq} = \overline{q}\,\overline{p}, \tag{2.3}$$

which leaves the scalar part q_r unchanged. This leads to the *norm* of $q \in \mathbb{H}$

$$|q| = \sqrt{q\overline{q}} = \sqrt{q_r^2 + q_i^2 + q_j^2 + q_k^2}, \qquad |pq| = |p|\,|q|\,. \tag{2.4}$$

The part $\mathbf{V}(q) = q - q_r = \frac{1}{2}(q - \overline{q}) = q_i\boldsymbol{i} + q_j\boldsymbol{j} + q_k\boldsymbol{k}$ is called a *pure* quaternion, and it squares to the negative number $-(q_i^2 + q_j^2 + q_k^2)$. Every unit quaternion (*i.e.*,

---

[1] Compare Section 3.4, in particular Theorem 3.5.

[2] This may be important in generalisations of the QFT, such as to a space-time Fourier transform in [4], or a general two-sided Clifford–Fourier transform in [7].

$|q| = 1$) can be written as:

$$q = q_r + q_i\boldsymbol{i} + q_j\boldsymbol{j} + q_k\boldsymbol{k} = q_r + \sqrt{q_i^2 + q_j^2 + q_k^2}\,\boldsymbol{\mu}(q) \\ = \cos\alpha + \boldsymbol{\mu}(q)\sin\alpha = e^{\alpha\,\boldsymbol{\mu}(q)}, \tag{2.5}$$

where

$$\cos\alpha = q_r, \qquad \sin\alpha = \sqrt{q_i^2 + q_j^2 + q_k^2}, \\ \boldsymbol{\mu}(q) = \frac{\mathbf{V}(q)}{|q|} = \frac{q_i\boldsymbol{i} + q_j\boldsymbol{j} + q_k\boldsymbol{k}}{\sqrt{q_i^2 + q_j^2 + q_k^2}}, \qquad \text{and} \qquad \boldsymbol{\mu}(q)^2 = -1. \tag{2.6}$$

The *inverse* of a non-zero quaternion is

$$q^{-1} = \frac{\overline{q}}{|q|^2} = \frac{\overline{q}}{q\overline{q}}. \tag{2.7}$$

The *scalar part* of a quaternion is defined as

$$\mathrm{S}(q) = q_r = \frac{1}{2}(q + \overline{q}), \tag{2.8}$$

with *symmetries*

$$\mathrm{S}(pq) = \mathrm{S}(qp) = p_r q_r - p_i q_i - p_j q_j - p_k q_k, \quad \mathrm{S}(q) = \mathrm{S}(\overline{q}), \quad \forall p, q \in \mathbb{H}, \tag{2.9}$$

and *linearity*

$$\mathrm{S}(\alpha p + \beta q) = \alpha\,\mathrm{S}(p) + \beta\,\mathrm{S}(q) = \alpha p_r + \beta q_r, \quad \forall p, q \in \mathbb{H}, \ \alpha, \beta \in \mathbb{R}. \tag{2.10}$$

The scalar part and the quaternion conjugate allow the definition of the $\mathbb{R}^4$ *inner product*[3] of two quaternions p, q as

$$\mathrm{S}(p\overline{q}) = p_r q_r + p_i q_i + p_j q_j + p_k q_k \in \mathbb{R}. \tag{2.11}$$

Definition 2.1 (Orthogonality of quaternions). Two quaternions $p, q \in \mathbb{H}$ are *orthogonal* $p \perp q$, if and only if the inner product $\mathrm{S}(p\overline{q}) = 0$.

The *orthogonal*[4] *2D planes split* (OPS) of quaternions with respect to the orthonormal pure unit quaternions $\boldsymbol{i}, \boldsymbol{j}$ [4, 5] is defined by

$$q = q_+ + q_-, \quad q_\pm = \frac{1}{2}(q \pm \boldsymbol{i}q\boldsymbol{j}). \tag{2.12}$$

Explicitly in real components $q_r, q_i, q_j, q_k \in \mathbb{R}$ using (2.1) we get

$$q_\pm = \{q_r \pm q_k + \boldsymbol{i}(q_i \mp q_j)\}\frac{1 \pm \boldsymbol{k}}{2} = \frac{1 \pm \boldsymbol{k}}{2}\{q_r \pm q_k + \boldsymbol{j}(q_j \mp q_i)\}. \tag{2.13}$$

This leads to the following new Pythagorean *modulus identity* [5]

Lemma 2.2 (Modulus identity). *For* $q \in \mathbb{H}$

$$|q|^2 = |q_-|^2 + |q_+|^2. \tag{2.14}$$

---

[3]Note that we do not use the notation $p \cdot q$, which is unconventional for full quaternions.

[4]Compare Lemma 2.3.

Lemma 2.3 (Orthogonality of OPS split parts). *Given any two quaternions* $p, q \in \mathbb{H}$ *and applying the OPS of* (2.12) *the resulting parts are orthogonal*

$$\mathrm{S}(p_+\overline{q_-}) = 0, \qquad \mathrm{S}(p_-\overline{q_+}) = 0, \tag{2.15}$$

i.e., $p_+ \perp q_-$ *and* $p_- \perp q_+$.

In Lemma 2.3 (proved in [5]) the second identity follows from the first by $\mathrm{S}(\overline{x}) = \mathrm{S}(x)$, $\forall x \in \mathbb{H}$, and $\overline{p_-\overline{q_+}} = q_+\overline{p_-}$.

It is evident, that instead of $\boldsymbol{i}, \boldsymbol{j}$, any pair of orthonormal pure quaternions can be used to produce an analogous split. This is a first indication, that the OPS of (2.12) is in fact *steerable*. We observe, that $\boldsymbol{i}q\boldsymbol{j} = q_+ - q_-$, *i.e.*, under the map $\boldsymbol{i}(\,)\boldsymbol{j}$ the q_+ part is invariant, the q_- part changes sign. Both parts are according to (2.13) two-dimensional, and by Lemma 2.3 they span two completely orthogonal planes. The q_+-plane is spanned by the orthogonal quaternions $\{\boldsymbol{i} - \boldsymbol{j}, 1 + \boldsymbol{i}\boldsymbol{j} = 1 + \boldsymbol{k}\}$, whereas the q_--plane is, *e.g.*, spanned by $\{\boldsymbol{i} + \boldsymbol{j}, 1 - \boldsymbol{i}\boldsymbol{j} = 1 - \boldsymbol{k}\}$, *i.e.*, we have the two 2D subspace bases

$$q_+\text{-basis: } \{\boldsymbol{i} - \boldsymbol{j}, 1 + \boldsymbol{i}\boldsymbol{j} = 1 + \boldsymbol{k}\}, \qquad q_-\text{-basis: } \{\boldsymbol{i} + \boldsymbol{j}, 1 - \boldsymbol{i}\boldsymbol{j} = 1 - \boldsymbol{k}\}. \tag{2.16}$$

Note that all basis vectors of (2.16)

$$\{\boldsymbol{i} - \boldsymbol{j}, 1 + \boldsymbol{i}\boldsymbol{j}, \boldsymbol{i} + \boldsymbol{j}, 1 - \boldsymbol{i}\boldsymbol{j}\} \tag{2.17}$$

together form an orthogonal basis of $\mathbb{H}$ interpreted as $\mathbb{R}^4$.

The map $\boldsymbol{i}(\,)\boldsymbol{j}$ rotates the q_--plane by 180° around the 2D q_+ axis plane. Note that in agreement with its geometric interpretation, the map $\boldsymbol{i}(\,)\boldsymbol{j}$ is an *involution*, because applying it twice leads to identity

$$\boldsymbol{i}(\boldsymbol{i}q\boldsymbol{j})\boldsymbol{j} = \boldsymbol{i}^2 q \boldsymbol{j}^2 = (-1)^2 q = q. \tag{2.18}$$

3. General Orthogonal 2D Planes Split

We will study generalizations of the OPS split by replacing $\boldsymbol{i}, \boldsymbol{j}$ by arbitrary unit quaternions f, g. Even with this generalization, the map $f(\,)g$ continues to be an involution, because $f^2 q g^2 = (-1)^2 q = q$. For clarity we study the cases $f \neq \pm g$, and $f = g$ separately, though they have a lot in common, and do not always need to be distinguished in specific applications.

3.1. Orthogonal 2D Planes Split Using Two Linearly Independent Pure Unit Quaternions

Our result is now, that all these properties hold, even if in the above considerations the pair $\boldsymbol{i}, \boldsymbol{j}$ is replaced by an arbitrary pair of linearly independent nonorthogonal pure quaternions f, g, $f^2 = g^2 = -1, f \neq \pm g$. The OPS is then *re-defined* with respect to the linearly independent pure unit quaternions f, g as

$$q_\pm = \frac{1}{2}(q \pm f q g). \tag{3.1}$$

Equation (2.12) is a special case with $f = \boldsymbol{i}, g = \boldsymbol{j}$. We observe from (3.1), that $fqg = q_+ - q_-$, *i.e.*, under the map $f(\,)g$ the q_+ part is invariant, but the q_- part changes sign

$$fq_\pm g = \frac{1}{2}(fqg \pm f^2qg^2) = \frac{1}{2}(fqg \pm q) = \pm\frac{1}{2}(q \pm fqg) = \pm q_\pm. \tag{3.2}$$

We now show that even for (3.1) both parts are two-dimensional, and span two completely orthogonal planes. The q_+-plane is spanned[5] by the *orthogonal* pair of quaternions $\{f-g, 1+fg\}$:

$$\begin{aligned} &\mathrm{S}\Big((f-g)\overline{(1+fg)}\Big) = \mathrm{S}((f-g)(1+(-g)(-f))) \\ &= \mathrm{S}(f+fgf-g-g^2f) \overset{(2.9)}{=} \mathrm{S}(f+f^2g-g+f) = 2\,\mathrm{S}(f-g) = 0, \end{aligned} \tag{3.3}$$

whereas the q_--plane is, *e.g.*, spanned by $\{f+g, 1-fg\}$. The quaternions $f+g, 1-fg$ can be proved to be mutually *orthogonal* by simply replacing $g \to -g$ in (3.3). Note that we have

$$\begin{aligned} f(f-g)g &= f^2g - fg^2 = -g+f = f-g, \\ f(1+fg)g &= fg + f^2g^2 = fg+1 = 1+fg, \end{aligned} \tag{3.4}$$

as well as

$$\begin{aligned} f(f+g)g &= f^2g + fg^2 = -g-f = -(f+g), \\ f(1-fg)g &= fg - f^2g^2 = fg-1 = -(1-fg). \end{aligned} \tag{3.5}$$

We now want to generalize Lemma 2.3.

Lemma 3.1 (Orthogonality of two OPS planes). *Given any two quaternions $q, p \in \mathbb{H}$ and applying the OPS (3.1) with respect to two linearly independent pure unit quaternions f, g we get zero for the scalar part of the mixed products*

$$\mathrm{S}(p_+\overline{q_-}) = 0, \qquad \mathrm{S}(p_-\overline{q_+}) = 0. \tag{3.6}$$

We prove the first identity, the second follows from $\mathrm{S}(x) = \mathrm{S}(\overline{x})$.

$$\begin{aligned} \mathrm{S}(p_+\overline{q_-}) &= \frac{1}{4}\,\mathrm{S}((p+fpg)(\overline{q}-g\overline{q}f)) = \frac{1}{4}\,\mathrm{S}(p\overline{q} - fpgg\overline{q}f + fpg\overline{q} - pg\overline{q}f) \\ &\overset{(2.10),(2.9)}{=} \frac{1}{4}\,\mathrm{S}(p\overline{q} - p\overline{q} + pg\overline{q}f - pg\overline{q}f) = 0. \end{aligned} \tag{3.7}$$

Thus the set

$$\{f-g, 1+fg, f+g, 1-fg\} \tag{3.8}$$

forms a 4D orthogonal basis of $\mathbb{H}$ interpreted by (2.11) as $\mathbb{R}^4$, where we have for the orthogonal 2D planes the subspace bases:

$$q_+\text{-basis: } \{f-g, 1+fg\}, \qquad q_-\text{-basis: } \{f+g, 1-fg\}. \tag{3.9}$$

---

[5]For $f = \boldsymbol{i}, g = \boldsymbol{j}$ this is in agreement with (2.13) and (2.16)!

We can therefore use the following representation for every $q \in \mathbb{H}$ by means of four real coefficients $q_1, q_2, q_3, q_4 \in \mathbb{R}$

$$q = q_1(1 + fg) + q_2(f - g) + q_3(1 - fg) + q_4(f + g), \tag{3.10}$$

where

$$\begin{aligned} q_1 &= \mathrm{S}\big(q(1 + fg)^{-1}\big), & q_2 &= \mathrm{S}\big(q(f - g)^{-1}\big), \\ q_3 &= \mathrm{S}\big(q(1 - fg)^{-1}\big), & q_4 &= \mathrm{S}\big(q(f + g)^{-1}\big). \end{aligned} \tag{3.11}$$

As an example we have for $f = \boldsymbol{i}, g = \boldsymbol{j}$ according to (2.13) the coefficients for the decomposition with respect to the orthogonal basis (3.8)

$$q_1 = \frac{1}{2}(q_r + q_k), \quad q_2 = \frac{1}{2}(q_i - q_j), \quad q_3 = \frac{1}{2}(q_r - q_k), \quad q_4 = \frac{1}{2}(q_i + q_j). \tag{3.12}$$

Moreover, using

$$f - g = f(1 + fg) = (1 + fg)(-g), \quad f + g = f(1 - fg) = (1 - fg)g, \tag{3.13}$$

we have the following left and right factoring properties

$$\begin{aligned} q_+ = q_1(1 + fg) + q_2(f - g) &= (q_1 + q_2 f)(1 + fg) \\ &= (1 + fg)(q_1 - q_2 g), \end{aligned} \tag{3.14}$$

$$\begin{aligned} q_- = q_3(1 - fg) + q_4(f + g) &= (q_3 + q_4 f)(1 - fg) \\ &= (1 - fg)(q_3 + q_4 g). \end{aligned} \tag{3.15}$$

Equations (3.4) and (3.5) further show that the map $f(\,)g$ rotates the q_--plane by $180°$ around the q_+ axis plane. We found that our interpretation of the map $f(\,)g$ is in perfect agreement with Coxeter's notion of *half-turn* in [1]. This opens the way for new types of QFTs, where the pair of square roots of -1 involved does not necessarily need to be orthogonal.

Before suggesting a generalization of the QFT, we will establish a new set of very useful algebraic identities. Based on (3.14) and (3.15) we get for $\alpha, \beta \in \mathbb{R}$

$$\begin{aligned} e^{\alpha f} q e^{\beta g} &= e^{\alpha f} q_+ e^{\beta g} + e^{\alpha f} q_- e^{\beta g}, \\ e^{\alpha f} q_+ e^{\beta g} &= (q_1 + q_2 f) e^{\alpha f}(1 + fg) e^{\beta g} = e^{\alpha f}(1 + fg) e^{\beta g}(q_1 - q_2 g), \\ e^{\alpha f} q_- e^{\beta g} &= (q_3 + q_4 f) e^{\alpha f}(1 - fg) e^{\beta g} = e^{\alpha f}(1 - fg) e^{\beta g}(q_3 + q_4 g). \end{aligned} \tag{3.16}$$

Using (3.14) again we obtain

$$\begin{aligned} e^{\alpha f}(1 + fg) &= (\cos\alpha + f \sin\alpha)(1 + fg) \\ &\overset{(3.14)}{=} (1 + fg)(\cos\alpha - g\sin\alpha) = (1 + fg)e^{-\alpha g}, \end{aligned} \tag{3.17}$$

where we set $q_1 = \cos\alpha$, $q_2 = \sin\alpha$ for applying (3.14). Replacing in (3.17) $-\alpha \to \beta$ we get

$$e^{-\beta f}(1 + fg) = (1 + fg)e^{\beta g}, \tag{3.18}$$

Furthermore, replacing in (3.17) $g \to -g$ and subsequently $\alpha \to \beta$ we get

$$\begin{aligned} e^{\alpha f}(1-fg) &= (1-fg)e^{\alpha g}, \\ e^{\beta f}(1-fg) &= (1-fg)e^{\beta g}. \end{aligned} \tag{3.19}$$

Applying (3.14), (3.16), (3.17) and (3.18) we can rewrite

$$\begin{aligned} e^{\alpha f}q_+e^{\beta g} &\overset{(3.16)}{=} (q_1+q_2f)e^{\alpha f}(1+fg)e^{\beta g} \overset{(3.17)}{=} (q_1+q_2f)(1+fg)e^{(\beta-\alpha)g} \\ &\overset{(3.14)}{=} q_+e^{(\beta-\alpha)g}, \end{aligned} \tag{3.20}$$

or equivalently as

$$\begin{aligned} e^{\alpha f}q_+e^{\beta g} &\overset{(3.16)}{=} e^{\alpha f}(1+fg)e^{\beta g}(q_1-q_2g) \overset{(3.18)}{=} e^{(\alpha-\beta)f}(1+fg)(q_1-q_2g) \\ &\overset{(3.14)}{=} e^{(\alpha-\beta)f}q_+. \end{aligned} \tag{3.21}$$

In the same way by changing $g \to -g, \beta \to -\beta$ in (3.20) and (3.21) we can rewrite

$$e^{\alpha f}q_-e^{\beta g} = e^{(\alpha+\beta)f}q_- = q_-e^{(\alpha+\beta)g}. \tag{3.22}$$

The result is therefore

$$e^{\alpha f}q_\pm e^{\beta g} = q_\pm e^{(\beta\mp\alpha)g} = e^{(\alpha\mp\beta)f}q_\pm. \tag{3.23}$$

3.2. Orthogonal 2D Planes Split Using One Pure Unit Quaternion

We now treat the case for $g = f, f^2 = -1$. We then have the map $f(\,)f$, and the OPS split with respect to $f \in \mathbb{H}, f^2 = -1$,

$$q_\pm = \frac{1}{2}(q \pm fqf). \tag{3.24}$$

The pure quaternion $\boldsymbol{i}$ can be rotated by $R = \boldsymbol{i}(\boldsymbol{i}+f)$, see (3.27), into the quaternion unit f and back. Therefore studying the map $\boldsymbol{i}(\,)\boldsymbol{i}$ is, up to the constant rotation between $\boldsymbol{i}$ and f, the same as studying $f(\,)f$. This gives

$$\boldsymbol{i}q\boldsymbol{i} = \boldsymbol{i}(q_r + q_i\boldsymbol{i} + q_j\boldsymbol{j} + q_k\boldsymbol{k})\boldsymbol{i} = -q_r - q_i\boldsymbol{i} + q_j\boldsymbol{j} + q_k\boldsymbol{k}. \tag{3.25}$$

The OPS with respect to $f = g = \boldsymbol{i}$ gives

$$q_\pm = \frac{1}{2}(q \pm \boldsymbol{i}q\boldsymbol{i}), \quad q_+ = q_j\boldsymbol{j} + q_k\boldsymbol{k} = (q_j + q_k\boldsymbol{i})\boldsymbol{j}, \quad q_- = q_r + q_i\boldsymbol{i}, \tag{3.26}$$

where the q_+-plane is two-dimensional and manifestly orthogonal to the 2D q_--plane. This form (3.26) of the OPS is therefore identical to the quaternionic simplex/perplex split of [3].

For $g = f$ the q_--plane is always spanned by $\{1, f\}$. The rotation operator $R = \boldsymbol{i}(\boldsymbol{i}+f)$, with squared norm $|R|^2 = |\boldsymbol{i}(\boldsymbol{i}+f)|^2 = |(\boldsymbol{i}+f)|^2 = -(\boldsymbol{i}+f)^2$, rotates $\boldsymbol{i}$ into f according to

$$\begin{aligned} R^{-1}\boldsymbol{i}R = \frac{\overline{R}}{|R|^2}\boldsymbol{i}R &= \frac{(\boldsymbol{i}+f)\boldsymbol{i}\boldsymbol{i}\boldsymbol{i}(\boldsymbol{i}+f)}{-(\boldsymbol{i}+f)^2} = \frac{(\boldsymbol{i}+f)\boldsymbol{i}(\boldsymbol{i}(-f)+1)f}{(\boldsymbol{i}+f)^2} \\ &= \frac{(\boldsymbol{i}+f)(f+\boldsymbol{i})f}{(\boldsymbol{i}+f)^2} = f. \end{aligned} \tag{3.27}$$

The rotation R leaves 1 invariant and thus rotates the whole $\{1, \boldsymbol{i}\}$ plane into the q_--plane spanned by $\{1, f\}$. Consequently R also rotates the $\{\boldsymbol{j}, \boldsymbol{k}\}$ plane into the q_+-plane spanned by $\{\boldsymbol{j}' = R^{-1}\boldsymbol{j}R,\ \boldsymbol{k}' = R^{-1}\boldsymbol{k}R\}$. We thus constructively obtain the fully *orthonormal* 4D basis of $\mathbb{H}$ as

$$\{1, f, \boldsymbol{j}', \boldsymbol{k}'\} = R^{-1}\{1, \boldsymbol{i}, \boldsymbol{j}, \boldsymbol{k}\}R, \qquad R = \boldsymbol{i}(\boldsymbol{i} + f), \tag{3.28}$$

for any chosen pure unit quaternion f. We further have, for the orthogonal 2D planes created in (3.24) the subspace bases:

$$q_+\text{-basis: } \{\boldsymbol{j}', \boldsymbol{k}'\}, \qquad q_-\text{-basis: } \{1, f\}. \tag{3.29}$$

The rotation R (an orthogonal transformation!) of (3.27) preserves the fundamental quaternionic orthonormality and the anticommutation relations

$$f\boldsymbol{j}' = \boldsymbol{k}' = -\boldsymbol{j}'f, \qquad \boldsymbol{k}'f = \boldsymbol{j}' = -f\boldsymbol{k}' \qquad \boldsymbol{j}'\boldsymbol{k}' = f = -\boldsymbol{k}'\boldsymbol{j}'. \tag{3.30}$$

Hence

$$fqf = f(q_+ + q_-)f = q_+ - q_-, \quad i.e., \quad fq_\pm f = \pm q_\pm, \tag{3.31}$$

which represents again a half-turn by 180° in the 2D q_--plane around the 2D q_+-plane (as axis).

Figures 1 and 2 illustrate this decomposition for the case where f is the unit pure quaternion $\frac{1}{\sqrt{3}}(\boldsymbol{i} + \boldsymbol{j} + \boldsymbol{k})$. This decomposition corresponds (for pure quaternions) to the classical *luminance-chrominance* decomposition used in colour image processing, as illustrated, for example, in [3, Figure 2]. Three hundred unit quaternions randomly oriented in 4-space were decomposed. Figure 1 shows the three hundred points in 4-space, projected onto the six orthogonal planes $\{e, \boldsymbol{i}'\}, \{e, \boldsymbol{j}'\}, \{e, \boldsymbol{k}'\}, \{\boldsymbol{i}', \boldsymbol{j}'\}, \{\boldsymbol{j}', \boldsymbol{k}'\}, \{\boldsymbol{k}', \boldsymbol{i}'\}$ where $e = 1$ and $\boldsymbol{i}' = f$, as given in (3.28). The six views at the top show the q_+-plane, and the six below show the q_--plane.

Figure 2 shows the vector parts of the decomposed quaternions. The basis for the plot is $\{\boldsymbol{i}', \boldsymbol{j}', \boldsymbol{k}'\}$, where $\boldsymbol{i}' = f$ as given in (3.28). The green circles show the components in the $\{1, f\}$ plane, which intersects the 3-space of the vector part only along the line f (which is the *luminance* or *grey line* of colour image pixels). The red line on the figure corresponds to f. The blue circles show the components in the $\{\boldsymbol{j}', \boldsymbol{k}'\}$ plane, which is entirely within the 3-space. It is orthogonal to f and corresponds to the *chrominance plane* of colour image processing.

The next question is the influence the current OPS (3.24) has for left and right exponential factors of the form

$$e^{\alpha f} q_\pm e^{\beta f}. \tag{3.32}$$

We learn from (3.30) that

$$e^{\alpha f} q_\pm e^{\beta f} = e^{(\alpha \mp \beta) f} q_\pm = q_\pm e^{(\beta \mp \alpha) f}, \tag{3.33}$$

which is identical to (3.23), if we insert $g = f$ in (3.23).

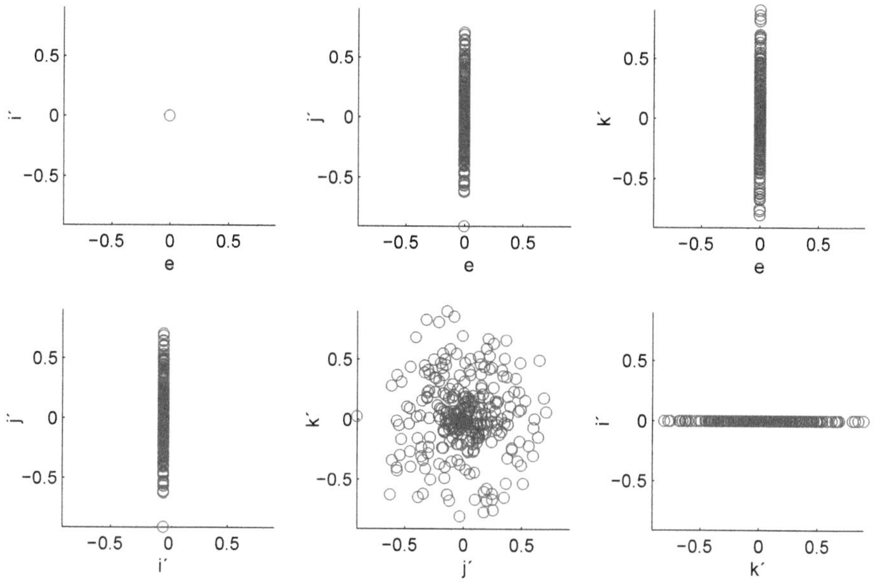

(a) q_+ component

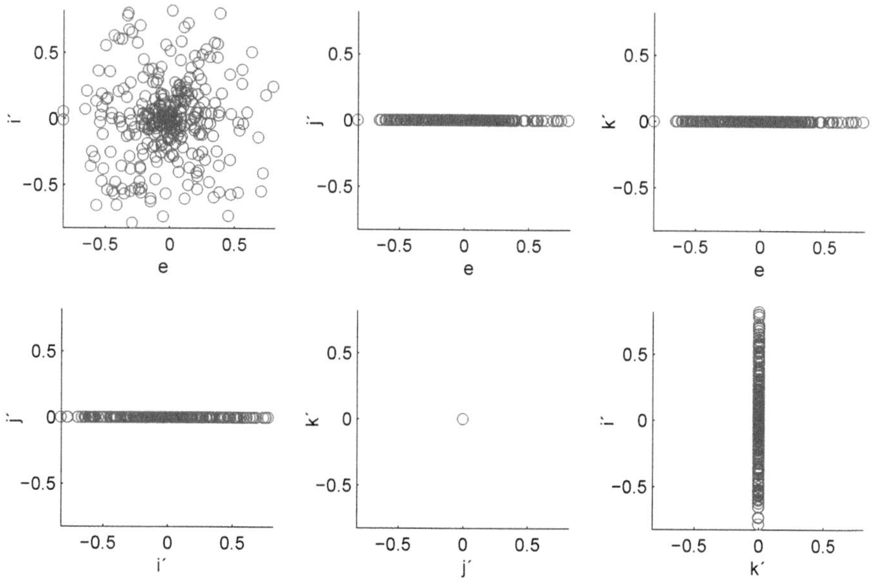

(b) q_- component

FIGURE 1. 4D scatter plot of quaternions decomposed using the orthogonal planes split of (3.24) with one unit pure quaternion $f = \boldsymbol{i}' = \frac{1}{\sqrt{3}}(\boldsymbol{i} + \boldsymbol{j} + \boldsymbol{k}) = g$.

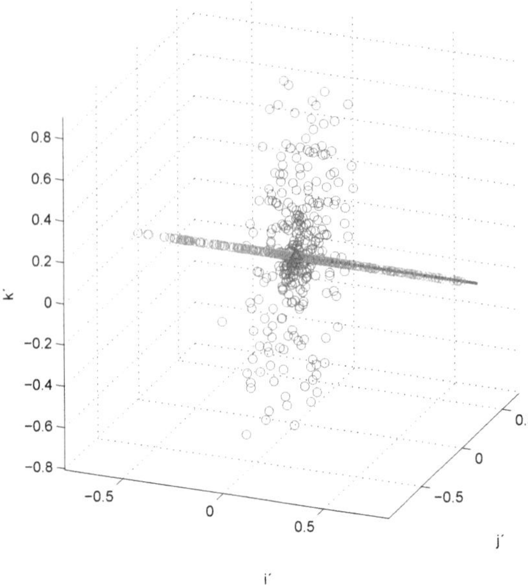

FIGURE 2. Scatter plot of vector parts of quaternions decomposed using the orthogonal planes split of (3.24) with one pure unit quaternion $f = \boldsymbol{i}' = \frac{1}{\sqrt{3}}(\boldsymbol{i}+\boldsymbol{j}+\boldsymbol{k}) = g$. The red line corresponds to the direction of f.

Next, we consider $g = -f, f^2 = -1$. We then have the map $f(\,)(-f)$, and the OPS split with respect to $f, -f \in \mathbb{H}, f^2 = -1$,

$$q_\pm = \frac{1}{2}(q \pm fq(-f)) = \frac{1}{2}(q \mp fqf). \tag{3.34}$$

Again we can study $f = \boldsymbol{i}$ first, because for general pure unit quaternions f the unit quaternion $\boldsymbol{i}$ can be rotated by (3.27) into the quaternion unit f and back. Therefore studying the map $\boldsymbol{i}(\,)(-\boldsymbol{i})$ is up to the constant rotation R of (3.27) the same as studying $f(\,)(-f)$. This gives the map

$$\boldsymbol{i}q(-\boldsymbol{i}) = \boldsymbol{i}(q_r + q_i\boldsymbol{i} + q_j\boldsymbol{j} + q_k\boldsymbol{k})(-\boldsymbol{i}) = q_r + q_i\boldsymbol{i} - q_j\boldsymbol{j} - q_k\boldsymbol{k}. \tag{3.35}$$

The OPS with respect to $f = \boldsymbol{i}, g = -\boldsymbol{i}$ gives

$$q_\pm = \frac{1}{2}(q \pm \boldsymbol{i}q(-\boldsymbol{i})), \quad q_- = q_j\boldsymbol{j} + q_k\boldsymbol{k} = (q_j + q_k\boldsymbol{i})\boldsymbol{j}, \quad q_+ = q_r + q_i\boldsymbol{i}, \tag{3.36}$$

where, compared to $f = g = \boldsymbol{i}$, the 2D q_+-plane and the 2D q_--planes appear interchanged. The form (3.36) of the OPS is again identical to the quaternionic simplex/perplex split of [3], but the simplex and perplex parts appear interchanged.

For $g = -f$ the q_+-plane is always spanned by $\{1, f\}$. The rotation R of (3.27) rotates $\boldsymbol{i}$ into f and leaves 1 invariant and thus rotates the whole $\{1, \boldsymbol{i}\}$

plane into the q_+-plane spanned by $\{1, f\}$. Consequently, R of (3.27) also rotates the $\{\boldsymbol{j}, \boldsymbol{k}\}$ plane into the q_--plane spanned by $\{\boldsymbol{j}' = R^{-1}\boldsymbol{j}R,\ \boldsymbol{k}' = R^{-1}\boldsymbol{k}R\}$.

We therefore have for the orthogonal 2D planes created in (3.34) the subspace bases:

$$q_+\text{-basis: } \{1, f\}, \qquad q_-\text{-basis: } \{\boldsymbol{j}', \boldsymbol{k}'\}. \tag{3.37}$$

We again obtain the fully *orthonormal* 4D basis (3.28) of $\mathbb{H}$, preserving the fundamental quaternionic orthonormality and the anticommutation relations (3.30).

Hence for (3.34)

$$fq(-f) = f(q_+ + q_-)(-f) = q_+ - q_-, \quad i.e., \quad fq_\pm(-f) = \pm q_\pm, \tag{3.38}$$

which represents again a half-turn by 180° in the 2D q_--plane around the 2D q_+-plane (as axis).

The remaining question is the influence the current OPS (3.34) has for left and right exponential factors of the form

$$e^{\alpha f} q_\pm e^{-\beta f}. \tag{3.39}$$

We learn from (3.30) that

$$e^{\alpha f} q_\pm e^{-\beta f} = e^{(\alpha \mp \beta) f} q_\pm = q_\pm e^{-(\beta \mp \alpha) f}, \tag{3.40}$$

which is identical to (3.23), if we insert $g = -f$ in (3.23).

For (3.23) therefore, we do not any longer need to distinguish the cases $f \neq \pm g$ and $f = \pm g$. This motivates us to a general OPS definition for any pair of pure quaternions f, g, and we get a general lemma.

Definition 3.2 (General orthogonal 2D planes split). Let $f, g \in \mathbb{H}$ be an arbitrary pair of pure quaternions f, g, $f^2 = g^2 = -1$, including the cases $f = \pm g$. The general OPS is then defined with respect to the two pure unit quaternions f, g as

$$q_\pm = \frac{1}{2}(q \pm fqg). \tag{3.41}$$

Remark 3.3. The three generalized OPS (3.1), (3.24), and (3.34) are formally identical and are now subsumed in (3.41) of Definition 3.2, where the values $g = \pm f$ are explicitly included, *i.e.*, any pair of pure unit quaternions $f, g \in \mathbb{H}$, $f^2 = g^2 = -1$, is admissible.

Lemma 3.4. *With respect to the general OPS of Definition* 3.2 *we have for left and right exponential factors the identity*

$$e^{\alpha f} q_\pm e^{\beta g} = q_\pm e^{(\beta \mp \alpha) g} = e^{(\alpha \mp \beta) f} q_\pm. \tag{3.42}$$

3.3. Geometric Interpretation of Left and Right Exponential Factors in f, g

We obtain the following general *geometric interpretation*. The map $f(\,)g$ always represents a rotation by angle π in the q_--plane (around the q_+-plane), the map $f^t(\,)g^t$, $t \in \mathbb{R}$, similarly represents a rotation by angle $t\pi$ in the q_--plane (around the q_+-plane as axis). Replacing[6] $g \to -g$ in the map $f(\,)g$ we further find that

$$fq_\pm(-g) = \mp q_\pm. \tag{3.43}$$

Therefore the map $f(\,)(-g) = f(\,)g^{-1}$, because $g^{-1} = -g$, represents a rotation by angle π in the q_+-plane (around the q_--plane), exchanging the roles of 2D rotation plane and 2D rotation axis. Similarly, the map $f^s(\,)g^{-s}$, $s \in \mathbb{R}$, represents a rotation by angle $s\pi$ in the q_+-plane (around the q_--plane as axis).

The product of these two rotations gives

$$\begin{aligned} f^{t+s}qg^{t-s} &= e^{(t+s)\frac{\pi}{2}f}qe^{(t-s)\frac{\pi}{2}g} = e^{\alpha f}qe^{\beta g}, \\ \alpha &= (t+s)\frac{\pi}{2}, \qquad \beta = (t-s)\frac{\pi}{2}, \end{aligned} \tag{3.44}$$

where based on (2.5) we used the identities $f = e^{\frac{\pi}{2}f}$ and $g = e^{\frac{\pi}{2}g}$.

The *geometric interpretation* of (3.44) is a rotation by angle $\alpha + \beta$ in the q_--plane (around the q_+-plane), and a second rotation by angle $\alpha - \beta$ in the q_+-plane (around the q_--plane). For $\alpha = \beta = \pi/2$ we recover the map $f(\,)g$, and for $\alpha = -\beta = \pi/2$ we recover the map $f(\,)g^{-1}$.

3.4. Determination of f, g for Given Steerable Pair of Orthogonal 2D Planes

Equations (3.9), (3.29), and (3.37) tell us how the pair of pure unit quaternions $f, g \in \mathbb{H}$ used in the general OPS of Definition 3.2, leads to an explicit basis for the resulting two orthogonal 2D planes, the q_+-plane and the q_--plane. We now ask the *opposite* question: how can we determine from a given steerable pair of orthogonal 2D planes in $\mathbb{H}$ the pair of pure unit quaternions $f, g \in \mathbb{H}$, which splits $\mathbb{H}$ exactly into this given pair of orthogonal 2D planes?

To answer this question, we first observe that in a 4D space it is sufficient to know only one 2D plane explicitly, specified, *e.g.*, by a pair of orthogonal unit quaternions $a, b \in \mathbb{H}$, $|a| = |b| = 1$, and without restriction of generality $b^2 = -1$, *i.e.*, b can be a pure unit quaternion $b = \boldsymbol{\mu}(b)$. But for $a = \cos\alpha + \boldsymbol{\mu}(a)\sin\alpha$, compare (2.6), we must distinguish $\mathrm{S}(a) = \cos\alpha \neq 0$ and $\mathrm{S}(a) = \cos\alpha = 0$, *i.e.*, of a also being a pure quaternion with $a^2 = -1$. The second orthogonal 2D plane is then simply the *orthogonal complement* in $\mathbb{H}$ to the a, b-plane.

Let us first treat the case $\mathrm{S}(a) = \cos\alpha \neq 0$. We set

$$f := ab, \qquad g := \overline{a}b. \tag{3.45}$$

[6] Alternatively and equivalently we could replace $f \to -f$ instead of $g \to -g$.

With this setting we get for the basis of the q_--plane

$$\begin{aligned} f+g &= ab+\overline{a}b = 2\,\mathrm{S}(a)\,b, \\ 1-fg &= 1-ab\overline{a}b = 1-a^2b^2 = 1+a^2 \\ &= 1+\cos^2\alpha - \sin^2\alpha + 2\boldsymbol{\mu}(a)\cos\alpha\sin\alpha \\ &= 2\cos\alpha(\cos\alpha+\boldsymbol{\mu}(a)\sin\alpha) = 2\,\mathrm{S}(a)\,a. \end{aligned} \tag{3.46}$$

For the equality $ab\overline{a}b = a^2b^2$ we used the orthogonality of a, b, which means that the vector part of a must be orthogonal to the pure unit quaternion b, *i.e.*, it must anticommute with b

$$ab = b\overline{a}, \qquad ba = \overline{a}b. \tag{3.47}$$

Comparing (3.9) and (3.46), the plane spanned by the two orthogonal unit quaternions $a, b \in \mathbb{H}$ is indeed the q_--plane for $\mathrm{S}(a) = \cos\alpha \neq 0$. The orthogonal q_+-plane is simply given by its basis vectors (3.9), inserting (3.45). This leads to the pair of orthogonal unit quaternions c, d for the q_+-plane as

$$c = \frac{f-g}{|f-g|} = \frac{ab-\overline{a}b}{|(a-\overline{a})b|} = \frac{a-\overline{a}}{|a-\overline{a}|}b = \boldsymbol{\mu}(a)b, \tag{3.48}$$

$$d = \frac{1+fg}{|1+fg|} = \frac{f-g}{|f-g|}g = cg = \boldsymbol{\mu}(a)bg = \boldsymbol{\mu}(a)b\overline{a}b = \boldsymbol{\mu}(a)ab^2 = -\boldsymbol{\mu}(a)a, \tag{3.49}$$

where we have used (3.13) for the second, and (3.47) for the sixth equality in (3.49).

Let us also verify that f, g of (3.45) are both pure unit quaternions using (3.47)

$$f^2 = abab = (a\overline{a})bb = -1, \qquad g^2 = \overline{a}b\overline{a}b = (\overline{a}a)bb = -1. \tag{3.50}$$

Note, that if we would set in (3.45) for $g := -\overline{a}b$, then the a, b-plane would have become the q_+-plane instead of the q_--plane. We can therefore determine by the sign in the definition of g, which of the two OPS planes the a, b-plane is to represent.

For both a and b being two pure orthogonal quaternions, we can again set

$$f := ab \Rightarrow f^2 = abab = -a^2b^2 = -1, \qquad g := \overline{a}b = -ab = -f, \tag{3.51}$$

where due to the orthogonality of the pure unit quaternions a, b we were able to use $ba = -ab$. In this case $f = ab$ is thus also shown to be a pure unit quaternion. Now the q_--plane of the corresponding OPS (3.34) is spanned by $\{a, b\}$, whereas the q_+-plane is spanned by $\{1, f\}$. Setting instead $g := -\overline{a}b = ab = f$, the q_--plane of the corresponding OPS (3.24) is spanned by $\{1, f\}$, wheras the q_+-plane is spanned by $\{a, b\}$.

We summarize our results in the following theorem.

Theorem 3.5. (Determination of f, g from given steerable 2D plane) *Given any 2D plane in $\mathbb{H}$ in terms of two unit quaternions a, b, where b is without restriction* *of generality pure, i.e.,* $b^2 = -1$, *we can make the given plane the* q_-*-plane of the OPS* $q_\pm = \frac{1}{2}(q \pm fqg)$, *by setting*

$$f := ab, \qquad g := \bar{a}b. \tag{3.52}$$

For $\mathrm{S}(a) \neq 0$ *the orthogonal* q_+*-plane is fully determined by the orthogonal unit quaternions*

$$c = \boldsymbol{\mu}(a)b, \qquad d = -\boldsymbol{\mu}(a)a. \tag{3.53}$$

where $\boldsymbol{\mu}(a)$ *is as defined in* (2.6). *For* $\mathrm{S}(a) = 0$ *the orthogonal* q_+*-plane with basis* $\{1, f\}$ *is instead fully determined by* $f = -g = ab$.

Setting alternatively

$$f := ab, \qquad g := -\bar{a}b. \tag{3.54}$$

makes the given a,b*-plane the* q_+*-plane instead. For* $\mathrm{S}(a) \neq 0$ *the orthogonal* q_-*-plane is then fully determined by* (3.54) *and* (3.9), *with the same orthogonal unit quaternions* $c = \boldsymbol{\mu}(a)b, d = -\boldsymbol{\mu}(a)a$ *as in* (3.53). *For* $\mathrm{S}(a) = 0$ *the orthogonal* q_-*-plane with basis* $\{1, f\}$ *is then instead fully determined by* $f = g = ab$.

An illustration of the decomposition is given in Figures 3 and 4. Again, three hundred unit pure quaternions randomly oriented in 4-space have been decomposed into two sets using the decomposition of Definition 3.2 and two unit pure quaternions f and g computed as in Theorem 3.5. b was the pure unit quaternion $\frac{1}{\sqrt{3}}(\boldsymbol{i} + \boldsymbol{j} + \boldsymbol{k})$ and a was the full unit quaternion $\frac{1}{\sqrt{2}} + \frac{1}{2}(\boldsymbol{i} - \boldsymbol{j})$. c and d were computed by (3.53) as $c = \boldsymbol{\mu}(a)b$ and $d = -\boldsymbol{\mu}(a)a$.

Figure 3 shows the three hundred points in 4-space, projected onto the six orthogonal planes $\{c,d\}, \{c,b\}, \{c,a\}, \{d,b\}, \{b,a\}, \{a,d\}$ where the orthonormal 4-space basis $\{c,d,b,a\} = \{(f-g)/|f-g|, (1+fg)/|1+fg|, (f+g)/|f+g|, (1-fg)/|1-fg|\}$. The six views at the top show the q_+-plane, and the six below show the q_--plane. Figure 4 shows the vector parts of the decomposed quaternions.

4. New QFT Forms: OPS-QFTs with Two Pure Unit Quaternions f, g

4.1. Generalized OPS Leads to New Steerable Type of QFT

We begin with a straightforward generalization of the (double-sided form of the) QFT [4, 5] in $\mathbb{H}$ by replacing $\boldsymbol{i}$ with f and $\boldsymbol{j}$ with g defined as

Definition 4.1. (QFT with respect to two pure unit quaternions f, g)
Let $f, g \in \mathbb{H}$, $f^2 = g^2 = -1$, be any two pure unit quaternions. The quaternion Fourier transform with respect to f, g is

$$\mathcal{F}^{f,g}\{h\}(\boldsymbol{\omega}) = \int_{\mathbb{R}^2} e^{-fx_1\omega_1} h(\boldsymbol{x})\, e^{-gx_2\omega_2} d^2\boldsymbol{x}, \tag{4.1}$$

where $h \in L^1(\mathbb{R}^2, \mathbb{H})$, $d^2\boldsymbol{x} = dx_1 dx_2$ and $\boldsymbol{x}, \boldsymbol{\omega} \in \mathbb{R}^2$.

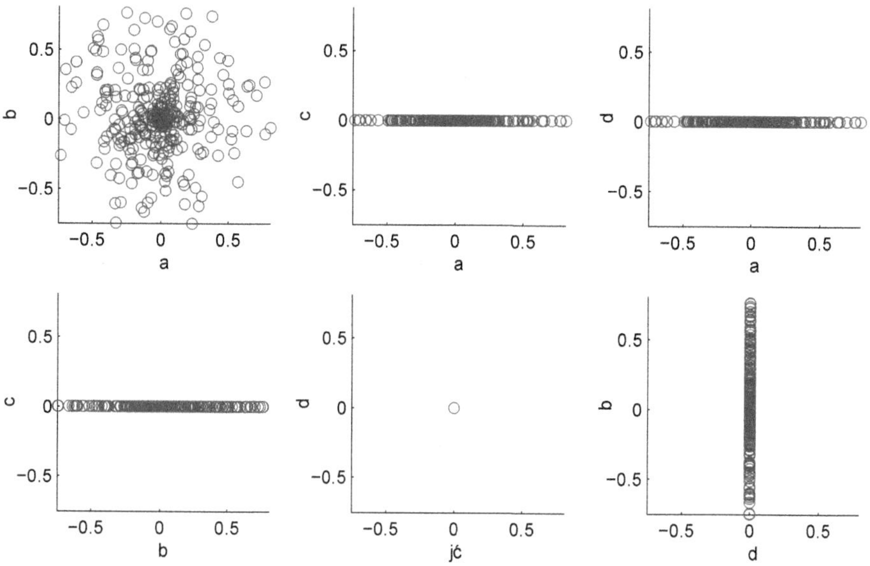

(a) q_+ component

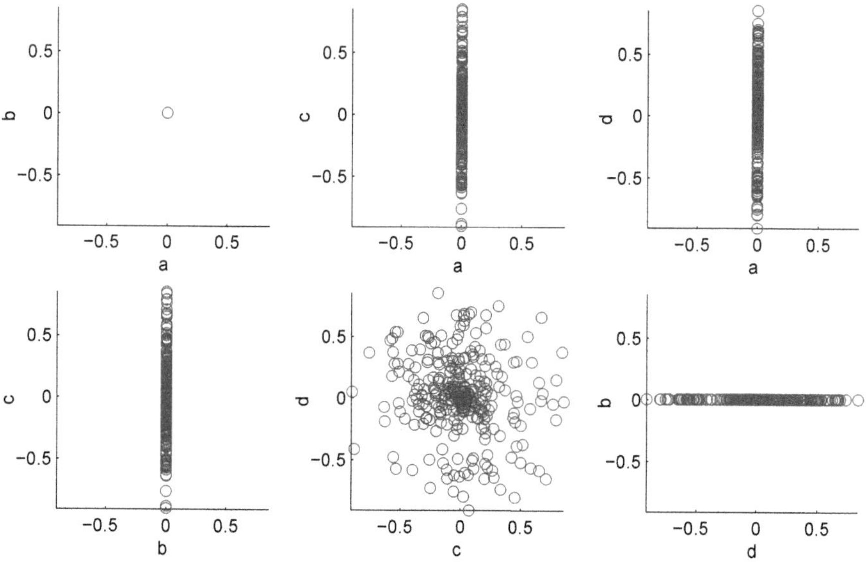

(b) q_- component

FIGURE 3. 4D scatter plot of quaternions decomposed using the orthogonal planes split of Definition 3.2.

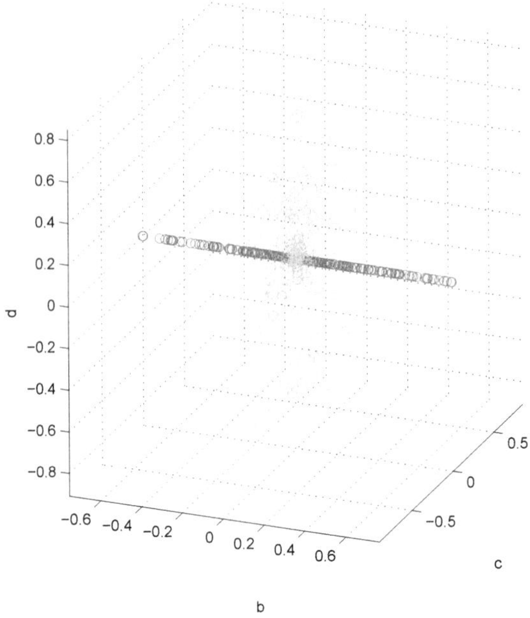

FIGURE 4. Scatter plot of vector parts of quaternions decomposed using the orthogonal planes split of Definition 3.2.

Note, that the pure unit quaternions f, g in Definition 4.1 do not need to be orthogonal, and that the cases $f = \pm g$ are fully included.

Linearity of the integral (4.1) allows us to use the OPS split $h = h_- + h_+$

$$\begin{aligned} \mathcal{F}^{f,g}\{h\}(\boldsymbol{\omega}) &= \mathcal{F}^{f,g}\{h_-\}(\boldsymbol{\omega}) + \mathcal{F}^{f,g}\{h_+\}(\boldsymbol{\omega}) \\ &= \mathcal{F}_-^{f,g}\{h\}(\boldsymbol{\omega}) + \mathcal{F}_+^{f,g}\{h\}(\boldsymbol{\omega}), \end{aligned} \tag{4.2}$$

since by their construction the operators of the Fourier transformation $\mathcal{F}^{f,g}$, and of the OPS with respect to f, g commute. From Lemma 3.4 follows

Theorem 4.2 (QFT of $h_\pm$). *The QFT of the $h_\pm$ OPS split parts, with respect to two unit quaternions f, g, of a quaternion module function $h \in L^1(\mathbb{R}^2, \mathbb{H})$ have the quasi-complex forms*

$$\begin{aligned} \mathcal{F}_\pm^{f,g}\{h\} = \mathcal{F}^{f,g}\{h_\pm\} &= \int_{\mathbb{R}^2} h_\pm e^{-g(x_2\omega_2 \mp x_1\omega_1)} d^2x \\ &= \int_{\mathbb{R}^2} e^{-f(x_1\omega_1 \mp x_2\omega_2)} h_\pm d^2x \ . \end{aligned} \tag{4.3}$$

Remark 4.3. The quasi-complex forms in Theorem 4.2 allow us to establish *discretized* and *fast* versions of the QFT of Definition 4.1 as sums of two complex discretized and fast Fourier transformations (FFT), respectively.

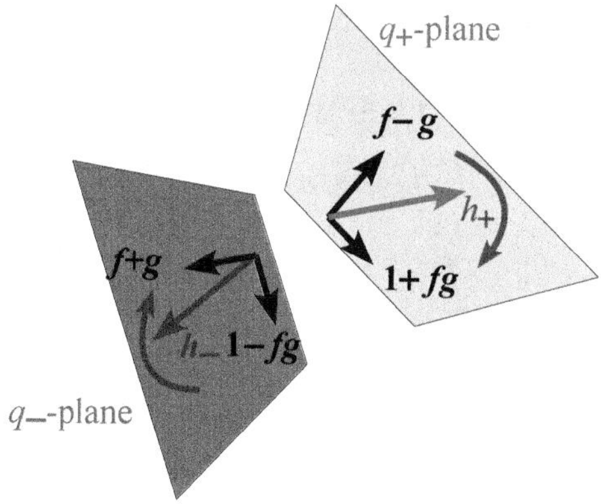

FIGURE 5. Geometric interpretation of integrand of $\text{QFT}^{f,g}$ in Definition 4.1 in terms of local phase rotations in $q_\pm$-planes.

We can now give a *geometric interpretation* of the integrand of the $\text{QFT}^{f,g}$ in Definition 4.1 in terms of local phase rotations, compare Section 3.3. The integrand product

$$e^{-fx_1\omega_1}h(\boldsymbol{x})\,e^{-gx_2\omega_2} \tag{4.4}$$

represents a *local rotation* by the phase angle $-(x_1\omega_1 + x_2\omega_2)$ in the q_--plane, and by the phase angle $-(x_1\omega_1 - x_2\omega_2) = x_2\omega_2 - x_1\omega_1$ in the orthogonal q_+-plane, compare Figure 5, which depicts two completely orthogonal planes in four dimensions.

Based on Theorem 3.5 the two phase rotation planes (analysis planes) can be freely *steered* by defining the two pure unit quaternions f, g used in Definition 4.1 according to (3.52) or (3.54).

4.2. Two Phase Angle Version of QFT

The above newly gained geometric understanding motivates us to propose a further new version of the $\text{QFT}^{f,g}$, with a straightforward two *phase angle* interpretation.

Definition 4.4. (Phase angle QFT with respect to f, g)
Let $f, g \in \mathbb{H}$, $f^2 = g^2 = -1$, be any two pure unit quaternions. The phase angle quaternion Fourier transform with respect to f, g is

$$\mathcal{F}_D^{f,g}\{h\}(\boldsymbol{\omega}) = \int_{\mathbb{R}^2} e^{-f\frac{1}{2}(x_1\omega_1 + x_2\omega_2)}h(\boldsymbol{x})\,e^{-g\frac{1}{2}(x_1\omega_1 - x_2\omega_2)}d^2\boldsymbol{x}. \tag{4.5}$$

where again $h \in L^1(\mathbb{R}^2, \mathbb{H})$, $d^2\boldsymbol{x} = dx_1dx_2$ and $\boldsymbol{x}, \boldsymbol{\omega} \in \mathbb{R}^2$.

The *geometric interpretation* of the integrand of (4.5) is a local phase rotation by angle $-(x_1\omega_1 + x_2\omega_2)/2 - (x_1\omega_1 - x_2\omega_2)/2 = -x_1\omega_1$ in the q_--plane, and a second local phase rotation by angle $-(x_1\omega_1 + x_2\omega_2)/2 + (x_1\omega_1 - x_2\omega_2)/2 = -x_2\omega_2$ in the q_+-plane, compare Section 3.3.

If we apply the $\text{OPS}^{f,g}$ split to (4.5) we obtain the following theorem.

Theorem 4.5 (Phase angle QFT of $h_\pm$). *The phase angle QFT of Definition 4.4 applied to the $h_\pm$ OPS split parts, with respect to two pure unit quaternions f, g, of a quaternion module function $h \in L^1(\mathbb{R}^2, \mathbb{H})$ leads to the quasi-complex expressions*

$$\mathcal{F}_{D+}^{f,g}\{h\} = \mathcal{F}_{D}^{f,g}\{h_+\} = \int_{\mathbb{R}^2} h_+ e^{+gx_2\omega_2} d^2x = \int_{\mathbb{R}^2} e^{-fx_2\omega_2} h_+ d^2x\ , \tag{4.6}$$

$$\mathcal{F}_{D-}^{f,g}\{h\} = \mathcal{F}_{D}^{f,g}\{h_-\} = \int_{\mathbb{R}^2} h_- e^{-gx_1\omega_1} d^2x = \int_{\mathbb{R}^2} e^{-fx_1\omega_1} h_- d^2x. \tag{4.7}$$

Note that based on Theorem 3.5 the two phase rotation planes (analysis planes) are again freely steerable.

Theorem 4.5 allows us to establish *discretized* and *fast* versions of the phase angle QFT of Definition 4.4 as sums of two complex discretized and fast Fourier transformations (FFT), respectively.

The maps $f(\,)g$ considered so far did not involve *quaternion conjugation* $q \rightarrow \overline{q}$. In the following we investigate maps which additionally conjugate the argument, *i.e.*, of type $f\overline{(\,)}g$, which are also *involutions*.

5. Involutions and QFTs Involving Quaternion Conjugation

5.1. Involutions Involving Quaternion Conjugations

The simplest case is quaternion conjugation itself

$$q \rightarrow \overline{q} = q_r - q_i\boldsymbol{i} - q_j\boldsymbol{j} - q_k\boldsymbol{k}, \tag{5.1}$$

which can be interpreted as a *reflection at the real line* q_r. The real line through the origin remains pointwise invariant, while every other point in the 3D hyperplane of pure quaternions is reflected to the opposite side of the real line. The related involution

$$q \rightarrow -\overline{q} = -q_r + q_i\boldsymbol{i} + q_j\boldsymbol{j} + q_k\boldsymbol{k}, \tag{5.2}$$

is the *reflection at the* 3D *hyperplane* of pure quaternions (which stay invariant), *i.e.*, only the real line is changed into its negative $q_r \rightarrow -q_r$.

Similarly any pure unit quaternion factor like $\boldsymbol{i}$ in the map

$$q \rightarrow \boldsymbol{i}\overline{q}\boldsymbol{i} = -q_r + q_i\boldsymbol{i} - q_j\boldsymbol{j} - q_k\boldsymbol{k}, \tag{5.3}$$

leads to a reflection at the (pointwise invariant) line through the origin with direction $\boldsymbol{i}$, while the map

$$q \rightarrow -\boldsymbol{i}\overline{q}\boldsymbol{i} = q_r - q_i\boldsymbol{i} + q_j\boldsymbol{j} + q_k\boldsymbol{k}, \tag{5.4}$$

leads to a reflection at the invariant 3D hyperplane orthogonal to the line through the origin with direction $\boldsymbol{i}$. The map

$$q \rightarrow f\overline{q}f, \tag{5.5}$$

leads to a reflection at the (pointwise invariant) line with direction f through the origin, while the map

$$q \rightarrow -f\overline{q}f, \tag{5.6}$$

leads to a reflection at the invariant 3D hyperplane orthogonal to the line with direction f through the origin.

Next we turn to a map of the type

$$q \to -e^{\alpha f}\overline{q}e^{\alpha f}. \tag{5.7}$$

Its set of pointwise invariants is given by

$$\begin{aligned} q = -e^{\alpha f}\overline{q}e^{\alpha f} &\Leftrightarrow\ e^{-\alpha f}q = -\overline{q}e^{\alpha f} \ \Leftrightarrow\ e^{-\alpha f}q + \overline{q}e^{\alpha f} = 0 \\ &\Leftrightarrow\ \mathrm{S}(\overline{q}e^{\alpha f}) = 0 \ \Leftrightarrow\ q \perp e^{\alpha f}. \end{aligned} \tag{5.8}$$

We further observe that

$$e^{\alpha f} \to -e^{\alpha f}e^{-\alpha f}e^{\alpha f} = -e^{\alpha f}. \tag{5.9}$$

The map $-a\overline{(\,)}a$, with unit quaternion $a = e^{\alpha f}$, therefore represents a reflection at the invariant 3D hyperplane orthogonal to the line through the origin with direction a.

Similarly, the map $a\overline{(\,)}a$, with unit quaternion $a = e^{\alpha f}$, then represents a reflection at the (pointwise invariant) line with direction a through the origin.

The combination of two such reflections (both at 3D hyperplanes, or both at lines), given by unit quaternions a, b, leads to a rotation

$$\begin{aligned} -b\overline{-a\overline{q}a}b = b\overline{a\overline{q}a}b &= b\overline{a}q\overline{a}b = rqs, \\ r = b\overline{a}, \qquad s = \overline{a}b, \qquad & |r| = |b|\,|a| = 1 = |s|, \end{aligned} \tag{5.10}$$

in two orthogonal planes, exactly as studied in Section 3.3.

The combination of three reflections at 3D hyperplanes, given by unit quaternions a, b, c, leads to

$$-c\overline{[-b\overline{-a\overline{q}a}b]}c = d\,\overline{q}t, \quad d = -c\overline{b}a, \ \ t = a\overline{b}c, \ \ |d| = |c|\,|b|\,|a| = |t| = 1. \tag{5.11}$$

The product of the reflection map $-\overline{q}$ of (5.2) with $d\,\overline{q}t$ leads to $-dqt$, a *double rotation* as studied in Section 4. Therefore $d\,\overline{(\,)}\,t$ represents a *rotary reflection* (rotation reflection). The three reflections $-a\overline{q}a$, $-b\overline{q}b$, $-c\overline{q}c$ have the intersection of the three 3D hyperplanes as a resulting common pointwise invariant line, which is $d + t$, because

$$d\,\overline{(d+t)}\,t = d\,\overline{t}t + d\,\overline{d}t = d + t. \tag{5.12}$$

In the remaining 3D hyperplane, orthogonal to the pointwise invariant line through the origin in direction $d + t$, the axis of the rotary reflection is

$$d\,\overline{(d-t)}\,t = -d\,\overline{t}t + d\,\overline{d}t = -d + t = -(d - t). \tag{5.13}$$

We now also understand that a sign change of $d \to -d$ (compare three reflections at three 3D hyperplanes $-c\overline{[-b\overline{(-a\overline{q}a)}b]}c$ with three reflections at three lines $+c\overline{[+b\overline{(+a\overline{q}a)}b]}c$) simply exchanges the roles of pointwise invariant line $d + t$ and rotary reflection axis $d - t$.

Next, we seek for an explicit description of the rotation plane of the rotary reflection $d\overline{(\,)}t$. We find that for the unit quaternions $d = e^{\alpha g}, t = e^{\beta f}$ the commutator

$$[d,t] = dt - td = e^{\alpha g}e^{\beta f} - e^{\beta f}e^{\alpha g} = (gf - fg)\sin\alpha\sin\beta, \tag{5.14}$$

is a pure quaternion, because

$$\overline{gf - fg} = fg - gf = -(gf - fg). \tag{5.15}$$

Moreover, $[d,t]$ is orthogonal to d and t, and therefore orthogonal to the plane spanned by the pointwise invariant line $d + t$ and the rotary reflection axis $d - t$, because

$$\mathrm{S}([d,t]\bar{d}) = \mathrm{S}(dt\bar{d} - td\bar{d}) = 0, \qquad \mathrm{S}([d,t]\bar{t}) = 0. \tag{5.16}$$

We obtain a second quaternion in the plane orthogonal to $d + t$, and $d - t$, by applying the rotary reflection to $[d,t]$

$$d\,\overline{[d,t]}\,t = -d[d,t]t = -[d,t]\bar{d}t, \tag{5.17}$$

because d is orthogonal to the pure quaternion $[d,t]$. We can construct an *orthogonal basis of the plane of the rotary reflection* $d\overline{(\,)}t$ by computing the pair of orthogonal quaternions

$$v_{1,2} = [d,t] \mp d\,\overline{[d,t]}\,t = [d,t] \pm [d,t]\bar{d}t = [d,t](1 \pm \bar{d}t). \tag{5.18}$$

For finally computing the rotation angle, we need to know the relative length of the two orthogonal quaternions v_1, v_2 of (5.18). For this it helps to represent the unit quaternion $\bar{d}t$ as

$$\bar{d}t = e^{\gamma u}, \qquad \gamma \in \mathbb{R},\ u \in \mathbb{H},\ u^2 = -1. \tag{5.19}$$

We then obtain for the length ratio

$$\begin{aligned} r^2 = \frac{|v_1|^2}{|v_2|^2} = \frac{|1+\bar{d}t|^2}{|1-\bar{d}t|^2} = \frac{(1+\bar{d}t)(1+\bar{t}d)}{(1-\bar{d}t)(1-\bar{t}d)} &= \frac{1+\bar{d}t\bar{t}d+\bar{d}t+\bar{t}d}{1+\bar{d}t\bar{t}d-\bar{d}t-\bar{t}d} \\ = \frac{2+2\cos\gamma}{2-2\cos\gamma} &= \frac{1+\cos\gamma}{1-\cos\gamma}. \end{aligned} \tag{5.20}$$

By applying the rotary reflection $d\overline{(\,)}t$ to v_1 and decomposing the result with respect to the pair of orthogonal quaternions in the rotary reflection plane (5.18) we can compute the rotation angle. Applying the rotary reflection to v_1 gives

$$\begin{aligned} d\overline{v_1}t &= d\,\overline{[d,t] - d\overline{[d,t]}t}\,t = d\,\overline{[d,t](1+\bar{d}t)}t = d(1+\bar{t}d)\overline{[d,t]}t \\ &= d(1+\bar{t}d)(-[d,t])t = -[d,t](\bar{d}t + \bar{d}t\bar{d}t). \end{aligned} \tag{5.21}$$

The square of $\bar{d}t$ is

$$\begin{aligned} (\bar{d}t)^2 = (\cos\gamma + u\sin\gamma)^2 &= -1 + 2\cos\gamma\,[\cos\gamma + u\sin\gamma] \\ &= -1 + 2\cos\gamma\,\bar{d}t. \end{aligned} \tag{5.22}$$

We therefore get

$$\begin{aligned} d\,\overline{v_1}\,t = -[d,t](\overline{d}t - 1 + 2\cos\gamma\overline{d}t) &= [d,t](1-(1+2\cos\gamma)\overline{d}t) \\ &= a_1 v_1 + a_2 r v_2, \end{aligned} \tag{5.23}$$

and need to solve

$$1-(1+2\cos\gamma)\overline{d}t = a_1(1+\overline{d}t) + a_2 r(1-\overline{d}t), \tag{5.24}$$

which leads to

$$d\,\overline{v_1}\,t = -\cos\gamma v_1 + \sin\gamma r v_2 = \cos(\pi-\gamma)v_1 + \sin(\pi-\gamma) r v_2. \tag{5.25}$$

The rotation angle of the rotary reflection $d\,\overline{(\,)}\,t$ in its rotation plane v_1, v_2 is therefore

$$\Gamma = \pi - \gamma, \qquad \gamma = \arccos \mathrm{S}(\overline{d}t)\,. \tag{5.26}$$

In terms of $d = e^{\alpha g}, t = e^{\beta f}$ we get

$$\overline{d}t = \cos\alpha\cos\beta - g\sin\alpha\cos\beta + f\cos\alpha\sin\beta - gf\sin\alpha\sin\beta. \tag{5.27}$$

And with the angle ω between g and f

$$\begin{aligned} gf &= \frac{1}{2}(gf+fg) + \frac{1}{2}(gf-fg) = \mathrm{S}(gf) + \frac{1}{2}[g,f] \\ &= -\cos\omega - \sin\omega\frac{[g,f]}{|[g,f]|}, \end{aligned} \tag{5.28}$$

we finally obtain for γ the scalar part $\mathrm{S}(\overline{d}t)$ as

$$\begin{aligned} \mathrm{S}(\overline{d}t) &= \cos\gamma = \cos\alpha\cos\beta + \cos\omega\sin\alpha\sin\beta \\ &= \cos\alpha\cos\beta - \mathrm{S}(gf)\sin\alpha\sin\beta. \end{aligned} \tag{5.29}$$

In the special case of $g = \pm f$, $\mathrm{S}(gf) = \mp 1$, *i.e.*, for $\omega = 0, \pi$, we get from (5.29) that

$$\begin{aligned} \mathrm{S}(\overline{d}t) = \cos\alpha\cos\beta \pm \sin\alpha\sin\beta &= \cos\alpha\cos\beta + \sin(\pm\alpha)\sin\beta \\ &= \cos(\pm\alpha - \beta), \end{aligned} \tag{5.30}$$

and thus using (5.26) the rotation angle would become

$$\Gamma = \pi - (\pm\alpha - \beta) = \pi \mp \alpha + \beta. \tag{5.31}$$

Yet (5.26) was derived assuming $[d,t] \neq 0$. But direct inspection shows that (5.31) is indeed correct: For $g = \pm f$ the plane $d+t, d-t$ is identical to the $1, f$ plane. The rotation plane is thus a plane of pure quaternions orthogonal to the $1, f$ plane. The quaternion conjugation in $q \mapsto d\,\overline{q}\,t$ leads to a rotation by π and the left and right factors lead to further rotations by $\mp\alpha$ and β, respectively. Thus (5.31) is verified as a special case of (5.26) for $g = \pm f$.

By substituting in Lemma 3.4 $(\alpha, \beta) \to (-\beta, -\alpha)$, and by taking the quaternion conjugate we obtain the following lemma.

Lemma 5.1. *Let* $q_\pm = \frac{1}{2}(q \pm fqg)$ *be the OPS of Definition* 3.2. *For left and right exponential factors we have the identity*

$$e^{\alpha g}\,\overline{q_\pm} e^{\beta f} = \overline{q_\pm} e^{(\beta\mp\alpha)f} = e^{(\alpha\mp\beta)g}\,\overline{q_\pm}. \tag{5.32}$$

5.2. New Steerable QFTs with Quaternion Conjugation and Two Pure Unit Quaternions f, g

We therefore consider now the following new variant of the (double-sided form of the) QFT [4, 5] in $\mathbb{H}$ (replacing both $\boldsymbol{i}$ with g and $\boldsymbol{j}$ with f, and using quaternion conjugation). It is essentially the quaternion conjugate of the new QFT of Definition 4.1, but because of its distinct local transformation geometry it deserves separate treatment.

Definition 5.2. (QFT with respect to f, g, including quaternion conjugation) Let $f, g \in \mathbb{H}$, $f^2 = g^2 = -1$, be any two pure unit quaternions. The quaternion Fourier transform with respect to f, g, involving quaternion conjugation, is

$$\mathcal{F}_c^{g,f}\{h\}(\boldsymbol{\omega}) = \overline{\mathcal{F}^{f,g}\{h\}(-\boldsymbol{\omega})} = \int_{\mathbb{R}^2} e^{-gx_1\omega_1}\overline{h(\boldsymbol{x})}\,e^{-fx_2\omega_2}d^2\boldsymbol{x}, \tag{5.33}$$

where $h \in L^1(\mathbb{R}^2, \mathbb{H})$, $d^2\boldsymbol{x} = dx_1dx_2$ and $\boldsymbol{x}, \boldsymbol{\omega} \in \mathbb{R}^2$.

Linearity of the integral in (5.33) of Definition 5.2 leads to the following corollary to Theorem 4.2.

Corollary 5.3 (QFT $\mathcal{F}_c^{g,f}$ of $h_\pm$). *The QFT* $\mathcal{F}_c^{g,f}$ (5.33) *of the* $h_\pm = \frac{1}{2}(h \pm fhg)$ *OPS split parts, with respect to any two unit quaternions f, g, of a quaternion module function* $h \in L^1(\mathbb{R}^2, \mathbb{H})$ *have the quasi-complex forms*

$$\begin{aligned}\mathcal{F}_c^{g,f}\{h_\pm\}(\boldsymbol{\omega}) = \overline{\mathcal{F}^{f,g}\{h_\pm\}(-\boldsymbol{\omega})} &= \int_{\mathbb{R}^2} \overline{h_\pm} e^{-f(x_2\omega_2\mp x_1\omega_1)}d^2x \\ &= \int_{\mathbb{R}^2} e^{-g(x_1\omega_1\mp x_2\omega_2)}\overline{h_\pm}d^2x\ .\end{aligned} \tag{5.34}$$

Note, that the pure unit quaternions f, g in Definition 5.2 and Corollary 5.3 do not need to be orthogonal, and that the cases $f = \pm g$ are fully included. Corollary 5.3 leads to discretized and fast versions of the QFT with quaternion conjugation of Definition 5.2.

It is important to note that the roles (sides) of f, g appear exchanged in (5.33) of Definition 5.2 and in Corollary 5.3, although the same OPS of Definition 3.2 is applied to the signal h as in Sections 3 and 4. This role change is due to the presence of quaternion conjugation in Definition 5.2. Note that it is possible to first apply (5.33) to h, and subsequently split the integral with the $\mathrm{OPS}^{g,f}$ $\mathcal{F}_{c,\pm} = \frac{1}{2}(\mathcal{F}_c \pm g\mathcal{F}_c f)$, where the particular order of g from the left and f from the right is due to the application of conjugation in (5.34) to $h_\pm$ *after* h is split with (3.41) into h_+ and h_-.

5.3. Local Geometric Interpretation of the QFT with Quaternion Conjugation

Regarding the *local geometric interpretation* of the QFT with quaternion conjugation of Definition 5.2 we need to distinguish the following cases, depending on $[d,t]$ and on whether the left and right phase factors

$$d = e^{-gx_1\omega_1}, \qquad t = e^{-fx_2\omega_2}, \tag{5.35}$$

attain scalar values ± 1 or not.

Let us first assume that $[d,t] \neq 0$, which by (5.14) is equivalent to $g \neq \pm f$, and $\sin(x_1\omega_1) \neq 0$, and $\sin(x_2\omega_2) \neq 0$. Then we have the generic case of a local rotary reflection with pointwise invariant line of direction

$$d + t = e^{-gx_1\omega_1} + e^{-fx_2\omega_2}, \tag{5.36}$$

rotation axis in direction

$$d - t = e^{-gx_1\omega_1} - e^{-fx_2\omega_2}, \tag{5.37}$$

rotation plane with basis (5.18), and by (5.26) and (5.29) the general rotation angle

$$\Gamma = \pi - \arccos \mathrm{S}(\overline{d}t)\,,$$
$$\mathrm{S}(\overline{d}t) = \cos(x_1\omega_1)\cos(x_2\omega_2) - \mathrm{S}(gf)\sin(x_1\omega_1)\sin(x_2\omega_2). \tag{5.38}$$

Whenever $g = \pm f$, or when $\sin(x_1\omega_1) = 0$ ($x_1\omega_1 = 0, \pi[\mathrm{mod}\, 2\pi]$, *i.e.*, $d = \pm 1$), we get for the pointwise invariant line in direction $d + t$ the simpler unit quaternion direction expression $e^{-\frac{1}{2}(\pm x_1\omega_1 + x_2\omega_2)f}$, because we can apply

$$e^{\alpha f} + e^{\beta f} = e^{\frac{1}{2}(\alpha+\beta)f}\big(e^{\frac{1}{2}(\alpha-\beta)f} + e^{\frac{1}{2}(\beta-\alpha)f}\big) = e^{\frac{1}{2}(\alpha+\beta)f}\, 2\cos\frac{\alpha-\beta}{2}, \tag{5.39}$$

and similarly for the rotation axis $d - t$ we obtain the direction expression

$$e^{-\frac{1}{2}(\pm x_1\omega_1 + x_2\omega_2 + \pi)f},$$

whereas the rotation angle is by (5.31) simply

$$\Gamma = \pi \pm x_1\omega_1 - x_2\omega_2. \tag{5.40}$$

For $\sin(x_2\omega_2) = 0$ ($x_2\omega_2 = 0, \pi[\mathrm{mod}\, 2\pi]$, *i.e.*, $t = \pm 1$), the pointwise invariant line in direction $d+t$ simplifies by (5.39) to $e^{-\frac{1}{2}(x_1\omega_1 + x_2\omega_2)g}$, and the rotation axis with direction $d-t$ simplifies to $e^{-\frac{1}{2}(x_1\omega_1 + x_2\omega_2 + \pi)g}$, whereas the angle of rotation is by (5.31) simply

$$\Gamma = \pi + x_1\omega_1 - x_2\omega_2. \tag{5.41}$$

5.4. Phase Angle QFT with Respect to f, g, Including Quaternion Conjugation

Even when quaternion conjugation is applied to the signal h we can propose a further new version of the $\mathrm{QFT}_c^{g,f}$, with a straightforward two *phase angle* interpretation. The following definition to some degree ignores the resulting local rotary reflection effect of combining quaternion conjugation and left and right phase factors of Section 5.3, but depending on the application context, it may nevertheless be of interest in its own right.

Definition 5.4 (Phase angle QFTwith respect to f, g, including quaternion conjugation). Let $f, g \in \mathbb{H}$, $f^2 = g^2 = -1$, be any two pure unit quaternions. The phase angle quaternion Fourier transform with respect to f, g, involving quaternion conjugation, is

$$\mathcal{F}_{cD}^{g,f}\{h\}(\boldsymbol{\omega}) = \overline{\mathcal{F}_D^{f,g}\{h\}(-\omega_1, \omega_2)} = \int_{\mathbb{R}^2} e^{-g\frac{1}{2}(x_1\omega_1 + x_2\omega_2)}\overline{h(\boldsymbol{x})}\, e^{-f\frac{1}{2}(x_1\omega_1 - x_2\omega_2)} d^2\boldsymbol{x}. \tag{5.42}$$

where again $h \in L^1(\mathbb{R}^2, \mathbb{H})$, $d^2\boldsymbol{x} = dx_1 dx_2$ and $\boldsymbol{x}, \boldsymbol{\omega} \in \mathbb{R}^2$.

Based on Lemma 5.1, one possible *geometric interpretation* of the integrand of (5.42)) is a local phase rotation of $\overline{h_+}$ by angle $-(x_1\omega_1 - x_2\omega_2)/2 + (x_1\omega_1 + x_2\omega_2)/2 = +x_2\omega_2$ in the $\overline{q_+}$ plane, and a second local phase rotation of $\overline{h_-}$ by angle $-(x_1\omega_1 - x_2\omega_2)/2 - (x_1\omega_1 + x_2\omega_2)/2 = -x_1\omega_1$ in the $\overline{q_-}$ plane. This is expressed in the following corollary to Theorem 4.5.

Corollary 5.5 (Phase angle QFT of $h_\pm$, involving quaternion conjugation). *The phase angle QFT with quaternion conjugation of Definition 5.4 applied to the $h_\pm$ OPS split parts, with respect to any two pure unit quaternions f, g, of a quaternion module function $h \in L^1(\mathbb{R}^2, \mathbb{H})$ leads to the quasi-complex expressions*

$$\mathcal{F}_{cD}^{g,f}\{h_+\}(\boldsymbol{\omega}) = \overline{\mathcal{F}_D^{g,f}\{h_+\}(-\omega_1, \omega_2)} = \int_{\mathbb{R}^2} \overline{h_+} e^{+fx_2\omega_2} d^2x = \int_{\mathbb{R}^2} e^{-gx_2\omega_2}\overline{h_+} d^2x\ , \tag{5.43}$$

$$\mathcal{F}_{cD}^{g,f}\{h_-\}(\boldsymbol{\omega}) = \overline{\mathcal{F}_D^{g,f}\{h_-\}(-\omega_1, \omega_2)} = \int_{\mathbb{R}^2} \overline{h_-} e^{\ fx_1\omega_1} d^2x = \int_{\mathbb{R}^2} e^{-gx_1\omega_1}\overline{h_-} d^2x\ . \tag{5.44}$$

Note that based on Theorem 3.5 the two phase rotation planes (analysis planes) are again freely steerable. Corollary 5.5 leads to discretized and fast versions of the phase angle QFT with quaternion conjugation of Definition 5.4.

6. Conclusion

The involution maps $\boldsymbol{i}(\ )\boldsymbol{j}$ and $f(\ )g$ have led us to explore a range of similar quaternionic maps $q \mapsto aqb$ and $q \mapsto a\bar{q}b$, where a, b are taken to be unit quaternions. Geometric interpretations of these maps as reflections, rotations, and rotary reflections in 4D can mostly be found in [1]. We have further developed these geometric interpretations to gain a *complete local transformation geometric understanding* of the integrands of the proposed new quaternion Fourier transformations (QFTs) applied to general quaternionic signals $h \in L^1(\mathbb{R}^2, \mathbb{H})$. This new geometric understanding is also valid for the special cases of the hitherto well-known left-sided, right-sided, and left- and right-sided (two-sided) QFTs of [2, 4, 8, 3] and numerous other references.

Our newly gained geometric understanding itself motivated us to propose *new types* of QFTs with specific geometric properties. The investigation of these new types of QFTs with the generalized form of the orthogonal 2D planes split of Definition 3.2 lead to important QFT split theorems, which allow the use of discrete and (complex) Fourier transform software for efficient discretized and fast numerical implementations.

Finally, we are convinced that our geometric interpretation of old and new QFTs paves the way for new applications, *e.g.*, regarding *steerable filter design* for specific tasks in image, colour image and signal processing, *etc.*

Acknowledgement. E.H. wishes to thank God for the joy of doing this research, his family, and S.J.S. for his great cooperation and hospitality.

References

[1] H.S.M. Coxeter. Quaternions and reflections. *The American Mathematical Monthly*, 53(3):136–146, Mar. 1946.

[2] T.A. Ell. Quaternion-Fourier transforms for analysis of 2-dimensional linear time-invariant partial-differential systems. In *Proceedings of the 32nd Conference on Decision and Control*, pages 1830–1841, San Antonio, Texas, USA, 15–17 December 1993. IEEE Control Systems Society.

[3] T.A. Ell and S.J. Sangwine. Hypercomplex Fourier transforms of color images. *IEEE Transactions on Image Processing*, 16(1):22–35, Jan. 2007.

[4] E. Hitzer. Quaternion Fourier transform on quaternion fields and generalizations. *Advances in Applied Clifford Algebras*, 17(3):497–517, May 2007.

[5] E. Hitzer. Directional uncertainty principle for quaternion Fourier transforms. *Advances in Applied Clifford Algebras*, 20(2):271–284, 2010.

[6] E. Hitzer. OPS-QFTs: A new type of quaternion Fourier transform based on the orthogonal planes split with one or two general pure quaternions. In *International Conference on Numerical Analysis and Applied Mathematics*, volume 1389 of *AIP Conference Proceedings*, pages 280–283, Halkidiki, Greece, 19–25 September 2011. American Institute of Physics.

[7] E. Hitzer. Two-sided Clifford Fourier transform with two square roots of -1 in $C\ell_{p,q}$. In *Proceedings of the 5th Conference on Applied Geometric Algebras in Computer Science and Engineering* (*AGACSE* 2012), La Rochelle, France, 2–4 July 2012.

[8] S.J. Sangwine. Fourier transforms of colour images using quaternion, or hypercomplex, numbers. *Electronics Letters*, 32(21):1979–1980, 10 Oct. 1996.

Eckhard Hitzer
College of Liberal Arts, Department of Material Science
International Christian University, 181-8585 Tokyo, Japan
e-mail: `hitzer@icu.ac.jp`

Stephen J. Sangwine
School of Computer Science and Electronic Engineering
University of Essex, Wivenhoe Park, Colchester, CO4 3SQ, UK
e-mail: `sjs@essex.ac.uk`

Quaternion and Clifford–Fourier Transforms and Wavelets
Trends in Mathematics, 41–56

3 Quaternionic Spectral Analysis of Non-Stationary Improper Complex Signals

Nicolas Le Bihan and Stephen J. Sangwine

Abstract. We consider the problem of the spectral content of a complex improper signal and the time-varying behaviour of this spectral content. The signals considered are one-dimensional (1D), complex-valued, with possible correlation between the real and imaginary parts, *i.e.*, *improper* complex signals. As a consequence, it is well known that the 'classical' (complex-valued) Fourier transform does not exhibit Hermitian symmetry and also that it is necessary to consider simultaneously the spectrum and the pseudo-spectrum to completely characterize such signals. Hence, an 'augmented' representation is necessary. However, this does not provide a *geometric* analysis of the complex improper signal.

We propose another approach for the analysis of improper complex signals based on the use of a 1D Quaternion Fourier Transform (QFT). In the case where complex signals are non-stationary, we investigate the extension of the well-known 'analytic signal' and introduce the quaternion-valued 'hyperanalytic signal'. As with the hypercomplex two-dimensional (2D) extension of the analytic signal proposed by Bülow in 2001, our extension of analytic signals for complex-valued signals can be obtained by the inverse QFT of the quaternion-valued spectrum after suppressing negative frequencies.

Analysis of the hyperanalytic signal reveals the time-varying frequency content of the corresponding complex signal. Using two different representations of quaternions, we show how modulus and quaternion angles of the hyperanalytic signal are linked to geometric features of the complex signal. This allows the definition of the angular velocity and the *complex envelope* of a complex signal. These concepts are illustrated on synthetic signal examples.

The hyperanalytic signal can be seen as the exact counterpart of the classical analytic signal, and should be thought of as the very first and simplest quaternionic time-frequency representation for improper non-stationary complex-valued signals.

Mathematics Subject Classification (2010). 65T50, 11R52.

Keywords. Quaternions, complex signals, Fourier transform, analytic signal.

1. Introduction

The analytic signal has been known since 1948 from the work of Ville [21] and Gabor [7]. It can be easily described, even though its theoretical ramifications are deep. Its use in non-stationary signal analysis is routine and it has been used in numerous applications. Simply put, given a real-valued signal $f(t)$, its analytic signal $a(t)$ is a complex signal with real part equal to $f(t)$, and imaginary part orthogonal to $f(t)$. The imaginary part is sometimes known as the quadrature signal – in the case where $f(t)$ is a sinusoid, the imaginary part of the analytic signal is in quadrature, that is with a phase difference of $-\pi/2$. The orthogonal signal is related to $f(t)$ by the Hilbert transform [9, 10]. The analytic signal has the interesting property that its modulus $|a(t)|$ is an envelope of the signal $f(t)$. The envelope is also known as the *instantaneous amplitude*. Thus if $f(t)$ is an amplitude-modulated sinusoid, the envelope $|a(t)|$, subject to certain conditions, is the modulating signal. The argument of the analytic signal, $\angle a(t)$ is known as the *instantaneous phase*. The analytic signal has a third interesting property: it has a one-sided Fourier transform. Thus a simple algorithm for constructing the analytic signal (algebraically or numerically) is to compute the Fourier transform of $f(t)$, multiply the Fourier transform by a unit step which is zero for negative frequencies, and then construct the inverse Fourier transform.

In this chapter we extend the concept of the analytic signal from the case of a real signal $f(t)$ with a complex analytic signal $a(t)$, to a complex signal $z(t)$ with a *hypercomplex* analytic signal $h(t)$, which we call the *hyperanalytic signal*. Just as the classical complex analytic signal contains both the original real signal (in the real part) and a real orthogonal signal (in the imaginary part), a hyperanalytic signal contains two complex signals: the original signal and an orthogonal signal. We have previously published partial results on this topic [19, 15, 14]. Here we develop a clear idea of how to generalise the classic case of amplitude modulation to a complex signal, and show for the first time that this leads to a correctly extended analytic signal concept in which the (*complex*) envelope and phase have clear interpretations.

The construction of an orthogonal signal alone would not constitute a full generalisation of the classical analytic signal to the complex case: it is also necessary to generalise the envelope and phase concepts, and this can only be done by interpreting the original and orthogonal complex signals as a pair. In this chapter we have only one way to do this: by representing the pair of complex signals as a quaternion signal. This arises naturally from the method above for creating an orthogonal signal, but also from the Cayley–Dickson construction of a quaternion as a complex number with complex real and imaginary parts (with different roots of -1 used in each of the two levels of complex number).

Extension of the analytic signal concept to 2D signals, that is images, with real, complex or quaternion-valued pixels is of interest, but outside the scope of this chapter. Some work has been done on this, notably by Bülow, Felsberg, Sommer and Hahn [1, 2, 6, 8]. The principal issue to be solved in the 2D case is to generalise the concept of a single-sided spectrum. Hahn considered a single quadrant or orthant spectrum, Sommer *et al.* considered a spectrum with support limited to half the complex plane, not necessarily confined to two quadrants, but still with real sample or pixel values.

Recently, Lilly and Olhede [16] have published a paper on bivariate analytic signal concepts without explicitly considering the complex signal case which we cover here. Their approach is linked to a specific signal model, the *modulated elliptical signal*, which they illustrate with the example of a drifting oceanographic float. The approach taken in the present chapter is more general and without reference to a specific signal model.

2. 1D Quaternion Fourier Transform

In this section, we will be concerned with the definition and properties of the quaternion Fourier transform (QFT) of complex-valued signals. Before introducing the main definitions, we give some prerequisites on quaternion-valued signals.

2.1. Preliminary Remarks

Quaternions were discovered by Sir W.R. Hamilton in 1843 [11]. Quaternions are 4D hypercomplex numbers that form a noncommutative division ring denoted $\mathbb{H}$. A quaternion $q \in \mathbb{H}$ has a Cartesian form: $q = q_0 + q_1\boldsymbol{i} + q_2\boldsymbol{j} + q_3\boldsymbol{k}$, with $q_0, q_1, q_2, q_3 \in \mathbb{R}$ and $\boldsymbol{i}, \boldsymbol{j}, \boldsymbol{k}$ roots of -1 satisfying $\boldsymbol{i}^2 = \boldsymbol{j}^2 = \boldsymbol{k}^2 = \boldsymbol{ijk} = -1$. The *scalar part* of q is: $\mathrm{S}(q) = q_0$. The *vector part* of q is: $\mathbf{V}(q) = q - \mathrm{S}(q)$. Quaternion multiplication is not commutative, so that in general $qp \neq pq$ for $p, q \in \mathbb{H}$. The conjugate of q is $\overline{q} = q_0 - q_1\boldsymbol{i} - q_2\boldsymbol{j} - q_3\boldsymbol{k}$. The norm of q is $\|q\| = |q|^2 = (q_0^2 + q_1^2 + q_2^2 + q_3^2) = q\overline{q}$. A quaternion with $q_0 = 0$ is called pure. If $|q| = 1$, then q is called a *unit* quaternion. The inverse of q is $q^{-1} = \overline{q}/\|q\|$. Pure unit quaternions are special quaternions, among which are $\boldsymbol{i}$, $\boldsymbol{j}$ and $\boldsymbol{k}$. Together with the identity of $\mathbb{H}$, they form a *quaternion basis*: $\{1, \boldsymbol{i}, \boldsymbol{j}, \boldsymbol{k}\}$. In fact, given any two unit pure quaternions, μ and ξ, which are orthogonal to each other (*i.e.*, $\mathrm{S}(\mu\xi) = 0$), then $\{1, \mu, \xi, \mu\xi\}$ is a quaternion basis.

Quaternions can also be viewed as complex numbers with complex components, *i.e.*, one can write $q = z_1 + \boldsymbol{j}z_2$ in the basis $\{1, \boldsymbol{i}, \boldsymbol{j}, \boldsymbol{k}\}$ with $z_1, z_2 \in \mathbb{C}^{\boldsymbol{i}}$, *i.e.*, $z_\alpha = \Re(z_\alpha) + \boldsymbol{i}\Im(z_\alpha)$ for $\alpha = 1, 2$[1]. This is called the Cayley–Dickson form.

Among the possible representations of q, two of them will be used in this chapter: the polar (Euler) form and the polar Cayley–Dickson form [20].

Polar form. Any non-zero quaternion q can be written:

$$q = |q|\, e^{\mu_q \phi_q} = |q|\left(\cos\phi_q + \mu_q \sin\phi_q\right),$$

[1]In the sequel, we denote by $\mathbb{C}^\mu$ the set of complex numbers with root of $-1 = \mu$. Note that these are degenerate quaternions, where all vector parts point in the same direction.

where μ_q is the axis of q and ϕ_q is the angle of q. For future use in this chapter, we give here their explicit expressions:

$$\begin{cases} \mu_q = \mathbf{V}(q) \, / \, |\mathbf{V}(q)| \, , \\ \phi_q = \arctan\left(|\mathbf{V}(q)| \, / \, \mathrm{S}(q) \right) . \end{cases} \tag{2.1}$$

The axis and angle are used in interpreting the hyperanalytic signal in Section 4.

Polar Cayley–Dickson form. Any quaternion q also has a unique polar Cayley–Dickson form [20] given by:

$$q = A_q \exp(B_q \boldsymbol{j}) = (a_0 + a_1 \boldsymbol{i}) \exp((b_0 + b_1 \boldsymbol{i})\boldsymbol{j}), \tag{2.2}$$

where $A_q = a_0 + a_1 \boldsymbol{i}$ is the complex modulus of q and $B_q = b_0 + b_1 \boldsymbol{i}$ its complex phase. It is proven in [20] that given a quaternion $p = (b_0 + b_1 \boldsymbol{i})\boldsymbol{j} = b_0 \boldsymbol{j} + b_1 \boldsymbol{k}$, its exponential is given as:

$$e^p = \cos|p| + \frac{p}{|p| \sin|p|} \tag{2.3}$$

$$\begin{aligned} &= \cos(\sqrt{b_0^2 + b_1^2}) + \boldsymbol{j} \frac{b_0}{\sqrt{b_0^2 + b_1^2}} \sin(\sqrt{b_0^2 + b_1^2}) \\ &\quad + \boldsymbol{k} \frac{b_1}{\sqrt{b_0^2 + b_1^2}} \sin(\sqrt{b_0^2 + b_1^2}). \end{aligned} \tag{2.4}$$

As a consequence, using the right-hand side expression in (2.2), the Cartesian coordinates of q can be linked to the polar Cayley–Dickson ones in the following way:

$$\begin{cases} A_q = \dfrac{q_0 + \boldsymbol{i} q_1}{\cos(\sqrt{q_2^2 + q_3^2})} \\ B_q = \arctan(\sqrt{q_2^2 + q_3^2}) \sqrt{q_2^2 + q_3^2} \left(\dfrac{q_0 q_3 + q_1 q_2}{q_0^2 + q_1^2} + \boldsymbol{i} \dfrac{q_0 q_3 - q_1 q_2}{q_0^2 + q_1^2} \right) . \end{cases} \tag{2.5}$$

In Section 4, we will make use of the complex modulus A_q for interpretation of the hyperanalytic signal.

2.2. 1D Quaternion Fourier Transform

In this chapter, we use a 1D version of the (right-side) QFT first defined in discrete-time form in [17]. Thus, it is necessary to specify the axis (a pure unit quaternion) of the transform. So, we will denote by $\mathcal{F}_\mu \,[\,]$ a QFT with transformation axis μ. For convenience below, we refer to this as a QFT_μ. In order to work with the classical quaternion basis, in the sequel we will use $\boldsymbol{j}$ as the transform axis. The only restriction on the transform axis is that it must be orthogonal to the original basis of the signal (here $\{1, \boldsymbol{i}\}$). We now present the definition and some properties of the transform used here.

Definition 1. Given a complex-valued signal $z(t) = z_r(t) + \boldsymbol{i} z_i(t)$, its quaternion Fourier transform with respect to axis $\boldsymbol{j}$ is:

$$Z_{\boldsymbol{j}}(\nu) = \mathcal{F}_{\boldsymbol{j}}\left[z(t)\right] = \int_{-\infty}^{+\infty} z(t) e^{-\boldsymbol{j}2\pi\nu t} \mathrm{d}t, \tag{2.6}$$

and the inverse transform is:

$$z(t) = \mathcal{F}_{\boldsymbol{j}}^{-1}\left[Z_{\boldsymbol{j}}(\nu)\right] = \int_{-\infty}^{+\infty} Z_{\boldsymbol{j}}(\nu) e^{\boldsymbol{j}2\pi\nu t} \mathrm{d}\nu. \tag{2.7}$$

Property 1. Given a complex signal $z(t) = z_r(t) + \boldsymbol{i} z_i(t)$ and its quaternion Fourier transform denoted $Z_{\boldsymbol{j}}(\nu)$, then the following properties hold:

- The even part of $z_r(t)$ is in $\Re(Z_{\boldsymbol{j}}(\nu))$,
- The odd part of $z_r(t)$ is in $\Im_{\boldsymbol{j}}(Z_{\boldsymbol{j}}(\nu))$,
- The even part of $z_i(t)$ is in $\Im_{\boldsymbol{i}}(Z_{\boldsymbol{j}}(\nu))$,
- The odd part of $z_i(t)$ is in $\Im_{\boldsymbol{k}}(Z_{\boldsymbol{j}}(\nu))$.

Proof. Expand (2.6) into real and imaginary parts with respect to $\boldsymbol{i}$, and expand the quaternion exponential into cosine and sine components:

$$\begin{aligned} Z_{\boldsymbol{j}}(\nu) &= \int_{-\infty}^{+\infty} \left[z_r(t) + \boldsymbol{i} z_i(t)\right] \left[\cos(2\pi\nu t) - \boldsymbol{j}\sin(2\pi\nu t)\right] dt \\ &= \int_{-\infty}^{+\infty} z_r(t)\cos(2\pi\nu t)\mathrm{d}t - \boldsymbol{j}\int_{-\infty}^{+\infty} z_r(t)\sin(2\pi\nu t)\mathrm{d}t \\ &\quad + \boldsymbol{i}\int_{-\infty}^{+\infty} z_i(t)\cos(2\pi\nu t)\mathrm{d}t - \boldsymbol{k}\int_{-\infty}^{+\infty} z_i(t)\sin(2\pi\nu t)\mathrm{d}t, \end{aligned}$$

from which the stated properties are evident. □

These properties are central to the justification of the use of the QFT to analyze a complex-valued signal carrying complementary but different information in its real and imaginary parts. Using the QFT, it is possible to have the odd and even parts of the real and imaginary parts of the signal in four different components in the transform domain. This idea was also the initial motivation of Bülow, Sommer and Felsberg when they developed the monogenic signal for images [1, 2, 6]. Note that the use of hypercomplex Fourier transforms was originally introduced in 2D Nuclear Magnetic Resonance image analysis [5, 3].

We now turn to the link between a complex signal and the quaternion signal that can be uniquely associated to it.

Property 2. A one-sided QFT $X(\nu)$, obtained from a complex signal as in Definition 1 by suppressing the negative frequencies, that is with $X(\nu) = 0$ for $\nu < 0$, corresponds to a full quaternion-valued signal in the time domain.

Proof. The proof is based on the symmetry properties, listed in Property 1, fulfilled by the $QFT_{\boldsymbol{j}}$ of complex-valued signals. From this, it is easily verified that, given the $QFT_{\boldsymbol{j}}$ of a complex signal $u(t)$ denoted by $U_{\boldsymbol{j}}(\nu)$, then $(1 + \text{sign}(\nu))U_{\boldsymbol{j}}(\nu)$ is right-sided (*i.e.*, it vanishes for all $\nu < 0$). Here $\text{sign}(\nu)$ is the classical sign function, which is equal to 1 for $\nu > 0$ and equal to -1 for $\nu < 0$. Second, the inverse transform, or $IQFT_{\boldsymbol{j}}$ of $(1 + \text{sign}(\nu))U_{\boldsymbol{j}}(\nu)$ can be decomposed as follows:

$$IQFT_{\boldsymbol{j}}\left[(1 + \text{sign}(\nu))U_{\boldsymbol{j}}(\nu)\right] = IQFT_{\boldsymbol{j}}\left[U_{\boldsymbol{j}}(\nu)\right] + IQFT_{\boldsymbol{j}}\left[\text{sign}(\nu)U_{\boldsymbol{j}}(\nu)\right]$$

By definition, $IQFT_{\boldsymbol{j}}\left[U_{\boldsymbol{j}}(\nu)\right]$ is a complex-valued signal with non-zero real and $\boldsymbol{i}$-imaginary parts, and null $\boldsymbol{j}$-imaginary and $\boldsymbol{k}$-imaginary parts. To complete the proof we show that $IQFT_{\boldsymbol{j}}\left[\text{sign}(\nu)U_{\boldsymbol{j}}(\nu)\right]$ has null real and $\boldsymbol{i}$-imaginary parts, and non-zero $\boldsymbol{j}$-imaginary and $\boldsymbol{k}$-imaginary parts. Consider the original complex signal $u(t) = u_r(t) + u_i(t)\boldsymbol{i}$, as composed of odd and even, real and imaginary parts (four components in total). Property 1 shows how these four components map to the four Cartesian components of $U_{\boldsymbol{j}}(\nu)$, namely:

- The even part of $u_r(t) \mapsto \Re(U_{\boldsymbol{j}}(\nu))$, which is even,
- The even part of $u_i(t) \mapsto \Im_{\boldsymbol{i}}(U_{\boldsymbol{j}}(\nu))$, which is even,
- The odd part of $u_r(t) \mapsto \Im_{\boldsymbol{j}}(U_{\boldsymbol{j}}(\nu))$, which is odd.
- The odd part of $u_i(t) \mapsto \Im_{\boldsymbol{k}}(U_{\boldsymbol{j}}(\nu))$, which is odd.

Multiplication by $\text{sign}(\nu)$ changes the parity to the following:

- $\text{sign}(\nu)\Re(U_{\boldsymbol{j}}(\nu))$ is odd,
- $\text{sign}(\nu)\Im_{\boldsymbol{i}}(U_{\boldsymbol{j}}(\nu))$, is odd,
- $\text{sign}(\nu)\Im_{\boldsymbol{j}}(U_{\boldsymbol{j}}(\nu))$, is even.
- $\text{sign}(\nu)\Im_{\boldsymbol{k}}(U_{\boldsymbol{j}}(\nu))$, is even.

and the inverse transform $IQFT_{\boldsymbol{j}}$ maps these components to:

- $\text{sign}(\nu)\Re(U_{\boldsymbol{j}}(\nu)) \mapsto \boldsymbol{j}u_1(t)$,
- $\text{sign}(\nu)\Im_{\boldsymbol{i}}(U_{\boldsymbol{j}}(\nu)) \mapsto \boldsymbol{ij}u_2(t) = \boldsymbol{k}u_2(t)$,
- $\text{sign}(\nu)\Im_{\boldsymbol{j}}(U_{\boldsymbol{j}}(\nu)) \mapsto \boldsymbol{j}u_3(t)$,
- $\text{sign}(\nu)\Im_{\boldsymbol{k}}(U_{\boldsymbol{j}}(\nu)) \mapsto \boldsymbol{k}u_4(t)$,

where $u_x(t), x = 1, 2, 3, 4$ are real functions of t. Hence $IQFT_{\boldsymbol{j}}\left[\text{sign}(\nu)U_{\boldsymbol{j}}(\nu)\right]$ has null real and null $\boldsymbol{i}$-imaginary parts, but non-zero $\boldsymbol{j}$-imaginary and $\boldsymbol{k}$-imaginary parts as previously stated. □

Property 3. Given a complex signal $x(t)$, one can associate to it a unique canonical pair corresponding to a modulus and phase. These modulus and phase are uniquely defined through the hyperanalytic signal, which is quaternion valued.

Proof. Cancelling the negative frequencies of the QFT leads to a quaternion signal in the time domain. Then, any quaternion signal has a modulus and phase defined using its Cayley–Dickson polar form. □

2.3. Convolution

We consider the special case of convolution of a complex signal by a real signal. Consider g and f such that: $g : \mathbb{R}^+ \to \mathbb{C}$ and $f : \mathbb{R}^+ \to \mathbb{R}$. Now, consider the $\mathrm{QFT}_{\boldsymbol{j}}$ of their convolution:

$$\begin{aligned}
\mathcal{F}_{\boldsymbol{j}}\left[(g * f)(t)\right] &= \int_{-\infty}^{+\infty}\int_{-\infty}^{+\infty} g(\tau)f(t-\tau)\mathrm{d}\tau e^{-2\boldsymbol{j}\pi\nu t}\mathrm{d}t \\
&= \int_{-\infty}^{+\infty}\int_{-\infty}^{+\infty} g(\tau)e^{-\boldsymbol{j}2\pi\nu(t'+\tau)}f(t')d\tau\mathrm{d}t' \\
&= \int_{-\infty}^{+\infty} g(\tau)e^{-\boldsymbol{j}2\pi\nu\tau}\mathrm{d}\tau \int_{-\infty}^{+\infty} f(t')e^{-\boldsymbol{j}2\pi\nu t'}\mathrm{d}t' \\
&= \mathcal{F}_{\boldsymbol{j}}\left[g(t)\right]\mathcal{F}_{\boldsymbol{j}}\left[f(t)\right].
\end{aligned} \tag{2.8}$$

Thus, the definition used for the QFT here verifies the convolution theorem in the considered case. This specific case will be of use in our definition of the hyperanalytic signal.

2.4. The Quaternion Fourier Transform of the Hilbert Transform

It is straightforward to verify that the quaternion Fourier transform of a real signal $x(t) = 1/\pi t$ is $-\boldsymbol{j}\,\mathrm{sign}(\nu)$, where $\boldsymbol{j}$ is the axis of the transform. Substituting $x(t)$ into (2.6), we get:

$$\mathcal{F}_{\boldsymbol{j}}\left[\frac{1}{\pi t}\right] = \frac{1}{\pi}\int_{-\infty}^{+\infty} \frac{e^{-\boldsymbol{j}2\pi\nu t}}{t}\mathrm{d}t,$$

and this is clearly isomorphic to the classical complex case. The solution in the classical case is $-i\,\mathrm{sign}(\nu)$, and hence in the quaternion case must be as stated above.

It is also straightforward to see that, given an arbitrary real signal $y(t)$, subject only to the constraint that its classical Hilbert transform $\mathcal{H}\left[y(t)\right]$ exists, then one can easily show that the classical Hilbert transform of the signal may be obtained using a quaternion Fourier transform as follows:

$$\mathcal{H}\left[y(t)\right] = \mathcal{F}_{\boldsymbol{j}}^{-1}\left[-\boldsymbol{j}\,\mathrm{sign}(\nu)Y_{\boldsymbol{j}}(\nu)\right] \tag{2.9}$$

where $Y_{\boldsymbol{j}}(\nu) = \mathcal{F}_{\boldsymbol{j}}\left[y(t)\right]$. This result follows from the isomorphism between the quaternion and complex Fourier transforms when operating on a real signal, and it may be seen to be the result of a convolution between the signal $y(t)$ and the quaternion Fourier transform of $x(t) = 1/\pi t$. Note that $\boldsymbol{j}$ and $Y_{\boldsymbol{j}}(\nu)$ commute as a consequence of $y(t)$ being real.

3. The Hypercomplex Analytic Signal

We define the hyperanalytic signal $z_+(t)$ by a similar approach to that originally developed by Ville [21]. The following definitions give the details of the construction of this signal. *Note that the signal $z(t)$ is considered to be non-analytic, or* improper, *in the classical (complex) sense, that is its real and imaginary parts are* not *orthogonal. However, the following definitions are valid if $z(t)$ is analytic, as it can be considered as a degenerate case of the more general non-analytic case.*

Definition 2. Consider a complex signal $z(t) = z_r(t) + \boldsymbol{i}z_i(t)$ and its quaternion Fourier transform $Z_{\boldsymbol{j}}(\nu)$ as defined in Definition 1. Then, the hypercomplex analogue of the Hilbert transform of $z(t)$, is as follows:

$$\mathcal{H}_{\boldsymbol{j}}\left[z(t)\right] = \mathcal{F}_{\boldsymbol{j}}^{-1}\left[-\boldsymbol{j}\,\mathrm{sign}(\nu)Z_{\boldsymbol{j}}(\nu)\right], \tag{3.1}$$

where the Hilbert transform is thought of as: $\mathcal{H}_{\boldsymbol{j}}\left[z(t)\right] = p.v.\left(z * (1/\pi t)\right)$, where the principal value (p.v.) is understood in its classical way. This result follows from equation (2.9) and the linearity of the quaternion Fourier transform. To extract the imaginary part, the vector part of the quaternion signal must be multiplied by $-\boldsymbol{i}$. An alternative is to take the scalar or inner product of the vector part with $\boldsymbol{i}$. Note that $\boldsymbol{j}$ and $Z_{\boldsymbol{j}}(\nu)$ anticommute because $\boldsymbol{j}$ is orthogonal to $\boldsymbol{i}$, the axis of $Z_{\boldsymbol{j}}(\nu)$. Therefore the ordering is not arbitrary, but changing it simply conjugates the result.

Definition 3. Given a complex-valued signal $z(t)$ that can be expressed in the form of a quaternion as $z(t) = z_r(t) + \boldsymbol{i}z_i(t)$, then the hypercomplex analytic signal of $z(t)$, denoted $z_+(t)$ is given by:

$$z_+(t) = z(t) + \boldsymbol{j}\mathcal{H}_{\boldsymbol{j}}\left[z(t)\right], \tag{3.2}$$

where $\mathcal{H}_{\boldsymbol{j}}\left[z(t)\right]$ is the hypercomplex analogue of the Hilbert transform of $z(t)$ defined in the preceding definition. The quaternion Fourier transform of this hypercomplex analytic signal is thus:

$$\begin{aligned} Z_+(\nu) &= Z_{\boldsymbol{j}}(\nu) - \boldsymbol{j}^2\,\mathrm{sign}(\nu)Z_{\boldsymbol{j}}(\nu) \\ &= \left[1 + \mathrm{sign}(\nu)\right]Z_{\boldsymbol{j}}(\nu) \\ &= 2U(\nu)Z_{\boldsymbol{j}}(\nu), \end{aligned}$$

where $U(\nu)$ is the classical unit step function.

This result is unique to the quaternion Fourier transform representation of the hypercomplex analytic signal – the hypercomplex analytic signal has a one-sided quaternion Fourier spectrum. This means that the hypercomplex analytic signal may be constructed from a complex signal $z(t)$ in exactly the same way that an analytic signal may be constructed from a real signal $x(t)$, by suppression of negative frequencies in the Fourier spectrum. The only difference is that in the hypercomplex analytic case, a quaternion rather than a complex Fourier transform must be used, and of course the complex signal must be put in the form $z(t) =$

$z_r(t) + \boldsymbol{i}z_i(t)$ which, although a quaternion signal, is isomorphic to the original complex signal.

A second important property of the hypercomplex analytic signal is that it maintains a separation between the different even and odd parts of the original signal.

Property 4. The original signal $z(t)$ is the *simplex* part [4, Theorem 1], [18, § 13.1.3] of its corresponding hypercomplex analytic signal. The perplex part is the orthogonal or 'quadrature' component, $o(t)$. They are obtained by:

$$z(t) = \frac{1}{2}\left(z_+(t) - \boldsymbol{i}z_+(t)\boldsymbol{i}\right), \tag{3.3}$$

$$o(t) = \frac{1}{2}\left(z_+(t) + \boldsymbol{i}z_+(t)\boldsymbol{i}\right). \tag{3.4}$$

Proof. This follows from equation (3.2). Writing this in full by substituting the orthogonal signal for $\mathcal{H}_{\boldsymbol{j}}\left[z(t)\right]$:

$$z_+(t) = z(t) + \boldsymbol{j}o(t) = z_r(t) + \boldsymbol{i}z_i(t) + \boldsymbol{j}o_r(t) - \boldsymbol{k}o_i(t),$$

and substituting this into equation (3.3), we get:

$$z(t) = \frac{1}{2}\begin{pmatrix} z_r(t) + \boldsymbol{i}z_i(t) + \boldsymbol{j}o_r(t) - \boldsymbol{k}o_i(t) \\ -\boldsymbol{i}\left[z_r(t) + \boldsymbol{i}z_i(t) + \boldsymbol{j}o_r(t) - \boldsymbol{k}o_i(t)\right]\boldsymbol{i} \end{pmatrix},$$

and since $\boldsymbol{i}$ and $\boldsymbol{j}$ are orthogonal unit pure quaternions, $\boldsymbol{ij} = -\boldsymbol{ji}$:

$$= \frac{1}{2}\begin{pmatrix} & z_r(t) + \boldsymbol{i}z_i(t) + \boldsymbol{j}o_r(t) - \boldsymbol{k}o_i(t) \\ + & z_r(t) + \boldsymbol{i}z_i(t) - \boldsymbol{j}o_r(t) + \boldsymbol{k}o_i(t) \end{pmatrix},$$

from which the first part of the result follows. Equation (3.4) differs only in the sign of the second term, and it is straightforward to see that if $z_+(t)$ is substituted, $z(t)$ cancels out, leaving $o(t)$. □

4. Geometric Instantaneous Amplitude and Phase

Using the hypercomplex analytic signal, we now present the geometric features that can be obtained thanks to two representations of $z_+(t)$. First, we notice that the hypercomplex analytic signal has a polar form given as:

$$z_+(t) = \rho_{\mathbf{R}}^+(t)e^{\mu^+(t)\phi^+(t)}, \tag{4.1}$$

where $\rho_{\mathbf{R}}^+ = |z_+(t)|$ is the *real modulus* of $z_+(t)$, $\phi_+(t)$ is its *argument* and $\mu_+(t)$ its *axis*. The real modulus, or *real envelope*, is not very informative, as it is real valued and so does not provide a 2D description of the slowly varying part (envelope) of $z(t)$. Nonetheless, a complex modulus can be defined using the modulus of the polar Cayley–Dickson representation that is more informative on the geometric features of the original improper signal $z(t)$.

4.1. Complex Envelope

The complex modulus of the hypercomplex analytic signal has properties very similar to the 'classical' case. It is defined as the modulus of the Cayley–Dickson polar representation of $z_+(t)$. Considering the Cayley–Dickson polar form of the hypercomplex analytic signal given as:

$$z_+(t) = \rho_{\mathbf{C}}^+(t) e^{\Phi^+(t)\boldsymbol{j}}, \tag{4.2}$$

then the *complex envelope* of $z(t)$ is simply $\rho_{\mathbf{C}}^+(t)$. An illustration of a complex envelope of a complex improper signal is given in Figures 1 and 3. In Figure 1, an improper complex signal made of a low frequency envelope modulating a high frequency linear frequency sweep is presented. The complex envelope obtained through the hyperanalytic signal is displayed in red (and in blue a negated version to show how the envelope encompass the signal). In Figure 3, a similar signal, but with a quadratic sweep, is displayed. Again, the complex envelope fits the slowly varying pattern of the signal.

Note that we do not make use of the phase of the Cayley–Dickson polar form in the sequel. It was demonstrated in [15] that this phase can reveal information on the 'modulation' part of the improper complex signal. Here we are looking for geometric descriptors of the improper complex signal through the hyperanalytic signal. For this purpose, we now show how the phase of the polar form of $z_+(t)$ can be linked with the angular velocity concept.

4.2. Angular Velocity

The *phase* of a hyperanalytic signal is slightly different from the well-known concept of phase for a complex-valued signal. First, it is a three-dimensional quantity made of an axis and a phase (*i.e.*, a pure unit quaternion and a scalar). Using the polar form of $z_+(t)$, its normalized part, denoted $\tilde{z}_+(t)$, is simply:

$$\tilde{z}_+(t) = \frac{z_+(t)}{\rho_{\mathbf{R}}^+(t)}, \tag{4.3}$$

where it is assumed that $\rho_{\mathbf{R}}^+(t) \neq 0$ for all t. In the case where $\rho_{\mathbf{R}}^+(t) = 0$, it simply means that the original signal $z(t) = 0$ and so is $z_+(t)$. If not, then $\tilde{z}_+(t)$ is a unit quaternion, *i.e.*, an element of $\mathcal{S}^3$. Without loss of generality, one can write $\tilde{z}_+(t)$ as:

$$\tilde{z}_+(t) = \exp\left[\theta(t)\mathbf{v}(t)\right], \tag{4.4}$$

where $\theta(t)$ is scalar valued and $\mathbf{v}(t)$ is a pure unit quaternion. Now, it is well known [13, 12] from quaternion formulations of kinematics that the instantaneous frequency is given by:

$$\phi(t) = \frac{\mathrm{d}\theta(t)}{\mathrm{d}t} = \arg\left(\frac{\mathrm{d}\tilde{z}_+(t)}{\mathrm{d}t}\right), \tag{4.5}$$

where the time derivative of $\tilde{z}_+(t)$ is understood as:

$$\frac{\mathrm{d}\tilde{z}_+(t)}{\mathrm{d}t} = \lim_{\Delta t \to 0} \frac{\tilde{z}_+(t+\Delta t)\tilde{z}_+(t)^{-1}}{\Delta t}. \tag{4.6}$$

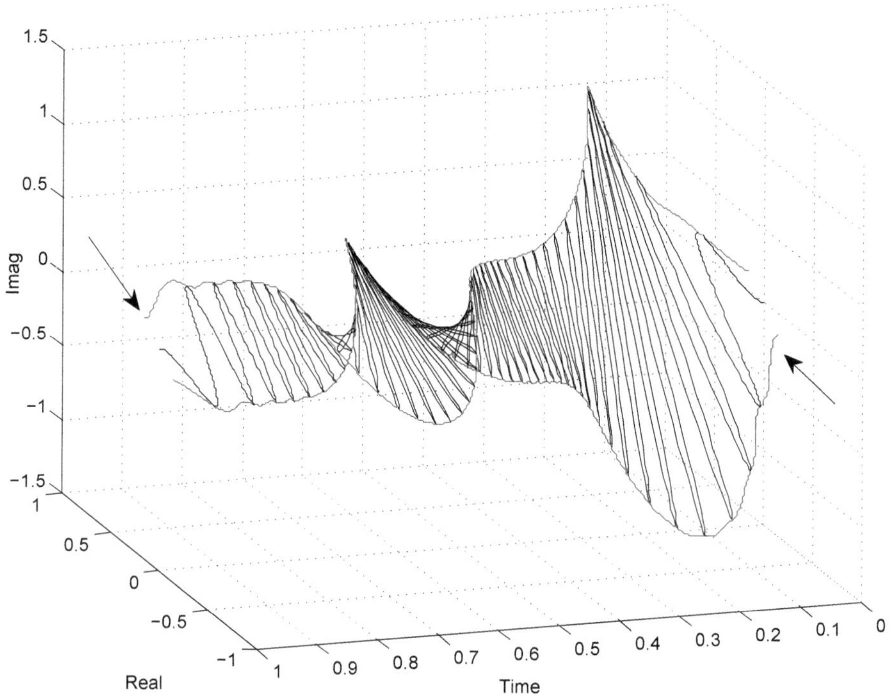

FIGURE 1. Improper complex signal consisting of a modulating low frequency envelope and a high frequency **linear** frequency sweep. The envelope $\rho_{\mathbb{C}}^{+}(t)$ is indicated by the arrow at left and $-\rho_{\mathbb{C}}^{+}(t)$ by the arrow at right.

This is in fact the multiplicative increment (a unit pure quaternion) between the hyperanalytic signal at time t and time $t + \Delta t$. The argument in (4.5) is thus the amount of angle per unit of time by which the signal has been rotated (in rad/s). The axis of this increment gives the direction of this rotation. This is an *angular velocity*, which corresponds, for a complex improper signal, to its *instantaneous frequency*. This *angular velocity* is a geometrical and spectral local information on the signal $z(t)$, as it gives locally the frequency content and the geometrical orientation and behaviour of the signal with time.

In Figures 1 to 4, we illustrate this concept of instantaneous frequency for improper complex signals. In Figure 1, an improper complex signal made of a linear sweep (single frequency signal with frequency linearly changing with time) and a complex envelope is presented. Its *complex envelope* is presented in red (the inverse of the envelope is also plotted, in blue, for visualization purposes), which fits the 'low frequency' behaviour of the signal. In Figure 2, the angular velocity of the hyperanalytic signal is compared to the frequency sweep used to generate the

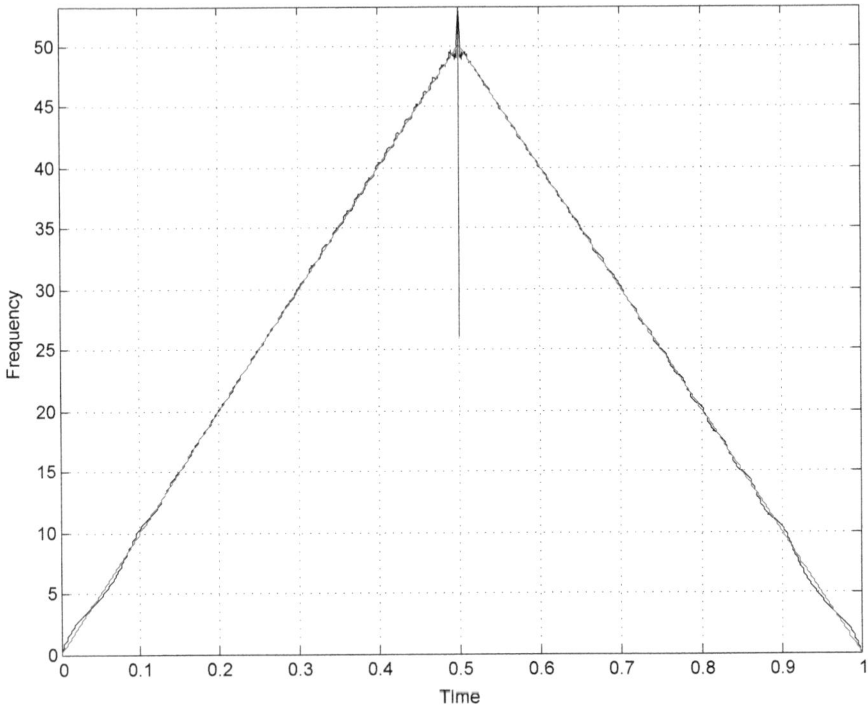

FIGURE 2. Linear frequency sweep estimated as the angular velocity of the hyperanalytic signal (with ripples and discontinuity), and original linear frequency sweep used to generate the original improper complex signal (without ripples). The discontinuity in the estimate is due to the fact that the angular velocity obtained from the hyperanalytic signal is not differentiable at $t = 0.5$.

signal, showing good agreement, except at a singularity point (where the phase of the hyperanalytic signal is not differentiable). A similar procedure is carried out in Figures 3 and 4, where the 'high frequency' content is now a quadratic sweep (frequency varying quadratically with time). The same conclusions can be drawn about the complex envelope and the angular velocity.

The complex envelope and the angular velocity are the extensions, for complex improper signals, of the well-known envelope and instantaneous frequency for real signals. They allow a local description of the geometrical and spectral behaviour of the signal and thus consist in the simplest time-frequency representation for improper complex signals. As in the classical case, these local descriptors have their limitations. Here, the number of frequency components must be kept to one for the angular velocity to be able to recover the instantaneous frequency. This limitation makes impossible the identification of local spectral contents in the case of mixtures of signals or wide-band signals. Such signals require more sophisticated

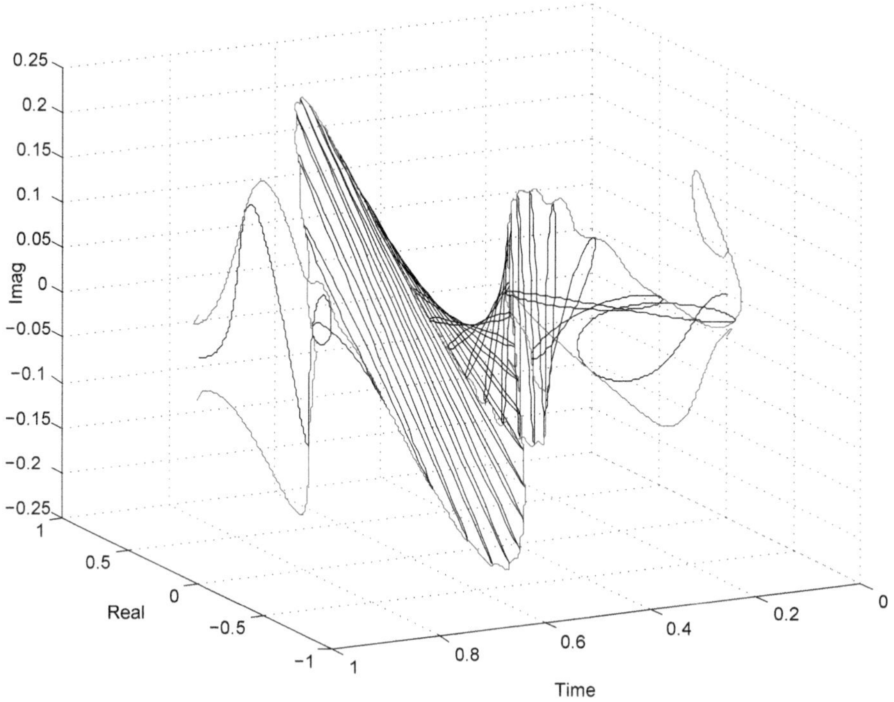

FIGURE 3. Improper complex signal consisting in a modulating low frequency envelope and a high frequency *quadratic* frequency sweep.

time-frequency representations that could be developed, based on the quaternion FT. The study of such representations is a natural step after the work presented.

5. Conclusions

We have shown in this chapter how the classical analytic signal concept may be extended to the case of the hyperanalytic signal of an original complex signal. The quaternion based approach yields an interpretation of the hyperanalytic signal as a quaternion signal which leads naturally to the definition of the complex envelope. Also, the use of the polar form of the hyperanalytic signal allows the derivation of an angular velocity, which is indeed an instantaneous frequency. Both the envelope and the angular velocity allow a local and spectral description of a complex improper signal, leading to the first geometrical time-frequency representation for complex improper signals.

Future work will consist in developing other time-frequency representations for improper complex signals with more diverse spectral behaviour.

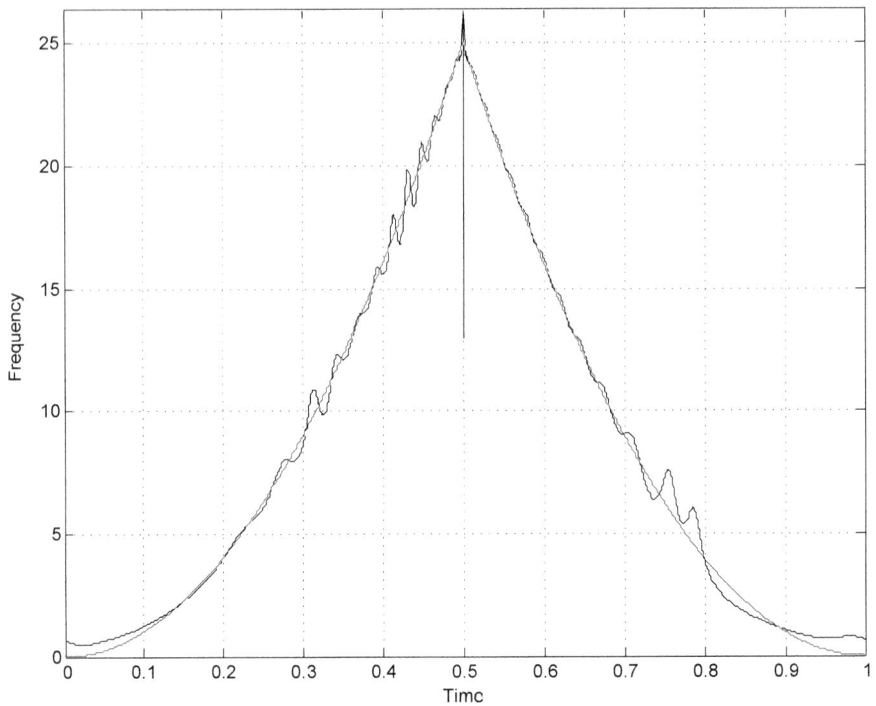

FIGURE 4. Quadratic frequency sweep estimated as the angular velocity of the hyperanalytic signal (with ripples), and original quadratic frequency sweep used to generate the original improper complex signal (without ripples). The discontinuity at $t = 0.5$ is due to the fact that the angular velocity obtained from the hyperanalytic signal is not differentiable at this point.

Acknowledgement

We thank Eckhard Hitzer for suggesting the proof of Property 2.

References

[1] T. Bülow. *Hypercomplex Spectral Signal Representations for the Processing and Analysis of Images.* PhD thesis, University of Kiel, Germany, Institut für Informatik und Praktische Mathematik, Aug. 1999.

[2] T. Bülow and G. Sommer. Hypercomplex signals – a novel extension of the analytic signal to the multidimensional case. *IEEE Transactions on Signal Processing*, 49(11):2844–2852, Nov. 2001.

[3] M.A. Delsuc. Spectral representation of 2D NMR spectra by hypercomplex numbers. *Journal of magnetic resonance*, 77(1):119–124, Mar. 1988.

[4] T.A. Ell and S.J. Sangwine. Hypercomplex Wiener-Khintchine theorem with application to color image correlation. In *IEEE International Conference on Image Processing* (*ICIP* 2000), volume II, pages 792–795, Vancouver, Canada, 11–14 Sept. 2000. IEEE.

[5] R.R. Ernst, G. Bodenhausen, and A. Wokaun. *Principles of Nuclear Magnetic Resonance in One and Two Dimensions*. International Series of Monographs on Chemistry. Oxford University Press, 1987.

[6] M. Felsberg and G. Sommer. The monogenic signal. *IEEE Transactions on Signal Processing*, 49(12):3136–3144, Dec. 2001.

[7] D. Gabor. Theory of communication. *Journal of the Institution of Electrical Engineers*, 93(26):429–457, 1946. Part III.

[8] S.L. Hahn. Multidimensional complex signals with single-orthant spectra. *Proceedings of the IEEE*, 80(8):1287–1300, Aug. 1992.

[9] S.L. Hahn. Hilbert transforms. In A.D. Poularikas, editor, *The transforms and applications handbook*, chapter 7, pages 463–629. CRC Press, Boca Raton, 1996. A CRC handbook published in cooperation with IEEE press.

[10] S.L. Hahn. *Hilbert transforms in signal processing*. Artech House signal processing library. Artech House, Boston, London, 1996.

[11] W.R. Hamilton. *Lectures on Quaternions*. Hodges and Smith, Dublin, 1853. Available online at Cornell University Library: `http://historical.library.cornell.edu/math/`.

[12] A.J. Hanson. *Visualizing Quaternions*. The Morgan Kaufmann Series in Interactive 3D Technology. Elsevier/Morgan Kaufmann, San Francisco, 2006.

[13] J.B. Kuipers. *Quaternions and Rotation Sequences*. Princeton University Press, Princeton, New Jersey, 1999.

[14] N. Le Bihan and S.J. Sangwine. About the extension of the 1D analytic signal to improper complex valued signals. In *Eighth International Conference on Mathematics in Signal Processing*, page 45, The Royal Agricultural College, Cirencester, UK, 16–18 December 2008.

[15] N. Le Bihan and S.J. Sangwine. The H-analytic signal. In *Proceedings of EUSIPCO* 2008*, 16th European Signal Processing Conference*, page 5, Lausanne, Switzerland, 25–29 Aug. 2008. European Association for Signal Processing.

[16] J.M. Lilly and S.C. Olhede. Bivariate instantaneous frequency and bandwidth. *IEEE Transactions on Signal Processing*, 58(2):591–603, Feb. 2010.

[17] S.J. Sangwine and T.A. Ell. The discrete Fourier transform of a colour image. In J.M. Blackledge and M.J. Turner, editors, *Image Processing II Mathematical Methods, Algorithms and Applications*, pages 430–441, Chichester, 2000. Horwood Publishing for Institute of Mathematics and its Applications. Proceedings Second IMA Conference on Image Processing, De Montfort University, Leicester, UK, September 1998.

[18] S.J. Sangwine, T.A. Ell, and N. Le Bihan. Hypercomplex models and processing of vector images. In C. Collet, J. Chanussot, and K. Chehdi, editors, *Multivariate Image Processing*, Digital Signal and Image Processing Series, chapter 13, pages 407–436. ISTE Ltd, and John Wiley, London, and Hoboken, NJ, 2010.

[19] S.J. Sangwine and N. Le Bihan. Hypercomplex analytic signals: Extension of the analytic signal concept to complex signals. In *Proceedings of EUSIPCO* 2007*, 15th*

European Signal Processing Conference, pages 621–4, Poznan, Poland, 3–7 Sept. 2007. European Association for Signal Processing.

[20] S.J. Sangwine and N. Le Bihan. Quaternion polar representation with a complex modulus and complex argument inspired by the Cayley–Dickson form. *Advances in Applied Clifford Algebras*, 20(1):111–120, Mar. 2010. Published online 22 August 2008.

[21] J. Ville. Théorie et applications de la notion de signal analytique. *Cables et Transmission*, 2A:61–74, 1948.

Nicolas Le Bihan
GIPSA-Lab/CNRS
11 Rue des mathématiques
F-38402 Saint Martin d'Hères, France
e-mail: nicolas.le-bihan@gipsa-lab.grenoble-inp.fr

Stephen J. Sangwine
School of Computer Science and Electronic Engineering
University of Essex
Wivenhoe Park
Colchester CO4 3SQ, UK
e-mail: sjs@essex.ac.uk

Quaternion and Clifford–Fourier Transforms and Wavelets
Trends in Mathematics, 57–83

4 Quaternionic Local Phase for Low-level Image Processing Using Atomic Functions

E. Ulises Moya-Sánchez and E. Bayro-Corrochano

Abstract. In this work we address the topic of image processing using an atomic function (AF) in a representation of quaternionic algebra. Our approach is based on the most important AF, the $\mathrm{up}(x)$ function. The main reason to use the atomic function $\mathrm{up}(x)$ is that this function can express analytically multiple operations commonly used in image processing such as low-pass filtering, derivatives, local phase, and multiscale and steering filters. Therefore, the modelling process in low level-processing becomes easy using this function. The quaternionic algebra can be used in image analysis because lines (even), edges (odd) and the symmetry of some geometric objects in $\mathbb{R}^2$ are enhanced. The applications show an example of how $\mathrm{up}(x)$ can be applied in some basic operations in image processing and for quaternionic phase computation.

Mathematics Subject Classification (2010). Primary 11R52; secondary 65D18.
Keywords. Quaternionic phase.

1. Introduction

The visual system is the most advanced of our senses. Therefore, it is easy to understand that the processing of images plays an important role in human perception and computer vision [3, 9]. In this chapter we address the topic of image processing using geometric algebra (GA) for computer vision applications in low-level and mid-level (geometric feature extraction and analysis) processing, which belong to the first layers of a bottom-up computer vision system.

Complex and hypercomplex numbers play an important role in signal processing [9, 5], especially in order to obtain local features such as the frequency and phase. In 1D signal analysis, it is usual to compute *via* the FFT a local magnitude and phase using the *analytic signal*. The entire potential of the local phase

This work has been supported by CINVESTAV and CONACYT.

information of the images is shown when constraints or invariants are required to be found. In other words, we use the quaternionic (local) phase because the local phase can be used to link the low-level image processing with the upper layers. The phase information is invariant to illumination changes and can be used to detect low-level geometric characteristics of lines or edges [9, 3, 13]. The phase can also be used to measure the local decomposition of the image according to its symmetries [6, 9].

This work has two basic goals: to present a new approach based on an atomic function (AF) up(x) in a representation of geometric algebra (GA) and to apply the phase information to reduce the gap between low-level processing and image analysis. Our approach is based on the most important AF, the up(x) function [12], quaternionic algebra, and multiscale and steering filters. The up(x) function has good locality properties (compact support), the derivative of any order can be expressed easily, the up(x) and dup(x) can mimic the simple cells of the mammalian visual processing system [14], and the approximation to other functions (polynomial) is relatively simple [12]. As we show in this work, the atomic function up(x) can be used as a building block to build multiple operations commonly used in image processing, such as low-pass filtering, nth-order derivatives, local phase, *etc.*

The applications presented show an example of how the AF can be applied to quaternionic phase analysis. It is based on line and edge detection and symmetry measurement using the phase. As a result, we can reduce the gap between the low-level processing and the computer vision applications without abandoning the geometric algebra framework.

2. Atomic Functions

The atomic functions were first developed in the 1970s, jointly by V.L. and V.A. Rvachev. By definition, the AF are compactly supported, infinitely differentiable solutions of differential functional equations (see (2.1)) with a shifted argument [12], that is

$$Lf(x) = \lambda \sum_{k=1}^{M} c(k) f(ax - b(k)), \quad |a| > 1, \quad b, c, \lambda \in N, \tag{2.1}$$

where $L = \frac{\mathrm{d}^n}{\mathrm{d}x^n} + a_1 \frac{\mathrm{d}^{n-1}}{\mathrm{d}x^{n-1}} + \cdots + a_n$ is a linear differential operator with constant coefficients. In the AF class, the function up(x) is the simplest and, at the same time, the most useful primitive function to generate other kinds of AFs [12].

2.1. Mother Atomic Function up(x)

In general, the atomic function up(x) is generated by infinite convolutions of rectangular impulses. The function up(x) has the following representation in terms of

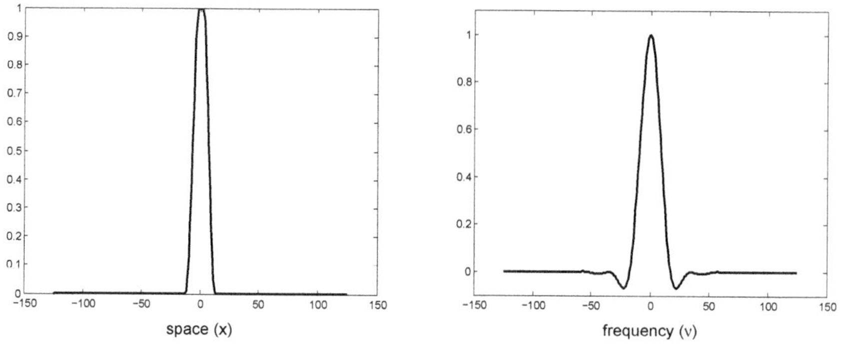

FIGURE 1. Left: atomic function $\mathrm{up}(x)$; right: Fourier transform $\widehat{\mathrm{up}}(\nu)$.

the Fourier transform [12, 8]:

$$\mathrm{up}(x) = \frac{1}{2\pi}\int_{-\infty}^{\infty}\prod_{k=1}^{\infty}\frac{\sin(\nu 2^{-k})}{\nu 2^{-k}}e^{i\nu x}\mathrm{d}\nu. \tag{2.2}$$

$$= \frac{1}{2\pi}\int_{-\infty}^{\infty}\widehat{\mathrm{up}}(\nu)e^{i\nu x}\mathrm{d}\nu \tag{2.3}$$

Figure 1 shows the $\mathrm{up}(x)$ and its Fourier transform $\widehat{\mathrm{up}}(\nu)$.

The AF windows were compared with classic windows by means of parameters such as the equivalent noise bandwidth, the 50% overlapping region correlation, the parasitic modulation amplitude, the maximum conversion losses (in decibels), the maximum side lobe level (in decibels), the asymptotic decay rate of the side lobes (in decibels per octave), the window width at the six- decibel level, the coherent gain, *etc.* All atomic windows exceed classic ones in terms of the asymptotic decay rate [12].

However the main reasons that we found to use the AF $\mathrm{up}(x)$ instead of other kernels such as Gauss, Gabor, and log-Gabor is that many operations commonly used in image processing can be expressed analytically, in contrast using the Gauss or the log-Gabor this could be impossible. Additionally the following useful properties of the AF have been reported in [12, 10]

- There are explicit equations for the values of the moments and Fourier transform (see (2.3)). The even moments of $\mathrm{up}(x)$ are

$$a_n = \int_{-1}^{1} x^n\,\mathrm{up}(x)\mathrm{d}x \tag{2.4}$$

$$a_{2n} = \frac{(2n)!}{2^{2^n}-1}\sum_{k=1}^{n}\frac{a_{2n-2k}}{(2n-2k)!(2k+1)!} \tag{2.5}$$

where $a_0 = 1$ and $a_{2n+1} = 0$. The odd-order moments are

$$b_n = \int_0^1 x^n \operatorname{up}(x) dx \tag{2.6}$$

$$b_{2n+1} = \frac{1}{(n+1)2^{2n+1}} \sum_{k=1}^{n+1} a_{2n-2k+2} \binom{2k}{2n+2k} \tag{2.7}$$

where $b_{2n} = a_{2n}$. A similar relation for Gauss or Log-Gabor does not exist.

- The $\operatorname{up}(x)$ function is a compactly supported function in the spatial domain. Therefore, we can obtain good local characteristics. In addition to compactly supported functions and integrable functions, functions that have a sufficiently rapid decay at infinity can also be convolved. The Gauss and Log-Gabor functions do not have such compact support.
- Translations with smaller steps yield polynomials (x^n) of any degree, *i.e.*,

$$\sum_{k=-\infty}^{\infty} c_k \operatorname{up}(x - (k2^n)) \equiv x^n \quad c_k, x \in \mathbb{R}. \tag{2.8}$$

- Since derivatives of any order can be represented in terms of simple shifts, we can easily represent any derivative operator or nth-order derivative:

$$\mathrm{d}^{(n)} \operatorname{up}(x) = 2^{n(n+1)/2} \sum_{k=1}^{2^n} \delta_k \operatorname{up}(2^n x + 2^n + 1 - 2k), \tag{2.9}$$

where $\delta_{2k} = -\delta_k$, $\delta_{2k-1} = \delta_k$, $\delta_{2k} = 1$.
- The AFs are infinitely differentiable (C^∞). As a result, the AFs and their Fourier transforms are rapidly decreasing functions. (Their Fourier transforms decrease on the real axis faster than any power function.)

A natural extension of the $\operatorname{up}(x)$ function to the case of many variables is based on the usual tensor product of 1D $\operatorname{up}(x)$ [10]. As a result, we have

$$\operatorname{up}(x, y) = \operatorname{up}(x) \operatorname{up}(y). \tag{2.10}$$

In Figure 2, we show a 2D atomic function in the spatial and frequency domains.

2.2. The $\operatorname{dup}(x)$ Function

There are some mask operators, including the Sobel, Prewitt, and Kirsh, that are used to extract edges from images. A common drawback of these operators is that it is impossible to ensure that they adapt to the intrinsic signal parameters over a wide range of the working band, *i.e.*, they are truncated and discrete versions of a differential operator. [8]. This means that adaptation of the differential operator to the behavior of the input signal by broadening or narrowing its band is desirable, in order to ensure a maximum signal-to-noise ratio [8].

This problem reduces to the synthesis of infinitely differentiable finite functions with a small wide bandwidth that are used for constructing the weighting windows [8]. One of the most effective solutions is obtained with the help of the

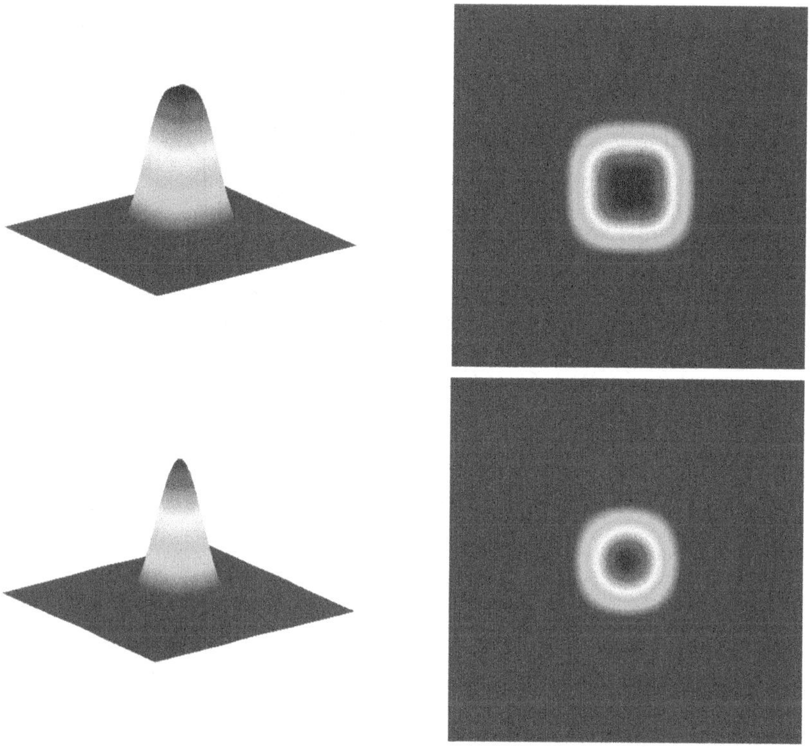

FIGURE 2. Top: Two views of $\mathrm{up}(x, y)$. Bottom: Fourier transform $\widehat{\mathrm{up}}(\nu, \upsilon)$.

atomic functions [8]. The AF can be used in two ways: construction of a window in the frequency region to obtain the required improvement in properties of the pulse characteristic; or direct synthesis based on (2.1) [8]. Therefore, the function $\mathrm{up}(x)$ satisfies (2.1) as follows:

$$\mathrm{dup}(x) = 2\,\mathrm{up}(2x+1) - 2\,\mathrm{up}(2x-1) \tag{2.11}$$

Figure 3 shows the $\mathrm{dup}(x)$ function and its Fourier transform $\widehat{dup}(\nu)$. If we compute the Fourier transform of (2.11), we obtain

$$i\nu F[\mathrm{up}(2x)] = \left(e^{i\nu} - e^{-i\nu}\right) F[\mathrm{up}(2x)]) \tag{2.12}$$

$$F(\mathrm{dup}(x)) = 2\boldsymbol{i}\sin(\nu) F(\mathrm{up}(2x)) \tag{2.13}$$

By differentiating (2.1) term by term, we obtain [8]

$$\mathrm{d}^{(n)}\,\mathrm{up}(x) = 2^{n(n+1)/2} \sum_{k=1}^{2^n} \delta_k\,\mathrm{up}(2^n x + 2^n + 1 - 2k), \tag{2.14}$$

where $\delta_{2k} = -\delta_k$, $\delta_{2k-1} = \delta_k$, and $\delta_{2k} = 1$. The function $\mathrm{dup}(x)$ provides a good window in the spatial frequency regions because the side lobe has been completely

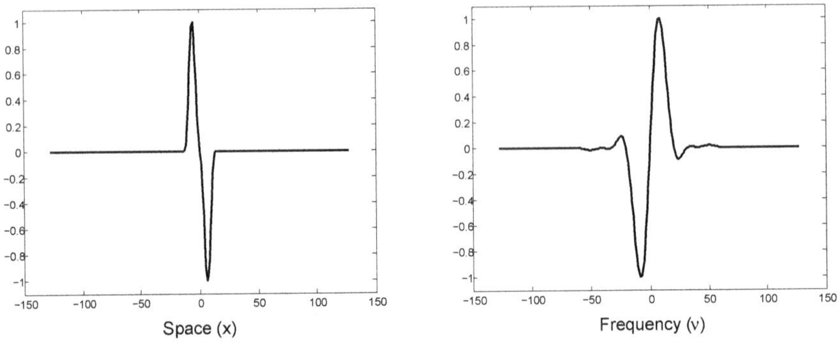

FIGURE 3. Left: derivative of atomic function $\mathrm{dup}(x)$; Right: Fourier transform $\widehat{\mathrm{dup}}(\nu)$.

eliminated [8]. Similarly to (2.10), we can get a 2D expression of each derivative:

$$\mathrm{dup}(x,y)_x = \mathrm{dup}(x)\,\mathrm{up}(y) \tag{2.15}$$

$$\mathrm{dup}(x,y)_y = \mathrm{up}(x)\,\mathrm{dup}(y) \tag{2.16}$$

$$\mathrm{dup}(x,y)_{x,y} = \mathrm{dup}(x)\,\mathrm{dup}(y) \tag{2.17}$$

Figure 4 shows two graphics of $\mathrm{dup}(x,y)_y$, $\mathrm{dup}(x,y)_{x,y}$ in the spatial domain.

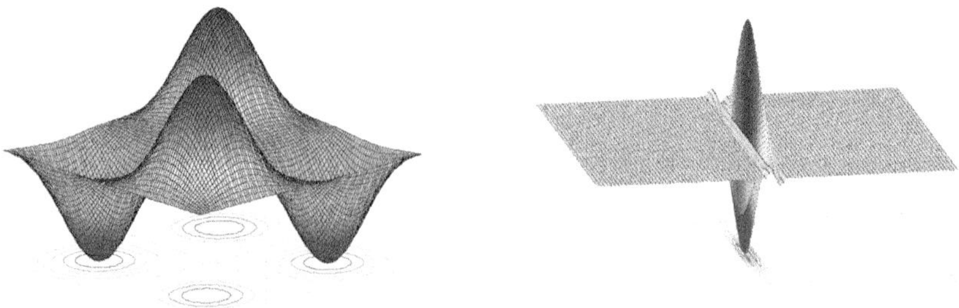

FIGURE 4. Left: $\mathrm{dup}(x,y)_{x,y}$; right: $\mathrm{dup}(x,y)_y$.

3. Why the Use of the Atomic Function?

In this section we justify our introduction of the atomic function $\mathrm{up}(x)$ as a kernel. In this regard, we discuss the role of the Gauss, LogGauss and the $\mathrm{up}(x)$ kernels in low-level image processing. First, we analyse their characteristics from a mathematical perspective and then we carry out an experimental test to show their strengths in the processing of real images. We have to stress that we are interested in the application of the $\mathrm{up}(x)$ kernel not just for filtering in scale space. This up(x) kernel is promising because it has a range of multiple possible applications which is wider than the possible applications of the Gauss or LogGauss kernels.

3.1. Dyadic Shifts and the Riesz Transform

As was explained in Section 2, atomic functions are compactly supported, infinitely differentiable solutions of differential equations with a shifted argument. Consequently an atomic function can be seen as an appropriate building block of linear shift invariant (LSI) operators to implement complex operators for image processing. In contrast the Gauss and LogGauss functions require derivatives and convolutions with a function of series of impulses for building complex operators. We believe that atomic functions are promising to develop operators based on linear combination of shifted atomic functions. Section 2.2 illustrates the differentiator atomic function dup(x). In 2002 Petermichl [16] showed that the Hilbert transform lies in the closed convex hull of dyadic singular operators, thus the Hilbert transform can be represented as an average of dyadic shifts. Petermichl *et al.* [17] show that the same is true for $\mathbb{R}^n$, therefore the Riesz transforms can also be obtained as the results of averaging of dyadic shifts. Thus we claim that the Riesz transforms can be obtained by averaging dyadic shifts of atomic functions.

3.2. Monogenic Signals and the Atomic Function

The monogenic signal was introduced by Felsberg, Bülow, and Sommer [7]. We outline briefly the concept of the monogenic signal and then explain the use of the atomic-function based monogenic signal. If we embed $\mathbb{R}^3$ into a subspace of $\mathbb{H}$ spanned by just $\{1, \boldsymbol{i}, \boldsymbol{j}\}$ according to

$$q = (\boldsymbol{i}, \boldsymbol{j}, 1)\boldsymbol{x} = x_3 + x_1\boldsymbol{i} + x_2\boldsymbol{j}, \tag{3.1}$$

and further embed the vector field $\boldsymbol{g}$ as follows:

$$g_{\mathbb{H}} = (-\boldsymbol{i}, -\boldsymbol{j}, 1)\boldsymbol{g} = g_3 - g_1\boldsymbol{i} - g_2\boldsymbol{j} \tag{3.2}$$

then $\nabla \times g(\boldsymbol{x}) = 0$ and $\nabla \cdot g(\boldsymbol{x}) = \langle \nabla, g(\boldsymbol{x})\rangle = 0$ are equivalent to the generalized Cauchy–Riemann equations from Clifford analysis [4, 15]. All functions that fulfill these equations are known as *monogenic* functions. Using the same embedding, the monogenic signal can be defined in the frequency domain as follows:

$$\begin{aligned} F_M(u) &= G_3(u_1, u_2, 0) - \boldsymbol{i}G_1(u_1, u_2, 0) - \boldsymbol{j}G_2(u_1, u_2, 0) \\ &= F(\boldsymbol{u}) - (\boldsymbol{i}, \boldsymbol{j})\boldsymbol{F}_R(\boldsymbol{u}) = \frac{|\boldsymbol{u}| + (1, \boldsymbol{k})\boldsymbol{u}}{|\boldsymbol{u}|} F(\boldsymbol{u}), \end{aligned} \tag{3.3}$$

where the inverse Fourier transform of $F_M(u)$ is given by

$$f_M(\boldsymbol{x}) = f(\boldsymbol{x}) - (\boldsymbol{i}, \boldsymbol{j})\boldsymbol{f}_R(\boldsymbol{x}) = f(\boldsymbol{x}) + h_R \star f(\boldsymbol{x}), \tag{3.4}$$

where $\boldsymbol{f}_R(\boldsymbol{x})$ stands for the Riesz transform obtained by taking the inverse transform of $\boldsymbol{F}_R(\boldsymbol{u})$ as follows:

$$\begin{aligned} \boldsymbol{F}_R(\boldsymbol{u}) = \frac{i\boldsymbol{u}}{|\boldsymbol{u}|}F(\boldsymbol{u}) = \boldsymbol{H}_R(\boldsymbol{u})F(\boldsymbol{u}) \longleftrightarrow \boldsymbol{f}_R(\boldsymbol{x}) &= -\frac{\boldsymbol{x}}{2\pi\,|\boldsymbol{x}|^2} \star f(\boldsymbol{x}) \\ &= h_R(\boldsymbol{x}) \star f(\boldsymbol{x}), \end{aligned} \tag{3.5}$$

where $\star$ stands for the convolution operation. Note that here the Riesz transform is the generalization of the 1D Hilbert transform. Using the fundamental solution of the 3D Laplace equation restricted to the open half-space $z > 0$ with boundary condition, the solution is defined as

$$\begin{aligned} f_M(x,y,z) &= h_P \star f(x,y,z) + h_P \star h_R \star f(x,y,z) \\ &= h_P \star (1 + h_R\star) f(x,y,z), \end{aligned} \tag{3.6}$$

where h_P stands for the 2D Poisson kernel. Setting in $f_M(x,y,z)$ the variable z equal to zero, we obtain the so-called monogenic signal. Some authors have used the Gauss kernel instead of the Poisson kernel, because the Poisson kernel establishes a linear scale space similar to the Gaussian scale space.

The atomic function is also an LSI operator; therefore, it appears that its use ensures a computation in a linear scale space as well.

The monogenic functions are the solutions of the generalized Cauchy–Riemann equations or Laplace-type equations. The atomic function can be used to compute compactly supported solutions of functional differential equations, for example, (2.1). Conditions under which the type of equations (2.1) have solutions with compact support and an explicit form were obtained by Ravachev [12]. Compactly supported solutions of equations of the type (2.1) are called *atomic functions*.

Now, for the case of 2D signal processing, we can apply the wavelet steerability and the Riesz transform. In this regard we will utilize the quaternion wavelet atomic function, which will be discussed in more depth below. For the scale space filtering, authors use the Gauss or Poisson kernels for the Riesz transformation [6, 7]. It is known that the Poisson kernel is the fundamental solution of the 3D Laplace equation, however there are authors who use instead the Poisson, Gauss or LogGauss function. Felsberg [6] showed that the spatial extent of the Poisson kernel is greater than that of the Gaussian kernel. In addition the uncertainty of the Poisson kernel for 1D signals is slightly worse than that of the Gaussian kernel (by a factor of $\sqrt{2}$). Nowadays many authors are now using the LogGauss kernel for implementing monogenic signals, because it has better properties than both the Gauss and Poisson kernels. Thus, we can infer that, in contrast, the atomic function as a spatially-compact kernel guarantees an analytic and closed solution of Laplace type equations. In addition, as we said above, the Riesz transforms can be obtained by averaging dyadic operators, thus the use of a compact atomic function avoids increased truncation errors. However, by the case of the noncompact Poisson, Gauss and LogGauss kernels, in practice, their larger spatial extents require either larger filter masks otherwise they cause increased truncation errors.

4. Quaternion Algebra $\mathbb{H}$

The even subalgebra $\mathcal{G}^+_{3,0,0}$ (bivector basis) is isomorphic to the quaternion algebra $\mathbb{H}$, which is an associative, non-commutative, four-dimensional algebra that consists of one real element and three imaginary elements.

$$q = a + b\boldsymbol{i} + c\boldsymbol{j} + d\boldsymbol{k}, \quad a, b, c, d \in \mathbb{R} \tag{4.1}$$

The units $\boldsymbol{i}, \boldsymbol{j}$ obey the relations $\boldsymbol{i}^2 = \boldsymbol{j}^2 = -1, \boldsymbol{ij} = \boldsymbol{k}$. $\mathbb{H}$ is geometrically inspired, and the imaginary components can be described in terms of the basis of $\mathbb{R}^3$ space, $\boldsymbol{i} \to \boldsymbol{e}_{23}, \boldsymbol{j} \to \boldsymbol{e}_{12}, \boldsymbol{k} \to \boldsymbol{e}_{31}$. Another important property of $\mathbb{H}$ is the phase concept. A polar representation of q is

$$q = |q|\, e^{\boldsymbol{i}\phi} e^{\boldsymbol{k}\psi} e^{\boldsymbol{j}\theta}, \tag{4.2}$$

where $|q| = \sqrt{q\bar{q}}$ where $\bar{q}$ is a conjugate of $q = a - b\boldsymbol{i} - c\boldsymbol{j} - d\boldsymbol{k}$ and the angles (ϕ, θ, ψ) represent the three quaternionic phases [5].

4.1. Quaternionic Atomic Function $\mathrm{qup}(x, y)$

Since a 2D signal can be split into even (e) and odd (o) parts [5],

$$f(x, y) = f_{\mathrm{ee}}(x, y) + f_{\mathrm{oe}}(x, y) + f_{\mathrm{eo}}(x, y) + f_{\mathrm{oo}}(x, y), \tag{4.3}$$

we can separate the four components of $\mathrm{up}(x, y)$ and represent it as a quaternion as follows [13, 14]:

$$\begin{aligned} \mathrm{qup}(x, y) &= \mathrm{up}(x, y)[\cos(w_x)\cos(w_y) \\ &\quad + \boldsymbol{i}\sin(w_x)\cos(w_y) + \boldsymbol{j}\cos(w_x)\sin(w_y) + \boldsymbol{k}\sin(w_x)\sin(w_y)] \\ &= \mathrm{qup}_{\mathrm{ee}}(x, y) + \boldsymbol{i}\,\mathrm{qup}_{\mathrm{oe}}(x, y) + \boldsymbol{j}\,\mathrm{qup}_{\mathrm{eo}}(x, y) + \boldsymbol{k}\,\mathrm{qup}_{\mathrm{oo}}(x, y). \end{aligned} \tag{4.4}$$

Figure 5 shows the quaternion atomic function $\mathrm{qup}(x, y)$ in the spatial domain with its four components: the real part $\mathrm{qup}_{\mathrm{ee}}(x, y)$ and the imaginary parts $\mathrm{qup}_{\mathrm{eo}}(x, y)$, $\mathrm{qup}_{\mathrm{oe}}(x, y)$, and $\mathrm{qup}_{\mathrm{oo}}(x, y)$. We can see even and odd symmetries in the horizontal, vertical, and diagonal axes.

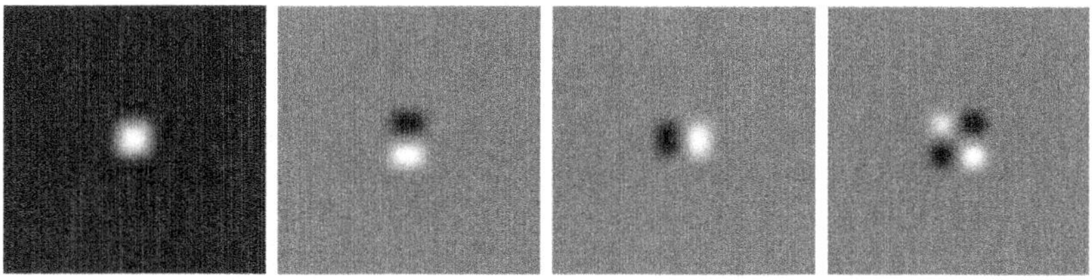

FIGURE 5. Atomic function $\mathrm{qup}(x, y)$. From left to right: $\mathrm{qup}_{\mathrm{ee}}(x, y)$, $\mathrm{qup}_{\mathrm{oe}}(x, y)$, $\mathrm{qup}_{\mathrm{eo}}(x, y)$, and $\mathrm{qup}_{\mathrm{oo}}(x, y)$.

4.2. Quaternionic Atomic Wavelet

The importance of wavelet transforms (real, complex, hypercomplex) has been discussed in many references [1, 18]. In this work, we use the quaternionic wavelet transform (QWT). The QWT can be seen as an extension of the complex wavelet transform (CWT) [1]. The multiresolution analysis applied in RWT and CWT can be straightforwardly extended to the QWT [1].

Our approach is based on the quaternionic Fourier transform (QFT), (4.5), which is also the basis for the quaternionic analytic function defined by Bülow [5]. The kernel of the 2D QFT is given by

$$e^{-i2\pi vx} f(x,y) e^{-j2\pi\nu y}, \tag{4.5}$$

where the real part is $\cos(2\pi\nu x)\cos(2\pi v y)$; the imaginary parts $(\boldsymbol{i},\boldsymbol{j},\boldsymbol{k})$ are its partial Hilbert transform $\cos(2\pi\nu x)\sin(2\pi v y)$, $\sin(2\pi\nu x)\cos(2\pi v y)$ (horizontal and vertical); and total Hilbert transform (diagonal) $\sin(2\pi\nu x)\sin(2\pi v y)$. Using the QFT basis, we can obtain the 2D phase information that satisfies the definition of the quaternionic analytic signal. To compute the multiscale approach, we use a 2D separable implementation. We independently apply two sets of h and g wavelet filters.

$$\begin{aligned} h &= \exp\left(\boldsymbol{i}\frac{c_1 u_1 x}{\sigma_1}\right) \mathrm{up}(x,y,\sigma_1) \exp\left(\boldsymbol{j}\frac{s_1 \omega v_1 y}{\sigma_1}\right) \\ &= h_{\mathrm{ee}} + \boldsymbol{i}h_{\mathrm{oe}} + \boldsymbol{j}h_{\mathrm{eo}} + \boldsymbol{k}h_{\mathrm{oo}} \end{aligned} \tag{4.6}$$

$$\begin{aligned} g &= \exp\left(\boldsymbol{i}\frac{c_2 w_2 x}{\sigma_2}\right) \mathrm{up}(x,y,\sigma_1) \exp\left(\boldsymbol{j}\frac{s_2 \omega v_2 y}{\sigma_2}\right) \\ &= g_{\mathrm{ee}} + \boldsymbol{i}g_{\mathrm{oe}} + \boldsymbol{j}g_{\mathrm{eo}} + \boldsymbol{k}g_{\mathrm{oo}}, \end{aligned} \tag{4.7}$$

where the extra τ parameter in the up function stands for the filter width. The procedure for quaternionic wavelet multiresolution analysis depicted partially in Figure 6 is as follows [1]:

1. Convolve the image (2D signal) at level n with the scale and wave filters h and g along the rows of the 2D signal. The latter filters are the discrete versions of those filters given in (4.6) and (4.7).
2. The h and g filters are convolved with the columns of the previous responses of the filters h and g.
3. Subsample the responses of these filters by a factor of 2 ($\uparrow 2$).
4. The real part of the approximation at level $n+1$ is taken as input at the next level. This process continues through all levels, repeating the steps just stated.

As can be seen in Figure 6, the low level of the pyramid is the highest level of resolution, and as you move up, the resolution decreases.

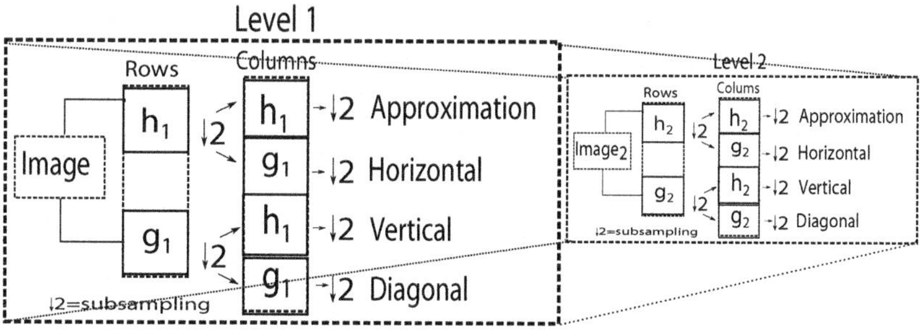

FIGURE 6. Multiresolution approach.

4.3. Steerable Quaternionic Filter

Due to our approach not being invariant to rotations [6], we require a filter bank in order to get different line or edge orientations. Figure 7 shows different orientation filters and real and imaginary parts. We steer the mother atomic function wavelet through a multiresolution pyramid in order to detect quaternionic phase changes, which can be used for the feature detection of lines, edges, centroids, and orientation in geometric structures.

5. Quaternionic Local Phase Information

In this chapter, we refer to the phase information as the local phase in order to separate the structure or geometric information and the amplitude in a certain part of the signal. Moreover, the phase information permits us to obtain invariant or equivariant[1] response. For instance, it has been shown that the phase has an invariant response to changes in image brightness and the phase can be used to measure the symmetry or asymmetry of objects [11, 2, 9]. These invariant and equivariant responses are the key part to link the low-level processing with the image analysis and the upper layers in computer vision applications.

The local phase means the computation of the phase at a certain position in a real signal. In 1D signals, the analytic signal based on the Hilbert transform $(f_{\boldsymbol{H}}(x))$ [5] is given by

$$f_A(f(x)) = f(x) + \boldsymbol{i} f_{\boldsymbol{H}}(x), \tag{5.1}$$

$$f_A(f(x)) = |A|\, e^{\boldsymbol{i}\theta}, \tag{5.2}$$

where $|A| = \sqrt{f(x)^2 + f_{\boldsymbol{H}}(x)^2}$ and $\theta = \arctan\left(\frac{f(x)}{f_{\boldsymbol{H}}(x)}\right)$ permits us to extract the magnitude and phase independently. In 2D signals, the Hilbert transform is not enough to compute the magnitude and phase independently in any direction [6]. In order to solve this, the quaternionic analytic (see (5.3)) signal and the monogenic

[1]Equivariance: monotonic dependency of value or parameter under some transformation.

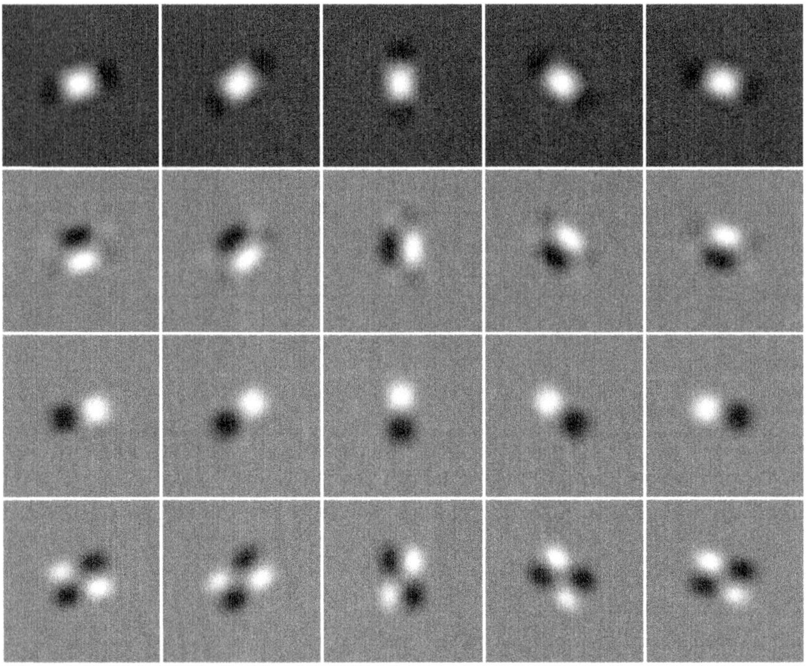

FIGURE 7. From left to right: qup(15°), qup(40°), qup(90°), qup(140°), and qup(155°)

signal have been proposed by Bülow [5] and Felsberg [6], respectively. Until now, we have used an approximation of the quaternionic analytic signal based on the basis of QFT to extract some oriented axis symmetries.

Figure 8 contains, at the top, an image with lines and edges, at the centre an image profile, and at the bottom the profiles of the three quaternionic phases. In the phase profiles, we can distinguish between a line (even) and an edge (odd) using the phase (θ). These results are similar to the results reported by Granlund [9] using only a complex phase, because they used an image that changes in one direction.

5.1. Quaternionic Analytic Signal

The quaternionic analytic signal in the spatial domain is defined as [5]:

$$f_A^q(x,y) = f(x,y) + \boldsymbol{i} f_{\boldsymbol{Hi}}(x,y) + \boldsymbol{j} f_{\boldsymbol{Hj}}(x,y) + \boldsymbol{k} f_{\boldsymbol{Hk}}(x,y), \tag{5.3}$$

where $f_{\boldsymbol{Hi}}(x,y) = f(x,y) * (\delta(y)/\pi x)$ and $f_{\boldsymbol{Hj}}(x,y) = f(x,y) * (\delta(x)/\pi y)$ are the partial Hilbert transforms and $f_{\boldsymbol{Hk}}(x,y) = f(x,y) * \big(\delta(x,y)/\pi^2 xy\big)$ is the total Hilbert transform. Bülow has shown that the QFT kernel is expressed in terms of the Hilbert transforms. The phases can be computed easily using a 3D rotation matrix $\mathcal{M}$, which can be factored into three rotations, $R = R_x(2\phi), R_z(2\psi), R_y(2\theta)$,

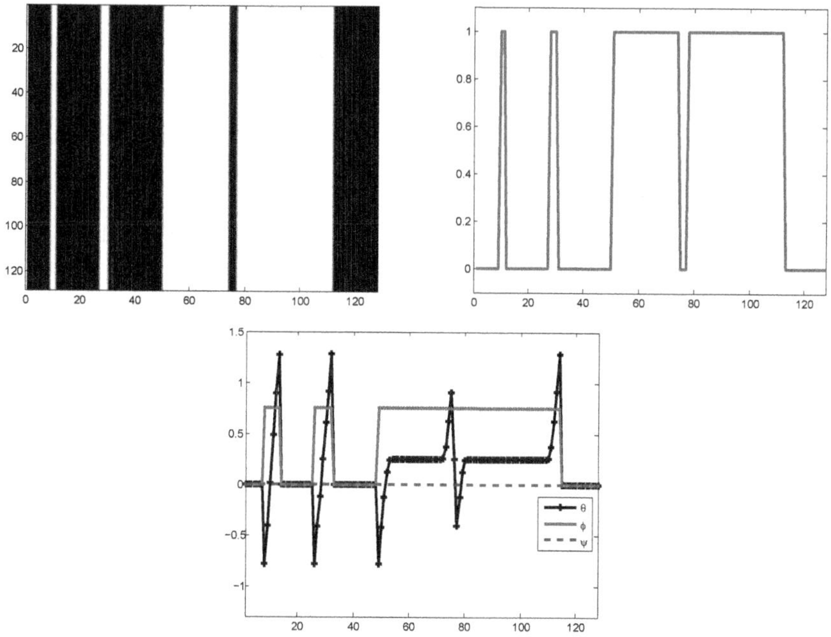

FIGURE 8. Image (top left), image profile (top right), and three quaternionic phases: profile, line, and edge (bottom).

in the coordinate axes [5], *i.e.*,

$$\mathcal{M}(q) = \mathcal{M}(q_1)\mathcal{M}(q_2)\mathcal{M}(q_3) \tag{5.4}$$

$$q_1 = e^{\boldsymbol{i}\phi}, \quad q_2 = e^{\boldsymbol{j}\theta}, \quad q_3 = e^{\boldsymbol{k}\psi} \tag{5.5}$$

$$\mathcal{M}(q) = \begin{pmatrix} a^2+b^2-c^2-d^2 & 2(bc-ad) & 2(bd+ac) \\ 2(bc+ad) & a^2-b^2+c^2-d^2 & 2(cd-ab) \\ 2(bd-ac) & 2(cd+ab) & a^2-b^2-c^2+d^2 \end{pmatrix} \tag{5.6}$$

$$R = \begin{pmatrix} \cos(2\psi)\cos(2\theta) & -\sin(2\psi) & \cos(2\psi)\sin(2\theta) \\ \begin{matrix}\cos(2\phi)\sin(2\psi)\cos(2\theta) \\ +\sin(2\phi)\sin(2\theta)\end{matrix} & \cos(2\phi)\cos(2\psi) & \begin{matrix}\cos(2\phi)\sin(2\psi)\sin(2\theta) \\ -\sin(2\phi)\cos(2\theta)\end{matrix} \\ \begin{matrix}\sin(2\phi)\sin(2\psi)\cos(2\theta) \\ -\cos(2\phi)\sin(2\theta)\end{matrix} & \sin(2\phi)\cos(2\psi) & \begin{matrix}\sin(2\phi)\sin(2\psi)\sin(2\theta) \\ +\cos(2\phi)\cos(2\theta)\end{matrix} \end{pmatrix}. \tag{5.7}$$

The quaternionic phases are expressed by the following rules:

$$\psi = -\frac{1}{2}\arcsin(2(bc-ad)). \tag{5.8}$$

- If $\psi \in \left]-\frac{\pi}{4}, \frac{\pi}{4}\right[$, then $\phi = \frac{1}{2}\arg_{\boldsymbol{i}}\left(q\mathcal{T}_{\boldsymbol{j}}\left(\overline{q}\right)\right)$ and $\theta = \frac{1}{2}\arg_{\boldsymbol{j}}\left(\mathcal{T}_{\boldsymbol{i}}\left(\overline{q}\right)q\right)$.
- If $\psi = \pm\frac{\pi}{4}$, then select either $\phi = 0$ and $\theta = \frac{1}{2}\arg_{\boldsymbol{j}}\left(\mathcal{T}_{\boldsymbol{k}}\left(\overline{q}\right)q\right)$ or $\theta = 0$ and $\phi = \frac{1}{2}\arg_{\boldsymbol{i}}\left(q\mathcal{T}_{\boldsymbol{k}}\left(\overline{q}\right)\right)$.
- If $e^{\boldsymbol{i}\phi}e^{\boldsymbol{k}\psi}e^{\boldsymbol{j}\theta} = -q$ and $\phi \geq 0$, then $\phi \to \phi - \pi$.
- If $e^{\boldsymbol{i}\phi}e^{\boldsymbol{k}\psi}e^{\boldsymbol{j}\theta} = -q$ and $\phi < 0$, then $\phi \to \phi + \pi$.

The phase ranges are $(\phi, \theta, \psi) \in [-\pi, \pi[\times \left[-\frac{\pi}{2}, \frac{\pi}{2}\right[\times \left[-\frac{\pi}{4}, \frac{\pi}{4}\right]$. The applications of the quaternionic analytic signal in image processing have to be limited in a narrow band, and Bülow used a Gauss window with the QFT kernel (which can be seen as Gabor) to approximate the quaternionic analytic function. In our work, instead of a Gauss window, we use a compact support window, the $\mathrm{up}(x, y)$ function.

5.2. Quaternionic Phase Analysis

There are many points of view to see the phase information. The local phase can be used as a measure of the symmetry in the 2D signals of the object [11, 9]. The symmetry is related to middle-level properties if it remains invariant under some transformation. A symmetric analysis related to the phase was proposed by P. Kovesi, using the even $(e_n(x))$ and odd $(o_n(x))$ responses of wavelets at scale n [11]:

$$\mathrm{Sym}(x) = \frac{\sum_n |e_n(x)| - |o_n(x)|}{\sum_n A_n(x)} \tag{5.9}$$

$$\mathrm{Asym}(x) = \frac{\sum_n |o_n(x)| - |e_n(x)|}{\sum_n A_n(x)} \tag{5.10}$$

where $A_n(x) = \sqrt{e_n(x)^2 + o_n(x)^2}$. Since the phase information has the capability to decode the geometrical information into even or odd symmetries (see Figure 8), we use the phase information to do a geometric analysis of the image. Note that the quaternionic phase analysis helps to reduce the gap between the low-level and the middle-level processing.

5.3. Hilbert Transform Using AF

The Hilbert transform and the derivative are closely related, and the Hilbert transform can actually be computed using a derivative and some convolution properties [19]:

$$f \star (g \star h) = (f \star g) \star h \tag{5.11}$$

$$\nabla (f \star g) = \nabla f \star g = f \star \nabla g, \tag{5.12}$$

where

$$f(x,y), g(x,y), h(x,y) \in \mathbb{R}^2 \quad \text{and} \quad \nabla = \boldsymbol{e}_1 \frac{\partial}{\partial x} + \boldsymbol{e}_2 \frac{\partial}{\partial y}.$$

If $g(x,y) = -(1/\pi)\log|x|\log|y|$ [2], and if we use the convolution distribution properties, we can express the Hilbert transform and the partial Hilbert transform (see (5.3)) as

$$f_{\boldsymbol{H}_i}(x,y) = \frac{\partial f(x,y)}{\partial x} \star -\frac{1}{\pi}\log|x| \tag{5.13}$$

$$f_{\boldsymbol{H}_j}(x,y) = \frac{\partial f(x,y)}{\partial y} \star -\frac{1}{\pi}\log|y| \tag{5.14}$$

$$f_{\boldsymbol{H}_k}(x,y) = \frac{\partial^2 f(x,y)}{\partial x \partial y} \star -\frac{1}{\pi^2}\log|x|\log|y| \tag{5.15}$$

and we can use the convolution association property to get the equation of a certain part of the signal in terms of dup(x):

$$f_{\boldsymbol{H}_i}(x,y) = f(x,y) \star \left(\operatorname{dup}(x,y)_x \star -\frac{1}{\pi}\log|x|\right) \tag{5.16}$$

$$f_{\boldsymbol{H}_j}(x,y) = f(x,y) \star \left(\operatorname{dup}(x,y)_y \star -\frac{1}{\pi}\log|y|\right) \tag{5.17}$$

$$f_{\boldsymbol{H}_k}(x,y) = f(x,y) \star \left(\operatorname{dup}(x,y)_{xy} \star -\frac{1}{\pi^2}\log|x|\log|y|\right). \tag{5.18}$$

Moreover, it has been shown by Petermichl *et al.* [17] that the Hilbert and Riesz transforms can be implemented with the average of dyadic shifts, and the dyadic shift operations appear naturally using the AF and its derivatives.

6. Applications

6.1. Convolution $I \star \operatorname{qup}(x,y)$

Figure 9 shows the convolution of the qup (real and imaginary parts) filter on a shadow chessboard image. The real part corresponds to a low-pass filtering of the image. In addition, we can see a selective detection of lines in the horizontal and

[2] $\log|x|$ is the fundamental solution of the Laplace equation.

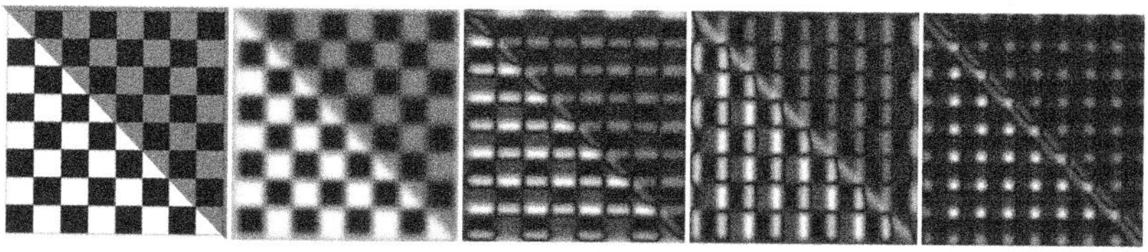

FIGURE 9. Convolution of qup(x,y) with chessboard image. From left to right: original image, real part, $\boldsymbol{i}$-part, $\boldsymbol{j}$-part, and $\boldsymbol{k}$-part.

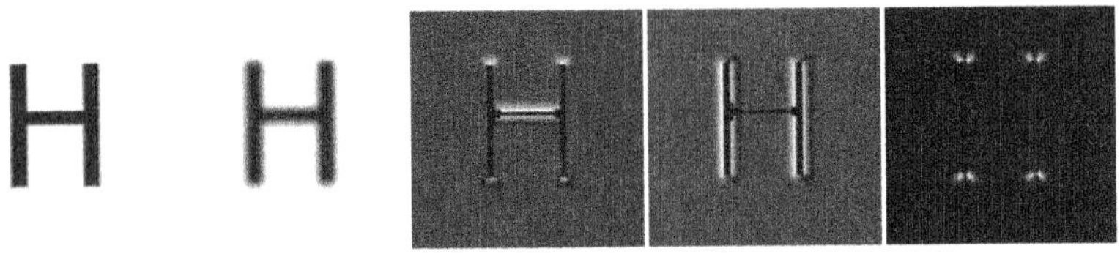

FIGURE 10. Convolution of $\mathrm{qup}(x, y)$ with image (letter H). From left to right: original image, real part, $\boldsymbol{i}$-part, $\boldsymbol{j}$-part, and $\boldsymbol{k}$-part.

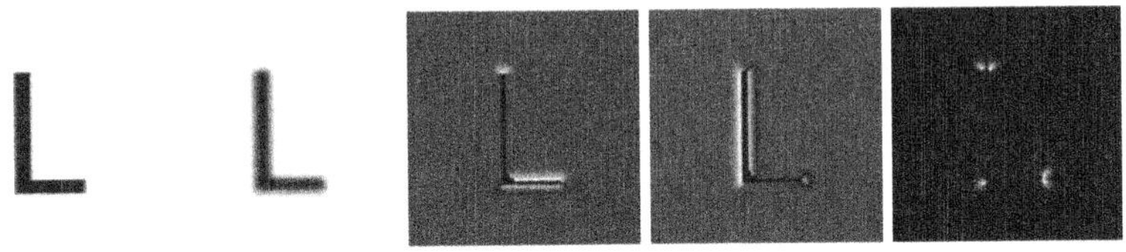

FIGURE 11. Convolution of $\mathrm{qup}(x, y)$ with image (letter L). From left to right: original image, real part, $\boldsymbol{i}$-part, $\boldsymbol{j}$-part, and $\boldsymbol{k}$-part are shown.

vertical orientations, particularly in the $\boldsymbol{i}$-part and $\boldsymbol{j}$-part. On the other hand, the $\boldsymbol{k}$-part can be used to detect the corners as a diagonal response. We can see how the response in the light part is more intensive than that in the shadowed part. We show in Figures 10 and 11 how the $\mathrm{qup}(x, y)$ filters respond to other images We can see a similar behaviour in Figure 9. We can also notice that the direct convolution of $\mathrm{qup}(x, y)$ with the image is sensitive to the contrast.

6.2. Derivative of $\mathrm{up}(x)$

Figure 12 illustrates the convolution of the first derivative, $\mathrm{dup}(x, y)$, and a shadow chessboard. The convolution with dup can be used as an oriented change detector with a simple rotation. Similarly to Figure 9, the convolved image has a low response in the shadow part.

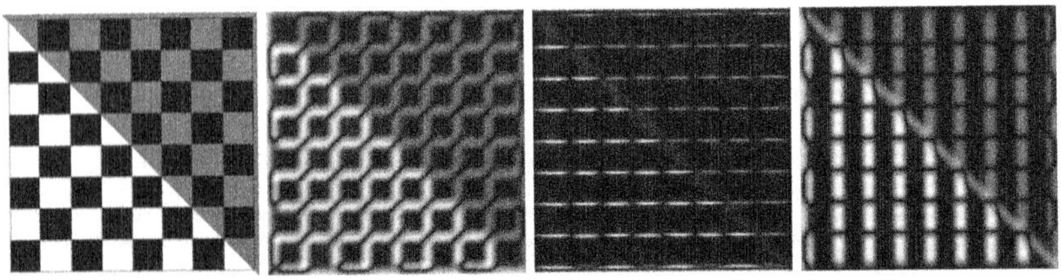

FIGURE 12. Convolution of $\mathrm{dup}(x, y)$ with the chess image. (a) Test image; (b) result of the convolution of the image with $\mathrm{dup}(x, y, 0°)$; (c) $\mathrm{dup}(x, y, 45°)$; (d) $\mathrm{dup}(x, y, 135°)$.

6.3. Image Processing Using Monogenic Signals

We use these three kernels to implement monogenic signals whereby the aperture of the filters are varied at three scales of a multiresolution pyramid. The same reference frequency and the same aperture rate variation are used for the three monogenic signals. In Figures 13, 14 and 15, you can appreciate at these scales the better consistency of the up(x) kernel particularly to detect *via* the phase concept

(a) Upper row: Gauss, Hx and Hy frequency; lower row: energy, θ, ψ

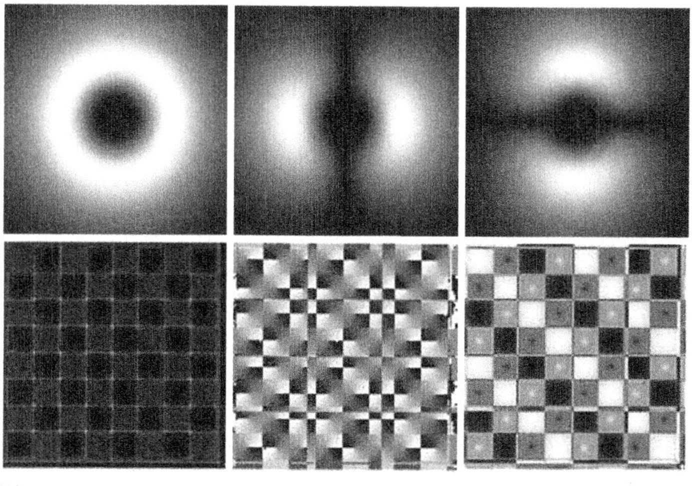

(b) Upper row: LogGabor, Hx and Hy frequency; lower row: energy, θ, ψ

FIGURE 13. Monogenic signal and response at the first level of filtering ($s = 1$). (a) Gabor monogenic; (b) LogGabor monogenic signal.

(c) Upper row: up, Hx and Hy frequency; lower row: energy, θ, ψ

FIGURE 13. Monogenic signal and response at the first level of filtering ($s = 1$). (c) up monogenic.

the corners and borders diminishing the expected blurring by a checkerboard image. We suspect that this an effect of the compactness in space of the up(x) kernel, the Moire effects are milder especially if you observe the images at scales 2 and 3, whereas in both the Gauss and LogGauss kernel-based monogenic phases there are noticeable artifacts in the form of bubbles or circular spots. These are often a result of Moire effects. In contrast the phases of the up(x) based monogenic signal still preserves the scales and edges of the checkerboard.

6.4. Quaternionic Local Phase

As we have mentioned before, the (local) phase information can be used to extract the structure information (line and edges) independently of illumination changes, and we can measure the symmetry (or asymmetry) with the phase information as a middle-level property. The performance of qup to detect edges or lines using the quaternionic phases in different images is shown in this section.

In Figure 17, we present different geometric figures with lines that represent the profiles (top, centre, and bottom). Similarly to Figure 8, the profiles of the images are shown as well as the quaternionic phase profile. This figure illustrates how the three phases have a different symmetric or antisymmetric response on the top, center, and bottom profile. These results motivate us to use the quaternionic phases as a measure of symmetry using the even or odd response of the quaternionic phases in a horizontal or vertical direction. In others words, the quaternionic phase information decomposes the image into oriented edges or lines.

As an example of other possible applications of the phase information in image analysis of geometric figures, we show Figure 18. The original image, the magnitude, and the three phases are shown. The first row of Figure 18 shows four squares with a different illumination and area; one of the squares is rotated, and this is the only square that can be seen in the ψ phase. In the second row, the four squares have been rotated 15°, and the ψ phase responds to the four squares, whereas in the third row, the four figures are rotated 45°, and the ψ phase only detects three of the four squares. In this example, the ψ phase only responds to the orientation of the square (a geometric transformation), independently of the size, contrast, or position of the image.

The magnitude and the quaternionic phases for a circle with lines are shown in Figure 19. In this image, we tuned the filter parameters in order to obtain a response (positive and negative) to diagonal lines. Again, in this image, the shadow or the size of the lines in the image does not change the phase response. The black lines correspond to a cone beam with a 45° orientation (±15°) and the white lines are oriented 135° (±15°).

In Figures 19 and 18, lines or geometric objects have been shown separately. Figure 20 illustrates an image with lines and geometric objects. In this case, we can use the quaternionic phases to extract different oriented edges of geometric objects or different oriented line textures. The θ and ϕ phases detect lines or edges in some vertical and horizontal directions, while the ψ phase detects the diagonal response and the corners in squares or geometric figures. Even if the geometric objects have internal lines, different illuminations, or positions, the edges of each square can

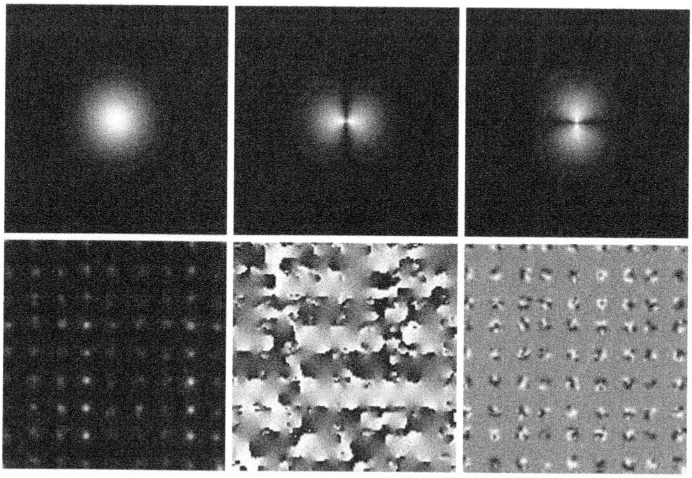

(a) Upper row: Gauss, Hx and Hy frequency; lower row: energy, θ, ψ

FIGURE 14. Monogenic signal and response at the second level of filtering ($s = 2$). (a) Gabor filters.

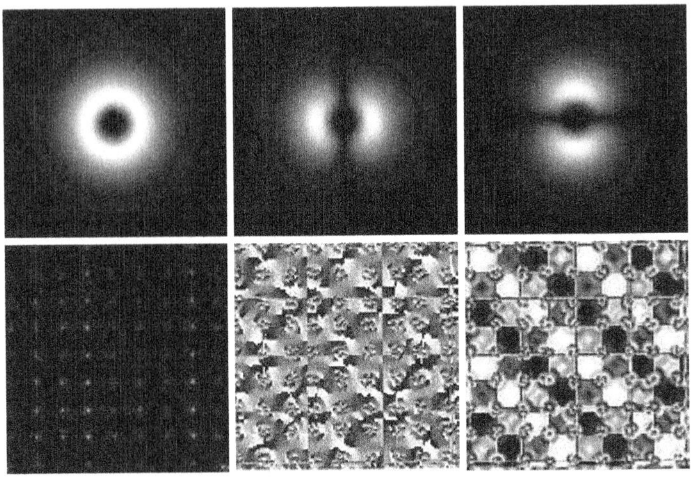

(b) Upper row: LogGabor, Hx and Hy frequency; lower row: energy, θ, ψ

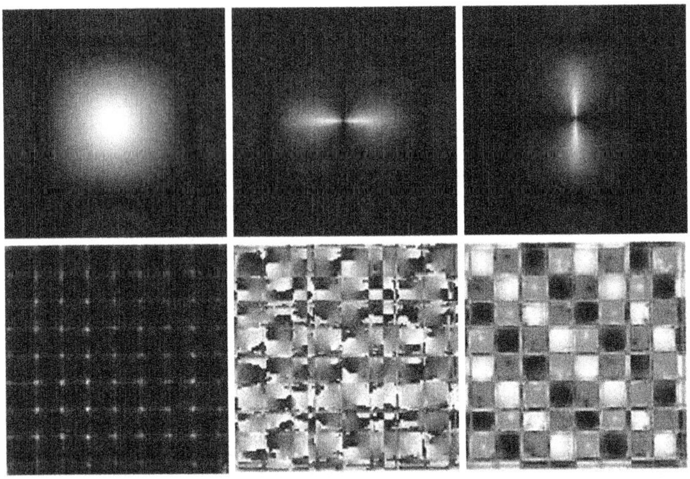

(c) Upper row: up, Hx and Hy frequency; lower row: energy, θ, ψ

FIGURE 14. Monogenic signal and response at the second level of filtering ($s = 2$). (b) LogGabor and (c) up filters.

be detected in the ϕ and θ images. Furthermore, in the ϕ phase, the horizontal lines are highlighted, whereas the vertical lines do not appear. The θ phases show a similar result, but in this case the vertical edges and lines are highlighted. In the ψ phase, the vertical lines or edges are highlighted.

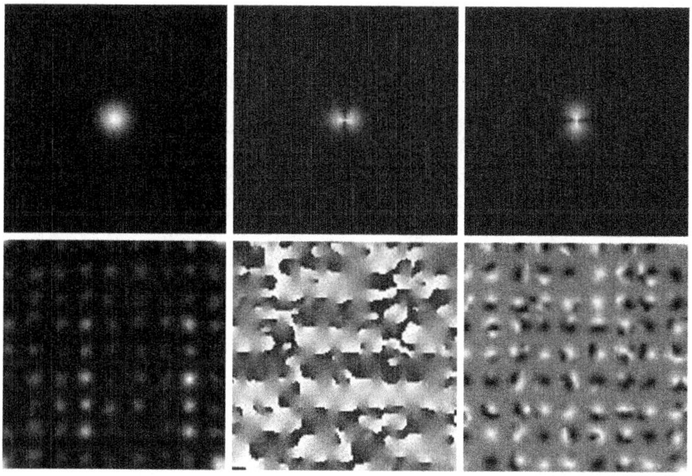

(a) Upper row: Gauss, Hx and Hy frequency; lower row: energy, θ, ψ

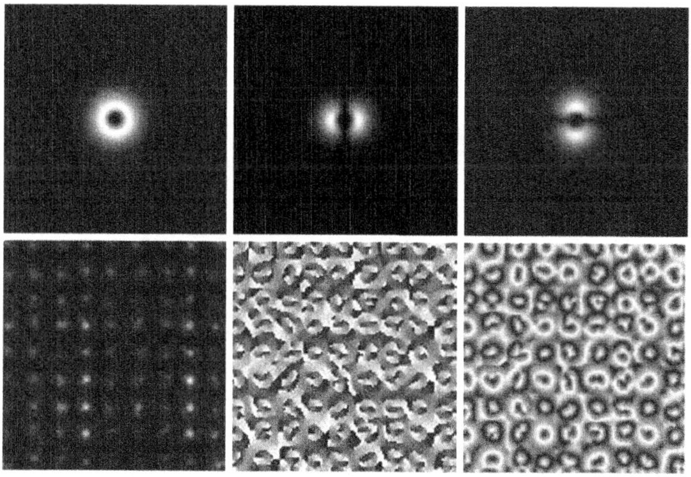

(b) Upper row: LogGabor, Hx and Hy frequency; lower row: energy, θ, ψ

FIGURE 15. Monogenic signal and response at the third level of filtering ($s = 3$). (a) Gabor and (b) LogGabor filters.

6.5. Phase Symmetry

As an example of many possible applications, we show the computation of the main axis in one orientation using the θ phase. We use the information of Figure 21 to determine a threshold. A reflection operation can be used to measure the symmetry.

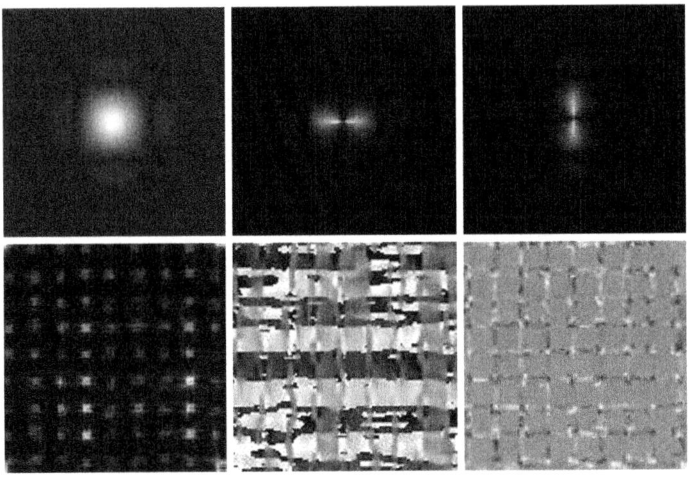

(c) Upper row: up, Hx and Hy frequency; lower row: energy, θ, ψ

FIGURE 15. Monogenic signal and response at the third level of filtering ($s = 3$). (c) up filters.

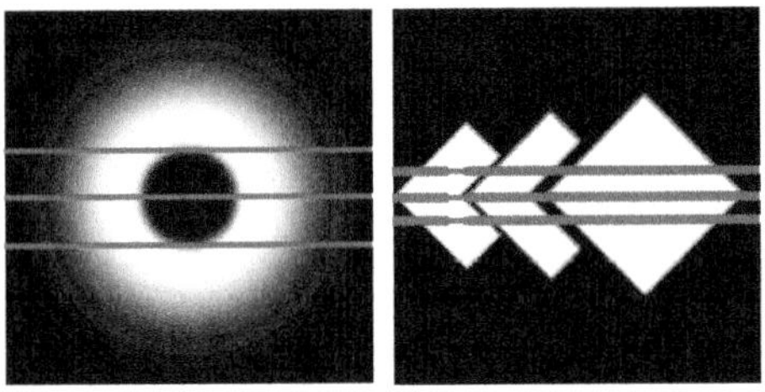

FIGURE 16. From left to right: circle; edges.

6.6. Multiresolution and Steerable Filters: $\mathrm{qup}(x, y)$

In this section, we present multiscale and steerable filter results. Figure 22 shows a circle and its multiscale processing. In the left column, we see a multiscale filtering. In this figure, the ψ phase responds better to diagonal lines, and we can see coarse to fine details. In the right group of the figure, we can see the effect of different orientations of our steerable filters.

7. Conclusions

In this work, we have presented image processing and analysis using an atomic function and the quaternionic phase concept. As we have shown, the atomic function up(x) has shown potential in image processing as a building block to build

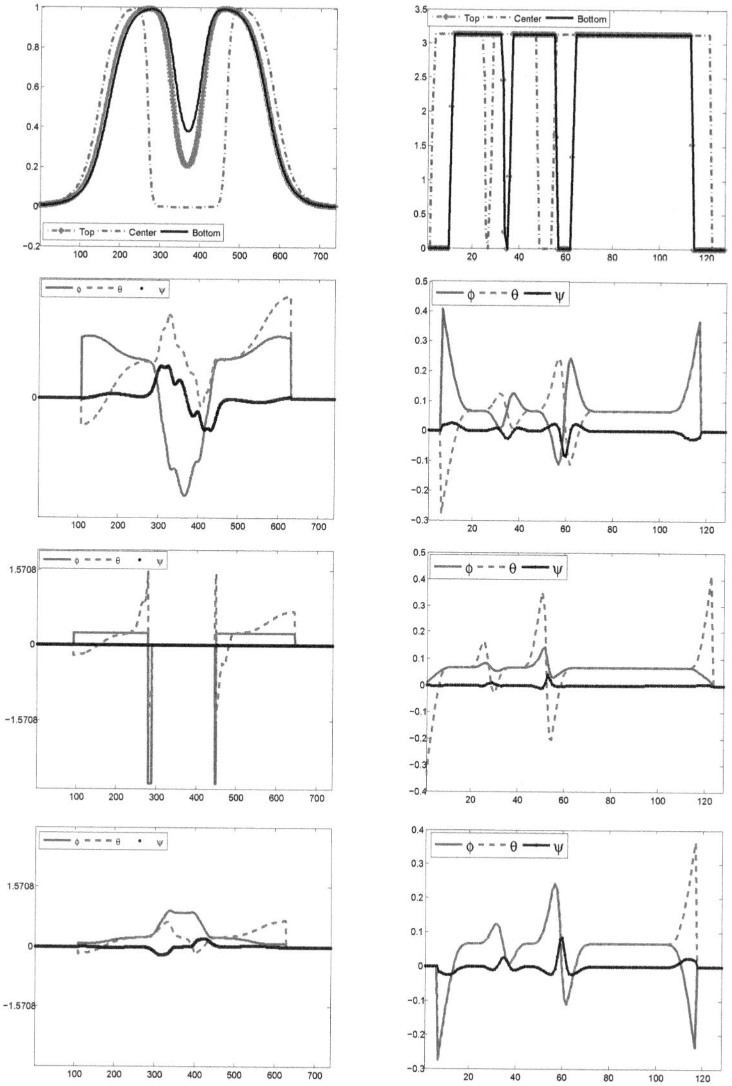

FIGURE 17. Left: quaternionic phase profiles of a circle. Top, centre and bottom phase profiles are shown. Right: quaternionic phase profiles of edge images. Top, centre and bottom phase profiles are shown.

multiple operations that can be done analytically such as low-pass filtering, derivatives, local phase, and multiscale and steering filters. We have shown that the function $\mathrm{qup}(x, y)$ is useful to detect lines or edges in a specific orientation using the quaternionic phase concept. Additionally, an oriented texture can be chosen

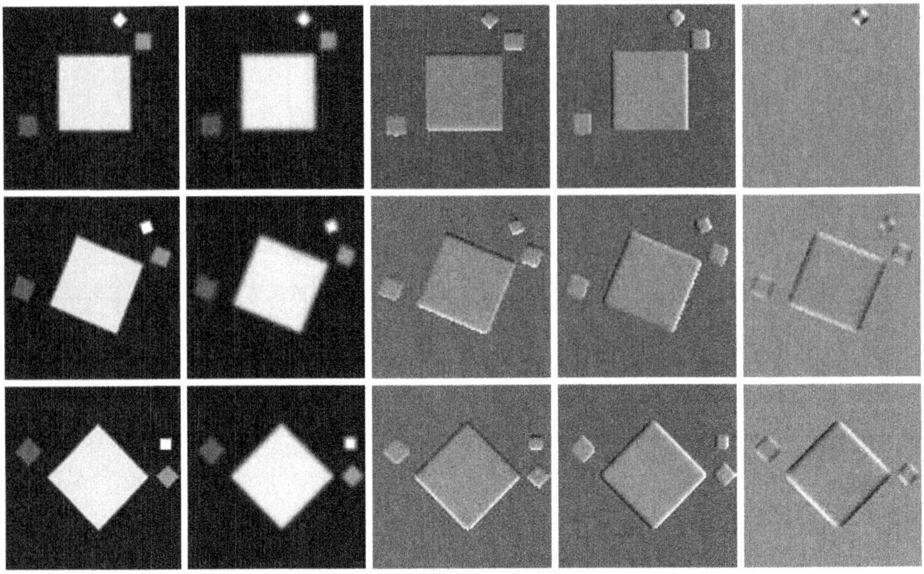

FIGURE 18. From left to right: image, magnitude, and the quaternionic phases ϕ, θ and ψ. We can see how the ψ phase responds to the rotated object.

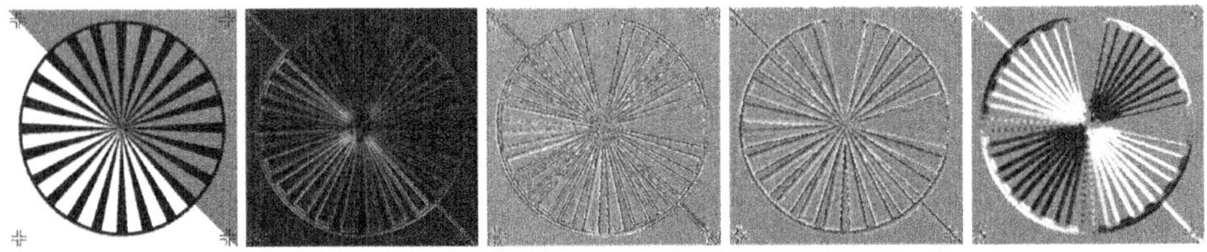

FIGURE 19. From left to right: image, magnitude, and the quaternionic phases. We can see how the ψ phase responds to the rotated object.

using the quaternionic phases. As an initial step, we have shown how to do the image analysis of geometric objects in $\mathbb{R}^2$ using the symmetry response of the phase. As in other applications of geometric algebra we can take advantage of the constraints. Since the information from the three phases is independent of illumination changes, algorithms using the quaternionic atomic function can be less sensitive than other methods based on the illumination changes. These results motivated us to find other invariants such as rotation invariants using the Riesz transform. In future work, we expect to develop a complete computer vision approach based on geometric algebra.

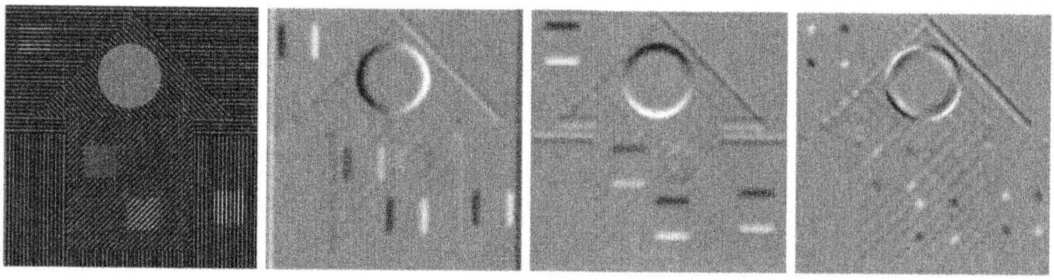

FIGURE 20. Image (left) and 2D quaternionic phases (left to right: ϕ, θ, ψ). A texture based on lines can be detected or discriminated, and at the same time, the phase information can highlight the edges.

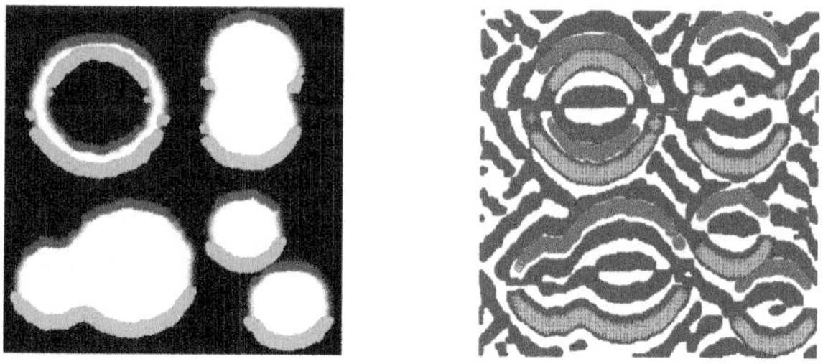

FIGURE 21. Circle image. The symmetry behaviour of the phase is shown.

References

[1] E. Bayro-Corrochano. The theory and use of the quaternion wavelet transform. *Journal of Mathematical Imaging and Vision*, 24:19–35, 2006.

[2] J. Bernd. *Digital Image Processing.* Springer-Verlag, New York, 1993.

[3] J. Bigun. *Vision with Direction.* Springer, 2006.

[4] F. Brackx, R. Delanghe, and F. Sommen. *Clifford Analysis*, volume 76. Pitman, Boston, 1982.

[5] T. Bülow. *Hypercomplex Spectral Signal Representations for the Processing and Analysis of Images.* PhD thesis, University of Kiel, Germany, Institut für Informatik und Praktische Mathematik, Aug. 1999.

[6] M. Felsberg. *Low-Level Image Processing with the Structure Multivector.* PhD thesis, Christian-Albrechts-Universität, Institut für Informatik und Praktische Mathematik, Kiel, 2002.

[7] M. Felsberg and G. Sommer. The monogenic signal. *IEEE Transactions on Signal Processing*, 49(12):3136–3144, Dec. 2001.

[8] A.S. Gorshkov, V.F. Kravchenko, and V.A. Rvachev. Estimation of the discrete derivative of a signal on the basis of atomic functions. *Izmeritelnaya Tekhnika*, 1(8):10, 1992.

[9] G.H. Granlund and H. Knutsson. *Signal Processing for Computer Vision.* Kluwer, Dordrecht, 1995.

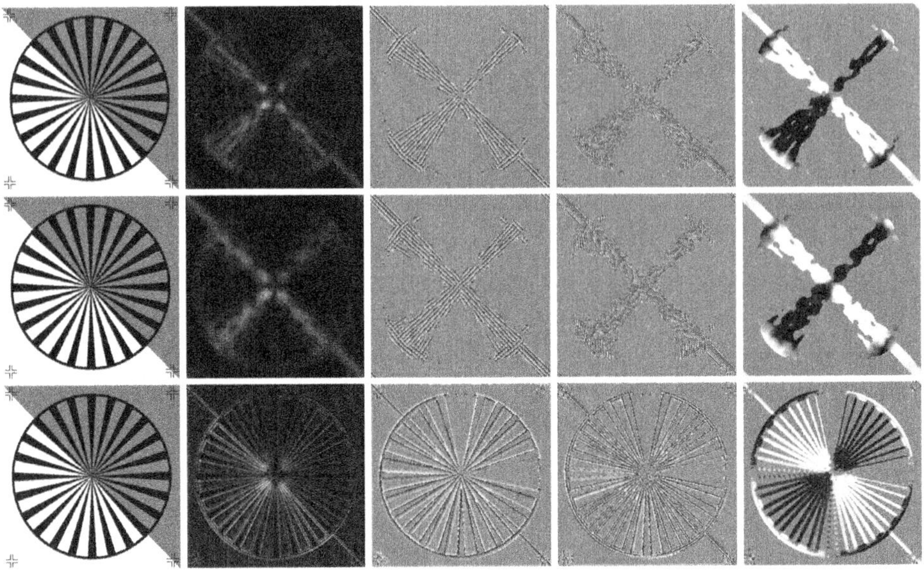

(a)

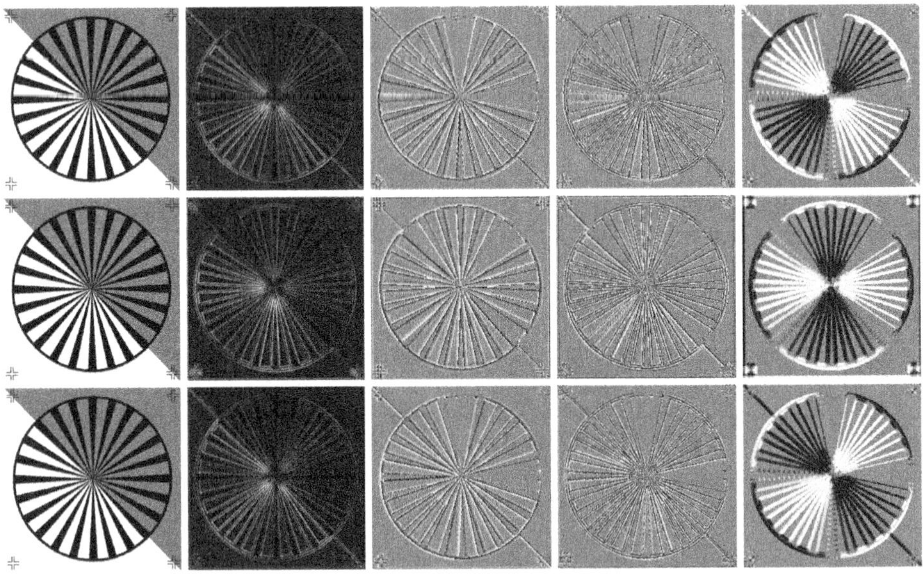

(b)

FIGURE 22. Multiresolution and steerable filters. Columns left to right: image, magnitude, quaternionic phases ϕ, θ, ψ. (a) multiscale approach. (b) steering filters top to bottom $0^\circ, 45^\circ, 90^\circ$ convolved with the image.

[10] V.M. Kolodyazhnya and V.A. Rvachov. Atomic functions: Generalization to the multivariable case and promising applications. *Cybernetics and Systems Analysis*, 46(6), 2007.

[11] P. Kovesi. *Invariant Measures of Image Features from Phase Information.* PhD thesis, University of Western Australia, Australia, 1996.

[12] V. Kravchenko, V. Ponomaryov, and H. Perez-Meana. *Adaptive digital processing of multidimensional signals with applications.* Moscow Fizmatlit, Moscow, 2010.

[13] E. Moya-Sánchez and E. Bayro-Corrochano. Quaternion atomic function wavelet for applications in image processing. In I. Bloch and R. Cesar, editors, *Progress in Pattern Recognition, Image Analysis, Computer Vision, and Applications*, volume 6419 of *Lecture Notes in Computer Science*, pages 346–353. Springer, 2010.

[14] E. Moya-Sánchez and E. Vázquez-Santacruz. A geometric bio-inspired model for recognition of low-level structures. In T. Honkela, W. Duch, M. Girolami, and S. Kaski, editors, *Artificial Neural Networks and Machine Learning – ICANN* 2011, volume 6792 of *Lecture Notes in Computer Science*, pages 429–436. Springer, 2011.

[15] M.N. Nabighian. Toward a three-dimensional automatic interpretation of potential field data via generalized Hilbert transforms: Fundamental relations. *Geophysics*, 49(6):780–786, June 1982.

[16] S. Petermichl. Dyadic shifts and logarithmic estimate for Hankel operators with matrix symbol. *Comptes Rendus de l'Académie des Sciences*, 330(6):455–460, 2000.

[17] S. Petermichl, S. Treil, and A. Volberg. Why are the Riesz transforms averages of the dyadic shifts? *Publicacions matematiques*, 46(Extra 1):209–228, 2002. Proceedings of the 6th International Conference on Harmonic Analysis and Partial Differential Equations, El Escorial (Madrid), 2002.

[18] I.W. Selesnick, R.G. Baraniuk, and N.G. Kingsbury. The dual-tree complex wavelet transform. *IEEE Signal Processing Magazine*, 22(6):123–151, Nov. 2005.

[19] B. Svensson. *A Multidimensional Filtering Framework with Applications to Local Structure Analysis and Image Enhancement.* PhD thesis, Linköping University, Linkoöping, Sweden, 2008.

E. Ulises Moya-Sánchez and E. Bayro-Corrochano
CINVESTAV
Campus Guadalajara, Av. del Bosque 1145
Colonia El Bajio, CP 45019
Zapopan, Jalisco, Mexico
e-mail: `emoya@gdl.cinvestav.mx`
`edb@gdl.cinvestav.mx`

Quaternion and Clifford–Fourier Transforms and Wavelets
Trends in Mathematics, 85–104

5 Bochner's Theorems in the Framework of Quaternion Analysis

S. Georgiev and J. Morais

Abstract. Let $\sigma(x)$ be a nondecreasing function, such that $\sigma(-\infty) = 0$, $\sigma(\infty) = 1$ and let us denote by $\mathcal{B}$ the class of functions which can be represented by a Fourier–Stieltjes integral $f(t) = \int_{-\infty}^{\infty} e^{itx} d\sigma(x)$. The purpose of this chapter is to give a characterization of the class $\mathcal{B}$ and to give a generalization of the classical theorem of Bochner in the framework of quaternion analysis.

Mathematics Subject Classification (2010). Primary 30G35; secondary 42A38.

Keywords. Quaternion analysis, quaternion Fourier transform, quaternion Fourier–Stieltjes integral, Bochner theorem.

1. Introduction

In a recent paper [5], we discussed special properties of the asymptotic behaviour of the quaternion Fourier transform (QFT) and provided a straightforward generalization of the classical Bochner–Minlos theorem to the framework of quaternion analysis. The main objective of the present chapter is to extend, using similar techniques, the theorem of Bochner on Fourier integral transforms of complex-valued functions of positive type to functions with values in the Hamiltonian quaternion algebra in which the exponential function is replaced by a (noncommutative) quaternion exponential product. The Fourier–Stieltjes transform (FST) is a well-known generalization of the standard Fourier transform, and is frequently applied in certain areas of theoretical and applied probability and stochastic processes contexts. The present study aims to develop further numerical integration methods for solving partial differential equations in the quaternion analysis setting.

The chapter is organized as follows. Section 2 recalls the classical Bochner theorem on the Fourier integral transforms of functions of positive type and collects some basic concepts in quaternion analysis. Section 3 defines and analyzes different types of quaternion Fourier–Stieltjes transforms (QFST) and establishes a number of their important properties. The underlying signals are continuous functions of bounded variation defined in $\mathbb{R}^2$ and taking values on the quaternion algebra. We proceed by proving the uniform continuity on this transform. Then, we describe the interplay between uniform continuity and quaternion distribution. Section 4 contains the main result of the chapter – the counterpart of the Bochner theorem for the noncommutative structure of quaternion functions (see Theorem 4.3 below). To prove our Bochner theorem, we introduce the notion of positive-type quaternion function and deduce some of its characteristics, and rely on the asymptotic behaviour and other general properties of the QFST. To the best of our knowledge this is done here for the first time. In the interests of simplicity of presentation, we have not extended this work to its most general form. Further investigations are in progress and will be reported in full in a forthcoming paper.

2. Preliminaries

2.1. Bochner Theorem

In this subsection, we review the classical Bochner theorem on Fourier integral transforms, which can be found, *e.g.*, in [1, 2].

A complex-valued function $f(t)$ defined on the interval $(-\infty,\infty)$ is said to be *positive definite* if it satisfies the following conditions:

1. f is bounded and continuous on $(-\infty,\infty)$;
2. $f(-t)=\overline{f(t)}$, for all t;
3. for any set of real numbers $t_1,\dots,t_s$, complex numbers $a_1,\dots,a_s$, and any positive integer s, the inequality $\sum_{m=1}^{s}\sum_{n=1}^{s} a_m\overline{a_n}f(t_m-t_n)\geq 0$ is satisfied.

The following theorem is due to Bochner [1].

Theorem 2.1 (Bochner). *If $\sigma(x)$ is a nondecreasing bounded function on the interval $(-\infty,\infty)$, and if $f(t)$ is defined by the Stieltjes integral*

$$f(t)=\int_{-\infty}^{\infty} e^{itx}d\sigma(x), \qquad -\infty<t<\infty, \tag{2.1}$$

then $f(t)$ is a continuous function of the positive type.

It is of interest to remark at this point that the Fourier–Stieltjes transform of nondecreasing bounded functions can be easily seen continuous functions of positive type. Conversely, if $f(t)$ is measurable on $(-\infty,\infty)$, and f is of the positive type, then there exists a nondecreasing bounded function $\sigma(x)$ such that $f(t)$ is given by (2.1) for almost all x, $-\infty<x<\infty$. In the converse part, Bochner assumed $f(t)$ to be continuous, and showed that $\sigma(x)$ is such that (2.1) is true for all t. Riesz, on the other hand, succeeded to prove that the measurability of f was sufficient in the converse.

2.2. Quaternion Analysis

The present subsection collects some basic facts about quaternions and the (left-sided) QFT, which will be needed throughout the text.

In all that follows let

$$\mathbb{H} := \{z = a + b\boldsymbol{i} + c\boldsymbol{j} + d\boldsymbol{k} : \ a, b, c, d \in \mathbb{R}\} \tag{2.2}$$

denote the Hamiltonian skew field, *i.e.*, quaternions, where the imaginary units $\boldsymbol{i}$, $\boldsymbol{j}$, and $\boldsymbol{k}$ are subject to the multiplication rules:

$$\begin{gathered} \boldsymbol{i}^2 = \boldsymbol{j}^2 = \boldsymbol{k}^2 = -1, \\ \boldsymbol{ij} = \boldsymbol{k} = -\boldsymbol{ji}, \qquad \boldsymbol{jk} = \boldsymbol{i} = -\boldsymbol{kj}, \qquad \boldsymbol{ki} = \boldsymbol{j} = -\boldsymbol{ik}. \end{gathered} \tag{2.3}$$

Like in the complex case, $\mathrm{S}(z) = a$ and $\mathbf{V}(z) = b\boldsymbol{i} + c\boldsymbol{j} + d\boldsymbol{k}$ define the scalar and vector parts of z. The conjugate of z is $\overline{z} = a - b\boldsymbol{i} - c\boldsymbol{j} - d\boldsymbol{k}$, and the norm of z is defined by

$$|z| = \sqrt{z\overline{z}} = \sqrt{\overline{z}z} = \sqrt{a^2 + b^2 + c^2 + d^2}, \tag{2.4}$$

which coincides with the corresponding Euclidean norm of z as a vector in $\mathbb{R}^4$. For $\mathbf{x} := (x_1, x_2) \in \mathbb{R}^2$ we consider $\mathbb{H}$-valued functions defined in $\mathbb{R}^2$, *i.e.*, functions of the form

$$\begin{gathered} f(\mathbf{x}) := [f(\mathbf{x})]_0 + [f(\mathbf{x})]_1\boldsymbol{i} + [f(\mathbf{x})]_2\boldsymbol{j} + [f(\mathbf{x})]_3\boldsymbol{k}, \\ [f]_l : \mathbb{R}^2 \longrightarrow \mathbb{R} \qquad (l = 0, 1, 2, 3). \end{gathered} \tag{2.5}$$

Properties (like integrability, continuity or differentiability) that are ascribed to f have to be fulfilled by all its components $[f]_l$.

Let $L^1(\mathbb{R}^2; \mathbb{H})$ denote the linear space of integrable $\mathbb{H}$-valued functions defined in $\mathbb{R}^2$. The *left-sided QFT* of $f \in L^1(\mathbb{R}^2; \mathbb{H})$ is given by [6]

$$\mathcal{F}(f) : \mathbb{R}^2 \to \mathbb{H}, \qquad \mathcal{F}(f)(\boldsymbol{\omega}) := \int_{\mathbb{R}^2} \mathbf{e}(\boldsymbol{\omega}, \mathbf{x})\, f(\mathbf{x})\, d^2\mathbf{x}, \tag{2.6}$$

where the kernel function

$$\mathbf{e} : \mathbb{R}^2 \times \mathbb{R}^2 \to \mathbb{H}, \qquad \mathbf{e}(\boldsymbol{\omega}, \mathbf{x}) := e^{-\boldsymbol{j}\omega_2 x_2} e^{-\boldsymbol{i}\omega_1 x_1}, \tag{2.7}$$

is called the (left-sided) quaternion Fourier kernel. For $i = 1, 2$, x_i will denote the space coordinates and ω_i the angular frequencies. The previous definition of the QFT varies from the original one only in the fact that we use 2D vectors instead of scalars and that it is defined to be 2D. It is of interest to remark at this point that the product in (2.6) has to be performed in a fixed order since, in general, $\mathbf{e}(\boldsymbol{\omega}, \mathbf{x})$ does not commute with every element of the algebra.

Under suitable conditions, the original signal f can be reconstructed from $\mathcal{F}(f)$ by the inverse transform. The *(left-sided) inverse QFT* of $g \in L^1(\mathbb{R}^2; \mathbb{H})$ is given by

$$\mathcal{F}^{-1}(g) : \mathbb{R}^2 \to \mathbb{H}, \qquad \mathcal{F}^{-1}(g)(\mathbf{x}) = \frac{1}{(2\pi)^2} \int_{\mathbb{R}^2} \overline{\mathbf{e}(\boldsymbol{\omega}, \mathbf{x})}\, g(\boldsymbol{\omega})\, d^2\boldsymbol{\omega}\,, \tag{2.8}$$

where $\overline{\mathbf{e}(\boldsymbol{\omega}, \mathbf{x})}$ is called the inverse (left-sided) quaternion Fourier kernel.

3. Quaternion Fourier–Stieltjes Transform and its Properties

This section generalizes the classical FST to Hamilton's quaternion algebra. Using the 4D analogue of the (complex-valued) function $\sigma(x)$ described before, we extend the FST to the QFST. As we shall see later, some properties of the FST can be extended in this context.

3.1. Quaternion Fourier–Stieltjes Transform

Partially motivated by the results from Bülow [3], the idea behind the construction of a quaternion counterpart of the Stieltjes integral is to replace the exponential function in (2.1) by a suitable (noncommutative) quaternion exponential product.

In the sequel, we consider the functions $\sigma^1, \sigma^2 : \mathbb{R} \to \mathbb{H}$ of the quaternion form (2.5), such that $|\sigma^i| \leq \delta_i$ for real numbers $\delta_i < \infty$ $(i = 1, 2)$. Due to the noncommutativity of the quaternions, we shall define three different types of QFST.

Definition 3.1. The *QFST* $\mathcal{FS}(\sigma^1, \sigma^2) : \mathbb{R}^2 \to \mathbb{H}$ of $\sigma^1(x_1)$ and $\sigma^2(x_2)$ is defined as the Stieltjes integrals:

1. Right-sided QFST:
$$\mathcal{FS}_r(\sigma^1, \sigma^2)(\omega_1, \omega_2) := \int_{\mathbb{R}^2} d\sigma^1(x_1) d\sigma^2(x_2) e^{i\omega_1 x_1} e^{j\omega_2 x_2}. \tag{3.1}$$
2. Left-sided QFST:
$$\mathcal{FS}_l(\sigma^1, \sigma^2)(\omega_1, \omega_2) := \int_{\mathbb{R}^2} e^{j\omega_2 x_2} e^{i\omega_1 x_1} d\sigma^1(x_1) d\sigma^2(x_2). \tag{3.2}$$
3. Two-sided QFST:
$$\mathcal{FS}_s(\sigma^1, \sigma^2)(\omega_1, \omega_2) := \int_{\mathbb{R}^2} e^{i\omega_1 x_1} d\sigma^1(x_1) d\sigma^2(x_2) e^{j\omega_2 x_2}. \tag{3.3}$$

Remark 3.2. We remind the reader that, the order of the exponentials in (3.1)–(3.3) is fixed because of the noncommutativity of the quaternion product. It is of interest to remark at this point that in case that $d\sigma^1 d\sigma^2 = f(\mathbf{x}) d^2\mathbf{x}$, the formulae above reduce to the usual definitions for the right-, left- and two-sided QFT [4, 7, 6], which only differ by the signs in the exponential kernel terms.

It is significant to note that, in practice, the integrals (3.1)–(3.3) will always exist. For example, for any two real variables ω_1, ω_2, and for real constants a, b it holds that

$$\begin{aligned} |\mathcal{FS}_r(\sigma^1, \sigma^2)| \leq & \left| \int_{-\infty}^{a} \left(\int_{-\infty}^{b} + \int_{b}^{\infty} \right) d\sigma^1(x_1) d\sigma^2(x_2) e^{i\omega_1 x_1} e^{j\omega_2 x_2} \right| \\ & + \left| \int_{a}^{\infty} \left(\int_{-\infty}^{b} + \int_{b}^{\infty} \right) d\sigma^1(x_1) d\sigma^2(x_2) e^{i\omega_1 x_1} e^{j\omega_2 x_2} \right| \\ \leq & \left(|\sigma^1(\infty) - \sigma^1(b)| + |\sigma^1(b) - \sigma^1(-\infty)| \right) \\ & \times \left(|\sigma^2(\infty) - \sigma^2(a)| + |\sigma^2(a) - \sigma^2(-\infty)| \right). \end{aligned} \tag{3.4}$$

Remark 3.3. Throughout this text we will only investigate the integral (3.1), which we denote for simplicity by $\mathcal{FS}(\sigma^1, \sigma^2)$. Nevertheless, all computations can be easily adapted to (3.2) and (3.3). In view of (3.1) and (3.2), a straightforward calculation shows that

$$\begin{aligned} \mathcal{FS}_r(\sigma^1, \sigma^2)(\omega_1, \omega_2) &= \overline{\int_{\mathbb{R}^2} e^{-j\omega_2 x_2} e^{-i\omega_1 x_1} \overline{d\sigma^1(x_1) d\sigma^2(x_2)}} \\ &= \overline{\mathcal{FS}_l(\overline{\sigma^2}, \overline{\sigma^1})}(-\omega_1, -\omega_2). \end{aligned} \tag{3.5}$$

For the sake of further simplicity, in the considerations to follow we will often omit the subscript and, additionally, write only $\mathcal{FS}(\omega_1,\omega_2)$ instead of $\mathcal{FS}(\sigma^1,\sigma^2)(\omega_1,\omega_2)$. We can restate the definition of right-sided QFST (3.1) in equivalent terms as follows.

Lemma 3.4. *The (right-sided) QFST has the closed-form representation*

$$\mathcal{FS}(\omega_1, \omega_2) := [\Phi(\omega_1, \omega_2)]_0 + [\Phi(\omega_1, \omega_2)]_1 + [\Phi(\omega_1, \omega_2)]_2 + [\Phi(\omega_1, \omega_2)]_3, \tag{3.6}$$

where we used the integrals

$$[\Phi(\omega_1, \omega_2)]_0 = \int_{\mathbb{R}^2} d\sigma^1(x_1) d\sigma^2(x_2) \cos(\omega_1 x_1) \cos(\omega_2 x_2), \tag{3.7}$$

$$[\Phi(\omega_1, \omega_2)]_1 = \int_{\mathbb{R}^2} d\sigma^1(x_1) d\sigma^2(x_2) \boldsymbol{i} \sin(\omega_1 x_1) \cos(\omega_2 x_2), \tag{3.8}$$

$$[\Phi(\omega_1, \omega_2)]_2 = \int_{\mathbb{R}^2} d\sigma^1(x_1) d\sigma^2(x_2) \boldsymbol{j} \cos(\omega_1 x_1) \sin(\omega_2 x_2), \tag{3.9}$$

$$[\Phi(\omega_1, \omega_2)]_3 = \int_{\mathbb{R}^2} d\sigma^1(x_1) d\sigma^2(x_2) \boldsymbol{k} \sin(\omega_1 x_1) \sin(\omega_2 x_2). \tag{3.10}$$

Corollary 3.5. *The (right-sided) QFST satisfies the following identities:*

$$\mathcal{FS}(\omega_1, \omega_2) + \mathcal{FS}(\omega_1, -\omega_2) = 2\left(\Phi_0(\omega_1, \omega_2) + \Phi_1(\omega_1, \omega_2)\right), \tag{3.11}$$

$$\mathcal{FS}(\omega_1, \omega_2) - \mathcal{FS}(\omega_1, -\omega_2) = 2\left(\Phi_2(\omega_1, \omega_2) + \Phi_3(\omega_1, \omega_2)\right), \tag{3.12}$$

$$\mathcal{FS}(\omega_1, \omega_2) + \mathcal{FS}(-\omega_1, \omega_2) = 2\left(\Phi_0((\omega_1, \omega_2) + \Phi_2(\omega_1, \omega_2)\right), \tag{3.13}$$

$$\mathcal{FS}(\omega_1, \omega_2) - \mathcal{FS}(-\omega_1, \omega_2) = 2\left(\Phi_1(\omega_1, \omega_2) + \Phi_3(\omega_1, \omega_2)\right), \tag{3.14}$$

$$\mathcal{FS}(\omega_1, \omega_2) + \mathcal{FS}(-\omega_1, -\omega_2) = 2\left(\Phi_0(\omega_1, \omega_2) + \Phi_3(\omega_1, \omega_2)\right), \tag{3.15}$$

$$\mathcal{FS}(\omega_1, \omega_2) - \mathcal{FS}(-\omega_1, -\omega_2) = 2\left(\Phi_1(\omega_1, \omega_2) + \Phi_2(\omega_1, \omega_2)\right), \tag{3.16}$$

$$\mathcal{FS}(\omega_1, -\omega_2) + \mathcal{FS}(-\omega_1, -\omega_2) = 2\left(\Phi_0(\omega_1, \omega_2) - \Phi_2(\omega_1, \omega_2)\right), \tag{3.17}$$

$$\mathcal{FS}(\omega_1, -\omega_2) - \mathcal{FS}(-\omega_1, -\omega_2) = 2\left(\Phi_1(\omega_1, \omega_2) - \Phi_3(\omega_1, \omega_2)\right), \tag{3.18}$$

$$\mathcal{FS}(-\omega_1, \omega_2) + \mathcal{FS}(-\omega_1, -\omega_2) = 2\left(\Phi_0(\omega_1, \omega_2) - \Phi_1(\omega_1, \omega_2)\right), \tag{3.19}$$

$$\mathcal{FS}(-\omega_1, \omega_2) - \mathcal{FS}(-\omega_1, -\omega_2) = 2\left(\Phi_2(\omega_1, \omega_2) - \Phi_3(\omega_1, \omega_2)\right). \tag{3.20}$$

3.2. Properties

This subsection presents certain properties of the asymptotic behaviour of the QFST, and establishes some of its basic properties.

We begin with the notions of monotonic increasing, bounded variation and distribution in the context of quaternion analysis.

Definition 3.6. A function $\sigma : \mathbb{R} \to \mathbb{H}$ is called *monotonic increasing*, if all of its components $[\sigma]_l$ $(l = 0, 1, 2, 3)$ are monotonic increasing functions.

Definition 3.7. A function $\sigma : \mathbb{R} \to \mathbb{H}$ is called *with bounded variation* on $\mathbb{R}$, if there exists a real number $M < \infty$ such that $\int_{\mathbb{R}} |d\sigma(x)| < M$.

Definition 3.8. A function $\sigma : \mathbb{R} \to \mathbb{H}$ is said to be a *quaternion distribution*, if it is of bounded variation and monotonic increasing, and if the following limits exist

$$\lim_{x \longrightarrow y+} \sigma(x) = \sigma(y+0), \quad \text{and} \quad \lim_{x \longrightarrow y-} \sigma(x) = \sigma(y-0), \tag{3.21}$$

(taken over all directions) for which

$$\sigma(y) = \frac{1}{2}\left[\sigma(y+0) + \sigma(y-0)\right] \tag{3.22}$$

holds almost everywhere on $\mathbb{R}$.

To proceed with, it is significant to note that, for every two functions $\sigma^1, \sigma^2 : \mathbb{R} \to \mathbb{H}$ of the quaternion form (2.5), the study of the properties of the distribution

$$\mathcal{FS}(\omega_1, \omega_2) := \int_{\mathbb{R}^2} d\sigma^1(x_1) d\sigma^2(x_2) e^{i\omega_1 x_1} e^{j\omega_2 x_2} \tag{3.23}$$

is reduced to the separate study of each component

$$\int_{\mathbb{R}^2} \left[d\sigma^1(x_1)\right]_l \left[d\sigma^2(x_2)\right]_m e^{i\omega_1 x_1} e^{j\omega_2 x_2} \qquad (l, m = 0, 1, 2, 3). \tag{3.24}$$

From now on, we denote the class of functions which can be represented as (3.23) by $\mathcal{B}$. Functions in $\mathcal{B}$ are called (*right*) *quaternion Bochner functions* and $\mathcal{B}$ will be referred to as the (*right*) *quaternion Bochner set*. It follows that all members of $\mathcal{B}$ are entire functions of the real variables ω_1, ω_2.

Firstly, we shall note the following property of $\mathcal{B}$:

Proposition 3.9. *$\mathcal{B}$ is a linear space.*

Proof. Let $f, g \in \mathcal{B}$ and z_1, z_2, z_3, z_4 be quaternion numbers. It is easily seen that

$$z_1 f z_2 + z_3 g z_4 \in \mathcal{B}. \tag{3.25}$$

□

The foregoing discussion suggests the computation of certain equalities for quaternion Bochner functions. If $f \in \mathcal{B}$ is given, then

$$f(0, \omega_2) = \left(\sigma^1(\infty) - \sigma^1(-\infty)\right) \int_{\mathbb{R}} d\sigma^2(x_2) e^{j\omega_2 x_2}, \tag{3.26}$$

$$f(\omega_1, 0) = \int_{\mathbb{R}} d\sigma^1(x_1) \left(\sigma^2(\infty) - \sigma^2(-\infty)\right) e^{i\omega_1 x_1}, \tag{3.27}$$

$$f(0,0) = \left(\sigma^1(\infty) - \sigma^1(-\infty)\right)\left(\sigma^2(\infty) - \sigma^2(-\infty)\right). \tag{3.28}$$

In particular, a simple argument gives

$$\begin{aligned}
f(-\omega_1,-\omega_2) &= \int_{\mathbb{R}^2} d\sigma^1(x_1)d\sigma^2(x_2)e^{-i\omega_1 x_1}e^{-j\omega_2 x_2}\\
&= \overline{\int_{\mathbb{R}^2} e^{j\omega_2 x_2}e^{i\omega_1 x_1}\overline{d\sigma^1(x_1)d\sigma^2(x_2)}}\\
&= \overline{g(\omega_1,\omega_2)},
\end{aligned} \tag{3.29}$$

where g is any function which can be represented as $\overline{\mathcal{FS}_l(\overline{\sigma^2},\overline{\sigma^1})}(\omega_1,\omega_2)$. We now have the following property:

Proposition 3.10. *Every element of $\mathcal{B}$ is a continuous bounded function.*

Proof. Let f be any function in $\mathcal{B}$. The continuity of f is obvious. For the boundedness a direct computation shows that

$$\begin{aligned}
|f(\omega_1,\omega_2)| &\le \int_{\mathbb{R}^2} |d\sigma^1(x_1)d\sigma^2(x_2)|\,|e^{i\omega_1 x_1}e^{j\omega_2 x_2}| \qquad (3.30)\\
&\le |\sigma^1(\infty)-\sigma^1(-\infty)|\,|\sigma^2(\infty)-\sigma^2(-\infty)|\,. \qquad \square
\end{aligned}$$

Suppose now that $\sigma^1,\sigma^2:\mathbb{R}\to\mathbb{H}$ are fixed, and $f\in\mathcal{B}$ does not vanish identically. Question: Depending on the parity of $\sigma^1(x_1)$ and $\sigma^2(x_2)$, what can we say about the representation of f? If there is any similarity or difference in the parity of these functions, does this lead to some particular functions $f\in\mathcal{B}$? The answers are given by the following proposition:

Proposition 3.11. *Let $\sigma^1,\sigma^2:\mathbb{R}\to\mathbb{H}$, and $f\in\mathcal{B}$ be given.*

1. *If $\sigma^1(x_1)$ is an odd function then*

$$\begin{aligned}
&f(\omega_1,\omega_2) = 2\int_{-\infty}^{\infty}\int_0^{\infty} d\sigma^1(x_1)d\sigma^2(x_2)\boldsymbol{i}\sin(\omega_1 x_1)e^{\boldsymbol{j}\omega_2 x_2},\\
&f(0,\omega_2)=0.
\end{aligned} \tag{3.31}$$

2. *If $\sigma^1(x_1)$ is an even function then*

$$\begin{aligned}
&f(\omega_1,\omega_2) = 2\int_{-\infty}^{\infty}\int_0^{\infty} d\sigma^1(x_1)d\sigma^2(x_2)\cos(\omega_1 x_1)e^{\boldsymbol{j}\omega_2 x_2},\\
&f(0,0) = 2\left(\sigma^1(\infty)-\sigma^1(0)\right)\left(\sigma^2(\infty)-\sigma^2(-\infty)\right).
\end{aligned} \tag{3.32}$$

3. *If $\sigma^2(x_2)$ is an odd function then*

$$\begin{aligned}
&f(\omega_1,\omega_2) = 2\int_0^{\infty}\int_{-\infty}^{\infty} d\sigma^1(x_1)d\sigma^2(x_2)e^{\boldsymbol{i}\omega_1 x_1}\sin(\omega_2 x_2)\boldsymbol{j},\\
&f(\omega_1,0)=0.
\end{aligned} \tag{3.33}$$

4. *If $\sigma^2(x_2)$ is an even function then*

$$f(\omega_1,\omega_2) = 2\int_0^\infty\int_{-\infty}^\infty d\sigma^1(x_1)d\sigma^2(x_2)e^{i\omega_1x_1}\cos(\omega_2x_2),$$
$$f(0,0) = 2\left(\sigma^1(\infty)-\sigma^1(-\infty)\right)\left(\sigma^2(\infty)-\sigma^2(0)\right). \tag{3.34}$$

5. *If $\sigma^1(x_1)$ and $\sigma^2(x_2)$ are both odd functions then*

$$f(\omega_1,\omega_2) = 4\int_0^\infty\int_0^\infty d\sigma^1(x_1)d\sigma^2(x_2)\boldsymbol{k}\sin(\omega_1x_1)\sin(\omega_2x_2),$$
$$f(\omega_1,0) = f(0,\omega_2) = 0. \tag{3.35}$$

6. *If $\sigma^1(x_1)$ is an odd function and $\sigma^2(x_2)$ an even function then*

$$f(\omega_1,\omega_2) = 4\int_0^\infty\int_0^\infty d\sigma^1(x_1)d\sigma^2(x_2)\boldsymbol{i}\sin(\omega_1x_1)\cos(\omega_2x_2),$$
$$f(0,\omega_2) = 0. \tag{3.36}$$

7. *If $\sigma^1(x_1)$ is an even function and $\sigma^2(x_2)$ an odd function then*

$$f(\omega_1,\omega_2) = 4\int_0^\infty\int_0^\infty d\sigma^1(x_1)d\sigma^2(x_2)\boldsymbol{j}\cos(\omega_1x_1)\sin(\omega_2x_2),$$
$$f(\omega_1,0) = 0. \tag{3.37}$$

8. *If $\sigma^1(x_1)$ and $\sigma^2(x_2)$ are both even functions then*

$$f(\omega_1,\omega_2) = 4\int_0^\infty\int_0^\infty d\sigma^1(x_1)d\sigma^2(x_2)\cos(\omega_1x_1)\cos(\omega_2x_2),$$
$$f(0,0) = 4\left(\sigma^1(\infty)-\sigma^1(0)\right)\left(\sigma^2(\infty)-\sigma^2(0)\right). \tag{3.38}$$

Proof. For sake of brevity we prove only the first statement, the proof of the remaining statements is similar. Let $\sigma^1(x_1)$ be an odd function. Using the trigonometric representation of the function $e^{i\omega_1x_1}$, a straightforward computation shows that

$$\begin{aligned}
f(\omega_1,\omega_2) &= \int_{-\infty}^\infty\int_{-\infty}^\infty d\sigma^1(x_1)d\sigma^2(x_2)\,e^{i\omega_1x_1}\,e^{j\omega_2x_2}\\
&= \int_{-\infty}^\infty\left(\int_{-\infty}^\infty d\sigma^1(x_1)\cos(\omega_1x_1)\right)d\sigma^2(x_2)\,e^{j\omega_2x_2}\\
&+ \int_{-\infty}^\infty\left(\int_{-\infty}^\infty d\sigma^1(x_1)\sin(\omega_1x_1)\right)d\sigma^2(x_2)\,\boldsymbol{i}\,e^{j\omega_2x_2} \qquad (3.39)\\
&= 2\int_{-\infty}^\infty\int_0^\infty d\sigma^1(x_1)d\sigma^2(x_2)\,\boldsymbol{i}\sin(\omega_1x_1)\,e^{j\omega_2x_2}.
\end{aligned}$$
□

3.3. Uniform Continuity

In this subsection we discuss uniform continuity and its relationship to the QFST. We begin by defining uniform continuity.

Definition 3.12. A quaternion function $f : \Omega \subset \mathbb{R}^2 \to \mathbb{H}$ is *uniformly continuous* on Ω, if and only if for all $\epsilon > 0$ there exists a $\delta > 0$, such that $|f(\omega_1) - f(\omega_2)| < \epsilon$ for all $\omega_1, \omega_2 \in \Omega$ whenever $|\omega_1 - \omega_2| < \delta$.

We now prove some results related to the asymptotic behaviour of the QFST.

Proposition 3.13. *Let f be an element of $\mathcal{B}$. For any natural number n, let $f^n : \mathbb{R} \times [-n, n] \to \mathbb{H}$ be the function given by*

$$f^n(\omega_1, \omega_2) = \int_{-n}^{n} \int_{-\infty}^{\infty} d\sigma^1(x_1) d\sigma^2(x_2) e^{i\omega_1 x_1} e^{j\omega_2 x_2}.$$

Then $f^n(\omega_1, \omega_2) \longrightarrow_{n\longrightarrow\infty} f(\omega_1, \omega_2)$ uniformly. Also, if the f^n are uniformly continuous functions, then f is also a uniformly continuous function.

Proof. A first straightforward computation shows that

$$\begin{aligned}
&|f(\omega_1, \omega_2) - f^n(\omega_1, \omega_2)| \\
&= \left| \int_{-\infty}^{\infty} \int_{-\infty}^{\infty} d\sigma^1(x_1) d\sigma^2(x_2) e^{i\omega_1 x_1} e^{j\omega_2 x_2} - \int_{-n}^{n} \int_{-\infty}^{\infty} d\sigma^1(x_1) d\sigma^2(x_2) e^{i\omega_1 x_1} e^{j\omega_2 x_2} \right| \\
&= \left| \int_{-\infty}^{-n} \int_{-\infty}^{\infty} d\sigma^1(x_1) d\sigma^2(x_2) e^{i\omega_1 x_1} e^{j\omega_2 x_2} + \int_{n}^{\infty} \int_{-\infty}^{\infty} d\sigma^1(x_1) d\sigma^2(x_2) e^{i\omega_1 x_1} e^{j\omega_2 x_2} \right| \\
&\leq |\sigma^1(\infty) - \sigma^1(-\infty)|\, |\sigma^2(-n) - \sigma^2(-\infty)| \\
&\quad + |\sigma^1(\infty) - \sigma^1(-\infty)|\, |\sigma^2(\infty) - \sigma^2(n)|.
\end{aligned} \tag{3.40}$$

Moreover, bearing in mind that

$$\sigma^2(-n) - \sigma^2(-\infty) \longrightarrow_{n\longrightarrow\infty} 0, \qquad \sigma^2(\infty) - \sigma^2(n) \longrightarrow_{n\longrightarrow\infty} 0, \tag{3.41}$$

and hence, $f^n(\omega_1, \omega_2) \longrightarrow_{n\longrightarrow\infty} f(\omega_1, \omega_2)$ uniformly. In addition, we claim that if all the $f^n(\omega_1, \omega_2)$ are uniformly continuous functions, it follows that $f(\omega_1, \omega_2)$ is also a uniformly continuous function. □

Likewise, we have the following analogous result.

Proposition 3.14. *Let f be an element of $\mathcal{B}$. For any natural number n, let $f^n : [-n, n] \times \mathbb{R} \to \mathbb{H}$ be the function given by*

$$f^n(\omega_1, \omega_2) = \int_{-\infty}^{\infty} \int_{-n}^{n} d\sigma^1(x_1) d\sigma^2(x_2)\, e^{i\omega_1 x_1}\, e^{j\omega_2 x_2}.$$

Then $f^n(\omega_1, \omega_2) \longrightarrow_{n\longrightarrow\infty} f(\omega_1, \omega_2)$ uniformly. Also, if all the f^n are uniformly continuous functions then f is also a uniformly continuous function.

Proposition 3.15. *Let f be an element of $\mathcal{B}$. For any natural number n, let $f^n : [-n,n]\times[-n,n] \to \mathbb{H}$ be the function given by*

$$f^n(\omega_1,\omega_2) = \int_{-n}^{n}\int_{-n}^{n} d\sigma^1(x_1)d\sigma^2(x_2)\, e^{i\omega_1 x_1}\, e^{j\omega_2 x_2}.$$

Then $f^n(\omega_1,\omega_2) \longrightarrow_{n\longrightarrow\infty} f(\omega_1,\omega_2)$ uniformly. Also, if all the f^n are uniformly continuous functions then f is also a uniformly continuous function.

Proof. We set $A := d\sigma^1(x_1)d\sigma^2(x_2)\, e^{i\omega_1 x_1}\, e^{j\omega_2 x_2}$. The proof follows from a simple observation:

$$\begin{aligned}\int_{-\infty}^{\infty}\int_{-\infty}^{\infty} A - \int_{-n}^{n}\int_{-n}^{n} A = {}& \int_{-\infty}^{-n}\int_{-\infty}^{-n} A + \int_{-\infty}^{-n}\int_{-n}^{n} A + \int_{-\infty}^{-n}\int_{n}^{\infty} A \\ &+ \int_{-n}^{n}\int_{-\infty}^{-n} A + \int_{-n}^{n}\int_{n}^{\infty} A + \int_{n}^{\infty}\int_{-\infty}^{-n} A \qquad (3.42)\\ &+ \int_{n}^{\infty}\int_{-n}^{n} A + \int_{n}^{\infty}\int_{n}^{\infty} A.\end{aligned}$$

□

We come now to the main theorem of this section.

Theorem 3.16. *Let $f \in \mathcal{B}$ be given, and $g : \mathbb{R}^2 \longrightarrow \mathbb{H}$ be a continuous and absolutely integrable function. For any $\sigma^1, \sigma^2 : \mathbb{R} \longrightarrow \mathbb{H}$ the following relations hold:*

1. $\displaystyle\int_{\mathbb{R}^2} f(t_1,t_2)g(\omega_1 - t_1, \eta - \omega_2)dt_1dt_2$

$$= \int_{\mathbb{R}^2} d\sigma^1(x_1)d\sigma^2(x_2) \int_{\mathbb{R}^2} e^{it_1x_1}e^{jt_2x_2} g(\omega_1 - t_1, \eta - \omega_2)dt_1dt_2; \qquad (3.43)$$

2. $\displaystyle\int_{\mathbb{R}^2} f(t_1,t_2)g(t_1,t_2)dt_1dt_2$

$$= (2\pi)^2 \int_{\mathbb{R}^2} d\sigma^1(x_1)d\sigma^2(x_2)\mathcal{F}^{-1}(g)(x_1,x_2). \qquad (3.44)$$

Proof. Assume $g : \mathbb{R}^2 \longrightarrow \mathbb{H}$ to be a continuous and absolutely integrable function. For any real variables ρ and χ we define the function

$$R(x_1,x_2,\rho,\chi) := \int_{-\chi}^{\chi}\int_{-\rho}^{\rho} e^{it_1x_1}e^{jt_2x_2} g(t_1,t_2)dt_1dt_2. \qquad (3.45)$$

We set $f^n(t_1,t_2) = \int_{-n}^{n}\int_{-n}^{n} d\sigma^1(x_1)d\sigma^2(x_2)e^{it_1x_1}e^{jt_2x_2}$. Using the fact that g is an absolutely integrable function, it follows that

$$\begin{aligned}&\int_{-\chi}^{\chi}\int_{-\rho}^{\rho} f^n(t_1,t_2)g(t_1,t_2)dt_1dt_2 \\ &\quad= \int_{-\chi}^{\chi}\int_{-\rho}^{\rho}\left(\int_{-n}^{n}\int_{-n}^{n} d\sigma^1(x_1)d\sigma^2(x_2)e^{it_1x_1}e^{jt_2x_2}\right) g(t_1,t_2)dt_1dt_2\end{aligned}$$

$$= \int_{-n}^{n}\int_{-n}^{n} d\sigma^1(x_1)d\sigma^2(x_2) \int_{-\chi}^{\chi}\int_{\rho}^{\rho} e^{it_1x_1}e^{jt_2x_2}g(t_1,t_2)dt_1dt_2$$

$$= \int_{-n}^{n}\int_{-n}^{n} d\sigma^1(x_1)d\sigma^2(x_2)R(x_1,x_2,\rho,\eta). \tag{3.46}$$

From the last proposition we know that $\lim\limits_{n\longrightarrow\infty} f^n(\omega_1,\omega_2) = f(\omega_1,\omega_2)$ converges uniformly. Moreover, since $g(t_1,t_2)$ is an absolutely integrable function, it follows that $R(x_1,x_2,\rho,\chi)$ is also a uniformly continuous function. Hence

$$\lim_{n\longrightarrow\infty} \int_{-\chi}^{\chi}\int_{-\rho}^{\rho} f^n(t_1,t_2)g(t_1,t_2)dt_1dt_2 = \int_{-\chi}^{\chi}\int_{-\rho}^{\rho} f(t_1,t_2)g(t_1,t_2)dt_1dt_2. \tag{3.47}$$

With this argument in hand, and based on (3.46) we conclude that

$$\lim_{n\longrightarrow\infty} \int_{-\chi}^{\chi}\int_{-\rho}^{\rho} f^n(t_1,t_2)g(t_1,t_2)dt_1dt_2 = \int_{\mathbb{R}^2} d\sigma^1(x_1)d\sigma^2(x_2)R(x_1,x_2,\rho,\eta). \tag{3.48}$$

From the last equality and from (3.47) we obtain

$$\int_{\mathbb{R}^2} d\sigma^1(x_1)d\sigma^2(x_2)R(x_1,x_2,\rho,\eta) = \int_{-\chi}^{\chi}\int_{\rho}^{\rho} f(t_1,t_2)g(t_1,t_2)dt_1dt_2. \tag{3.49}$$

In addition, we have

$$R(x_1,x_2,\rho,\chi) \underset{\substack{\rho\longrightarrow\infty\\ \chi\longrightarrow\infty}}{\longrightarrow} \int_{\mathbb{R}^2} e^{it_1x_1}\, e^{jt_2x_2}g(t_1,t_2)dt_1dt_2, \tag{3.50}$$

and hence, for any fixed $a,b>0$ it follows that

$$\int_{-a}^{a}\int_{-b}^{b} d\sigma^1(x_1)d\sigma^2(x_2)R(x_1,x_2,\rho,\eta) \tag{3.51}$$

$$\underset{\substack{\rho\longrightarrow\infty\\ \chi\longrightarrow\infty}}{\longrightarrow} \int_{-a}^{a}\int_{-b}^{b} d\sigma^1(x_1)d\sigma^2(x_2)\int_{\mathbb{R}^2} e^{it_1x_1}\, e^{jt_2x_2}g(t_1,t_2)dt_1dt_2.$$

For the sake of brevity, in the considerations to follow we will often omit the argument and write simply R instead of $R(x_1,x_2,\rho,\eta)$. Since R is uniformly bounded, there exists a positive constant M so that $|R|\le M$ for all $x_1,x_2,\rho,\eta\in\mathbb{R}$. We define

$$I := \left| \int_{-\infty}^{-a}\int_{-\infty}^{-b} d\sigma^1(x_1)d\sigma^2(x_2)R + \int_{-a}^{a}\int_{-\infty}^{-b} d\sigma^1(x_1)d\sigma^2(x_2)R \right.$$

$$+ \int_{a}^{\infty}\int_{-\infty}^{-b} d\sigma^1(x_1)d\sigma^2(x_2)R + \int_{-\infty}^{-a}\int_{-b}^{b} d\sigma^1(x_1)d\sigma^2(x_2)R$$

$$+\int_{a}^{\infty}\int_{-b}^{b} d\sigma^1(x_1)d\sigma^2(x_2)R + \int_{-\infty}^{-a}\int_{b}^{\infty} d\sigma^1(x_1)d\sigma^2(x_2)R$$
$$+\int_{-a}^{a}\int_{b}^{\infty} d\sigma^1(x_1)d\sigma^2(x_2)R + \int_{a}^{\infty}\int_{b}^{\infty} d\sigma^1(x_1)d\sigma^2(x_2)R \Bigg|. \tag{3.52}$$

Therefore, we obtain

$$\begin{aligned} I \leq M\,[&|\sigma^1(-b)-\sigma^1(-\infty)|\,|\sigma^2(-a)-\sigma^2(-\infty)| \\ &+ |\sigma^1(-b)-\sigma^1(-\infty)|\,|\sigma^2(a)-\sigma^2(-a)| \\ &+ |\sigma^1(-b)-\sigma^1(-\infty)|\,|\sigma^2(\infty)-\sigma^2(a)| \\ &+ |\sigma^1(b)-\sigma^1(-b)|\,|\sigma^2(-a)-\sigma^2(-\infty)| \\ &+ |\sigma^1(b)-\sigma^1(-b)|\,|\sigma^2(\infty)-\sigma^2(a)| \\ &+ |\sigma^1(\infty)-\sigma^1(b)|\,|\sigma^2(-a)-\sigma^2(-\infty)| \\ &+ |\sigma^1(\infty)-\sigma^1(b)|\,|\sigma^2(a)-\sigma^2(-a)| \\ &+ |\sigma^1(\infty)-\sigma^1(b)|\,|\sigma^2(\infty)-\sigma^2(a)|] \longrightarrow_{a,b\longrightarrow\infty} 0. \end{aligned} \tag{3.53}$$

Using the last inequality and (3.51) we get

$$\begin{aligned} \lim_{a,b\longrightarrow\infty} \int_{-a}^{a}\int_{-b}^{b} d\sigma^1(x_1)d\sigma^2(x_2)R(x_1,x_2,\rho,\eta) \\ = \int_{\mathbb{R}^2} d\sigma^1(x_1)d\sigma^2(x_2)\int_{\mathbb{R}^2} e^{it_1x_1}\, e^{jt_2x_2} g(t_1,t_2)dt_1dt_2. \end{aligned} \tag{3.54}$$

Based on this and with (3.48) we find

$$\begin{aligned} &\int_{\mathbb{R}^2} f(t_1,t_2)g(t_1,t_2)dt_1dt_2 \\ &\quad= \int_{\mathbb{R}^2} d\sigma^1(x_1)d\sigma^2(x_2)\int_{\mathbb{R}^2} e^{it_1x_1}\, e^{jt_2x_2} g(t_1,t_2)dt_1dt_2 \\ &\quad= (2\pi)^2\int_{\mathbb{R}^2} d\sigma^1(x_1)d\sigma^2(x_2)\,\mathcal{F}^{-1}(g)(x_1,x_2). \end{aligned} \tag{3.55}$$

Making the change of variables $t_1 \longrightarrow \omega_1 - t_1$, $t_2 \longrightarrow \eta - \omega_2$ in the definition of g, we finally find

$$\begin{aligned} &\int_{\mathbb{R}^2} f(t_1,t_2)g(\omega_1-t_1,\eta-\omega_2)dt_1dt_2 \\ &\quad= \int_{\mathbb{R}^2} d\sigma^1(x_1)d\sigma^2(x_2)\int_{\mathbb{R}^2} e^{it_1x_1}\, e^{jt_2x_2} g(\omega_1-t_1,\eta-\omega_2)dt_1dt_2. \end{aligned} \tag{3.56}$$

□

Theorem 3.17. *For any $\sigma^1,\sigma^2:\mathbb{R}\longrightarrow\mathbb{H}$, consider the functions*

$$g(\omega_1) = \int_{\mathbb{R}} d\sigma^1(x_1)\, e^{i\omega_1x_1}, \qquad h(\omega_2) = \int_{\mathbb{R}} d\sigma^2(x_2)\, e^{j\omega_2x_2}. \tag{3.57}$$

Then for any real number ρ the following equalities hold:

$$\sigma^1(\rho) - \sigma^1(0) = \frac{1}{2\pi}\int_{\mathbb{R}} g(\omega_1)\frac{e^{-\boldsymbol{i}\rho\omega_1} - 1}{-\boldsymbol{i}\omega_1}d\omega_1, \tag{3.58}$$

$$\sigma^2(\rho) - \sigma^2(0) = \frac{1}{2\pi}\int_{\mathbb{R}} h(\omega_2)\frac{e^{-\boldsymbol{j}\rho\omega_2} - 1}{-\boldsymbol{j}\omega_2}d\omega_2. \tag{3.59}$$

Proof. We begin with the following observation:

$$\begin{aligned}
&g(\omega_1)\left(e^{-\boldsymbol{i}\rho\omega_1} - 1\right)\\
&\quad= \int_{\mathbb{R}} d\sigma^1(x_1)\, e^{\boldsymbol{i}\omega_1 x_1}\left(e^{-\boldsymbol{i}\rho\omega_1} - 1\right)\\
&\quad= \int_{\mathbb{R}} d\sigma^1(x_1)\, e^{\boldsymbol{i}\omega_1(x_1-\rho)} - \int_{\mathbb{R}} d\sigma^1(x_1)\, e^{\boldsymbol{i}\omega_1 x_1}\\
&\quad= \int_{\mathbb{R}} d\sigma^1(x_1+\rho)\, e^{\boldsymbol{i}\omega_1 x_1} - \int_{\mathbb{R}} d\sigma^1(x_1)\, e^{\boldsymbol{i}\omega_1 x_1}\\
&\quad= \lim_{n\longrightarrow\infty}\left(\int_{-n}^{n} d\sigma^1(x_1+\rho)\, e^{\boldsymbol{i}\omega_1 x_1} - \int_{-n}^{n} d\sigma^1(x_1)\, e^{\boldsymbol{i}\omega_1 x_1}\right) \qquad (3.60)\\
&\quad= \lim_{n\longrightarrow\infty}\Big[\sigma^1(n+\rho)e^{\boldsymbol{i}n\omega_1} - \sigma^1(-n+\rho)e^{-\boldsymbol{i}n\omega_1} - \sigma^1(n)e^{\boldsymbol{i}n\omega_1}\\
&\quad+ \sigma^1(-n)e^{-\boldsymbol{i}n\omega_1} - \int_{-n}^{n}\left(\sigma^1(x_1+\rho) - \sigma^1(x_1)\right)dx_1\, e^{\boldsymbol{i}\omega_1 x_1}\boldsymbol{i}\omega_1\Big]\\
&\quad= -\int_{\mathbb{R}}\left(\sigma^1(x_1+\rho) - \sigma^1(x_1)\right)dx_1\, e^{\boldsymbol{i}\omega_1 x_1}\boldsymbol{i}\omega_1.
\end{aligned}$$

Therefore, it is easy to see that

$$g(\omega_1)\left(e^{-\boldsymbol{i}\rho\omega_1} - 1\right)\frac{1}{-\boldsymbol{i}\omega_1} = \int_{\mathbb{R}}\left(\sigma^1(x_1+\rho) - \sigma^1(x_1)\right)dx_1\, e^{\boldsymbol{i}\omega_1 x_1}. \tag{3.61}$$

From the last equality and from the inverse Fourier transform formula we find that

$$\sigma^1(x_1+\rho) - \sigma^1(x_1) = \frac{1}{2\pi}\int_{\mathbb{R}} g(\omega_1)\frac{e^{-\boldsymbol{i}\rho\omega_1} - 1}{-\boldsymbol{i}\omega_1}e^{-\boldsymbol{i}x_1\omega_1}d\omega_1, \tag{3.62}$$

and, in particular for $x_1 = 0$ we find that

$$\sigma^1(\rho) - \sigma^1(0) = \frac{1}{2\pi}\int_{\mathbb{R}} g(\omega_1)\frac{e^{-\boldsymbol{i}\rho\omega_1} - 1}{-\boldsymbol{i}\omega_1}d\omega_1. \tag{3.63}$$

In a similar way we may deduce that

$$\sigma^2(\rho) - \sigma^2(0) = \frac{1}{2\pi}\int_{\mathbb{R}} h(\omega_1)\frac{e^{-\boldsymbol{j}\rho\omega_1} - 1}{-\boldsymbol{j}\omega_1}d\omega_1\,, \tag{3.64}$$

which completes the proof. □

3.4. Connecting Uniform Continuity to Quaternion Distributions

In this subsection we describe the connection between uniform continuity and quaternion distributions.

Theorem 3.18. *Let the sequences of quaternion distributions*

$$\sigma^1(x_1), \sigma^2(x_1), \ldots, \sigma^n(x_1), \ldots, \qquad \tau^1(x_2), \tau^2(x_2), \ldots, \tau^n(x_2), \ldots \tag{3.65}$$

be convergent to the distributions $\sigma^0(x_1)$ *and* $\tau^0(x_2)$*, respectively. Assume that*

$$\lim_{n\longrightarrow\infty} \sigma^n(\pm\infty) = \sigma^0(\pm\infty), \qquad \lim_{n\longrightarrow\infty} \tau^n(\pm\infty) = \tau^0(\pm\infty), \tag{3.66}$$

where $f^n(\omega_1, \omega_2) \in \mathcal{B}$ *corresponds to the distributions* $\sigma^n(x_1)$ *and* $\tau^n(x_2)$*, and* $f^0(\omega_1, \omega_2)$ *corresponds to the distributions* $\sigma^0(x_1)$ *and* $\tau^0(x_2)$*, respectively. Then*

$$\lim_{n\longrightarrow\infty} f^n(\omega_1, \omega_2) = f^0(\omega_1, \omega_2).$$

Proof. To begin with, we choose $a > 0$ so that $\sigma^0(x_1)$ and $\tau^0(x_2)$ are continuous functions at $\pm a$. Then

$$\lim_{n\longrightarrow\infty} \sigma^n(\pm a) = \sigma^0(\pm a), \qquad \lim_{n\longrightarrow\infty} \tau^n(\pm a) = \tau^0(\pm a). \tag{3.67}$$

For brevity of exposition, in the sequel we omit the arguments and write simply $d\sigma^k$ and $d\tau^k$ instead of $d\sigma^k(x_1)$ and $d\tau^k(x_2)$, respectively. Now, we make use of the following representations for $f^n(\omega_1, \omega_2)$ and $f^0(\omega_1, \omega_2)$:

$$\begin{aligned}
f^n(\omega_1, \omega_2) &= \int_{\mathbb{R}^2} d\sigma^n d\tau^n \, e^{\boldsymbol{i}\omega_1 x_1} \, e^{\boldsymbol{j}\omega_2 x_2} \\
&= \int_{-a}^{a}\int_{-a}^{a} d\sigma^n d\tau^n \, e^{\boldsymbol{i}\omega_1 x_1} \, e^{\boldsymbol{j}\omega_2 x_2} + \int_{-\infty}^{-a}\int_{-\infty}^{-a} d\sigma^n d\tau^n \, e^{\boldsymbol{i}\omega_1 x_1} \, e^{\boldsymbol{j}\omega_2 x_2} \\
&\quad + \int_{-\infty}^{-a}\int_{-a}^{a} d\sigma^n d\tau^n \, e^{\boldsymbol{i}\omega_1 x_1} \, e^{\boldsymbol{j}\omega_2 x_2} + \int_{-\infty}^{-a}\int_{a}^{\infty} d\sigma^n d\tau^n \, e^{\boldsymbol{i}\omega_1 x_1} \, e^{\boldsymbol{j}\omega_2 x_2} \\
&\quad + \int_{-a}^{a}\int_{-\infty}^{-a} d\sigma^n d\tau^n \, e^{\boldsymbol{i}\omega_1 x_1} \, e^{\boldsymbol{j}\omega_2 x_2} + \int_{-a}^{a}\int_{a}^{\infty} d\sigma^n d\tau^n \, e^{\boldsymbol{i}\omega_1 x_1} \, e^{\boldsymbol{j}\omega_2 x_2} \\
&\quad + \int_{a}^{\infty}\int_{-\infty}^{-a} d\sigma^n d\tau^n \, e^{\boldsymbol{i}\omega_1 x_1} \, e^{\boldsymbol{j}\omega_2 x_2} + \int_{a}^{\infty}\int_{-a}^{a} d\sigma^n d\tau^n \, e^{\boldsymbol{i}\omega_1 x_1} \, e^{\boldsymbol{j}\omega_2 x_2} \\
&\quad + \int_{a}^{\infty}\int_{a}^{\infty} d\sigma^n d\tau^n \, e^{\boldsymbol{i}\omega_1 x_1} \, e^{\boldsymbol{j}\omega_2 x_2},
\end{aligned} \tag{3.68}$$

and,

$$\begin{aligned}
f^0(\omega_1, \omega_2) &= \int_{\mathbb{R}^2} d\sigma^0 d\tau^0 \, e^{\boldsymbol{i}\omega_1 x_1} \, e^{\boldsymbol{j}\omega_2 x_2} \\
&= \int_{-a}^{a}\int_{-a}^{a} d\sigma^0 d\tau^0 \, e^{\boldsymbol{i}\omega_1 x_1} \, e^{\boldsymbol{j}\omega_2 x_2} + \int_{-\infty}^{-a}\int_{-\infty}^{-a} d\sigma^0 d\tau^0 \, e^{\boldsymbol{i}\omega_1 x_1} \, e^{\boldsymbol{j}\omega_2 x_2} \\
&\quad + \int_{-\infty}^{-a}\int_{-a}^{a} d\sigma^0 d\tau^0 \, e^{\boldsymbol{i}\omega_1 x_1} \, e^{\boldsymbol{j}\omega_2 x_2} + \int_{-\infty}^{-a}\int_{a}^{\infty} d\sigma^0 d\tau^0 \, e^{\boldsymbol{i}\omega_1 x_1} \, e^{\boldsymbol{j}\omega_2 x_2}
\end{aligned}$$

$$+\int_{-a}^{a}\int_{-\infty}^{-a} d\sigma^0 d\tau^0\, e^{\boldsymbol{i}\omega_1 x_1}\, e^{\boldsymbol{j}\omega_2 x_2} + \int_{-a}^{a}\int_{a}^{\infty} d\sigma^0 d\tau^0\, e^{\boldsymbol{i}\omega_1 x_1}\, e^{\boldsymbol{j}\omega_2 x_2}$$
$$+\int_{a}^{\infty}\int_{-\infty}^{-a} d\sigma^0 d\tau^0\, e^{\boldsymbol{i}\omega_1 x_1}\, e^{\boldsymbol{j}\omega_2 x_2} + \int_{a}^{\infty}\int_{-a}^{a} d\sigma^0 d\tau^0\, e^{\boldsymbol{i}\omega_1 x_1}\, e^{\boldsymbol{j}\omega_2 x_2}$$
$$+\int_{a}^{\infty}\int_{a}^{\infty} d\sigma^0 d\tau^0\, e^{\boldsymbol{i}\omega_1 x_1}\, e^{\boldsymbol{j}\omega_2 x_2}. \tag{3.69}$$

Furthermore, we set

$$I^1(\omega_1,\omega_2,a) = \int_{-a}^{a}\int_{-a}^{a} d\sigma^0 d\tau^0\, e^{\boldsymbol{i}\omega_1 x_1}\, e^{\boldsymbol{j}\omega_2 x_2} - \int_{-a}^{a}\int_{-a}^{a} d\sigma^n d\tau^n\, e^{\boldsymbol{i}\omega_1 x_1}\, e^{\boldsymbol{j}\omega_2 x_2}, \tag{3.70}$$

$$I^2(\omega_1,\omega_2,a) = \int_{-\infty}^{-a}\int_{-\infty}^{-a} d\sigma^0 d\tau^0\, e^{\boldsymbol{i}\omega_1 x_1}\, e^{\boldsymbol{j}\omega_2 x_2} - \int_{-\infty}^{-a}\int_{-\infty}^{-a} d\sigma^n d\tau^n\, e^{\boldsymbol{i}\omega_1 x_1}\, e^{\boldsymbol{j}\omega_2 x_2}, \tag{3.71}$$

$$I^3(\omega_1,\omega_2,a) = \int_{-\infty}^{-a}\int_{-a}^{a} d\sigma^0 d\tau^0\, e^{\boldsymbol{i}\omega_1 x_1}\, e^{\boldsymbol{j}\omega_2 x_2} - \int_{-\infty}^{-a}\int_{-a}^{a} d\sigma^n d\tau^n\, e^{\boldsymbol{i}\omega_1 x_1}\, e^{\boldsymbol{j}\omega_2 x_2}, \tag{3.72}$$

$$I^4(\omega_1,\omega_2,a) = \int_{-\infty}^{-a}\int_{a}^{\infty} d\sigma^0 d\tau^0\, e^{\boldsymbol{i}\omega_1 x_1}\, e^{\boldsymbol{j}\omega_2 x_2} - \int_{-\infty}^{-a}\int_{a}^{\infty} d\sigma^n d\tau^n\, e^{\boldsymbol{i}\omega_1 x_1}\, e^{\boldsymbol{j}\omega_2 x_2}, \tag{3.73}$$

$$I^5(\omega_1,\omega_2,a) = \int_{-a}^{a}\int_{-\infty}^{-a} d\sigma^0 d\tau^0\, e^{\boldsymbol{i}\omega_1 x_1}\, e^{\boldsymbol{j}\omega_2 x_2} - \int_{-a}^{a}\int_{-\infty}^{-a} d\sigma^n d\tau^n\, e^{\boldsymbol{i}\omega_1 x_1}\, e^{\boldsymbol{j}\omega_2 x_2}, \tag{3.74}$$

$$I^6(\omega_1,\omega_2,a) = \int_{-a}^{a}\int_{a}^{\infty} d\sigma^0 d\tau^0\, e^{\boldsymbol{i}\omega_1 x_1}\, e^{\boldsymbol{j}\omega_2 x_2} - \int_{-a}^{a}\int_{a}^{\infty} d\sigma^n d\tau^n\, e^{\boldsymbol{i}\omega_1 x_1}\, e^{\boldsymbol{j}\omega_2 x_2}, \tag{3.75}$$

$$I^7(\omega_1,\omega_2,a) = \int_{a}^{\infty}\int_{-\infty}^{-a} d\sigma^0 d\tau^0\, e^{\boldsymbol{i}\omega_1 x_1}\, e^{\boldsymbol{j}\omega_2 x_2} - \int_{a}^{\infty}\int_{-\infty}^{-a} d\sigma^n d\tau^n\, e^{\boldsymbol{i}\omega_1 x_1}\, e^{\boldsymbol{j}\omega_2 x_2}, \tag{3.76}$$

$$I^8(\omega_1,\omega_2,a) = \int_{a}^{\infty}\int_{-a}^{a} d\sigma^0 d\tau^0\, e^{\boldsymbol{i}\omega_1 x_1}\, e^{\boldsymbol{j}\omega_2 x_2} - \int_{a}^{\infty}\int_{-a}^{a} d\sigma^n d\tau^n\, e^{\boldsymbol{i}\omega_1 x_1}\, e^{\boldsymbol{j}\omega_2 x_2}, \tag{3.77}$$

$$I^9(\omega_1,\omega_2,a) = \int_{a}^{\infty}\int_{a}^{\infty} d\sigma^0 d\tau^0\, e^{\boldsymbol{i}\omega_1 x_1}\, e^{\boldsymbol{j}\omega_2 x_2} - \int_{a}^{\infty}\int_{a}^{\infty} d\sigma^n d\tau^n\, e^{\boldsymbol{i}\omega_1 x_1}\, e^{\boldsymbol{j}\omega_2 x_2}. \tag{3.78}$$

We further define $A_l(\omega_1,\omega_2,a) = \overline{\lim}_{n\longrightarrow\infty} |I^l(\omega_1,a)|$ $(l = 1,2,3,4,5,6,7,8)$.

For $I^1(\omega_1, a)$ a straightforward computation shows that

$$
\begin{aligned}
&|I^1(\omega_1,\omega_2,a)| \\
&= \Bigg| \left(\sigma^n(a)e^{i\omega_1 a} - \sigma^n(-a)e^{-i\omega_1 a} - \int_{-a}^{a} \sigma^n(x_1) i\omega_1 e^{i\omega_1 x_1} dx_1\right) \\
&\quad \left(\tau^n(a)e^{j\omega_2 a} - \tau^n(-a)e^{-j\omega_2 a} - \int_{-a}^{a} \tau^n(x_2) i\omega_2 e^{j\omega_2 x_2} dx_2\right) \\
&\quad - \left(\sigma^0(a)e^{i\omega_1 a} - \sigma^0(-a)e^{-i\omega_1 a} - \int_{-a}^{a} \sigma^0(x_1) i\omega_1 e^{i\omega_1 x_1} dx_1\right) \\
&\quad \left(\tau^0(a)e^{j\omega_2 a} - \tau^n(-a)e^{-j\omega_2 a} - \int_{-a}^{a} \tau^0(x_2) j\omega_2 e^{j\omega_2 x_2} dx_2\right) \Bigg| . \qquad (3.79)
\end{aligned}
$$

Based on these results and on (3.67), and the convergence of the sequences $\sigma^n(x_1)$ and $\tau^n(x_2)$, it follows that $A_1(\omega_1, \omega_2, a) = 0$. Hence

$$
\begin{aligned}
|I^2(\omega_1, a)| &\leq \int_{-\infty}^{-a} \int_{-\infty}^{-a} d\sigma^0(x_1) d\tau^0(x_2) \left|e^{i\omega_1 x_1} e^{j\omega_2 x_2}\right| \\
&\quad + \int_{-\infty}^{-a} \int_{-\infty}^{-a} d\sigma^n(x_1) d\tau^n(x_2) \left|e^{i\omega_1 x_1} e^{j\omega_2 x_2}\right| \\
&< \left|\sigma^0(-a) - \sigma^0(-\infty)\right| \left|\tau^0(-a) - \tau^0(-\infty)\right| \\
&\quad + \left|\sigma^n(-a) - \sigma^n(-\infty)\right| \left|\tau^n(-a) - \tau^n(-\infty)\right| . \qquad (3.80)
\end{aligned}
$$

From the last inequality and from (3.66) it follows that for any $\epsilon > 0$ we can choose $a > 0$ large enough so that $A_2(\omega_1, \omega_2, a) < \epsilon$. In a similar way, we have

$$
A_l(\omega_1, \omega_2, a) < \epsilon, \qquad l = 3, 4, 5, 6, 7, 8, 9. \qquad (3.81)
$$

By combining these arguments, we finally obtain

$$
\lim_{n \longrightarrow \infty} f^n(\omega_1, \omega_2) = f^0(\omega_1, \omega_2), \qquad (3.82)
$$

which completes the proof. □

4. The Main Theorem

In this section we shall extend Bochner's result to the noncommutative structure of quaternion functions. We first define a few general properties of quaternion positive-type functions.

4.1. Positive Functions

Let us define the notion of positive definite measure in the context of quaternion analysis.

Definition 4.1. A function $f : \mathbb{R}^2 \longrightarrow \mathbb{H}$ is called *positive definite*, if it satisfies the following conditions:

1. f is bounded and continuous on $\mathbb{R}^2$.
2. $f(-\boldsymbol{\omega}) = \overline{g(\boldsymbol{\omega})}$ for all $\boldsymbol{\omega} \in \mathbb{R}^2$, where g is any function which can be represented as $\mathcal{FS}_l(\overline{\sigma^2}, \overline{\sigma^1})(\omega_1, \omega_2)$.
3. For any $\lambda_m = (\lambda_{1m}, \lambda_{2m}) \in \mathbb{R}^2$, quaternion numbers z_m $(m = 1, 2, \ldots, n)$, and any positive integer n the following inequality is satisfied:

$$\sum_{\substack{m,l=1\\ m\geq l}}^{n} f(\lambda_m - \lambda_l)\, z_m \overline{z_l} + \sum_{\substack{m,l=1\\ m>l}}^{n} z_m \overline{z_l}\,\overline{f(\lambda_l - \lambda_m)} \geq 0. \tag{4.1}$$

These parameters are measured such that:

(i) When $\lambda_m = \lambda_l$ for $m \neq l$ $(m, l = 1, 2, \ldots, n)$ it follows

$$\begin{aligned}
\sum_{\substack{m,l=1\\ m\leq l}}^{n} f(0,0) z_m \overline{z_l} + \sum_{\substack{m,l=1\\ m>l}}^{n} z_m \overline{z_l}\, \overline{f(0,0)} &= \sum_{\substack{m,l=1\\ m\leq l}}^{n} \left(f(0,0) z_m \overline{z_l} + z_l \overline{z_m}\, \overline{f(0,0)}\right)\\
&= \sum_{\substack{m,l=1\\ m\leq l}} \left(f(0,0) z_m \overline{z_l} + \overline{f(0,0) z_m \overline{z_l}}\right)\\
&= 2 \sum_{\substack{m,l=1\\ m\leq l}}^{n} \mathrm{S}(f(0,0) z_m \overline{z_l}) \geq 0.
\end{aligned} \tag{4.2}$$

(ii) When $\lambda_m = \lambda_l$, $z_m = z_l$ for $m \neq l$ $(m, l = 1, 2, \ldots, n)$, and by using the previous observation we get

$$2\,\mathrm{S}(f(0,0)) \sum_{m=1}^{n} |z_m|^2 \geq 0 \quad \Leftrightarrow \quad \mathrm{S}(f(0,0)) \geq 0. \tag{4.3}$$

(iii) When $n = 2$, $\lambda_1 = \boldsymbol{\omega}$, and $\lambda_2 = 0$, it follows that

$$\begin{aligned}
&f(\boldsymbol{\omega}) z_1 \overline{z_2} + f(0,0)\left(|z_1|^2 + |z_2|^2\right) + z_2 \overline{z_1}\, \overline{f(\boldsymbol{\omega})}\\
&\qquad = 2\,\mathrm{S}(f(\boldsymbol{\omega}) z_1 \overline{z_2}) + \left(|z_1|^2 + |z_2|^2\right) f(0,0).
\end{aligned} \tag{4.4}$$

In particular, when $z_1 = z_2$ we obtain

$$2\,|z_1|^2\,(\mathrm{S}(f(\boldsymbol{\omega})) + f(0,0)) \geq 0 \quad \Leftrightarrow \quad \mathrm{S}(f(\boldsymbol{\omega})) + f(0,0) \geq 0. \tag{4.5}$$

Remark 4.2. It should be noted that not all functions in $\mathcal{B}$ are positive definite. To verify this claim take, for example, the function $f(\omega_1, \omega_2) = -e^{-\boldsymbol{i}\omega_1}$. It is easy to see that expression (4.1) is not satisfied.

4.2. Bochner Theorem

Before proving the generalization of Bochner's theorem to quaternion functions, we consider the following set

$$\mathcal{C} := \left\{ f : \mathbb{R}^2 \longrightarrow \mathbb{H} : f = \int_{\mathbb{R}^2} d\left|\sigma^1(x_1)\right| d\left|\sigma^2(x_2)\right| e^{\boldsymbol{i}\omega_1 x_1} e^{\boldsymbol{j}\omega_2 x_2} \right\} \subset \mathcal{B}. \tag{4.6}$$

We are now ready for the classification theorem.

Theorem 4.3. *If $f \in \mathcal{C}$ then f is positive definite.*

Proof. Statement 1 of Definition 4.1 is proved in Proposition 3.10, and Statement 2 follows from (3.29). Let $\lambda_m = (\lambda_{1m}, \lambda_{2m}) \in \mathbb{R}^2$, $z_m \in \mathbb{H}$ $(m = 1, 2, \ldots, n)$. For any $\sigma^1, \sigma^2 : \mathbb{R} \to \mathbb{H}$, straightforward computations show that

$$
\begin{aligned}
&\sum_{\substack{m,l=1\\ m\le l}}^{n} f(\lambda_m - \lambda_l) z_m \overline{z_l} + \sum_{\substack{m,l=1\\ m>l}}^{n} z_m \overline{z_l}\, \overline{f(\lambda_l - \lambda_m)} \\
&= \sum_{\substack{m,l=1\\ m\le l}}^{n} \left(\int_{\mathbb{R}^2} d\left|\sigma^1(x_1)\right| d\left|\sigma^2(x_2)\right| e^{\boldsymbol{i}(\lambda_{1m}-\lambda_{1l})x_1} e^{\boldsymbol{j}(\lambda_{2m}-\lambda_{2l})x_2} \right) z_m \overline{z_l} \\
&\quad + \sum_{\substack{m,l=1\\ m>l}}^{n} z_m \overline{z_l}\, \overline{\int_{\mathbb{R}^2} d\left|\sigma^1(x_1)\right| d\left|\sigma^2(x_2)\right| e^{\boldsymbol{i}(\lambda_{1l}-\lambda_{1m})x_1} e^{\boldsymbol{j}(\lambda_{2m}-\lambda_{2l})x_2}} \\
&= \sum_{\substack{m,l=1\\ m<l}}^{n} \left(\int_{\mathbb{R}^2} d\left|\sigma^1(x_1)\right| d\left|\sigma^2(x_2)\right| e^{\boldsymbol{i}(\lambda_{1m}-\lambda_{1l})x_1} e^{\boldsymbol{j}(\lambda_{2m}-\lambda_{2l})x_2} \right) z_m \overline{z_l} \\
&\quad + \sum_{\substack{m,l=1\\ m>l}}^{n} z_m \overline{z_l} \int_{\mathbb{R}^2} e^{-\boldsymbol{j}(\lambda_{2m}-\lambda_{2l})x_2} e^{-\boldsymbol{i}(\lambda_{1l}-\lambda_{1m})x_1} d\left|\sigma^1(x_1)\right| d\left|\sigma^2(x_2)\right| \\
&= \sum_{m=1}^{n} |z_m|^2 \left(\left|\sigma^1(\infty)\right| - \left|\sigma^1(-\infty)\right| \right) \left(\left|\sigma^2(\infty)\right| - \left|\sigma^2(-\infty)\right| \right) \\
&\quad + \sum_{\substack{m,l=1\\ m<l}}^{n} \int_{\mathbb{R}^2} d\left|\sigma^1(x_1)\right| d\left|\sigma^2(x_2)\right| \begin{pmatrix} e^{\boldsymbol{i}(\lambda_{1m}-\lambda_{1l})x_1} e^{\boldsymbol{j}(\lambda_{2m}-\lambda_{2l})\beta} z_m \overline{z_l} \\ + \\ z_l \overline{z_m} e^{-\boldsymbol{j}(\lambda_{2m}-\lambda_{2l})x_2} e^{-\boldsymbol{i}(\lambda_{1m}-\lambda_{1l})x_1} \end{pmatrix} \\
&= \sum_{m=1}^{n} |z_m|^2 \left(\left|\sigma^1(\infty)\right| - \left|\sigma^1(-\infty)\right| \right) \left(\left|\sigma^2(\infty)\right| - \left|\sigma^2(-\infty)\right| \right) \\
&\quad + 2 \sum_{\substack{m,l=1\\ m<l}}^{n} \int_{\mathbb{R}^2} d\left|\sigma^1(x_1)\right| d\left|\sigma^2(x_2)\right| \mathrm{S}\left(e^{\boldsymbol{i}(\lambda_{1m}-\lambda_{1l})x_1} e^{\boldsymbol{j}(\lambda_{2m}-\lambda_{2l})x_2} z_m \overline{z_l} \right) \\
&\ge \sum_{m=1}^{n} |z_m|^2 \left(\left|\sigma^1(\infty)\right| - \left|\sigma^1(-\infty)\right| \right) \left(\left|\sigma^2(\infty)\right| - \left|\sigma^2(-\infty)\right| \right) \\
&\quad - 2 \sum_{\substack{m,l=1\\ m<l}}^{n} \int_{\mathbb{R}^2} |z_m||z_l|\, d\left|\sigma^1(x_1)\right| d\left|\sigma^2(x_2)\right|
\end{aligned}
$$

$$= \left(\sum_{m=1}^{n} |z_m|^2 - 2 \sum_{\substack{m,l=1 \\ m<l}}^{n} |z_m||z_l| \right) \left(|\sigma^1(\infty)| - |\sigma^1(-\infty)| \right) \left(|\sigma^2(\infty)| - |\sigma^2(-\infty)| \right)$$

$$= \left(\sum_{m=1}^{n} |z_m| \right)^2 \left(|\sigma^1(\infty)| - |\sigma^1(-\infty)| \right) \left(|\sigma^2(\infty)| - |\sigma^2(-\infty)| \right) \geq 0. \tag{4.7}$$

This proves that f is positive definite. □

The extension of Bochner's result to a much larger class of quaternion functions remains a challenge to future research.

Acknowledgement

Partial support from the Foundation for Science and Technology (FCT) *via* the grant DD-VU-02/90, Bulgaria, is acknowledged by the first named author. The second named author acknowledges financial support from the Foundation for Science and Technology (FCT) *via* the post-doctoral grant SFRH/ BPD/66342/2009. This work was supported by *FEDER* funds through the *COMPETE* – Operational Programme Factors of Competitiveness ('Programa Operacional Factores de Competitividade') and by Portuguese funds through the *Center for Research and Development in Mathematics and Applications* (University of Aveiro) and the Portuguese Foundation for Science and Technology ('FCT – Fundação para a Ciência e a Tecnologia'), within project PEst-C/MAT/UI4106/2011 with the COMPETE number FCOMP-01-0124-FEDER-022690.

References

[1] S. Bochner. Monotone funktionen, Stieltjessche integrate, und harmonische analyse. *Mathematische Annalen*, 108:378–410, 1933.

[2] S. Bochner. *Lectures on Fourier Integrals.* Princeton University Press, Princeton, New Jersey, 1959.

[3] T. Bülow. *Hypercomplex Spectral Signal Representations for the Processing and Analysis of Images.* PhD thesis, University of Kiel, Germany, Institut für Informatik und Praktische Mathematik, Aug. 1999.

[4] T. Bülow, M. Felsberg, and G. Sommer. Non-commutative hypercomplex Fourier transforms of multidimensional signals. In G. Sommer, editor, *Geometric computing with Clifford Algebras: Theoretical Foundations and Applications in Computer Vision and Robotics*, pages 187–207, Berlin, 2001. Springer.

[5] S. Georgiev, J. Morais, K.I. Kou, and W. Sprößig. Bochner–Minlos theorem and quaternion Fourier transform. To appear.

[6] B. Mawardi, E. Hitzer, A. Hayashi, and R. Ashino. An uncertainty principle for quaternion Fourier transform. *Computers and Mathematics with Applications*, 56(9):2411–2417, 2008.

[7] B. Mawardi and E.M.S. Hitzer. Clifford Fourier transformation and uncertainty principle for the Clifford algebra $C\ell_{3,0}$. *Advances in Applied Clifford Algebras*, 16(1):41–61, 2006.

S. Georgiev
Department of Differential Equations
University of Sofia
Sofia, Bulgaria
e-mail: sgg2000bg@yahoo.com

J. Morais
Centro de Investigação e Desenvolvimento
em Matemática e Aplicações (CIDMA)
Universidade de Aveiro
P3810-193 Aveiro, Portugal
e-mail: joao.pedro.morais@ua.pt

Quaternion and Clifford–Fourier Transforms and Wavelets
Trends in Mathematics, 105–120

6 Bochner–Minlos Theorem and Quaternion Fourier Transform

S. Georgiev, J. Morais, K.I. Kou and W. Sprößig

Abstract. There have been several attempts in the literature to generalize the classical Fourier transform by making use of the Hamiltonian quaternion algebra. The first part of this chapter features certain properties of the asymptotic behaviour of the quaternion Fourier transform. In the second part we introduce the quaternion Fourier transform of a probability measure, and we establish some of its basic properties. In the final analysis, we introduce the notion of positive definite measure, and we set out to extend the classical Bochner–Minlos theorem to the framework of quaternion analysis.

Mathematics Subject Classification (2010). Primary 30G35; secondary 42A38; tertiary 42A82.

Keywords. Quaternion analysis, quaternion Fourier transform, asymptotic behaviour, positive definitely measure, Bochner–Minlos theorem.

1. Introduction and Statement of Results

As is well known, the *classical Fourier transform* (FT) has wide applications in engineering, computer sciences, physics and applied mathematics. For instance, the FT can be used to provide signal analysis techniques where the signal from the original time domain is transformed to the frequency domain. Therefore there exists great interest and considerable effort to extend the FT to higher dimensions, and study its properties and interdependencies (see, *e.g.*, [1–7, 9–11, 17, 18, 20–23] and elsewhere). In view of numerous applications in physics and engineering problems, one is particularly interested in higher-dimensional analogues to $\mathbb{R}^n$, in particular, for $n = 4$. To this end, so far quaternion analysis offers the possibility of generalizing the underlying function theory in 2D to 4D, with the advantage of meeting exactly these goals. To aid the reader, see [15, 16, 19, 24, 25] for more complete accounts of this subject and related topics.

The first part of the present work is devoted to the study of the asymptotic behaviour of the *quaternion Fourier transform* (QFT). The QFT was first introduced by Ell in [10]. He proposed a two-sided QFT and studied some of its applications and properties. Later, Bülow [7] (*cf.* [8] and [1,20]) has also conducted a generalization of the real and complex FT using the quaternion algebra based on two complex variables, but, to the best of our knowledge, a detailed study of its asymptotic behaviour has not been carried out yet. The main motivation of the present study is to develop further general numerical methods for partial differential equations and to extend localization theorems for summation of Fourier series in the quaternion analysis setting. In a forthcoming article we shall describe these connections in more detail and illustrate them by some typical examples.

Due to the noncommutativity of the quaternions, there are three different types of QFT: a right-sided QFT, a left-sided QFT, and a two-sided QFT [23]. We will carry out the investigation of the following finite integral (defined from the time domain to the frequency domain)

$$\mathcal{F}_r(f)(\omega_1,\omega_2) := \int_a^b \int_a^b f(x_1,x_2)\, e^{-\boldsymbol{i}\omega_1 x_1}\, e^{-\boldsymbol{j}\omega_2 x_2}\, dx_1 dx_2, \tag{1.1}$$

where the signal $f : [a,b]\times[a,b] \subset \mathbb{R}^2 \longrightarrow \mathbb{H}$ will be taken to be

$$f(x_1,x_2) := [f(x_1,x_2)]_0 + [f(x_1,x_2)]_1\boldsymbol{i} + [f(x_1,x_2)]_2\boldsymbol{j} + [f(x_1,x_2)]_3\boldsymbol{k},$$

$$[f]_l : [a,b]\times[a,b] \longrightarrow \mathbb{R} \qquad (l = 0,1,2,3)$$

satisfying certain conditions, guaranteeing the convergence of the above integral. $\mathcal{F}_r(f)(\omega_1,\omega_2)$ is the (*finite*) *right-sided Fourier transform* [21] of the quaternion function $f(x_1,x_2)$, and it may be interpreted as a quaternionic extension of the classical FT; the exponential product $e^{-\boldsymbol{i}\omega_1 x_1}e^{-\boldsymbol{j}\omega_2 x_2}$ is called the (*right-sided*) *quaternion Fourier kernel*, and for $i = 1,2$; x_i will denote the *space* and ω_i the *angular frequency* variables. The previous definition of the QFT varies from the original one only in the fact that we use 2D vectors instead of scalars and that it is defined to be two-dimensional. Here $\boldsymbol{i}$, $\boldsymbol{j}$ and $\boldsymbol{k}$ are unit pure quaternions (*i.e.*, the quaternions with unit magnitude having no scalar part) that are orthogonal to each other. We point out that the product in (1.1) has to be written in a fixed order since, in general, $e^{-\boldsymbol{i}\omega_1 x_1}e^{-\boldsymbol{j}\omega_2 x_2}$ does not commute with every element of the algebra.

Remark 1.1. Throughout this text we investigate the integral (1.1) only that, for simplicity, we denote by $\mathcal{F}(f)$. Nevertheless, all results can be easily performed from the left-hand side:

$$\mathcal{F}_l(f)(\omega_1,\omega_2) := \int_a^b \int_a^b e^{-\boldsymbol{j}\omega_2 x_2}\, e^{-\boldsymbol{i}\omega_1 x_1} f(x_1,x_2)\, dx_1 dx_2,$$

since

$$\mathcal{F}_r(f)(-\omega_1,-\omega_2) = \int_a^b \int_a^b f(x_1,x_2) e^{\boldsymbol{i}\omega_1 x_1} e^{\boldsymbol{j}\omega_2 x_2} dx_1 dx_2$$

$$= \int_a^b \int_a^b f(x_1,x_2)\overline{e^{-j\omega_2 x_2} e^{-i\omega_1 x_1}} dx_1 dx_2$$

$$= \overline{\int_a^b \int_a^b e^{-j\omega_2 x_2} e^{-i\omega_1 x_1} \overline{f(x_1,x_2)} dx_1 dx_2} = \overline{\mathcal{F}_l(\overline{f})}(\omega_1,\omega_2).$$

Lemma 1.2. *The QFT of a 2D signal $f \in L^1([a,b]\times[a,b];\mathbb{H})$ has the closed-form representation:*

$$\mathcal{F}(f)(\omega_1,\omega_2) := \Phi_0(\omega_1,\omega_2) + \Phi_1(\omega_1,\omega_2) + \Phi_2(\omega_1,\omega_2) + \Phi_3(\omega_1,\omega_2),$$

where the integrals are

$$\begin{aligned}
\Phi_0(\omega_1,\omega_2) &= \int_a^b \int_a^b f(x_1,x_2)\cos(\omega_1 x_1)\cos(\omega_2 x_2) dx_1 dx_2,\\
\Phi_1(\omega_1,\omega_2) &= -\int_a^b \int_a^b f(x_1,x_2)\boldsymbol{i}\sin(\omega_1 x_1)\cos(\omega_2 x_2) dx_1 dx_2,\\
\Phi_2(\omega_1,\omega_2) &= -\int_a^b \int_a^b f(x_1,x_2)\boldsymbol{j}\cos(\omega_1 x_1)\sin(\omega_2 x_2) dx_1 dx_2,\\
\Phi_3(\omega_1,\omega_2) &= \int_a^b \int_a^b f(x_1,x_2)\boldsymbol{k}\sin(\omega_1 x_1)\sin(\omega_2 x_2) dx_1 dx_2.
\end{aligned}$$

For illustrative purposes, we have the following identities:

Corollary 1.3. *The QFT of a 2D signal $f \in L^1([a,b]\times[a,b];\mathbb{H})$ satisfies the following relations:*

$$\begin{aligned}
\mathcal{F}(f)(\omega_1,\omega_2) + \mathcal{F}(f)(\omega_1,-\omega_2) &= 2\left(\Phi_0(\omega_1,\omega_2) + \Phi_1(\omega_1,\omega_2)\right),\\
\mathcal{F}(f)(\omega_1,\omega_2) - \mathcal{F}(f)(\omega_1,-\omega_2) &= 2\left(\Phi_2(\omega_1,\omega_2) + \Phi_3(\omega_1,\omega_2)\right),\\
\mathcal{F}(f)(\omega_1,\omega_2) + \mathcal{F}(f)(-\omega_1,\omega_2) &= 2\left(\Phi_0(\omega_1,\omega_2) + \Phi_2(\omega_1,\omega_2)\right),\\
\mathcal{F}(f)(\omega_1,\omega_2) - \mathcal{F}(f)(-\omega_1,\omega_2) &= 2\left(\Phi_1(\omega_1,\omega_2) + \Phi_3(\omega_1,\omega_2)\right),\\
\mathcal{F}(f)(\omega_1,\omega_2) + \mathcal{F}(f)(-\omega_1,-\omega_2) &= 2\left(\Phi_0(\omega_1,\omega_2) + \Phi_3(\omega_1,\omega_2)\right),\\
\mathcal{F}(f)(\omega_1,\omega_2) - \mathcal{F}(f)(-\omega_1,-\omega_2) &= 2\left(\Phi_1(\omega_1,\omega_2) + \Phi_2(\omega_1,\omega_2)\right),\\
\mathcal{F}(f)(\omega_1,-\omega_2) + \mathcal{F}(f)(-\omega_1,-\omega_2) &= 2\left(\Phi_0(\omega_1,\omega_2) - \Phi_2(\omega_1,\omega_2)\right),\\
\mathcal{F}(f)(\omega_1,-\omega_2) - \mathcal{F}(f)(-\omega_1,-\omega_2) &= 2\left(\Phi_1(\omega_1,\omega_2) - \Phi_3(\omega_1,\omega_2)\right),\\
\mathcal{F}(f)(-\omega_1,\omega_2) + \mathcal{F}(f)(-\omega_1,-\omega_2) &= 2\left(\Phi_0(\omega_1,\omega_2) - \Phi_1(\omega_1,\omega_2)\right),\\
\mathcal{F}(f)(-\omega_1,\omega_2) - \mathcal{F}(f)(-\omega_1,-\omega_2) &= 2\left(\Phi_2(\omega_1,\omega_2) - \Phi_3(\omega_1,\omega_2)\right).
\end{aligned}$$

Under suitable conditions, the original signal f can be reconstructed from $\mathcal{F}(f)$ by the inverse transform (frequency to time domains).

Definition 1.4. The (*right-sided*) *inverse QFT* of $g \in L^1(\mathbb{R}^2;\mathbb{H})$ is given by

$$\mathcal{F}_r^{-1}(g) : \mathbb{R}^2 \longrightarrow \mathbb{H},$$

$$\mathcal{F}_r^{-1}(g)(x_1,x_2) := \frac{1}{(2\pi)^2}\int_{\mathbb{R}^2} g(\omega_1,\omega_2)\, e^{j\omega_2 x_2}\, e^{i\omega_1 x_1}\, d\omega_1 d\omega_2.$$

It has the closed-form representation:

$$\mathcal{F}_r^{-1}(g)(x_1,x_2) := \hat{\Phi}_0(x_1,x_2) + \hat{\Phi}_1(x_1,x_2) + \hat{\Phi}_2(x_1,x_2) + \hat{\Phi}_3(x_1,x_2),$$

where the integrals are

$$\begin{aligned}
\hat{\Phi}_0(x_1,x_2) &= \frac{1}{(2\pi)^2}\int_a^b\int_a^b g(\omega_1,\omega_2)\cos(\omega_1 x_1)\cos(\omega_2 x_2)d\omega_1 d\omega_2,\\
\hat{\Phi}_1(x_1,x_2) &= \frac{1}{(2\pi)^2}\int_a^b\int_a^b g(\omega_1,\omega_2)\boldsymbol{i}\sin(\omega_1 x_1)\cos(\omega_2 x_2)d\omega_1 d\omega_2,\\
\hat{\Phi}_2(x_1,x_2) &= \frac{1}{(2\pi)^2}\int_a^b\int_a^b g(\omega_1,\omega_2)\boldsymbol{j}\cos(\omega_1 x_1)\sin(\omega_2 x_2)d\omega_1 d\omega_2,\\
\hat{\Phi}_3(x_1,x_2) &= -\frac{1}{(2\pi)^2}\int_a^b\int_a^b g(\omega_1,\omega_2)\boldsymbol{k}\sin(\omega_1 x_1)\sin(\omega_2 x_2)d\omega_1 d\omega_2.
\end{aligned}$$

The quaternion exponential product $e^{\boldsymbol{j}x_2\omega_2}\, e^{\boldsymbol{i}\omega_1 x_1}$ is called the *inverse (right-sided) quaternion Fourier kernel.*

Remark 1.5. Again, all computations can easily be converted to other conventions, since

$$\begin{aligned}
\mathcal{F}_r^{-1}(g)(-x_1,-x_2) &= \frac{1}{(2\pi)^2}\int_a^b\int_a^b g(\omega_1,\omega_2)\overline{e^{\boldsymbol{i}\omega_1 x_1}\, e^{\boldsymbol{j}\omega_2 x_2}}d\omega_1 d\omega_2\\
&= \overline{\frac{1}{(2\pi)^2}\int_a^b\int_a^b e^{\boldsymbol{i}\omega_1 x_1}\, e^{\boldsymbol{j}\omega_2 x_2}\overline{g(\omega_1,\omega_2)}d\omega_1 d\omega_2}\\
&:= \overline{\mathcal{F}_l^{-1}(\overline{g})(x_1,x_2)}.
\end{aligned}$$

For convenience, below we will denote $\mathcal{F}_r^{-1}$ as $\mathcal{F}^{-1}$.

The present chapter has two main goals. The first consists in studying the asymptotic behaviour of the integral (1.1) under the assumption that f belongs to $L^1((a,b)\times(c,d);\mathbb{H})$ where a, b, c, d can be both finite and infinite points. We shall be interested in the connection between the function $f(x_1,x_2)$, and the behaviour of its Fourier transform $\mathcal{F}(f)(\omega_1,\omega_2)$ at infinity. These properties have an interest on their own for further applications to number theory, combinatorics, signal processing, imaging, computer vision and numerical analysis. The complexity of the underlying computations will need some attention. Central to this viewpoint are certain Fourier transform techniques, which as in the complex case, would be familiar to the reader. Our second goal consists in extending the classical Bochner–Minlos theorem to a noncommutative structure as in the case of quaternion functions. The resulting theorem guarantees the existence and uniqueness of the corresponding probability measure defined on a dual space. This will be done using the concept of *quaternion Fourier transform of a probability measure*. For the reader's convenience and for the sake of easy reference, the chapter is motivated by the results presented in [12, 13].

Although the presented results can be extended to generalized Clifford algebras as well, we will focus the discussion on the QFT for conciseness here. The proof of its generalization to higher dimensions is possible but needs more complicated calculations, which exceed the scope of this manuscript. These results are still under further investigation and will be reported in a forthcoming paper.

2. Preliminaries

At this stage we briefly recall basic algebraic facts about *quaternions* necessary for the sequel. Let $\mathbb{H} := \{p = a + b\boldsymbol{i} + c\boldsymbol{j} + d\boldsymbol{k} : a, b, c, d \in \mathbb{R}\}$ be a four-dimensional associative and noncommutative algebra, where the imaginary units $\boldsymbol{i}$, $\boldsymbol{j}$, and $\boldsymbol{k}$ are subject to the Hamiltonian multiplication rules

$$\boldsymbol{i}^2 = \boldsymbol{j}^2 = \boldsymbol{k}^2 = -1;$$

$$\boldsymbol{ij} = \boldsymbol{k} = -\boldsymbol{ji}, \qquad \boldsymbol{jk} = \boldsymbol{i} = -\boldsymbol{kj}, \qquad \boldsymbol{ki} = \boldsymbol{j} = -\boldsymbol{ik}.$$

The *scalar* and *vector parts* of p, $\mathrm{S}(p)$ and $\mathbf{V}(p)$, are defined as the a and $b\boldsymbol{i}+c\boldsymbol{j}+d\boldsymbol{k}$ terms, respectively. For the scalar part the cyclic product rule $\mathrm{S}(pqr) = \mathrm{S}(qrp)$ is valid. Further step is *quaternion conjugation* introduced similarly to that of the complex numbers $\overline{p} = a - b\boldsymbol{i} - c\boldsymbol{j} - d\boldsymbol{k}$. The quaternion conjugation is an anti-linear involution

$$\overline{\overline{p}} = p, \qquad \overline{p+q} = \overline{p} + \overline{q}, \qquad \overline{qp} = \overline{p}\,\overline{q}, \qquad \overline{\lambda p} = \lambda \overline{p} \quad (\forall\, \lambda \in \mathbb{R}).$$

The *norm* of p is defined by $|p| = \sqrt{p\overline{p}} = \sqrt{\overline{p}p} = \sqrt{a^2 + b^2 + c^2 + d^2}$, and it coincides with its corresponding Euclidean norm as a vector in $\mathbb{R}^4$. For every two quaternions p and q the triangle inequalities hold $|p + q| \le |p| + |q|$, and $|\,|p| - |q|\,| \le |p \pm q|$. Also, we have $|\mathrm{S}(p)| \le |p|$ and $|\mathbf{V}(p)| \le |p|$.

In the sequel, a *quaternion sequence* is a collection of real quaternions p_0, p_1, p_2, ... 'labelled' by nonnegative integers. We shall denote such a sequence by $\{p_n\}$ where $n = 0, 1, 2, \ldots$ and $p_n = p_{0,n} + p_{1,n}\boldsymbol{i} + p_{2,n}\boldsymbol{j} + p_{3,n}\boldsymbol{k}$ are the elements of the sequence, $p_{l,n} \in \mathbb{R}$ $(l = 0, 1, 2, 3)$. To supplement our investigations, we recall the key notion of convergence of a quaternion sequence.

Definition 2.1. The quaternion sequence $\{p_n\}$ is called *convergent* to the quaternion $p = a + b\boldsymbol{i} + c\boldsymbol{j} + d\boldsymbol{k}$ if $\lim_{n \longrightarrow \infty} |p_n - p| = 0$. We will use the traditional notation: $\lim_{n \longrightarrow \infty} p_n = p$.

Lemma 2.2. *Let $\{p_n\}$ and $\{q_n\}$ be two quaternion sequences for which $\lim_{n \longrightarrow \infty} p_n = p$ and $\lim_{n \longrightarrow \infty} q_n = q$, for $p, q \in \mathbb{H}$. Then*

1. $\lim_{n \longrightarrow \infty} (p_n \pm q_n) = p \pm q$;
2. $\lim_{n \longrightarrow \infty} (p_n q_n) = pq$;
3. $\lim_{n \longrightarrow \infty} (\alpha p_n) = \alpha p, \quad \alpha \in \mathbb{H}$.

Now, let s be the space of sequences

$$s := \{\{p_n\} : \lim_{n\longrightarrow\infty} n^t p_n = 0, \ \forall\, t \in \mathbb{N}_0\},$$

where $p_n \in \mathbb{H}$, and let

$$s_m := \{\{p_n\} : \|p\|_m^2 := \sum_{n=0}^{\infty}(1+n^2)^m |p_n|^2 < +\infty\}, \quad m \in \mathbb{Z}.$$

Proposition 2.3. *We have* $s = \cap_{m\in\mathbb{Z}} s_m$.

Proof. Let $p \in s$ be arbitrarily chosen and fixed, and $t \in \mathbb{N}_0$. From the definition of the space s we have that $\lim_{n\longrightarrow\infty} n^t p_n = 0$. Then for $C > 0$ a natural number $N = N(C)$ can be found so that for every $n > N$ the following holds

$$n^t |p_n| \le C.$$

Hence, it follows

$$\sum_{n=0}^{\infty} (1+n^2)^{\left[\frac{t}{2}\right]} |p_n|^2 \le |p_0|^2 + C^2 \sum_{n=1}^{\infty} (1+n^2)^{\left[\frac{t}{4}\right]} \frac{1}{n^{2t}} < +\infty,$$

and therefore $p \in s_{\left[\frac{t}{4}\right]}$. Since $t \in \mathbb{N}_0$ is arbitrary we conclude that $p \in s_m$ for every $m \in \mathbb{N}_0$, from where $p \in \cap_{m\in\mathbb{N}} s_m$. Since $p \in s$ was arbitrarily chosen it follows that $s \subset \cap_{m\in\mathbb{N}} s_m$. Now, let $p \in \cap_{m\in\mathbb{Z}} s_m$. Then for every $m \in \mathbb{N}_0$ we have that

$$\sum_{n=0}^{\infty} (1+n^2)^m |p_n|^2 < +\infty,$$

from where $\lim_{n\longrightarrow\infty} (1+n^2)^m |p_n|^2 = 0$. Therefore $\lim_{n\longrightarrow\infty} n^t p_n = 0$ for every $t \le 2m$. Since m was arbitrarily chosen then $\lim_{n\longrightarrow\infty} n^t p_n = 0$ for every $t \in \mathbb{N}_0$. Consequently, $p \in s$ and since $p \in \cap_{m\in\mathbb{N}} s_m$ was arbitrarily chosen we conclude that $\cap_{m\in\mathbb{N}} s_m \subset s$. □

In the sequel, consider the countable family of semi-norms on s

$$\|p\|_m^2 = \sum_{n=0}^{\infty} (1+n^2)^m |p_n|^2 .$$

Lemma 2.4. *s is completely a Hausdorff space.*

Proof. If $p \in s$ and $\|p\|_m = 0$ we get

$$\sum_{n=0}^{\infty}(1+n^2)^m |p_n|^2 = 0.$$

Whence, $p_n = 0$ for every natural n (including zero), and consequently $p = 0$. It follows that s is a Hausdorff space. Now, let $\{p_n^k\} \longrightarrow_{k\longrightarrow\infty} \{p_n\}$, *i.e.*, $\lim_{k\longrightarrow\infty} p_n^k = p_n$ for every $n \in \mathbb{N}_0$. Then

$$\left\|p^k - p\right\|_m \longrightarrow_{k\longrightarrow\infty} 0.$$

It follows that s is completely a space. □

Let us define the metric

$$\rho(p,q) = \sum_{m=0}^{\infty} \frac{\|p-q\|_m}{1+\|p-q\|_m} 2^{-m}$$

for $p, q \in s$. Evidently, the above metric has the translation property. In other words, from the countable family of seminorms we can define a metric with the translation property. From this follows the result that

Lemma 2.5. *s is a Fréchet space.*

Now, let s' denote the topological dual space to s given by $s' = \cup_{m\in\mathbb{Z}} s_m$. We will denote the set of all sequences by $\mathbb{R}^{\mathbb{N}_0}$, and we equip the space s' with cylindrical topology. Every element $p' \in s'$ acts on each element $p \in s$ as follows

$$\langle p', p\rangle = \sum_{n=0}^{\infty} p'_n p_n; \qquad \lim_{n\longrightarrow\infty} p'_n = p', \ \lim_{n\longrightarrow\infty} p_n = p.$$

3. Asymptotic Behaviour of the QFT

There are numerous questions that remain untouched about the behaviour of the QFT. We will fill in some of these gaps and discuss some rudiments of the asymptotic behaviour of (1.1).

We begin by proving the following result, which provides an interesting and efficient convergence characterization of the QFT.

Theorem 3.1. *Let $a, b, c, d \in \mathbb{R}$ and $f \in L^1((a,b)\times(c,d);\mathbb{H})$, then $\mathcal{F}(f)$ is uniformly continuous and bounded. Moreover, it satisfies*

$$\lim_{\omega_1\longrightarrow\pm\infty} \mathcal{F}(f)(\omega_1,\omega_2) = 0$$

uniformly in ω_2, and

$$\lim_{\omega_2\longrightarrow\pm\infty} \mathcal{F}(f)(\omega_1,\omega_2) = 0$$

uniformly in ω_1. The results hold with the same proof when the region $(a,b)\times(c,d)$ is replaced by its topological closure $\overline{(a,b)\times(c,d)}$ (a, b, c, d can be $\pm\infty$) or any region G (or $\overline{G}$) in the space $\mathbb{R}^2$.

Proof. It is easy to show that

$$\begin{aligned} |\mathcal{F}(f)(\omega_1,\omega_2)| &\le \int_a^b \int_c^d |f(x_1,x_2)|\, dx_1 dx_2 \\ &= \|f\|_{L^1((a,b)\times(c,d);\mathbb{H})}\,. \end{aligned} \tag{3.1}$$

So, the transform $\mathcal{F}(f)$ is bounded.

To prove the limit assertions, we use a density argument as in the classical case (Riemann–Lebesgue Lemma). We first assume that both f and $\partial_{x_1} f$ are continuous with compact support. Obviously, such functions form a dense subspace in $L^1((a,b)\times(c,d);\mathbb{H})$. By using integration by parts to the x_1-variable, with the iterated integration, we have

$$\begin{aligned}\mathcal{F}(f)(\omega_1,\omega_2) &= \int_a^b \left(f(b,x_2)\,\frac{1}{\mathbf{i}\omega_1}\,e^{-\mathbf{i}\omega_1 b}\,e^{-\mathbf{j}\omega_2 x_2} - f(a,x_2)\,\frac{1}{\mathbf{i}\omega_1}\,e^{-\mathbf{i}\omega_1 a}\,e^{-\mathbf{j}\omega_2 x_2}\right) dx_2 \\ &\quad - \int_a^b \int_a^b \partial_{x_1} f(x_1,x_2)\,\frac{1}{\mathbf{i}\omega_1}\,e^{-\mathbf{i}\omega_1 x_1}\,e^{-\mathbf{j}\omega_2 x_2}\,dx_1 dx_2 \\ &= -\int_a^b \int_c^d \partial_{x_1} f(x_1,x_2)\,\frac{1}{\mathbf{i}\omega_1}\,e^{-\mathbf{i}\omega_1 x_1}\,e^{-\mathbf{j}\omega_2 x_2}\,dx_1 dx_2.\end{aligned}$$

Therefore,

$$|\mathcal{F}(f)(\omega_1,\omega_2)| \le \frac{1}{|\omega_1|}\,\|\partial_{x_1} f\|_{L^1((a,b)\times(c,d);\mathbb{H})}\,. \tag{3.2}$$

So,

$$\lim_{\omega_1\longrightarrow\pm\infty} \mathcal{F}(f)(\omega_1,\omega_2) = 0$$

uniformly in ω_2. Similarly we can prove

$$\lim_{\omega_2\longrightarrow\pm\infty} \mathcal{F}(f)(\omega_1,\omega_2) = 0$$

uniformly in ω_1.

Now assume $f \in L^1((a,b)\times(c,d);\mathbb{H})$. For any given $\varepsilon > 0$, there exists a function f_ε in the above-mentioned dense class, such that

$$\|f - f_\varepsilon\|_{L^1((a,b)\times(c,d);\mathbb{H})} < \varepsilon.$$

By (3.1), we have

$$\begin{aligned}|\mathcal{F}(f)(\omega_1,\omega_2)| &\le |\mathcal{F}(f_\varepsilon)(\omega_1,\omega_2)| + \|f - f_\varepsilon\|_{L^1((a,b)\times(c,d);\mathbb{H})} \\ &\le |\mathcal{F}(f_\varepsilon)(\omega_1,\omega_2)| + \varepsilon.\end{aligned}$$

By using the result just proved for the density class, we have

$$\lim_{\omega_1\to\pm\infty} |\mathcal{F}(f)(\omega_1,\omega_2)| \le \varepsilon,$$

uniformly in ω_2. Since ε is arbitrary, we have

$$\lim_{\omega_1\to\pm\infty} \mathcal{F}(f)(\omega_1,\omega_2) = 0 \tag{3.3}$$

uniformly in ω_2. Similarly,

$$\lim_{\omega_2\to\pm\infty} \mathcal{F}(f)(\omega_1,\omega_2) = 0, \tag{3.4}$$

uniformly in ω_1.

Now we show the uniform continuity of $\mathcal{F}(f)(\omega_1,\omega_2)$. Given (3.3) and (3.4), since continuous functions are uniform continuous in compact sets, it suffices to show that $\mathcal{F}(f)$ is continuous at every point (ω_1, ω_2). In fact,

$$\begin{aligned}&\mathcal{F}(f)(\omega_1+\rho_1,\omega_2+\rho_2)-\mathcal{F}(f)(\omega_1,\omega_2)\\&=\int_a^b\int_c^d f(x_1,x_2)\left[e^{-ix_1(\omega_1+\rho_1)}e^{-jx_2(\omega_2+\rho_2)}-e^{-ix_2\omega_1}e^{-jx_2\omega_2}\right]dx_1dx_2.\end{aligned}$$

For any $\rho_1, \rho_2 > 0$, the integrand is dominated by a constant multiple of $|f(x_1, x_2)|$. Since the factor in the straight brackets tends to zero, by the Lebesgue Dominated Convergence Theorem, we have

$$\lim_{\rho_1,\rho_2\to 0}\mathcal{F}(f)(\omega_1+\rho_1,\omega_2+\rho_2)-\mathcal{F}(f)(\omega_1,\omega_2)=0,$$

proving the desired continuity. □

4. Bochner–Minlos Theorem

In this section we extend the classical Bochner–Minlos theorem (named after Bochner and Adol'fovich Minlos) to the framework of quaternion analysis. The resulting theorem guarantees the existence of the corresponding probability measure defined on a dual space. In particular, some interesting properties of the underlying measure are extended in this setting.

Definition 4.1. Let μ be a finite positive measure on $\mathbb{R}^2$. The QFT of μ is the function $\mathcal{F}_{ij}(\mu) : \mathbb{R}^2 \longrightarrow \mathbb{H}$ given by

$$\mathcal{F}_{ij}(\mu)(\omega_1,\omega_2):=\int_{\mathbb{R}^2} e^{-i\omega_1x_1}e^{-j\omega_2x_2}d\mu(x_1,x_2),$$

or the function $\mathcal{F}_{ji}(\mu) : \mathbb{R}^2 \longrightarrow \mathbb{H}$ given by

$$\mathcal{F}_{ji}(\mu)(\omega_1,\omega_2):=\int_{\mathbb{R}^2} e^{-j\omega_2x_2}e^{-i\omega_1x_1}d\mu(x_1,x_2).$$

Proposition 4.2. *Let μ be a finite positive measure on $\mathbb{R}^2$. The functionals $\mathcal{F}_{ij}(\mu)$ and $\mathcal{F}_{ji}(\mu)$ satisfy the following basic properties:*

1. $\mathcal{F}_{ij}(\mu)(0,0) = \mathcal{F}_{ji}(\mu)(0,0) = 1$;
2. $\mathcal{F}_{ij}(\mu)(-\omega_1,-\omega_2) = \overline{\mathcal{F}_{ji}(\mu)(\omega_1,\omega_2)}$;
3. $\mathcal{F}_{ji}(\mu)(-\omega_1,-\omega_2) = \overline{\mathcal{F}_{ij}(\mu)(\omega_1,\omega_2)}$;
4. $\mathcal{F}_{ij}(\mu)(-\omega_1,-\omega_2) + \mathcal{F}_{ij}(\mu)(\omega_1,\omega_2)$
$$= 2\int_{\mathbb{R}^2}(\cos(\omega_1x_1)\cos(\omega_2x_2)+\boldsymbol{k}\sin(\omega_1x_1)\sin(\omega_2x_2))\,d\mu(x_1,x_2);$$
5. $\mathcal{F}_{ij}(\mu)(\omega_1,\omega_2) + \mathcal{F}_{ji}(\mu)(-\omega_1,-\omega_2) = 2\int_{\mathbb{R}^2}\cos(\omega_1x_1)\cos(\omega_2x_2)d\mu(x_1,x_2)$.

Proof. For the first statement, note that

$$\mathcal{F}_{ij}(\mu)(0,0)=\int_{\mathbb{R}^2}d\mu(x_1,x_2)=\mu(\mathbb{R}^2).$$

For Statement 2 a straightforward computation shows that

$$\begin{aligned}\mathcal{F}_{\boldsymbol{ij}}(\mu)(-\omega_1,-\omega_2) &= \int_{\mathbb{R}^2} e^{\boldsymbol{i}\omega_1 x_1} e^{\boldsymbol{j}\omega_2 x_2} d\mu(x_1,x_2) \quad = \int_{\mathbb{R}^2} \overline{e^{-\boldsymbol{j}\omega_2 x_2} e^{-\boldsymbol{i}\omega_1 x_1}} d\mu(x_1,x_2) \\ &= \overline{\int_{\mathbb{R}^2} e^{-\boldsymbol{j}\omega_2 x_2} e^{-\boldsymbol{i}\omega_1 x_1} d\mu(x_1,x_2)} = \overline{\mathcal{F}_{\boldsymbol{ji}}(\mu)(\omega_1,\omega_2)}.\end{aligned}$$

The proof of Statement 3 will be omitted, being similar to the previous one. Now, taking into account that

$$\begin{aligned}e^{-\boldsymbol{i}\omega_1 x_1} e^{-\boldsymbol{j}\omega_2 x_2} = & \quad \cos(\omega_1 x_1)\cos(\omega_2 x_2) - \boldsymbol{i}\sin(\omega_1 x_1)\cos(\omega_2 x_2) \\ & - \boldsymbol{j}\cos(\omega_1 x_1)\sin(\omega_2 x_2) + \boldsymbol{k}\sin(\omega_1 x_1)\sin(\omega_2 x_2),\end{aligned}$$

and

$$\begin{aligned}e^{\boldsymbol{i}\omega_1 x_1} e^{\boldsymbol{j}\omega_2 x_2} = & \quad \cos(\omega_1 x_1)\cos(\omega_2 x_2) + \boldsymbol{i}\sin(\omega_1 x_1)\cos(\omega_2 x_2) \\ & + \boldsymbol{j}\cos(\omega_1 x_1)\sin(\omega_2 x_2) + \boldsymbol{k}\sin(\omega_1 x_1)\sin(\omega_2 x_2),\end{aligned}$$

we obtain

$$\begin{aligned}&\mathcal{F}_{\boldsymbol{ij}}(\mu)(\omega_1,\omega_2) + \mathcal{F}_{\boldsymbol{ij}}(\mu)(-\omega_1,-\omega_2) \\ &\qquad = \int_{\mathbb{R}^2} \left(e^{-\boldsymbol{i}\omega_1 x_1} e^{-\boldsymbol{j}\omega_2 x_2} + e^{\boldsymbol{i}\omega_1 x_1} e^{\boldsymbol{j}\omega_2 x_2}\right) d\mu(x_1,x_2) \\ &\qquad = 2\int_{\mathbb{R}^2} \left(\cos(\omega_1 x_1)\cos(\omega_2 x_2) + \boldsymbol{k}\sin(\omega_1 x_1)\sin(\omega_2 x_2)\right) d\mu(x_1,x_2).\end{aligned}$$

For the last statement we use the relation

$$\begin{aligned}e^{\boldsymbol{j}\omega_2 x_2} e^{\boldsymbol{i}\omega_1 x_1} = & \quad \cos(\omega_1 x_1)\cos(\omega_2 x_2) + \boldsymbol{i}\sin(\omega_1 x_1)\cos(\omega_2 x_2) \\ & + \boldsymbol{j}\cos(\omega_1 x_1)\sin(\omega_2 x_2) - \boldsymbol{k}\sin(\omega_1 x_1)\sin(\omega_2 x_2).\end{aligned}$$

Therefore, it follows

$$\begin{aligned}&\mathcal{F}_{\boldsymbol{ij}}(\mu)(\omega_1,\omega_2) + \mathcal{F}_{\boldsymbol{ji}}(\mu)(-\omega_1,-\omega_2) \\ &\qquad = \int_{\mathbb{R}^2} \left(e^{-\boldsymbol{i}\omega_1 x_1} e^{-\boldsymbol{j}\omega_2 x_2} + e^{\boldsymbol{j}\omega_2 x_2} e^{\boldsymbol{i}\omega_1 x_1}\right) d\mu(x_1,x_2) \\ &\qquad = 2\int_{\mathbb{R}^2} \cos(\omega_1 x_1)\cos(\omega_2 x_2) d\mu(x_1,x_2).\end{aligned}$$

□

In the sequel, let us denote by $\mathbb{S}(\mathbb{R}^2)$ the Schwartz space of smooth quaternion functions on $\mathbb{R}^2$. We formulate a first result.

Proposition 4.3. *Let $\phi \in \mathbb{S}(\mathbb{R}^2)$ and μ be a finite positive measure on $\mathbb{R}^2$. Then*

1. $\displaystyle\int_{\mathbb{R}^2} \mathcal{F}_{\boldsymbol{ij}}(\mu)(\omega_1,\omega_2)\phi(\omega_1,\omega_2)d\omega_1 d\omega_2 = \int_{\mathbb{R}^2} \mathcal{F}_{\boldsymbol{ij}}(\phi)(x_1,x_2)d\mu(x_1,x_2);$
2. $\displaystyle\int_{\mathbb{R}^2} \mathcal{F}_{\boldsymbol{ji}}(\mu)(\omega_1,\omega_2)\phi(\omega_1,\omega_2)d\omega_1 d\omega_2 = \int_{\mathbb{R}^2} \mathcal{F}_{\boldsymbol{ji}}(\phi)(x_1,x_2)d\mu(x_1,x_2).$

Proof. For simplicity we just present the computations of the first equality. The proof of the second one is similar. A direct computation shows that

$$\begin{aligned}
&\int_{\mathbb{R}^2} \mathcal{F}_{ij}(\mu)(\omega_1,\omega_2)\phi(\omega_1,\omega_2)d\omega_1 d\omega_2 \\
&\qquad = \int_{\mathbb{R}^2}\int_{\mathbb{R}^2} e^{-i\omega_1 x_1}e^{-j\omega_2 x_2}d\mu(x_1,x_2)\phi(\omega_1,\omega_2)d\omega_1 d\omega_2 \\
&\qquad = \int_{\mathbb{R}^2}\int_{\mathbb{R}^2} e^{-i\omega_1 x_1}e^{-j\omega_2 x_2}\phi(\omega_1,\omega_2)d\omega_1 d\omega_2 d\mu(x_1,x_2) \\
&\qquad = \int_{\mathbb{R}^2} \mathcal{F}_{ij}(\phi)(x_1,x_2)d\mu(x_1,x_2).
\end{aligned}$$

□

We now analyze some key properties of the above-mentioned functionals.

Proposition 4.4. *Let μ and ν be finite positive measures on $\mathbb{R}^2$. The functionals $\mathcal{F}_{ij}(\mu)$ and $\mathcal{F}_{ji}(\mu)$ are linear, i.e., for every $c, d \in \mathbb{H}$ it holds:*

$$\begin{aligned}
\mathcal{F}_{ij}(\mu c + \nu d) &= \mathcal{F}_{ij}(\mu)c + \mathcal{F}_{ij}(\nu)d, \\
\mathcal{F}_{ji}(\mu c + \nu d) &= \mathcal{F}_{ji}(\mu)c + \mathcal{F}_{ji}(\nu)d.
\end{aligned}$$

Proposition 4.5. *Let μ be a finite positive measure on $\mathbb{R}^2$. For any $a_i \in \mathbb{R}\setminus\{0\}$ $(i = 1, 2)$ the following conditions hold:*

1. $\mathcal{F}_{ij}(\mu(a_1x_1, a_2x_2)) = \mathcal{F}_{ij}(\mu(x_1,x_2))\left(\frac{\omega_1}{a_1}, \frac{\omega_2}{a_2}\right)$;
2. $\mathcal{F}_{ji}(\mu(a_1x_1, a_2x_2)) = \mathcal{F}_{ji}(\mu(x_1,x_2))\left(\frac{\omega_1}{a_1}, \frac{\omega_2}{a_2}\right)$.

Proof. For simplicity we just present the proof of the first condition. A straightforward computation shows that

$$\begin{aligned}
\mathcal{F}_{ij}(\mu(a_1x_1, a_2x_2)) &= \int_{\mathbb{R}^2} e^{-i\omega_1 x_1}e^{-j\omega_2 x_2}d\mu(a_1x_1,a_2x_2) \\
&= \int_{\mathbb{R}^2} e^{-i\frac{\omega_1}{a_1}(a_1x_1)}e^{-j\frac{\omega_2}{a_2}(a_2x_2)}d\mu(a_1x_1,a_2x_2) \\
&= \int_{\mathbb{R}^2} e^{-i\frac{\omega_1}{a_1}y_1}e^{-j\frac{\omega_2}{a_2}y_2}d\mu(y_1,y_2) \\
&= \mathcal{F}_{ij}(\mu(x_1,x_2))\left(\frac{\omega_1}{a_1}, \frac{\omega_2}{a_2}\right).
\end{aligned}$$

□

We proceed to define the notion of *positive definitely function* in the context of quaternion analysis.

Definition 4.6. Let f be a quaternion function on $\mathbb{R}^2$ that is continuous and bounded. For every finite positive measure μ on $\mathbb{R}^2$ the function f is said to be *positive definite* if

$$\sum_{k,l=1,k<l}^{N} z_k\overline{z_l}f(\lambda_k-\lambda_l) + \sum_{k,l=1,k<l}^{N} \overline{f(\lambda_k-\lambda_l)z_l\overline{z_k}} + \sum_{k=1}^{N}|z_k|^2\mu(\mathbb{R}^2) \geq 0$$

for every $\lambda_1, \lambda_2, \ldots, \lambda_N \in \mathbb{R}^2$, $z_1, z_2, \ldots, z_N \in \mathbb{H}$. These parameters are measured such that:

1. When $\lambda_1 = \lambda_2 = \cdots = \lambda_N$, and $z_1 = z_2 = \cdots = z_N$ it follows
$$2f(0,0) + \mu(\mathbb{R}^2) \geq 0;$$
2. When $N = 2$, $\lambda_1 = (x_1, x_2)$, and $\lambda_2 = (0,0)$, we have
$$z_1\overline{z}_2 f(x_1,x_2) + f(-x_1,-x_2)z_2\overline{z}_1 + \left(|z_1|^2 + |z_2|^2\right)\mu\left(\mathbb{R}^2\right) \geq 0,$$
which is valid if $f(-x_1,-x_2) = \overline{f(x_1,x_2)}$.

Proposition 4.7. *The functional $\mathcal{F}_{\boldsymbol{ij}}(\mu)$ is positive definite and bounded.*

Proof. Let $\lambda_1, \lambda_2, \ldots, \lambda_N \in \mathbb{R}^2$ such that $\lambda_i = (\lambda_{1k}, \lambda_{2k})$, and $z_1, z_2, \ldots, z_N \in \mathbb{H}$. Direct computations show that

$$\sum_{k,l,k<l}^{N} z_k\overline{z}_l \mathcal{F}_{\boldsymbol{ij}}(\mu)(\lambda_k - \lambda_l) + \sum_{k,l=1,k<l}^{N} \overline{\mathcal{F}_{\mathbf{ij}}(\mu)(\lambda_k - \lambda_l)z_l\overline{z_k}} + \sum_{k=1}^{N} |z_k|^2 \mu\left(\mathbb{R}^2\right)$$
$$= \int_{\mathbb{R}^2} \sum_{k,l=1,k<l}^{N} z_k\overline{z}_l e^{-\boldsymbol{i}(\lambda_{1k}-\lambda_{1l})x_1} e^{-\boldsymbol{j}(\lambda_{2k}-\lambda_{2l})x_2} d\mu(x_1,x_2)$$
$$+ \int_{\mathbb{R}^2} \sum_{k,l=1,k<l}^{N} e^{-\boldsymbol{j}(\lambda_{2k}-\lambda_{2l})x_2} e^{-\boldsymbol{i}(\lambda_{1k}-\lambda_{1l})x_1} z_l\overline{z_k} d\mu(x_1,x_2)$$
$$+ \int_{\mathbb{R}^2} \sum_{k=1}^{N} |z_k|^2 d\mu(x_1,x_2)$$
$$= \int_{\mathbb{R}^2} \sum_{k,l=1,k<l}^{N} z_k\overline{z}_l e^{-\boldsymbol{i}(\lambda_{1k}-\lambda_{1l})x_1} e^{-\boldsymbol{j}(\lambda_{2k}-\lambda_{2l})x_2} d\mu(x_1,x_2)$$
$$+ \int_{\mathbb{R}^2} \sum_{k,l=1,k<l}^{N} \overline{z_k\overline{z}_l e^{-\boldsymbol{i}(\lambda_{1k}-\lambda_{1l})x_1} e^{-\boldsymbol{j}(\lambda_{2k}-\lambda_{2l})x_2}} d\mu(x_1,x_2)$$
$$+ \int_{\mathbb{R}^2} \sum_{k=1}^{N} |z_k|^2 d\mu(x_1,x_2)$$
$$= \int_{\mathbb{R}^2} \left[\sum_{k=1}^{N} |z_k|^2 + 2\sum_{k,l=1,k<l}^{N} \mathrm{S}\left(z_k\overline{z}_l e^{-\boldsymbol{i}(\lambda_{1k}-\lambda_{1l})x_1} e^{-\boldsymbol{j}(\lambda_{2k}-\lambda_{2l})x_2}\right)\right] d\mu(x_1,x_2)$$
$$\geq \int_{\mathbb{R}^2} \left(\sum_{k=1}^{N} |z_k|^2 - 2\sum_{k,l=1,k<l}^{N} |z_k|\,|z_l|\right) d\mu(x_1,x_2) \geq 0.$$

Furthermore, it follows that

$$|\mathcal{F}_{\boldsymbol{ij}}(\mu)(\omega_1,\omega_2)| = \left|\int_{\mathbb{R}^2} e^{-\boldsymbol{i}\omega_1 x_1} e^{-\boldsymbol{j}\omega_2 x_2} d\mu(x_1,x_2)\right|$$

$$\leq \int_{\mathbb{R}^2} |e^{-i\omega_1 x_1} e^{-j\omega_2 x_2}|\, d\mu(x_1, x_2) = \mu\left(\mathbb{R}^2\right) < +\infty. \qquad \square$$

Likewise we can prove the following proposition.

Proposition 4.8. *The functional $\mathcal{F}_{ji}(\mu)$ is positive definite and bounded.*

Below we shall assume that μ is a probably measure on s'. For every elements $a \in s$ and $a' \in s'$ we define the functionals g_{ij} and g_{ji} on s as follows:

$$g_{ij} : s \longrightarrow s, \qquad a \longmapsto g_{ij}(a) := \int_{s'} e^{i\langle a',a\rangle} e^{j\langle a',a\rangle} d\mu(a')$$

and

$$g_{ji} : s \longrightarrow s, \qquad a \longmapsto g_{ji}(a) := \int_{s'} e^{j\langle a',a\rangle} e^{i\langle a',a\rangle} d\mu(a').$$

Next we present a generalization of the *classical Bochner–Minlos theorem* on positive definite functions to the case of quaternion functions.

Theorem 4.9 (Bochner–Minlos theorem). *The functional g_{ij} satisfies the following three conditions:*

1. *Normalization:* $g_{ij}(0) = 1$;
2. *Positivity:*
$$\sum_{k,l=1,k<l}^{n} z_k \overline{z}_l g_{ij}(a_k - a_l) + \sum_{k,l=1,k<l}^{n} \overline{g_{ij}(a_k - a_l) z_l \overline{z_k}} + \sum_k |z_k|^2 \geq 0;$$
3. *Continuity:* g_{ij} *is continuous in the sense of Fréchet topology.*

Proof. We begin the proof by noting that

$$g_{ij}(0) = \int_{s'} d\mu(a') = \mu(s') = 1.$$

For simplicity we will prove Statement 2 in the case $n = 2$ only, *i.e.*, we will prove that

$$z_1\overline{z}_2\, g_{ij}(a_1 - a_2) + \overline{g_{ij}(a_1 - a_2) z_2 \overline{z_1}} + |z_1|^2 + |z_2|^2 \geq 0.$$

For the sake of convenience we set $z = a_1 - a_2$. It follows that

$$\begin{aligned} &z_1\overline{z}_2\, g_{ij}(z) + \overline{g_{ij}(z) z_2 \overline{z_1}} + |z_1|^2 + |z_2|^2 \\ &= \int_{s'} \left(z_1\overline{z}_2 e^{i\langle a',z\rangle} e^{j\langle a',z\rangle} + \overline{z_1\overline{z}_2 e^{i\langle a',z\rangle} e^{j\langle a',z\rangle}} + |z_1|^2 + |z_2|^2 \right) d\mu(a') \\ &= \int_{s'} \left[2\,\mathrm{S}\!\left(z_1\overline{z_2} e^{i\langle a',z\rangle} e^{j\langle a',z\rangle} + |z_1|^2 + |z_2|^2 \right) \right] d\mu(a'). \end{aligned} \tag{4.1}$$

Notice that the last equality follows from the relation

$$z_1\overline{z}_2\, g_{ij}(z) = \overline{g_{ji}(-z) z_2 \overline{z}_1}.$$

Now, let $z_i = |z_i|\, e^{\frac{\mathbf{V}(z_i)}{|\mathbf{V}(z_i)|}\theta_i}$, with $\theta_i = \arg(z_i)$ $(i = 1, 2)$. Then

$$\mathrm{S}(z_1\overline{z}_2\, g_{ij}(z)) \geq -|z_1|\,|z_2|\,.$$

From here and (4.1) we obtain

$$z_1\overline{z}_2\, g_{ij}(z) + g_{ji}(-z)z_2\overline{z}_1 + |z_1|^2 + |z_2|^2$$
$$\geq \int_{s'} \left(|z_1|^2 + |z_2|^2 - 2\,|z_1|\,|z_2|\right) d\mu(a') \geq 0.$$

Using induction we may conclude that Statement 2 is valid for every natural n. For the proof of the remaining statement, let $\lim_{n\longrightarrow\infty} a_n = a$ be understood in the sense of the topology of Fréchet. Then $\lim_{n\longrightarrow\infty} a_n = a$ holds in the usual sense. From here and from the definition of the functional g_{ij} we conclude that $\lim_{n\longrightarrow\infty} g_{ij}(a_n) = g_{ij}(a)$. Therefore for every $\epsilon > 0$ a natural number $N = N(\epsilon) > 0$ can be found so that for $k > N$ the following holds

$$\|g(a_n) - g(a)\|_k < \epsilon.$$

Consequently, it follows that

$$\rho\,(g(a_n), g(a)) = \sum_{N+1}^{\infty} \frac{\|g(a_n) - g(a)\|_k}{1 + \|g(a_n) - g(a)\|_k} 2^{-k} \leq \epsilon \sum_{N+1}^{\infty} 2^{-k}. \qquad \square$$

Proposition 4.10. *For every element $a \in s$ the functionals g_{ij} and g_{ji} satisfy the additional properties:*

1. $g_{ij}(-a) = \overline{g_{ji}(a)}$;
2. $g_{ji}(-a) = \overline{g_{ij}(a)}$.

Though the significance of our approach to concrete applications, such as the characterization of measurement configurations for functional spaces, was the main reason for restricting ourselves to the quaternionic case, doubtless, the reduction in calculations for proving the results played an important role too. As was already mentioned, it is possible to perform an analogous study to generalized Clifford algebras following the same ideas. Further investigations will be presented in a forthcoming paper.

Acknowledgement

Partial support from the Foundation for Science and Technology (FCT) *via* the grant DD-VU-02/90, Bulgaria is acknowledged by the first named author. The second named author acknowledges financial support from the Foundation for Science and Technology (FCT) *via* the post-doctoral grant SFRH/ BPD/66342/2009. This work was supported by *FEDER* funds through *COM PETE* – Operational Programme Factors of Competitiveness ('Programa Operacional Factores de Competitividade') and by Portuguese funds through the *Center for Research and Development in Mathematics and Applications* (University of Aveiro) and the Portuguese Foundation for Science and Technology ('FCT – Fundação para a Ciência e a Tecnologia'), within project PEst-C/MAT/UI4106/2011 with COMPETE number FCOMP-01-0124-FEDER-022690. The third named author acknowledges financial support from the research grant of the University of Macau No. MYRG142(Y1-L2)-FST11-KKI.

References

[1] M. Bahri. Generalized Fourier transform in real clifford algebra $cl(0,n)$. *Far East Journal of Mathematical Sciences*, 48(1):11–24, Jan. 2011.

[2] P. Bas, N. Le Bihan, and J.M. Chassery. Color image watermarking using quaternion Fourier transform. In *Proceedings of the IEEE International Conference on Acoustics Speech and Signal Processing, ICASSP*, volume 3, pages 521–524. Hong-Kong, 2003.

[3] E. Bayro-Corrochano, N. Trujillo, and M. Naranjo. Quaternion Fourier descriptors for preprocessing and recognition of spoken words using images of spatiotemporal representations. *Mathematical Imaging and Vision*, 28(2):179–190, 2007.

[4] F. Brackx, N. De Schepper, and F. Sommen. The Clifford–Fourier transform. *Journal of Fourier Analysis and Applications*, 11(6):669–681, 2005.

[5] F. Brackx, N. De Schepper, and F. Sommen. The Fourier transform in Clifford analysis. *Advances in Imaging and Electron Physics*, 156:55–201, 2009.

[6] F. Brackx, R. Delanghe, and F. Sommen. *Clifford Analysis*, volume 76. Pitman, Boston, 1982.

[7] T. Bülow. *Hypercomplex Spectral Signal Representations for the Processing and Analysis of Images.* PhD thesis, University of Kiel, Germany, Institut für Informatik und Praktische Mathematik, Aug. 1999.

[8] T. Bülow, M. Felsberg, and G. Sommer. Non-commutative hypercomplex Fourier transforms of multidimensional signals. In G. Sommer, editor, *Geometric computing with Clifford Algebras: Theoretical Foundations and Applications in Computer Vision and Robotics*, pages 187–207, Berlin, 2001. Springer.

[9] J. Ebling and G. Scheuermann. Clifford Fourier transform on vector fields. *IEEE Transactions on Visualization and Computer Graphics*, 11(4):469–479, July/Aug. 2005.

[10] T.A. Ell. Quaternion-Fourier transforms for analysis of 2-dimensional linear time-invariant partial-differential systems. In *Proceedings of the 32nd Conference on Decision and Control*, pages 1830–1841, San Antonio, Texas, USA, 15–17 December 1993. IEEE Control Systems Society.

[11] T.A. Ell and S.J. Sangwine. Hypercomplex Fourier transforms of color images. *IEEE Transactions on Image Processing*, 16(1):22–35, Jan. 2007.

[12] S. Georgiev. Bochner–Minlos theorem and quaternion Fourier transform. In Gürlebeck [14].

[13] S. Georgiev, J. Morais, and W. Sprößig. Trigonometric integrals in the framework of quaternionic analysis. In Gürlebeck [14].

[14] K. Gürlebeck, editor. *9th International Conference on Clifford Algebras and their Applications*, Weimar, Germany, 15–20 July 2011.

[15] K. Gürlebeck and W. Sprößig. *Quaternionic Analysis and Elliptic Boundary Value Problems.* Berlin: Akademie-Verlag, Berlin, 1989.

[16] K. Gürlebeck and W. Sprößig. *Quaternionic and Clifford Calculus for Physicists and Engineers.* Wiley, Aug. 1997.

[17] E. Hitzer. Quaternion Fourier transform on quaternion fields and generalizations. *Advances in Applied Clifford Algebras*, 17(3):497–517, May 2007.

[18] E.M.S. Hitzer and B. Mawardi. Clifford Fourier transform on multivector fields and uncertainty principles for dimensions $n = 2(\mathrm{mod}\, 4)$ and $n = 3(\mathrm{mod}\, 4)$. *Advances in Applied Clifford Algebras*, 18(3-4):715–736, 2008.

[19] V.V. Kravchenko. *Applied Quaternionic Analysis*, volume 28 of *Research and Exposition in Mathematics*. Heldermann Verlag, Lemgo, Germany, 2003.

[20] B. Mawardi. Generalized Fourier transform in Clifford algebra $\mathcal{C}\ell_{0,3}$. *Far East Journal of Mathematical Sciences*, 44(2):143–154, 2010.

[21] B. Mawardi, E. Hitzer, A. Hayashi, and R. Ashino. An uncertainty principle for quaternion Fourier transform. *Computers and Mathematics with Applications*, 56(9):2411–2417, 2008.

[22] B. Mawardi and E.M.S. Hitzer. Clifford Fourier transformation and uncertainty principle for the Clifford algebra $C\ell_{3,0}$. *Advances in Applied Clifford Algebras*, 16(1):41–61, 2006.

[23] S.-C. Pei, J.-J. Ding, and J.-H. Chang. Efficient implementation of quaternion Fourier transform, convolution, and correlation by 2-D complex FFT. *IEEE Transactions on Signal Processing*, 49(11):2783–2797, Nov. 2001.

[24] M. Shapiro and N.L. Vasilevski. Quaternionic ψ-hyperholomorphic functions, singular integral operators and boundary value problems I. ψ-hyperholomorphic function theory. *Complex Variables, Theory and Application*, 27(1):17–46, 1995.

[25] M. Shapiro and N.L. Vasilevski. Quaternionic ψ-hyperholomorphic functions, singular integral operators and boundary value problems II. Algebras of singular integral operators and Riemann type boundary value problems. *Complex Variables, Theory Appl.*, 27(1):67–96, 1995.

S. Georgiev
Department of Differential Equations, University of Sofia
Sofia, Bulgaria
e-mail: sgg2000bg@yahoo.com

J. Morais
Centro de Investigação e Desenvolvimento em Matemática e Aplicações (CIDMA)
Universidade de Aveiro, 3810-193 Aveiro, Portugal
e-mail: joao.pedro.morais@ua.pt

K.I. Kou
Department of Mathematics, Faculty of Science and Technology
University of Macau, Macau
e-mail: kikououou@umac.mo

W. Sprößig
Freiberg University of Mining and Technology
Freiberg, Germany
e-mail: sproessig@math.tu-freiberg.de

Part II

Clifford Algebra

Quaternion and Clifford–Fourier Transforms and Wavelets
Trends in Mathematics, 123–153

7 Square Roots of −1 in Real Clifford Algebras

Eckhard Hitzer, Jacques Helmstetter and Rafał Abłamowicz

Abstract. It is well known that Clifford (geometric) algebra offers a geometric interpretation for square roots of -1 in the form of blades that square to minus 1. This extends to a geometric interpretation of quaternions as the side face bivectors of a unit cube. Systematic research has been done [33] on the biquaternion roots of -1, abandoning the restriction to blades. Biquaternions are isomorphic to the Clifford (geometric) algebra $C\ell_{3,0}$ of $\mathbb{R}^3$. Further research on general algebras $C\ell_{p,q}$ has explicitly derived the geometric roots of -1 for $p+q \leq 4$ [20]. The current research abandons this dimension limit and uses the Clifford algebra to matrix algebra isomorphisms in order to algebraically characterize the continuous manifolds of square roots of -1 found in the different types of Clifford algebras, depending on the type of associated ring ($\mathbb{R}$, $\mathbb{H}$, $\mathbb{R}^2$, $\mathbb{H}^2$, or $\mathbb{C}$). At the end of the chapter explicit computer generated tables of representative square roots of -1 are given for all Clifford algebras with $n = 5, 7$, and $s = 3 \,(\mathrm{mod}\, 4)$ with the associated ring $\mathbb{C}$. This includes, *e.g.*, $C\ell_{0,5}$ important in Clifford analysis, and $C\ell_{4,1}$ which in applications is at the foundation of conformal geometric algebra. All these roots of -1 are immediately useful in the construction of new types of geometric Clifford–Fourier transformations.

Mathematics Subject Classification (2010). Primary 15A66; secondary 11E88, 42A38, 30G35.

Keywords. Algebra automorphism, inner automorphism, center, centralizer, Clifford algebra, conjugacy class, determinant, primitive idempotent, trace.

1. Introduction

The young London Goldsmith professor of applied mathematics W.K. Clifford created his *geometric algebras*[1] in 1878 inspired by the works of Hamilton on

[1]In his original publication [11] Clifford first used the term *geometric algebras*. Subsequently in mathematics the new term *Clifford algebras* [28] has become the proper mathematical term. For emphasizing the *geometric* nature of the algebra, some researchers continue [9, 16, 17] to use the original term geometric algebra(s).

quaternions and by Grassmann's exterior algebra. Grassmann invented the anti-symmetric outer product of vectors, that regards the oriented parallelogram area spanned by two vectors as a new type of number, commonly called bivector. The bivector represents its own plane, because outer products with vectors in the plane vanish. In three dimensions the outer product of three linearly independent vectors defines a so-called trivector with the magnitude of the volume of the parallelepiped spanned by the vectors. Its orientation (sign) depends on the handedness of the three vectors.

In the Clifford algebra [16] of $\mathbb{R}^3$ the three bivector side faces of a unit cube $\{e_1e_2, e_2e_3, e_3e_1\}$ oriented along the three coordinate directions $\{e_1, e_2, e_3\}$ correspond to the three quaternion units $\boldsymbol{i}$, $\boldsymbol{j}$, and $\boldsymbol{k}$. Like quaternions, these three bivectors square to minus one and generate the rotations in their respective planes.

Beyond that Clifford algebra allows to extend complex numbers to higher dimensions [7, 17] and systematically generalize our knowledge of complex numbers, holomorphic functions and quaternions into the realm of Clifford analysis. It has found rich applications in symbolic computation, physics, robotics, computer graphics, *etc.* [8, 9, 12, 14, 27]. Since bivectors and trivectors in the Clifford algebras of Euclidean vector spaces square to minus one, we can use them to create new geometric kernels for Fourier transformations. This leads to a large variety of new Fourier transformations, which all deserve to be studied in their own right [5, 6, 9, 13, 18, 19, 22, 23, 26, 29–32].

In our current research we will treat square roots of -1 in Clifford algebras $C\ell_{p,q}$ of both Euclidean (positive definite metric) and non-Euclidean (indefinite metric) non-degenerate vector spaces, $\mathbb{R}^n = \mathbb{R}^{n,0}$ and $\mathbb{R}^{p,q}$, respectively. We know from Einstein's special theory of relativity that non-Euclidean vector spaces are of fundamental importance in nature [15]. They are further, *e.g.*, used in computer vision and robotics [12] and for general algebraic solutions to contact problems [27]. Therefore this chapter is about characterizing square roots of -1 in all Clifford algebras $C\ell_{p,q}$, extending previous limited research on $C\ell_{3,0}$ in [33] and $C\ell_{p,q}, n = p+q \leq 4$ in [20]. The manifolds of square roots of -1 in $C\ell_{p,q}$, $n = p+q = 2$, compare Table 1 of [20], are visualized in Figure 1.

First, we introduce necessary background knowledge of Clifford algebras and matrix ring isomorphisms and explain in more detail how we will characterize and classify the square roots of -1 in Clifford algebras in Section 2. Next, we treat section by section (in Sections 3 to 7) the square roots of -1 in Clifford algebras which are isomorphic to matrix algebras with associated rings $\mathbb{R}$, $\mathbb{H}$, $\mathbb{R}^2$, $\mathbb{H}^2$, and $\mathbb{C}$, respectively. The term *associated* means that the isomorphic matrices will only have matrix elements from the associated ring. The square roots of -1 in Section 7 with associated ring $\mathbb{C}$ are of particular interest, because of the existence of classes of *exceptional* square roots of -1, which all include a nontrivial term in the central element of the respective algebra different from the identity. Section 7 therefore includes a detailed discussion of all classes of square roots of -1 in the algebras $C\ell_{4,1}$, the isomorphic $C\ell_{0,5}$, and in $C\ell_{7,0}$. Finally, we add Appendix A with tables of square roots of -1 for all Clifford algebras with $n = 5, 7$, and $s = 3 \pmod 4$.

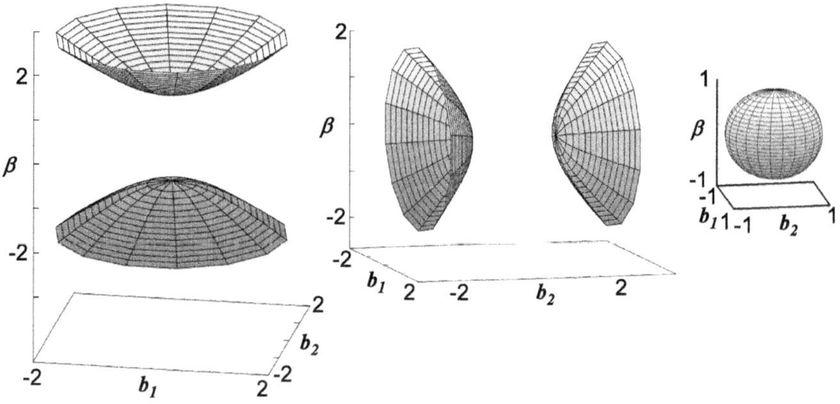

FIGURE 1. Manifolds of square roots f of -1 in $C\ell_{2,0}$ (left), $C\ell_{1,1}$ (center), and $C\ell_{0,2} \cong \mathbb{H}$ (right). The square roots are $f = \alpha + b_1 e_1 + b_2 e_2 + \beta e_{12}$, with $\alpha, b_1, b_2, \beta \in \mathbb{R}$, $\alpha = 0$, and $\beta^2 = b_1^2 e_2^2 + b_2^2 e_1^2 + e_1^2 e_2^2$.

The square roots of -1 in Section 7 and in Appendix A were all computed with the Maple package CLIFFORD [2], as explained in Appendix B.

2. Background and Problem Formulation

Let $C\ell_{p,q}$ be the algebra (associative with unit 1) generated over $\mathbb{R}$ by $p+q$ elements e_k (with $k = 1, 2, \ldots, p+q$) with the relations $e_k^2 = 1$ if $k \le p$, $e_k^2 = -1$ if $k > p$ and $e_h e_k + e_k e_h = 0$ whenever $h \neq k$, see [28]. We set the vector space dimension $n = p+q$ and the signature $s = p-q$. This algebra has dimension 2^n, and its even subalgebra $C\ell_0(p,q)$ has dimension 2^{n-1} (if $n > 0$). We are concerned with square roots of -1 contained in $C\ell_{p,q}$ or $C\ell_0(p,q)$. If the dimension of $C\ell_{p,q}$ or, $C\ell_0(p,q)$ is ≤ 2, it is isomorphic to $\mathbb{R} \cong C\ell_{0,0}$, $\mathbb{R}^2 \cong C\ell_{1,0}$, or $\mathbb{C} \cong C\ell_{0,1}$, and it is clear that there is no square root of -1 in $\mathbb{R}$ and $\mathbb{R}^2 = \mathbb{R} \times \mathbb{R}$, and that there are two squares roots i and $-i$ in $\mathbb{C}$. Therefore we only consider algebras of dimension ≥ 4. Square roots of -1 have been computed explicitly in [33] for $C\ell_{3,0}$, and in [20] for algebras of dimensions $2^n \le 16$.

An algebra $C\ell_{p,q}$ or $C\ell_0(p,q)$ of dimension ≥ 4 is isomorphic to one of the five matrix algebras: $\mathcal{M}(2d, \mathbb{R})$, $\mathcal{M}(d, \mathbb{H})$, $\mathcal{M}(2d, \mathbb{R}^2)$, $\mathcal{M}(d, \mathbb{H}^2)$ or $\mathcal{M}(2d, \mathbb{C})$. The integer d depends on n. According to the parity of n, it is either $2^{(n-2)/2}$ or $2^{(n-3)/2}$ for $C\ell_{p,q}$, and, either $2^{(n-4)/2}$ or $2^{(n-3)/2}$ for $C\ell_0(p,q)$. The associated ring (either $\mathbb{R}$, $\mathbb{H}$, $\mathbb{R}^2$, $\mathbb{H}^2$, or $\mathbb{C}$) depends on s in this way[2]:

[2]Compare Chapter 16 on *matrix representations and periodicity of* 8, as well as Table 1 on p. 217 of [28].

s mod 8	0	1	2	3	4	5	6	7
associated ring for $C\ell_{p,q}$	$\mathbb{R}$	$\mathbb{R}^2$	$\mathbb{R}$	$\mathbb{C}$	$\mathbb{H}$	$\mathbb{H}^2$	$\mathbb{H}$	$\mathbb{C}$
associated ring for $C\ell_0(p,q)$	$\mathbb{R}^2$	$\mathbb{R}$	$\mathbb{C}$	$\mathbb{H}$	$\mathbb{H}^2$	$\mathbb{H}$	$\mathbb{C}$	$\mathbb{R}$

Therefore we shall answer the following question: what can we say about the square roots of -1 in an algebra $\mathcal{A}$ that is isomorphic to $\mathcal{M}(2d,\mathbb{R})$, $\mathcal{M}(d,\mathbb{H})$, $\mathcal{M}(2d,\mathbb{R}^2)$, $\mathcal{M}(d,\mathbb{H}^2)$, or, $\mathcal{M}(2d,\mathbb{C})$? They constitute an algebraic submanifold in $\mathcal{A}$; how many connected components[3] (for the usual topology) does it contain? Which are their dimensions? This submanifold is invariant by the action of the *group* $\mathrm{Inn}(\mathcal{A})$ of *inner automorphisms*[4] of $\mathcal{A}$, *i.e.*, for every $r \in \mathcal{A}, r^2 = -1 \Rightarrow f(r)^2 = -1\ \forall f \in \mathrm{Inn}(\mathcal{A})$. The orbits of $\mathrm{Inn}(\mathcal{A})$ are called conjugacy classes[5]; how many conjugacy classes are there in this submanifold? If the associated ring is $\mathbb{R}^2$ or $\mathbb{H}^2$ or $\mathbb{C}$, the group $\mathrm{Aut}(\mathcal{A})$ of all automorphisms of $\mathcal{A}$ is larger than $\mathrm{Inn}(\mathcal{A})$, and the action of $\mathrm{Aut}(\mathcal{A})$ in this submanifold shall also be described.

We recall some properties of $\mathcal{A}$ that do not depend on the associated ring. The group $\mathrm{Inn}(\mathcal{A})$ contains as many connected components as the *group* $\mathrm{G}(\mathcal{A})$ of *invertible elements* in $\mathcal{A}$. We recall that this assertion is true for $\mathcal{M}(2d,\mathbb{R})$ but not for $\mathcal{M}(2d+1,\mathbb{R})$ which is not one of the relevant matrix algebras. If f is an element of $\mathcal{A}$, let $\mathrm{Cent}(f)$ be the centralizer of f, that is, the subalgebra of all $g \in \mathcal{A}$ such that $fg = gf$. The conjugacy class of f contains as many connected components[6] as $\mathrm{G}(\mathcal{A})$ if (and only if) $\mathrm{Cent}(f) \bigcap \mathrm{G}(\mathcal{A})$ is contained in the neutral[7] connected component of $\mathrm{G}(\mathcal{A})$, and the dimension of its conjugacy class is

$$\dim(\mathcal{A}) - \dim(\mathrm{Cent}(f)). \tag{2.1}$$

Note that for invertible $g \in \mathrm{Cent}(f)$ we have $g^{-1}fg = f$.

Besides, let $\mathrm{Z}(\mathcal{A})$ be the center of $\mathcal{A}$, and let $[\mathcal{A},\mathcal{A}]$ be the subspace spanned by all $[f,g] = fg - gf$. In all cases $\mathcal{A}$ is the direct sum of $\mathrm{Z}(\mathcal{A})$ and $[\mathcal{A},\mathcal{A}]$. For

[3]Two points are in the same connected component of a manifold, if they can be joined by a continuous path inside the manifold under consideration. (This applies to all topological spaces satisfying the property that each neighborhood of any point contains a neighborhood in which every pair of points can always be joined by a continuous path.)

[4]An inner automorphism f of $\mathcal{A}$ is defined as $f : \mathcal{A} \to \mathcal{A}, f(x) = a^{-1}xa, \forall x \in \mathcal{A}$, with given fixed $a \in \mathcal{A}$. The composition of two inner automorphisms $g(f(x)) = b^{-1}a^{-1}xab = (ab)^{-1}x(ab)$ is again an inner automorphism. With this operation the inner automorphisms form the group $\mathrm{Inn}(\mathcal{A})$, compare [35].

[5]The conjugacy class (similarity class) of a given $r \in \mathcal{A}, r^2 = -1$ is $\{f(r) : f \in \mathrm{Inn}(\mathcal{A})\}$, compare [34]. Conjugation is transitive, because the composition of inner automorphisms is again an inner automorphism.

[6]According to the general theory of groups acting on sets, the conjugacy class (as a topological space) of a square root f of -1 is isomorphic to the *quotient* of $\mathrm{G}(\mathcal{A})$ and $\mathrm{Cent}(f)$ (the subgroup of stability of f). Quotient means here the set of left handed classes modulo the subgroup. If the subgroup is contained in the neutral connected component of $\mathrm{G}(\mathcal{A})$, then the number of connected components is the same in the quotient as in $\mathrm{G}(\mathcal{A})$. See also [10].

[7]*Neutral* means to be connected to the identity element of $\mathcal{A}$.

example,[8] $Z(\mathcal{M}(2d,\mathbb{R})) = \{a\mathbf{1} \mid a \in \mathbb{R}\}$ and $Z(\mathcal{M}(2d,\mathbb{C})) = \{c\mathbf{1} \mid c \in \mathbb{C}\}$. If the associated ring is $\mathbb{R}$ or $\mathbb{H}$ (that is for even n), then $Z(\mathcal{A})$ is canonically isomorphic to $\mathbb{R}$, and from the projection $\mathcal{A} \to Z(\mathcal{A})$ we derive a linear form Scal : $\mathcal{A} \to \mathbb{R}$. When the associated ring[9] is $\mathbb{R}^2$ or $\mathbb{H}^2$ or $\mathbb{C}$, then $Z(\mathcal{A})$ is spanned by $\mathbf{1}$ (the unit matrix[10]) and some element ω such that $\omega^2 = \pm\mathbf{1}$. Thus, we get two linear forms Scal and Spec such that $\mathrm{Scal}(f)\mathbf{1}+\mathrm{Spec}(f)\omega$ is the projection of f in $Z(\mathcal{A})$ for every $f \in \mathcal{A}$. Instead of ω we may use $-\omega$ and replace Spec with −Spec. The following assertion holds for every $f \in \mathcal{A}$: The trace of each multiplication[11] $g \mapsto fg$ or $g \mapsto gf$ is equal to the product

$$\mathrm{tr}(f) = \dim(\mathcal{A})\,\mathrm{Scal}(f). \tag{2.2}$$

The word "trace" (when nothing more is specified) means a matrix trace in $\mathbb{R}$, which is the sum of its diagonal elements. For example, the matrix $M \in \mathcal{M}(2d,\mathbb{R})$ with elements $m_{kl} \in \mathbb{R}, 1 \le k,l \le 2d$ has the trace $\mathrm{tr}(M) = \sum_{k=1}^{2d} m_{kk}$ [24].

We shall prove that in all cases $\mathrm{Scal}(f) = 0$ for every square root of −1 in $\mathcal{A}$. Then, we may distinguish *ordinary* square roots of −1, and *exceptional* ones. In all cases the ordinary square roots of −1 constitute a unique[12] conjugacy class of dimension $\dim(\mathcal{A})/2$ which has as many connected components as $G(\mathcal{A})$, and they satisfy the equality $\mathrm{Spec}(f) = 0$ if the associated ring is $\mathbb{R}^2$ or $\mathbb{H}^2$ or $\mathbb{C}$. The exceptional square roots of −1 only exist[13] if $\mathcal{A} \cong \mathcal{M}(2d,\mathbb{C})$. In $\mathcal{M}(2d,\mathbb{C})$ there are $2d$ conjugacy classes of exceptional square roots of −1, each one characterized by an equality $\mathrm{Spec}(f) = k/d$ with $\pm k \in \{1,2,\ldots,d\}$ [see Section 7], and their dimensions are $< \dim(\mathcal{A})/2$ [see equation (7.5)]. For instance, ω (mentioned above) and $-\omega$ are central square roots of −1 in $\mathcal{M}(2d,\mathbb{C})$ which constitute two conjugacy classes of dimension 0. Obviously, $\mathrm{Spec}(\omega) = 1$.

For symbolic computer algebra systems (CAS), like MAPLE, there exist Clifford algebra packages, *e.g.*, CLIFFORD [2], which can compute idempotents [3] and square roots of −1. This will be of especial interest for the exceptional square roots of −1 in $\mathcal{M}(2d,\mathbb{C})$.

Regarding a square root r of −1, a Clifford algebra is the direct sum of the subspaces $\mathrm{Cent}(r)$ (all elements that commute with r) and the skew-centralizer

[8] A matrix algebra based proof is, *e.g.*, given in [4].
[9] This is the case for n (and s) odd. Then the pseudoscalar $\omega \in C\ell_{p,q}$ is also in $Z(C\ell_{p,q})$.
[10] The number 1 denotes the unit of the Clifford algebra $\mathcal{A}$, whereas the bold face $\mathbf{1}$ denotes the unit of the isomorphic matrix algebra $\mathcal{M}$.
[11] These multiplications are bilinear over the center of $\mathcal{A}$.
[12] Let $\mathcal{A}$ be an algebra $\mathcal{M}(m,\mathbb{K})$ where $\mathbb{K}$ is a division ring. Thus two elements f and g of $\mathcal{A}$ induce $\mathbb{K}$-linear endomorphisms f' and g' on $\mathbb{K}^m$; if $\mathbb{K}$ is not commutative, $\mathbb{K}$ operates on $\mathbb{K}^m$ on the right side. The matrices f and g are conjugate (or similar) if and only if there are two $\mathbb{K}$-bases B_1 and B_2 of $\mathbb{K}^m$ such that f' operates on B_1 in the same way as g' operates on B_2. This theorem allows us to recognize that in all cases but the last one (with exceptional square roots of $-\mathbf{1}$), two square roots of $-\mathbf{1}$ are always conjugate.
[13] The pseudoscalars of Clifford algebras whose isomorphic matrix algebra has ring $\mathbb{R}^2$ or $\mathbb{H}^2$ square to $\omega^2 = +1$.

SCent(r) (all elements that anticommute with r). Every Clifford algebra multivector has a unique split by this Lemma.

Lemma 2.1. *Every multivector $A \in C\ell_{p,q}$ has, with respect to a square root $r \in C\ell_{p,q}$ of -1, i.e., $r^{-1} = -r$, the unique decomposition*

$$A_{\pm} = \frac{1}{2}(A \pm r^{-1}Ar), \quad A = A_+ + A_-, \quad A_+ r = rA_+, \quad A_- r = -rA_-. \tag{2.3}$$

Proof. For $A \in C\ell_{p,q}$ and a square root $r \in C\ell_{p,q}$ of -1, we compute

$$\begin{aligned} A_{\pm}r &= \frac{1}{2}(A \pm r^{-1}Ar)r = \frac{1}{2}(Ar \pm r^{-1}A(-1)) \stackrel{r^{-1}=-r}{=} \frac{1}{2}(rr^{-1}Ar \pm rA) \\ &= \pm r\frac{1}{2}(A \pm r^{-1}Ar). \end{aligned}$$ □

For example, in Clifford algebras $C\ell_{n,0}$ [23] of dimensions $n = 2 \bmod 4$, Cent(r) is the even subalgebra $C\ell_0(n,0)$ for the unit pseudoscalar r, and the subspace $C\ell_1(n,0)$ spanned by all k-vectors of odd degree k, is SCent(r). The most interesting case is $\mathcal{M}(2d,\mathbb{C})$, where a whole range of conjugacy classes becomes available. These results will therefore be particularly relevant for constructing *Clifford–Fourier transformations* using the square roots of -1.

3. Square Roots of -1 in $\mathcal{M}(2d,\mathbb{R})$

Here $\mathcal{A} = \mathcal{M}(2d,\mathbb{R})$, whence $\dim(\mathcal{A}) = (2d)^2 = 4d^2$. The group $G(\mathcal{A})$ has *two* connected components determined by the inequalities $\det(g) > 0$ and $\det(g) < 0$.

For the case $d = 1$ we have, *e.g.*, the algebra $C\ell_{2,0}$ isomorphic to $\mathcal{M}(2,\mathbb{R})$. The basis $\{1, e_1, e_2, e_{12}\}$ of $C\ell_{2,0}$ is mapped to

$$\left\{ \begin{pmatrix} 1 & 0 \\ 0 & 1 \end{pmatrix}, \begin{pmatrix} 0 & 1 \\ 1 & 0 \end{pmatrix}, \begin{pmatrix} 1 & 0 \\ 0 & -1 \end{pmatrix}, \begin{pmatrix} 0 & -1 \\ 1 & 0 \end{pmatrix} \right\}.$$

The general element $\alpha + b_1e_1 + b_2e_2 + \beta e_{12} \in C\ell_{2,0}$ is thus mapped to

$$\begin{pmatrix} \alpha + b_2 & -\beta + b_1 \\ \beta + b_1 & \alpha - b_2 \end{pmatrix} \tag{3.1}$$

in $\mathcal{M}(2,\mathbb{R})$. Every element f of $\mathcal{A} = \mathcal{M}(2d,\mathbb{R})$ is treated as an $\mathbb{R}$-linear endomorphism of $V = \mathbb{R}^{2d}$. Thus, its scalar component and its trace (2.2) are related as follows: $\mathrm{tr}(f) = 2d\mathrm{Scal}(f)$. If f is a square root of $-\mathbf{1}$, it turns V into a vector space over $\mathbb{C}$ (if the complex number i operates like f on V). If $(e_1, e_2, \dots, e_d)$ is a $\mathbb{C}$-basis of V, then $(e_1, f(e_1), e_2, f(e_2), \dots, e_d, f(e_d))$ is an $\mathbb{R}$-basis of V, and the $2d \times 2d$ matrix of f in this basis is

$$\mathrm{diag}\Big(\underbrace{\begin{pmatrix} 0 & -1 \\ 1 & 0 \end{pmatrix}, \dots, \begin{pmatrix} 0 & -1 \\ 1 & 0 \end{pmatrix}}_{d}\Big) \tag{3.2}$$

Consequently all square roots of $-\mathbf{1}$ in $\mathcal{A}$ are conjugate. The centralizer of a square root f of $-\mathbf{1}$ is the algebra of all $\mathbb{C}$-linear endomorphisms g of V (since

i operates like f on V). Therefore, the $\mathbb{C}$-dimension of Cent(f) is d^2 and its $\mathbb{R}$-dimension is $2d^2$. Finally, the dimension (2.1) of the conjugacy class of f is $\dim(\mathcal{A}) - \dim(\mathrm{Cent}(f)) = 4d^2 - 2d^2 = 2d^2 = \dim(\mathcal{A})/2$. The two connected components of G($\mathcal{A}$) are determined by the sign of the determinant. Because of the next lemma, the $\mathbb{R}$-determinant of every element of Cent(f) is ≥ 0. Therefore, the intersection $\mathrm{Cent}(f) \bigcap \mathrm{G}(\mathcal{A})$ is contained in the neutral connected component of G($\mathcal{A}$) and, consequently, the conjugacy class of f has two connected components like G($\mathcal{A}$). Because of the next lemma, the $\mathbb{R}$-trace of f vanishes (indeed its $\mathbb{C}$-trace is di, because f is the multiplication by the scalar i: $f(v) = iv$ for all v) whence $\mathrm{Scal}(f) = 0$. This equality is corroborated by the matrix written above.

We conclude that the square roots of $-\mathbf{1}$ constitute one conjugacy class with two connected components of dimension $\dim(\mathcal{A})/2$ contained in the hyperplane defined by the equation

$$\mathrm{Scal}(f) = 0. \tag{3.3}$$

Before stating the lemma that here is so helpful, we show what happens in the easiest case $d = 1$. The square roots of $-\mathbf{1}$ in $\mathcal{M}(2,\mathbb{R})$ are the real matrices

$$\begin{pmatrix} a & c \\ b & -a \end{pmatrix} \text{ with } \begin{pmatrix} a & c \\ b & -a \end{pmatrix}\begin{pmatrix} a & c \\ b & -a \end{pmatrix} = (a^2 + bc)\,\mathbf{1} = -\mathbf{1}; \tag{3.4}$$

hence $a^2 + bc = -1$, a relation between a, b, c which is equivalent to $(b-c)^2 = (b+c)^2 + 4a^2 + 4 \Rightarrow (b-c)^2 \geq 4 \Rightarrow b - c \geq 2$ (one component) or $c - b \geq 2$ (second component). Thus, we recognize the two connected components of square roots of $-\mathbf{1}$: The inequality $b \geq c + 2$ holds in one connected component, and the inequality $c \geq b + 2$ in the other one, compare Figure 2.

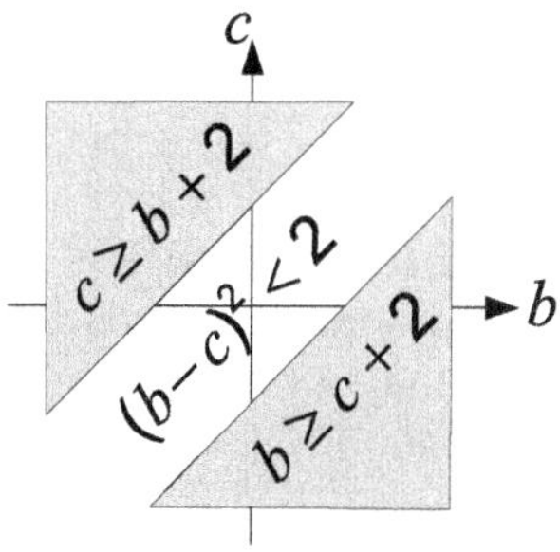

FIGURE 2. Two components of square roots of $-\mathbf{1}$ in $\mathcal{M}(2,\mathbb{R})$.

In terms of $C\ell_{2,0}$ coefficients (3.1) with $b - c = \beta + b_1 - (-\beta + b_1) = 2\beta$, we get the two component conditions simply as

$$\beta \geq 1 \quad \text{(one component)}, \qquad \beta \leq -1 \quad \text{(second component)}. \tag{3.5}$$

Rotations ($\det(g) = 1$) leave the pseudoscalar βe_{12} invariant (and thus preserve the two connected components of square roots of -1), but reflections ($\det(g') = -1$) change its sign $\beta e_{12} \to -\beta e_{12}$ (thus interchanging the two components).

Because of the previous argument involving a complex structure on the real space V, we conversely consider the complex space $\mathbb{C}^d$ with its structure of vector space over $\mathbb{R}$. If $(e_1, e_2, \ldots, e_d)$ is a $\mathbb{C}$-basis of $\mathbb{C}^d$, then $(e_1, ie_1, e_2, ie_2, \ldots, e_d, ie_d)$ is an $\mathbb{R}$-basis. Let g be a $\mathbb{C}$-linear endomorphism of $\mathbb{C}^d$ (*i.e.*, a complex $d \times d$ matrix), let $\mathrm{tr}_{\mathbb{C}}(g)$ and $\det_{\mathbb{C}}(g)$ be the trace and determinant of g in $\mathbb{C}$, and $\mathrm{tr}_{\mathbb{R}}(g)$ and $\det_{\mathbb{R}}(g)$ its trace and determinant for the real structure of $\mathbb{C}^d$.

Example. For $d = 1$ an endomorphism of $\mathbb{C}^1$ is given by a complex number $g = a + ib$, $a, b \in \mathbb{R}$. Its matrix representation is according to (3.2)

$$\begin{pmatrix} a & -b \\ b & a \end{pmatrix} \text{ with } \begin{pmatrix} a & -b \\ b & a \end{pmatrix}^2 = (a^2 - b^2)\begin{pmatrix} 1 & 0 \\ 0 & 1 \end{pmatrix} + 2ab\begin{pmatrix} 0 & -1 \\ 1 & 0 \end{pmatrix}. \tag{3.6}$$

Then we have $\mathrm{tr}_{\mathbb{C}}(g) = a + ib$, $\mathrm{tr}_{\mathbb{R}}\begin{pmatrix} a & -b \\ b & a \end{pmatrix} = 2a = 2\Re(\mathrm{tr}_{\mathbb{C}}(g))$ and $\det_{\mathbb{C}}(g) = a + ib$, $\det_{\mathbb{R}}\begin{pmatrix} a & -b \\ b & a \end{pmatrix} = a^2 + b^2 = |\det_{\mathbb{C}}(g)|^2 \geq 0$.

Lemma 3.1. *For every $\mathbb{C}$-linear endomorphism g we can write* $\mathrm{tr}_{\mathbb{R}}(g) = 2\Re(\mathrm{tr}_{\mathbb{C}}(g))$ *and* $\det_{\mathbb{R}}(g) = |\det_{\mathbb{C}}(g)|^2 \geq 0$.

Proof. There is a $\mathbb{C}$-basis in which the $\mathbb{C}$-matrix of g is triangular [then $\det_{\mathbb{C}}(g)$ is the product of the entries of g on the main diagonal]. We get the $\mathbb{R}$-matrix of g in the derived $\mathbb{R}$-basis by replacing every entry $a + bi$ of the $\mathbb{C}$-matrix with the elementary matrix $\begin{pmatrix} a & -b \\ b & a \end{pmatrix}$. The conclusion soon follows. The fact that the determinant of a block triangular matrix is the product of the determinants of the blocks on the main diagonal is used. □

4. Square Roots of -1 in $\mathcal{M}(2d, \mathbb{R}^2)$

Here $\mathcal{A} = \mathcal{M}(2d, \mathbb{R}^2) = \mathcal{M}(2d, \mathbb{R}) \times \mathcal{M}(2d, \mathbb{R})$, whence $\dim(\mathcal{A}) = 8d^2$. The group $\mathrm{G}(\mathcal{A})$ has four[14] connected components. Every element $(f, f') \in \mathcal{A}$ (with $f, f' \in \mathcal{M}(2d, \mathbb{R})$) has a determinant in $\mathbb{R}^2$ which is obviously $(\det(f), \det(f'))$, and the four connected components of $\mathrm{G}(\mathcal{A})$ are determined by the signs of the two components of $\det_{\mathbb{R}^2}(f, f')$.

The lowest-dimensional example ($d = 1$) is $C\ell_{2,1}$ isomorphic to $\mathcal{M}(2, \mathbb{R}^2)$. Here the pseudoscalar $\omega = e_{123}$ has square $\omega^2 = +1$. The center of the algebra is $\{1, \omega\}$ and includes the idempotents $\epsilon_\pm = (1 \pm \omega)/2$, $\epsilon_\pm^2 = \epsilon_\pm$, $\epsilon_+\epsilon_- = \epsilon_-\epsilon_+ = 0$. The basis of the algebra can thus be written as $\{\epsilon_+, e_1\epsilon_+, e_2\epsilon_+, e_{12}\epsilon_+, \epsilon_-, e_1\epsilon_-, e_2\epsilon_-, e_{12}\epsilon_-\}$, where the first (and the last) four elements form a basis of the

[14]In general, the number of connected components of $\mathrm{G}(\mathcal{A})$ is two if $\mathcal{A} = \mathcal{M}(m, \mathbb{R})$, and one if $\mathcal{A} = \mathcal{M}(m, \mathbb{C})$ or $\mathcal{A} = \mathcal{M}(m, \mathbb{H})$, because in all cases every matrix can be joined by a continuous path to a diagonal matrix with entries 1 or -1. When an algebra $\mathcal{A}$ is a direct product of two algebras $\mathcal{B}$ and $\mathcal{C}$, then $\mathrm{G}(\mathcal{A})$ is the direct product of $\mathrm{G}(\mathcal{B})$ and $\mathrm{G}(\mathcal{C})$, and the number of connected components of $\mathrm{G}(\mathcal{A})$ is the product of the numbers of connected components of $\mathrm{G}(\mathcal{B})$ and $\mathrm{G}(\mathcal{C})$.

subalgebra $C\ell_{2,0}$ isomorphic to $\mathcal{M}(2,\mathbb{R})$. In terms of matrices we have the identity matrix $(\mathbf{1},\mathbf{1})$ representing the scalar part, the idempotent matrices $(\mathbf{1},0)$, $(0,\mathbf{1})$, and the ω matrix $(\mathbf{1},-\mathbf{1})$, with $\mathbf{1}$ the unit matrix of $\mathcal{M}(2,\mathbb{R})$.

The square roots of $(-\mathbf{1},-\mathbf{1})$ in $\mathcal{A}$ are pairs of two square roots of $-\mathbf{1}$ in $\mathcal{M}(2d,\mathbb{R})$. Consequently they constitute a unique conjugacy class with four connected components of dimension $4d^2 = \dim(\mathcal{A})/2$. This number can be obtained in two ways. First, since every element $(f,f') \in \mathcal{A}$ (with $f,f' \in \mathcal{M}(2d,\mathbb{R})$) has twice the dimension of the components $f \in \mathcal{M}(2d,\mathbb{R})$ of Section 3, we get the component dimension $2\cdot 2d^2 = 4d^2$. Second, the centralizer $\mathrm{Cent}(f,f')$ has twice the dimension of $\mathrm{Cent}(f)$ of $\mathcal{M}(2d,\mathbb{R})$, therefore $\dim(\mathcal{A}) - \mathrm{Cent}(f,f') = 8d^2 - 4d^2 = 4d^2$. In the above example for $d=1$ the four components are characterized according to (3.5) by the values of the coefficients of $\beta e_{12}\epsilon_+$ and $\beta' e_{12}\epsilon_-$ as

$$\begin{aligned} c_1: &\quad \beta \geq 1, &\quad \beta' \geq 1,\\ c_2: &\quad \beta \geq 1, &\quad \beta' \leq -1,\\ c_3: &\quad \beta \leq -1, &\quad \beta' \geq 1,\\ c_4: &\quad \beta \leq -1, &\quad \beta' \leq -1. \end{aligned} \tag{4.1}$$

For every $(f,f') \in \mathcal{A}$ we can with (2.2) write $\mathrm{tr}(f) + \mathrm{tr}(f') = 2d\mathrm{Scal}(f,f')$ and

$$\mathrm{tr}(f) - \mathrm{tr}(f') = 2d\mathrm{Spec}(f,f') \quad \text{if} \quad \omega = (\mathbf{1},-\mathbf{1}); \tag{4.2}$$

whence $\mathrm{Scal}(f,f') = \mathrm{Spec}(f,f') = 0$ if (f,f') is a square root of $(-\mathbf{1},-\mathbf{1})$, compare (3.3).

The group $\mathrm{Aut}(\mathcal{A})$ is larger than $\mathrm{Inn}(\mathcal{A})$, because it contains the swap automorphism $(f,f') \mapsto (f',f)$ which maps the central element ω to $-\omega$, and interchanges the two idempotents ϵ_+ and ϵ_-. The group $\mathrm{Aut}(\mathcal{A})$ has eight connected components which permute the four connected components of the submanifold of square roots of $(-\mathbf{1},-\mathbf{1})$. The permutations induced by $\mathrm{Inn}(\mathcal{A})$ are the permutations of the *Klein group*. For example for $d=1$ of (4.1) we get the following $\mathrm{Inn}(\mathcal{M}(2,\mathbb{R}^2))$ permutations

$$\begin{aligned} \det(g) > 0, &\quad \det(g') > 0: &\quad \text{identity},\\ \det(g) > 0, &\quad \det(g') < 0: &\quad (c_1,c_2),(c_3,c_4),\\ \det(g) < 0, &\quad \det(g') > 0: &\quad (c_1,c_3),(c_2,c_4),\\ \det(g) < 0, &\quad \det(g') < 0: &\quad (c_1,c_4),(c_2,c_3). \end{aligned} \tag{4.3}$$

Beside the identity permutation, $\mathrm{Inn}(\mathcal{A})$ gives the three permutations that permute two elements and also the other two ones.

The automorphisms outside $\mathrm{Inn}(\mathcal{A})$ are

$$(f,f') \mapsto (gf'g^{-1}, g'fg'^{-1}) \quad \text{for some} \quad (g,g') \in \mathrm{G}(\mathcal{A}). \tag{4.4}$$

If $\det(g)$ and $\det(g')$ have opposite signs, it is easy to realize that this automorphism induces a circular permutation on the four connected components of square roots of $(-\mathbf{1},-\mathbf{1})$: If $\det(g)$ and $\det(g')$ have the same sign, this automorphism leaves globally invariant two connected components, and permutes the other two ones. For example, for $d = 1$ the automorphisms (4.4) outside $\mathrm{Inn}(\mathcal{A})$ permute the components (4.1) of square roots of $(-\mathbf{1}, -\mathbf{1})$ in $\mathcal{M}(2, \mathbb{R}^2)$ as follows

$$\begin{aligned} &\det(g) > 0, \quad \det(g') > 0: \quad (c_1), (c_2, c_3), (c_4), \\ &\det(g) > 0, \quad \det(g') < 0: \quad c_1 \to c_2 \to c_4 \to c_3 \to c_1, \\ &\det(g) < 0, \quad \det(g') > 0: \quad c_1 \to c_3 \to c_4 \to c_2 \to c_1, \\ &\det(g) < 0, \quad \det(g') < 0: \quad (c_1, c_4), (c_2), (c_3). \end{aligned} \tag{4.5}$$

Consequently, the quotient of the group $\mathrm{Aut}(\mathcal{A})$ by its neutral connected component is isomorphic to the group of isometries of a square in a Euclidean plane.

5. Square Roots of -1 in $\mathcal{M}(d, \mathbb{H})$

Let us first consider the easiest case $d = 1$, when $\mathcal{A} = \mathbb{H}$, *e.g.*, of $C\ell_{0,2}$. The square roots of -1 in $\mathbb{H}$ are the quaternions $ai + bj + cij$ with $a^2 + b^2 + c^2 = 1$. They constitute a compact and connected manifold of dimension 2. Every square root f of -1 is conjugate with i, *i.e.*, there exists $v \in \mathbb{H} : v^{-1} f v = i \Leftrightarrow fv = vi$. If we set $v = -fi + 1 = a + bij - cj + 1$ we have

$$fv = -f^2 i + f = f + i = (f(-i) + 1)i = vi.$$

v is invertible, except when $f = -i$. But i is conjugate with $-i$ because $ij = j(-i)$, hence, by transitivity f is also conjugate with $-i$.

Here $\mathcal{A} = \mathcal{M}(d, \mathbb{H})$, whence $\dim(A) = 4d^2$. The ring $\mathbb{H}$ is the algebra over $\mathbb{R}$ generated by two elements i and j such that $i^2 = j^2 = -1$ and $ji = -ij$. We identify $\mathbb{C}$ with the subalgebra generated by[15] i alone.

The group $\mathrm{G}(\mathcal{A})$ has only *one* connected component. We shall soon prove that every square root of $-\mathbf{1}$ in $\mathcal{A}$ is conjugate with $i\mathbf{1}$. Therefore, the submanifold of square roots of $-\mathbf{1}$ is a conjugacy class, and it is connected. The centralizer of $i\mathbf{1}$ in $\mathcal{A}$ is the subalgebra of all matrices with entries in $\mathbb{C}$. The $\mathbb{C}$-dimension of $\mathrm{Cent}(i\mathbf{1})$ is d^2, its $\mathbb{R}$-dimension is $2d^2$, and, consequently, the dimension (2.1) of the submanifold of square roots of $-\mathbf{1}$ is $4d^2 - 2d^2 = 2d^2 = \dim(\mathcal{A})/2$.

Here $V = \mathbb{H}^d$ is treated as a (unitary) *module* over $\mathbb{H}$ on the *right* side: The product of a line vector ${}^t v = (x_1, x_2, \ldots, x_d) \in V$ by $y \in \mathbb{H}$ is ${}^t v\, y = (x_1 y, x_2 y, \ldots, x_d y)$. Thus, every $f \in \mathcal{A}$ determines an $\mathbb{H}$-linear endomorphism of V: The matrix f multiplies the column vector $v = {}^t(x_1, x_2, \ldots, x_d)$ on the left side $v \mapsto fv$. Since $\mathbb{C}$ is a subring of $\mathbb{H}$, V is also a vector space of dimension $2d$ over $\mathbb{C}$. The scalar i always operates on the right side (like every scalar in $\mathbb{H}$). If $(e_1, e_2, \ldots, e_d)$ is an $\mathbb{H}$-basis of V, then $(e_1, e_1 j, e_2, e_2 j, \ldots, e_d, e_d j)$ is a $\mathbb{C}$-basis of V. Let f be a square root of $-\mathbf{1}$, then the eigenvalues of f in $\mathbb{C}$ are $+i$ or $-i$. If we treat V as a $2d$ vector space over $\mathbb{C}$, it is the direct ($\mathbb{C}$-linear) sum of the eigenspaces

$$V^+ = \{v \in V \mid f(v) = vi\} \quad \text{and} \quad V^- = \{v \in V \mid f(v) = -vi\}, \tag{5.1}$$

[15]This choice is usual and convenient.

representing f as a $2d \times 2d$ $\mathbb{C}$-matrix w.r.t. the $\mathbb{C}$-basis of V, with $\mathbb{C}$-scalar eigenvalues (multiplied from the right): $\lambda_\pm = \pm i$.

Since $ij = -ji$, the multiplication $v \mapsto vj$ permutes V^+ and V^-, as $f(v) = \pm vi$ is mapped to $f(v)j = \pm vij = \mp(vj)i$. Therefore, if $(e_1, e_2, \ldots, e_r)$ is a $\mathbb{C}$-basis of V^+, then $(e_1 j, e_2 j, \ldots, e_r j)$ is a $\mathbb{C}$-basis of V^-, consequently $(e_1,\ e_1 j,\ e_2,\ e_2 j,\ \ldots,\ e_r,\ e_r j)$ is a $\mathbb{C}$-basis of V, and $(e_1, e_2, \ldots, e_{r=d})$ is an $\mathbb{H}$-basis of V. Since f by $f(e_k) = e_k i$ for $k = 1, 2, \ldots, d$ operates on the $\mathbb{H}$-basis $(e_1, e_2, \ldots, e_d)$ in the same way as $i\mathbf{1}$ on the natural $\mathbb{H}$-basis of V, we conclude that f and $i\mathbf{1}$ are conjugate.

Besides, $\mathrm{Scal}(i\mathbf{1}) = 0$ because $2i\mathbf{1} = [j\mathbf{1}, ij\mathbf{1}] \in [\mathcal{A}, \mathcal{A}]$, thus $i\mathbf{1} \notin \mathrm{Z}(\mathcal{A})$. Whence,[16]

$$\mathrm{Scal}(f) = 0 \quad \text{for every square root of} \quad -\mathbf{1}. \tag{5.2}$$

These results are easily verified in the above example of $d = 1$ when $\mathcal{A} = \mathbb{H}$.

6. Square Roots of −1 in $\mathcal{M}(d, \mathbb{H}^2)$

Here, $\mathcal{A} = \mathcal{M}(d, \mathbb{H}^2) = \mathcal{M}(d, \mathbb{H}) \times \mathcal{M}(d, \mathbb{H})$, whence $\dim(A) = 8d^2$. The group $\mathrm{G}(\mathcal{A})$ has only one connected component (see Footnote 14).

The square roots of $(-\mathbf{1}, -\mathbf{1})$ in $\mathcal{A}$ are pairs of two square roots of $-\mathbf{1}$ in $\mathcal{M}(d, \mathbb{H})$. Consequently, they constitute a unique conjugacy class which is connected and its dimension is $2 \times 2d^2 = 4d^2 = \dim(\mathcal{A})/2$.

For every $(f, f') \in \mathcal{A}$ we can write $\mathrm{Scal}(f) + \mathrm{Scal}(f') = 2\,\mathrm{Scal}(f, f')$ and, similarly to (4.2),

$$\mathrm{Scal}(f) - \mathrm{Scal}(f') = 2\,\mathrm{Spec}(f, f') \quad \text{if} \quad \omega = (\mathbf{1}, -\mathbf{1}); \tag{6.1}$$

whence $\mathrm{Scal}(f, f') = \mathrm{Spec}(f, f') = 0$ if (f, f') is a square root of $(-\mathbf{1}, -\mathbf{1})$, compare with (5.2).

The group $\mathrm{Aut}(\mathcal{A})$ has two[17] connected components; the neutral component is $\mathrm{Inn}(\mathcal{A})$, and the other component contains the swap automorphism $(f, f') \mapsto (f', f)$.

The simplest example is $d = 1$, $\mathcal{A} = \mathbb{H}^2$, where we have the identity pair $(1, 1)$ representing the scalar part, the idempotents $(1, 0)$, $(0, 1)$, and ω as the pair $(1, -1)$.

$\mathcal{A} = \mathbb{H}^2$ is isomorphic to $C\ell_{0,3}$. The pseudoscalar $\omega = e_{123}$ has the square $\omega^2 = +1$. The center of the algebra is $\{1, \omega\}$, and includes the idempotents $\epsilon_\pm = \frac{1}{2}(1 \pm \omega)$, $\epsilon_\pm^2 = \epsilon_\pm$, $\epsilon_+\epsilon_- = \epsilon_-\epsilon_+ = 0$. The basis of the algebra can thus be written as $\{\epsilon_+, e_1\epsilon_+, e_2\epsilon_+, e_{12}\epsilon_+, \epsilon_-, e_1\epsilon_-, e_2\epsilon_-, e_{12}\epsilon_-\}$ where the first (and the last) four elements form a basis of the subalgebra $C\ell_{0,2}$ isomorphic to $\mathbb{H}$.

---

[16] Compare the definition of $\mathrm{Scal}(f)$ in Section 2, remembering that in the current section the associated ring is $\mathbb{H}$.

[17] Compare Footnote 14.

7. Square Roots of -1 in $\mathcal{M}(2d, \mathbb{C})$

The lowest-dimensional example for $d = 1$ is the Pauli matrix algebra $\mathcal{A} = \mathcal{M}(2, \mathbb{C})$ isomorphic to the geometric algebra $C\ell_{3,0}$ of the 3D Euclidean space and $C\ell_{1,2}$. The $C\ell_{3,0}$ vectors e_1, e_2, e_3 correspond one-to-one to the Pauli matrices

$$\sigma_1 = \begin{pmatrix} 0 & 1 \\ 1 & 0 \end{pmatrix}, \qquad \sigma_2 = \begin{pmatrix} 0 & -i \\ i & 0 \end{pmatrix}, \qquad \sigma_3 = \begin{pmatrix} 1 & 0 \\ 0 & -1 \end{pmatrix}, \tag{7.1}$$

with $\sigma_1\sigma_2 = i\sigma_3 = \begin{pmatrix} i & 0 \\ 0 & -i \end{pmatrix}$. The element $\omega = \sigma_1\sigma_2\sigma_3 = i\mathbf{1}$ represents the central pseudoscalar e_{123} of $C\ell_{3,0}$ with square $\omega^2 = -\mathbf{1}$. The Pauli algebra has the following idempotents

$$\epsilon_1 = \sigma_1^2 = \mathbf{1}, \qquad \epsilon_0 = (1/2)(\mathbf{1} + \sigma_3), \qquad \epsilon_{-1} = \mathbf{0}\,. \tag{7.2}$$

The idempotents correspond *via*

$$f = i(2\epsilon - \mathbf{1}), \tag{7.3}$$

to the square roots of $-\mathbf{1}$:

$$f_1 = i\mathbf{1} = \begin{pmatrix} i & 0 \\ 0 & i \end{pmatrix}, \; f_0 = i\sigma_3 = \begin{pmatrix} i & 0 \\ 0 & -i \end{pmatrix}, \; f_{-1} = -i\mathbf{1} = \begin{pmatrix} -i & 0 \\ 0 & -i \end{pmatrix}, \tag{7.4}$$

where by *complex* conjugation $f_{-1} = \overline{f_1}$. Let the idempotent $\epsilon_0' = \frac{1}{2}(\mathbf{1} - \sigma_3)$ correspond to the matrix $f_0' = -i\sigma_3$. We observe that f_0 is conjugate to $f_0' = \sigma_1^{-1} f_0 \sigma_1 = \sigma_1\sigma_2 = f_0$ using $\sigma_1^{-1} = \sigma_1$ but f_1 is not conjugate to f_{-1}. Therefore, only f_1, f_0, f_{-1} lead to three distinct conjugacy classes of square roots of $-\mathbf{1}$ in $\mathcal{M}(2, \mathbb{C})$. Compare Appendix B for the corresponding computations with CLIFFORD for Maple.

In general, if $\mathcal{A} = \mathcal{M}(2d, \mathbb{C})$, then $\dim(\mathcal{A}) = 8d^2$. The group $\mathrm{G}(\mathcal{A})$ has one connected component. The square roots of $-\mathbf{1}$ in $\mathcal{A}$ are in bijection with the idempotents ϵ [3] according to (7.3). According[18] to (7.3) and its inverse $\epsilon = \frac{1}{2}(\mathbf{1} - if)$ the square root of $-\mathbf{1}$ with $\mathrm{Spec}(f_-) = k/d = -1$, *i.e.*, $k = -d$ (see below), always corresponds to the trival idempotent $\epsilon_- = 0$, and the square root of $-\mathbf{1}$ with $\mathrm{Spec}(f_+) = k/d = +1$, $k = +d$, corresponds to the identity idempotent $\epsilon_+ = \mathbf{1}$.

If f is a square root of $-\mathbf{1}$, then $V = \mathbb{C}^{2d}$ is the direct sum of the eigenspaces[19] associated with the eigenvalues i and $-i$. There is an integer k such that the dimensions of the eigenspaces are respectively $d + k$ and $d - k$. Moreover, $-d \leq k \leq d$. Two square roots of $-\mathbf{1}$ are conjugate if and only if they give the same

[18] On the other hand it is clear that complex conjugation always leads to $f_- = \overline{f_+}$, where the overbar means complex conjugation in $\mathcal{M}(2d, \mathbb{C})$ and Clifford conjugation in the isomorphic Clifford algebra $C\ell_{p,q}$. So either the trivial idempotent $\epsilon_- = 0$ is included in the bijection (7.3) of idempotents and square roots of $-\mathbf{1}$, or alternatively the square root of $-\mathbf{1}$ with $\mathrm{Spec}(f_-) = -1$ is obtained from $f_- = \overline{f_+}$.

[19] The following theorem is sufficient for a matrix f in $\mathcal{M}(m, \mathbb{K})$, if $\mathbb{K}$ is a (commutative) field. The matrix f is diagonalizable if and only if $P(f) = 0$ for some polynomial P that has only simple roots, all of them in the field $\mathbb{K}$. (This implies that P is a multiple of the minimal polynomial, but we do not need to know whether P is or is not the minimal polynomial.)

integer k. Then, all elements of Cent(f) consist of diagonal block matrices with 2 square blocks of $(d+k)\times(d+k)$ matrices and $(d-k)\times(d-k)$ matrices. Therefore, the $\mathbb{C}$-dimension of Cent(f) is $(d+k)^2+(d-k)^2$. Hence the $\mathbb{R}$-dimension (2.1) of the conjugacy class of f:

$$8d^2 - 2(d+k)^2 - 2(d-k)^2 = 4(d^2 - k^2). \tag{7.5}$$

Also, from the equality $\mathrm{tr}(f) = (d+k)i - (d-k)i = 2ki$ we deduce that $\mathrm{Scal}(f) = 0$ and that $\mathrm{Spec}(f) = (2ki)/(2di) = k/d$ if $\omega = i\mathbf{1}$ (whence $\mathrm{tr}(\omega) = 2di$).

As announced on page 127, we consider that a square root of $-\mathbf{1}$ is *ordinary* if the associated integer k vanishes, and that it is *exceptional* if $k \neq 0$. Thus the following assertion is true in all cases: the ordinary square roots of $-\mathbf{1}$ in $\mathcal{A}$ constitute one conjugacy class of dimension $\dim(\mathcal{A})/2$ which has as many connected components as $\mathrm{G}(\mathcal{A})$, and the equality $\mathrm{Spec}(f) = 0$ holds for every ordinary square root of $-\mathbf{1}$ when the linear form Spec exists. All conjugacy classes of exceptional square roots of $-\mathbf{1}$ have a dimension $< \dim(\mathcal{A})/2$.

All square roots of $-\mathbf{1}$ in $\mathcal{M}(2d,\mathbb{C})$ constitute $(2d+1)$ conjugacy classes[20] which are also the connected components of the submanifold of square roots of $-\mathbf{1}$ because of the equality $\mathrm{Spec}(f) = k/d$, which is conjugacy class specific.

When $\mathcal{A} = \mathcal{M}(2d,\mathbb{C})$, the group $\mathrm{Aut}(\mathcal{A})$ is larger than $\mathrm{Inn}(\mathcal{A})$ since it contains the complex conjugation (that maps every entry of a matrix to the conjugate complex number). It is clear that the class of ordinary square roots of $-\mathbf{1}$ is invariant by complex conjugation. But the class associated with an integer k other than 0 is mapped by complex conjugation to the class associated with $-k$. In particular the complex conjugation maps the class $\{\omega\}$ (associated with $k = d$) to the class $\{-\omega\}$ associated with $k = -d$.

All these observations can easily verified for the above example of $d = 1$ of the Pauli matrix algebra $\mathcal{A} = \mathcal{M}(2,\mathbb{C})$. For $d = 2$ we have the isomorphism of $\mathcal{A} = \mathcal{M}(4,\mathbb{C})$ with $C\ell_{0,5}$, $C\ell_{2,3}$ and $C\ell_{4,1}$. While $C\ell_{0,5}$ is important in Clifford analysis, $C\ell_{4,1}$ is both the geometric algebra of the Lorentz space $\mathbb{R}^{4,1}$ and the conformal geometric algebra of 3D Euclidean geometry. Its set of square roots of $-\mathbf{1}$ is therefore of particular practical interest.

Example. Let $C\ell_{4,1} \cong \mathcal{A}$ where $\mathcal{A} = \mathcal{M}(4,\mathbb{C})$ for $d = 2$. The $C\ell_{4,1}$ 1-vectors can be represented[21] by the following matrices:

$$e_1 = \begin{pmatrix} 1 & 0 & 0 & 0 \\ 0 & -1 & 0 & 0 \\ 0 & 0 & -1 & 0 \\ 0 & 0 & 0 & 1 \end{pmatrix}, \quad e_2 = \begin{pmatrix} 0 & 1 & 0 & 0 \\ 1 & 0 & 0 & 0 \\ 0 & 0 & 0 & 1 \\ 0 & 0 & 1 & 0 \end{pmatrix}, \quad e_3 = \begin{pmatrix} 0 & -i & 0 & 0 \\ i & 0 & 0 & 0 \\ 0 & 0 & 0 & -i \\ 0 & 0 & i & 0 \end{pmatrix},$$

[20]Two conjugate (similar) matrices have the same eigenvalues and the same trace. This suffices to recognize that $2d+1$ conjugacy classes are obtained.

[21]For the computations of this example in the Maple package CLIFFORD we have used the identification $i = e_{23}$. Yet the results obtained for the square roots of $-\mathbf{1}$ are independent of this setting (we can alternatively use, *e.g.*, $i = e_{12345}$, or the imaginary unit $i \in \mathbb{C}$), as can easily be checked for f_1 of (7.7), f_0 of (7.8) and f_{-1} of (7.9) by only assuming the standard Clifford product rules for e_1 to e_5.

$$e_4 = \begin{pmatrix} 0 & 0 & 1 & 0 \\ 0 & 0 & 0 & -1 \\ 1 & 0 & 0 & 0 \\ 0 & -1 & 0 & 0 \end{pmatrix}, \; e_5 = \begin{pmatrix} 0 & 0 & -1 & 0 \\ 0 & 0 & 0 & 1 \\ 1 & 0 & 0 & 0 \\ 0 & -1 & 0 & 0 \end{pmatrix}. \tag{7.6}$$

We find five conjugacy classes of roots f_k of $-\mathbf{1}$ in $C\ell_{4,1}$ for $k \in \{0, \pm 1, \pm 2\}$: four exceptional and one ordinary. Since f_k is a root of $p(t) = t^2 + 1$ which factors over $\mathbb{C}$ into $(t-i)(t+i)$, the minimal polynomial $m_k(t)$ of f_k is one of the following: $t-i$, $t+i$, or $(t-i)(t+i)$. Respectively, there are three classes of characteristic polynomial $\Delta_k(t)$ of the matrix $\mathcal{F}_k$ in $\mathcal{M}(4,\mathbb{C})$ which corresponds to f_k, namely, $(t-i)^4$, $(t+i)^4$, and $(t-i)^{n_1}(t+i)^{n_2}$, where $n_1+n_2 = 2d = 4$ and $n_1 = d+k = 2+k$, $n_2 = d-k = 2-k$. As predicted by the above discussion, the ordinary root corresponds to $k=0$ whereas the exceptional roots correspond to $k \neq 0$.

1. For $k=2$, we have $\Delta_2(t) = (t-i)^4$, $m_2(t) = t-i$, and so $\mathcal{F}_2 = \text{diag}(i,i,i,i)$ which in the above representation (7.6) corresponds to the non-trivial central element $f_2 = \omega = e_{12345}$. Clearly, $\text{Spec}(f_2) = 1 = \frac{k}{d}$; $\text{Scal}(f_2) = 0$; the $\mathbb{C}$-dimension of the centralizer $\text{Cent}(f_2)$ is 16; and the $\mathbb{R}$-dimension of the conjugacy class of f_2 is zero as it contains only f_2 since $f_2 \in \text{Z}(\mathcal{A})$. Thus, the $\mathbb{R}$-dimension of the class is again zero in agreement with (7.5).
2. For $k=-2$, we have $\Delta_{-2}(t) = (t+i)^4$, $m_{-2}(t) = t+i$, and $\mathcal{F}_{-2} = \text{diag}(-i,-i,-i,-i)$ which corresponds to the central element $f_{-2} = -\omega = -e_{12345}$. Again, $\text{Spec}(f_{-2}) = -1 = \frac{k}{d}$; $\text{Scal}(f_{-2}) = 0$; the $\mathbb{C}$-dimension of the centralizer $\text{Cent}(f_{-2})$ is 16 and the conjugacy class of f_{-2} contains only f_{-2} since $f_{-2} \in \text{Z}(\mathcal{A})$. Thus, the $\mathbb{R}$-dimension of the class is again zero in agreement with (7.5).
3. For $k \neq \pm 2$, we consider three subcases when $k=1$, $k=0$, and $k=-1$. When $k=1$, then $\Delta_1(t) = (t-i)^3(t+i)$ and $m_1(t) = (t-i)(t+i)$. Then the root $\mathcal{F}_1 = \text{diag}(i,i,i,-i)$ corresponds to

$$f_1 = \frac{1}{2}(e_{23} + e_{123} - e_{2345} + e_{12345}). \tag{7.7}$$

Note that $\text{Spec}(f_1) = \frac{1}{2} = \frac{k}{d}$ so f_1 is an exceptional root of $-\mathbf{1}$.
When $k=0$, then $\Delta_0(t) = (t-i)^2(t+i)^2$ and $m_0(t) = (t-i)(t+i)$. Thus the root of $-\mathbf{1}$ in this case is $\mathcal{F}_0 = \text{diag}(i,i,-i,-i)$ which corresponds to just

$$f_0 = e_{123}. \tag{7.8}$$

Note that $\text{Spec}(f_0) = 0$ thus $f_0 = e_{123}$ is an ordinary root of $-\mathbf{1}$.
When $k=-1$, then $\Delta_{-1}(t) = (t-i)(t+i)^3$ and $m_{-1}(t) = (t-i)(t+i)$. Then, the root of $-\mathbf{1}$ in this case is $\mathcal{F}_{-1} = \text{diag}(i,-i,-i,-i)$ which corresponds to

$$f_{-1} = \frac{1}{2}(e_{23} + e_{123} + e_{2345} - e_{12345}). \tag{7.9}$$

Since $\text{Scal}(f_{-1}) = -\frac{1}{2} = \frac{k}{d}$, we gather that f_{-1} is an exceptional root.

As expected, we can also see that the roots ω and $-\omega$ are related *via* the grade involution whereas $f_1 = -\tilde{f}_{-1}$ where $\tilde{}$ denotes the reversion in $C\ell_{4,1}$.

Example. Let $C\ell_{0,5} \cong \mathcal{A}$ where $\mathcal{A} = \mathcal{M}(4, \mathbb{C})$ for $d = 2$. The $C\ell_{0,5}$ 1-vectors can be represented[22] by the following matrices:

$$e_1 = \begin{pmatrix} 0 & -1 & 0 & 0 \\ 1 & 0 & 0 & 0 \\ 0 & 0 & 0 & -1 \\ 0 & 0 & 1 & 0 \end{pmatrix}, e_2 = \begin{pmatrix} 0 & -i & 0 & 0 \\ -i & 0 & 0 & 0 \\ 0 & 0 & 0 & -i \\ 0 & 0 & -i & 0 \end{pmatrix}, e_3 = \begin{pmatrix} -i & 0 & 0 & 0 \\ 0 & i & 0 & 0 \\ 0 & 0 & i & 0 \\ 0 & 0 & 0 & -i \end{pmatrix},$$
$$e_4 = \begin{pmatrix} 0 & 0 & -1 & 0 \\ 0 & 0 & 0 & 1 \\ 1 & 0 & 0 & 0 \\ 0 & -1 & 0 & 0 \end{pmatrix}, e_5 = \begin{pmatrix} 0 & 0 & -i & 0 \\ 0 & 0 & 0 & i \\ -i & 0 & 0 & 0 \\ 0 & i & 0 & 0 \end{pmatrix}, \tag{7.10}$$

Like for $C\ell_{4,1}$, we have five conjugacy classes of the roots f_k of $-\mathbf{1}$ in $C\ell_{0,5}$ for $k \in \{0, \pm 1, \pm 2\}$: four exceptional and one ordinary. Using the same notation as in Example 7, we find the following representatives of the conjugacy classes.

1. For $k = 2$, we have $\Delta_2(t) = (t - i)^4$, $m_2(t) = t - i$, and $\mathcal{F}_2 = \operatorname{diag}(i, i, i, i)$ which in the above representation (7.10) corresponds to the non-trivial central element $f_2 = \omega = e_{12345}$. Then, $\operatorname{Spec}(f_2) = 1 = \frac{k}{d}$; $\operatorname{Scal}(f_2) = 0$; the $\mathbb{C}$-dimension of the centralizer $\operatorname{Cent}(f_2)$ is 16; and the $\mathbb{R}$-dimension of the conjugacy class of f_2 is zero as it contains only f_2 since $f_2 \in \mathrm{Z}(\mathcal{A})$. Thus, the $\mathbb{R}$-dimension of the class is again zero in agreement with (7.5).
2. For $k = -2$, we have $\Delta_{-2}(t) = (t + i)^4$, $m_{-2}(t) = t + i$, and $\mathcal{F}_{-2} = \operatorname{diag}(-i, -i, -i, -i)$ which corresponds to the central element $f_{-2} = -\omega = -e_{12345}$. Again, $\operatorname{Spec}(f_{-2}) = -1 = \frac{k}{d}$; $\operatorname{Scal}(f_{-2}) = 0$; the $\mathbb{C}$-dimension of the centralizer $\operatorname{Cent}(f_{-2})$ is 16 and the conjugacy class of f_{-2} contains only f_{-2} since $f_{-2} \in \mathrm{Z}(\mathcal{A})$. Thus, the $\mathbb{R}$-dimension of the class is again zero in agreement with (7.5).
3. For $k \neq \pm 2$, we consider three subcases when $k = 1$, $k = 0$, and $k = -1$. When $k = 1$, then $\Delta_1(t) = (t - i)^3(t + i)$ and $m_1(t) = (t - i)(t + i)$. Then the root $\mathcal{F}_1 = \operatorname{diag}(i, i, i, -i)$ corresponds to

$$f_1 = \frac{1}{2}(e_3 + e_{12} + e_{45} + e_{12345}). \tag{7.11}$$

Since $\operatorname{Spec}(f_1) = \frac{1}{2} = \frac{k}{d}$, f_1 is an exceptional root of $-\mathbf{1}$.

[22] For the computations of this example in the Maple package CLIFFORD we have used the identification $i = e_3$. Yet the results obtained for the square roots of $-\mathbf{1}$ are independent of this setting (we can alternatively use, *e.g.*, $i = e_{12345}$, or the imaginary unit $i \in \mathbb{C}$), as can easily be checked for f_1 of (7.11), f_0 of (7.12) and f_{-1} of (7.13) by only assuming the standard Clifford product rules for e_1 to e_5.

When $k = 0$, then $\Delta_0(t) = (t-i)^2(t+i)^2$ and $m_0(t) = (t-i)(t+i)$. Thus the root of $-\mathbf{1}$ is this case is $\mathcal{F}_0 = \operatorname{diag}(i, i, -i, -i)$ which corresponds to just

$$f_0 = e_{45}. \tag{7.12}$$

Note that $\operatorname{Spec}(f_0) = 0$ thus $f_0 = e_{45}$ is an ordinary root of $-\mathbf{1}$.
When $k = -1$, then $\Delta_{-1}(t) = (t-i)(t+i)^3$ and $m_{-1}(t) = (t-i)(t+i)$. Then, the root of $-\mathbf{1}$ in this case is $\mathcal{F}_{-1} = \operatorname{diag}(i, -i, -i, -i)$ which corresponds to

$$f_{-1} = \frac{1}{2}(-e_3 + e_{12} + e_{45} - e_{12345}). \tag{7.13}$$

Since $\operatorname{Scal}(f_{-1}) = -\frac{1}{2} = \frac{k}{d}$, we gather that f_{-1} is an exceptional root.

Again we can see that the roots f_2 and f_{-2} are related *via* the grade involution whereas $f_1 = -\tilde{f}_{-1}$ where $\tilde{}$ denotes the reversion in $C\ell_{0,5}$.

Example. Let $C\ell_{7,0} \cong \mathcal{A}$ where $\mathcal{A} = \mathcal{M}(8, \mathbb{C})$ for $d = 4$. We have nine conjugacy classes of roots f_k of $-\mathbf{1}$ for $k \in \{0, \pm 1, \pm 2 \pm 3 \pm 4\}$. Since f_k is a root of a polynomial $p(t) = t^2 + 1$ which factors over $\mathbb{C}$ into $(t-i)(t+i)$, its minimal polynomial $m(t)$ will be one of the following: $t - i$, $t + i$, or $(t-i)(t+i) = t^2 + 1$.

Respectively, each conjugacy class is characterized by a characteristic polynomial $\Delta_k(t)$ of the matrix $M_k \in \mathcal{M}(8, \mathbb{C})$ which represents f_k. Namely, we have

$$\Delta_k(t) = (t-i)^{n_1}(t+i)^{n_2},$$

where $n_1 + n_2 = 2d = 8$ and $n_1 = d + k = 4 + k$ and $n_2 = d - k = 4 - k$. The ordinary root of $-\mathbf{1}$ corresponds to $k = 0$ whereas the exceptional roots correspond to $k \neq 0$.

1. When $k = 4$, we have $\Delta_4(t) = (t-i)^8$, $m_4(t) = t - i$, and $\mathcal{F}_4 = \operatorname{diag}(\overbrace{i, \ldots, i}^{8})$ which in the representation used by CLIFFORD [2] corresponds to the non-trivial central element $f_4 = \omega = e_{1234567}$. Clearly, $\operatorname{Spec}(f_4) = 1 = \frac{k}{d}$; $\operatorname{Scal}(f_4) = 0$; the $\mathbb{C}$-dimension of the centralizer $\operatorname{Cent}(f_4)$ is 64; and the $\mathbb{R}$-dimension of the conjugacy class of f_4 is zero since $f_4 \in \mathrm{Z}(\mathcal{A})$. Thus, the $\mathbb{R}$-dimension of the class is again zero in agreement with (7.5).
2. When $k = -4$, we have $\Delta_{-4}(t) = (t+i)^8$, $m_{-4}(t) = t + i$, and $\mathcal{F}_{-4} = \operatorname{diag}(\overbrace{-i, \ldots, -i}^{8})$ which corresponds to $f_{-4} = -\omega = -e_{1234567}$. Again,
$$\operatorname{Spec}(f_{-4}) = -1 = \frac{k}{d}; \quad \operatorname{Scal}(f_{-4}) = 0;$$
the $\mathbb{C}$-dimension of the centralizer $\operatorname{Cent}(f)$ is 64 and the conjugacy class of f_{-4} contains only f_{-4} since $f_{-4} \in \mathrm{Z}(\mathcal{A})$. Thus, the $\mathbb{R}$-dimension of the class is again zero in agreement with (7.5).
3. When $k \neq \pm 4$, we consider seven subcases when $k = \pm 3$, $k = \pm 2$, $k = \pm 1$, and $k = 0$.

When $k=3$, then $\Delta_3(t)=(t-i)^7(t+i)$ and $m_3(t)=(t-i)(t+i)$. Then the root $\mathcal{F}_3=\mathrm{diag}(\overbrace{i,\ldots,i}^{7},-i)$ corresponds to

$$f_3=\frac{1}{4}(e_{23}-e_{45}+e_{67}-e_{123}+e_{145}-e_{167}+e_{234567}+3e_{1234567}). \qquad (7.14)$$

Since $\mathrm{Spec}(f_3)=\frac{3}{4}=\frac{k}{d}$, f_3 is an exceptional root of $-\mathbf{1}$.
When $k=2$, then $\Delta_2(t)=(t-i)^6(t+i)^2$ and $m_2(t)=(t-i)(t+i)$. Then the root $\mathcal{F}_2=\mathrm{diag}(\overbrace{i,\ldots,i}^{6},-i,-i)$ corresponds to

$$f_2=\frac{1}{2}(e_{67}-e_{45}-e_{123}+e_{1234567}). \qquad (7.15)$$

Since $\mathrm{Spec}(f_2)=\frac{1}{2}=\frac{k}{d}$, f_2 is also an exceptional root.
When $k=1$, then $\Delta_1(t)=(t-i)^5(t+i)^3$ and $m_1(t)=(t-i)(t+i)$. Then the root $\mathcal{F}_1=\mathrm{diag}(\overbrace{i,\ldots,i}^{5},-i,-i,-i)$ corresponds to

$$f_1=\frac{1}{4}(e_{23}-e_{45}+3e_{67}-e_{123}+e_{145}+e_{167}-e_{234567}+e_{1234567}). \qquad (7.16)$$

Since $\mathrm{Spec}(f_1)=\frac{1}{4}=\frac{k}{d}$, f_1 is another exceptional root.
When $k=0$, then $\Delta_0(t)=(t-i)^4(t+i)^4$ and $m_0(t)=(t-i)(t+i)$. Then the root $\mathcal{F}_0=\mathrm{diag}(i,i,i,i,-i,-i,-i,-i)$ corresponds to

$$f_0=\frac{1}{2}(e_{23}-e_{45}+e_{67}-e_{234567}). \qquad (7.17)$$

Since $\mathrm{Spec}(f_0)=0=\frac{k}{d}$, we see that f_0 is an ordinary root of $-\mathbf{1}$.
When $k=-1$, then $\Delta_{-1}(t)=(t-i)^3(t+i)^5$ and $m_{-1}(t)=(t-i)(t+i)$. Then the root $\mathcal{F}_{-1}=\mathrm{diag}(i,i,i,\overbrace{-i,\ldots,-i}^{5})$ corresponds to

$$f_{-1}=\frac{1}{4}(e_{23}-e_{45}+3e_{67}+e_{123}-e_{145}-e_{167}-e_{234567}-e_{1234567}). \qquad (7.18)$$

Thus, $\mathrm{Spec}(f_{-1})=-\frac{1}{4}=\frac{k}{d}$ and so f_{-1} is another exceptional root.
When $k=-2$, then $\Delta_{-2}(t)=(t-i)^2(t+i)^6$ and $m_{-2}(t)=(t-i)(t+i)$. Then the root $\mathcal{F}_{-2}=\mathrm{diag}(i,i,\overbrace{-i,\ldots,-i}^{6})$ corresponds to

$$f_{-2}=\frac{1}{2}(e_{67}-e_{45}+e_{123}-e_{1234567}). \qquad (7.19)$$

Since $\mathrm{Spec}(f_{-2})=-\frac{1}{2}=\frac{k}{d}$, we see that f_{-2} is also an exceptional root.
When $k=-3$, then $\Delta_{-3}(t)=(t-i)(t+i)^7$ and $m_{-3}(t)=(t-i)(t+i)$. Then the root $\mathcal{F}_{-3}=\mathrm{diag}(i,\overbrace{-i,\ldots,-i}^{7})$ corresponds to

$$f_{-3}=\frac{1}{4}(e_{23}-e_{45}+e_{67}+e_{123}-e_{145}+e_{167}+e_{234567}-3e_{1234567}). \qquad (7.20)$$

Again, $\mathrm{Spec}(f_{-3}) = -\frac{3}{4} = \frac{k}{d}$ and so f_{-3} is another exceptional root of $-\mathbf{1}$.

As expected, we can also see that the roots ω and $-\omega$ are related *via* the reversion whereas $f_3 = -\bar{f}_{-3}$, $f_2 = -\bar{f}_{-2}$, $f_1 = -\bar{f}_{-1}$ where $\bar{\ }$ denotes the conjugation in $C\ell_{7,0}$.

8. Conclusions

We proved that in all cases $\mathrm{Scal}(f) = 0$ for every square root of $-\mathbf{1}$ in $\mathcal{A}$ isomorphic to $C\ell_{p,q}$. We distinguished *ordinary* square roots of $-\mathbf{1}$, and *exceptional* ones.

In all cases the ordinary square roots f of $-\mathbf{1}$ constitute a unique conjugacy class of dimension $\dim(\mathcal{A})/2$ which has as many connected components as the group $\mathrm{G}(\mathcal{A})$ of invertible elements in $\mathcal{A}$. Furthermore, we have $\mathrm{Spec}(f) = 0$ (zero pseudoscalar part) if the associated ring is $\mathbb{R}^2$, $\mathbb{H}^2$, or $\mathbb{C}$. The exceptional square roots of $-\mathbf{1}$ *only* exist if $\mathcal{A} \cong \mathcal{M}(2d, \mathbb{C})$ (see Section 7).

For $\mathcal{A} = \mathcal{M}(2d, \mathbb{R})$ of Section 3, the centralizer and the conjugacy class of a square root f of $-\mathbf{1}$ both have $\mathbb{R}$-dimension $2d^2$ with two connected components, pictured in Figure 2 for $d = 1$.

For $\mathcal{A} = \mathcal{M}(2d, \mathbb{R}^2) = \mathcal{M}(2d, \mathbb{R}) \times \mathcal{M}(2d, \mathbb{R})$ of Section 4, the square roots of $(-\mathbf{1}, -\mathbf{1})$ are pairs of two square roots of $-\mathbf{1}$ in $\mathcal{M}(2d, \mathbb{R})$. They constitute a unique conjugacy class with four connected components, each of dimension $4d^2$. Regarding the four connected components, the group $\mathrm{Inn}(\mathcal{A})$ induces the permutations of the Klein group whereas the quotient group $\mathrm{Aut}(\mathcal{A})/\mathrm{Inn}(\mathcal{A})$ is isomorphic to the group of isometries of a Euclidean square in 2D.

For $\mathcal{A} = \mathcal{M}(d, \mathbb{H})$ of Section 5, the submanifold of the square roots f of $-\mathbf{1}$ is a single connected conjugacy class of $\mathbb{R}$-dimension $2d^2$ equal to the $\mathbb{R}$-dimension of the centralizer of every f. The easiest example is $\mathbb{H}$ itself for $d = 1$.

For $\mathcal{A} = \mathcal{M}(d, \mathbb{H}^2) = \mathcal{M}(2d, \mathbb{H}) \times \mathcal{M}(2d, \mathbb{H})$ of Section 6, the square roots of $(-\mathbf{1}, -\mathbf{1})$ are pairs of two square roots (f, f') of $-\mathbf{1}$ in $\mathcal{M}(2d, \mathbb{H})$ and constitute a unique connected conjugacy class of $\mathbb{R}$-dimension $4d^2$. The group $\mathrm{Aut}(\mathcal{A})$ has two connected components: the neutral component $\mathrm{Inn}(\mathcal{A})$ connected to the identity and the second component containing the swap automorphism $(f, f') \mapsto (f', f)$. The simplest case for $d = 1$ is $\mathbb{H}^2$ isomorphic to $C\ell_{0,3}$.

For $\mathcal{A} = \mathcal{M}(2d, \mathbb{C})$ of Section 7, the square roots of $-\mathbf{1}$ are in bijection to the idempotents. First, the ordinary square roots of $-\mathbf{1}$ (with $k = 0$) constitute a conjugacy class of $\mathbb{R}$-dimension $4d^2$ of a single connected component which is invariant under $\mathrm{Aut}(\mathcal{A})$. Second, there are $2d$ conjugacy classes of exceptional square roots of $-\mathbf{1}$, each composed of a single connected component, characterized by equality $\mathrm{Spec}(f) = k/d$ (the pseudoscalar coefficient) with $\pm k \in \{1, 2, \ldots, d\}$, and their $\mathbb{R}$-dimensions are $4(d^2 - k^2)$. The group $\mathrm{Aut}(\mathcal{A})$ includes conjugation of the pseudoscalar $\omega \mapsto -\omega$ which maps the conjugacy class associated with k to the class associated with $-k$. The simplest case for $d = 1$ is the Pauli matrix algebra isomorphic to the geometric algebra $C\ell_{3,0}$ of 3D Euclidean space $\mathbb{R}^3$, and to complex biquaternions [33].

Section 7 includes explicit examples for $d = 2$: $C\ell_{4,1}$ and $C\ell_{0,5}$, and for $d = 4$: $C\ell_{7,0}$. Appendix A summarizes the square roots of -1 in all $C\ell_{p,q} \cong \mathcal{M}(2d, \mathbb{C})$ for $d = 1, 2, 4$. Appendix B contains details on how square roots of -1 can be computed using the package CLIFFORD for Maple.

Among the many possible *applications* of this research, the possibility of *new integral transformations* in Clifford analysis is very promising. This field thus obtains essential algebraic information, which can, *e.g.*, be used to create *steerable* transformations, which may be steerable within a connected component of a submanifold of square roots of -1.

Appendix A. Summary of Roots of -1 in $C\ell_{p,q} \cong \mathcal{M}(2d, \mathbb{C})$ for $d = 1, 2, 4$

In this appendix we summarize roots of -1 for Clifford algebras $C\ell_{p,q} \cong \mathcal{M}(2d, \mathbb{C})$ for $d = 1, 2, 4$. These roots have been computed with CLIFFORD [2]. Maple [25] worksheets written to derive these roots are posted at [21].

TABLE 1. Square roots of -1 in $C\ell_{3,0} \cong \mathcal{M}(2, \mathbb{C})$, $d = 1$

k	f_k	$\Delta_k(t)$
1	$\omega = e_{123}$	$(t - i)^2$
0	e_{23}	$(t - i)(t + i)$
-1	$-\omega = -e_{123}$	$(t + i)^2$

TABLE 2. Square roots of -1 in $C\ell_{4,1} \cong \mathcal{M}(4, \mathbb{C})$, $d = 2$

k	f_k	$\Delta_k(t)$
2	$\omega = e_{12345}$	$(t - i)^4$
1	$\frac{1}{2}(e_{23} + e_{123} - e_{2345} + e_{12345})$	$(t - i)^3(t + i)$
0	e_{123}	$(t - i)^2(t + i)^2$
-1	$\frac{1}{2}(e_{23} + e_{123} + e_{2345} - e_{12345})$	$(t - i)(t + i)^3$
-2	$-\omega = -e_{12345}$	$(t + i)^4$

TABLE 3. Square roots of $-\mathbf{1}$ in $C\ell_{0,5} \cong \mathcal{M}(4,\mathbb{C})$, $d=2$

k	f_k	$\Delta_k(t)$
2	$\omega = e_{12345}$	$(t-i)^4$
1	$\frac{1}{2}(e_3 + e_{12} + e_{45} + e_{12345})$	$(t-i)^3(t+i)$
0	e_{45}	$(t-i)^2(t+i)^2$
-1	$\frac{1}{2}(-e_3 + e_{12} + e_{45} - e_{12345})$	$(t-i)(t+i)^3$
-2	$-\omega = -e_{12345}$	$(t+i)^4$

TABLE 4. Square roots of $-\mathbf{1}$ in $C\ell_{2,3} \cong \mathcal{M}(4,\mathbb{C})$, $d=2$

k	f_k	$\Delta_k(t)$
2	$\omega = e_{12345}$	$(t-i)^4$
1	$\frac{1}{2}(e_3 + e_{134} + e_{235} + \omega)$	$(t-i)^3(t+i)$
0	e_{134}	$(t-i)^2(t+i)^2$
-1	$\frac{1}{2}(-e_3 + e_{134} + e_{235} - \omega)$	$(t-i)(t+i)^3$
-2	$-\omega = -e_{12345}$	$(t+i)^4$

TABLE 5. Square roots of $-\mathbf{1}$ in $C\ell_{7,0} \cong \mathcal{M}(8,\mathbb{C})$, $d=4$

k	f_k	$\Delta_k(t)$
4	$\omega = e_{1234567}$	$(t-i)^8$
3	$\frac{1}{4}(e_{23} - e_{45} + e_{67} - e_{123} + e_{145} - e_{167} + e_{234567} + 3\omega)$	$(t-i)^7(t+i)$
2	$\frac{1}{2}(e_{67} - e_{45} - e_{123} + \omega)$	$(t-i)^6(t+i)^2$
1	$\frac{1}{4}(e_{23} - e_{45} + 3e_{67} - e_{123} + e_{145} + e_{167} - e_{234567} + \omega)$	$(t-i)^5(t+i)^3$
0	$\frac{1}{2}(e_{23} - e_{45} + e_{67} - e_{234567})$	$(t-i)^4(t+i)^4$
-1	$\frac{1}{4}(e_{23} - e_{45} + 3e_{67} + e_{123} - e_{145} - e_{167} - e_{234567} - \omega)$	$(t-i)^3(t+i)^5$
-2	$\frac{1}{2}(e_{67} - e_{45} + e_{123} - \omega)$	$(t-i)^2(t+i)^6$
-3	$\frac{1}{4}(e_{23} - e_{45} + e_{67} + e_{123} - e_{145} + e_{167} + e_{234567} - 3\omega)$	$(t-i)(t+i)^7$
-4	$-\omega = -e_{1234567}$	$(t+i)^8$

TABLE 6. Square roots of $-\mathbf{1}$ in $C\ell_{1,6} \cong \mathcal{M}(8,\mathbb{C})$, $d=4$

k	f_k	$\Delta_k(t)$
4	$\omega = e_{1234567}$	$(t-i)^8$
3	$\frac{1}{4}(e_4 - e_{23} - e_{56} + e_{1237} + e_{147} + e_{1567} - e_{23456} + 3\omega)$	$(t-i)^7(t+i)$
2	$\frac{1}{2}(-e_{23} - e_{56} + e_{147} + \omega)$	$(t-i)^6(t+i)^2$
1	$\frac{1}{4}(-e_4 - e_{23} - 3e_{56} - e_{1237} + e_{147} + e_{1567} - e_{23456} + \omega)$	$(t-i)^5(t+i)^3$
0	$\frac{1}{2}(e_4 + e_{23} + e_{56} + e_{23456})$	$(t-i)^4(t+i)^4$
−1	$\frac{1}{4}(-e_4 - e_{23} - 3e_{56} + e_{1237} - e_{147} - e_{1567} - e_{23456} - \omega)$	$(t-i)^3(t+i)^5$
−2	$\frac{1}{2}(-e_{23} - e_{56} - e_{147} - \omega)$	$(t-i)^2(t+i)^6$
−3	$\frac{1}{4}(e_4 - e_{23} - e_{56} - e_{1237} - e_{147} - e_{1567} - e_{23456} - 3\omega)$	$(t-i)(t+i)^7$
−4	$-\omega = -e_{1234567}$	$(t+i)^8$

TABLE 7. Square roots of $-\mathbf{1}$ in $C\ell_{3,4} \cong \mathcal{M}(8,\mathbb{C})$, $d=4$

k	f_k	$\Delta_k(t)$
4	$\omega = e_{1234567}$	$(t-i)^8$
3	$\frac{1}{4}(e_4 + e_{145} + e_{246} + e_{347} - e_{12456} - e_{13457} - e_{23467} + 3\omega)$	$(t-i)^7(t+i)$
2	$\frac{1}{2}(e_{145} - e_{12456} - e_{13457} + \omega)$	$(t-i)^6(t+i)^2$
1	$\frac{1}{4}(-e_4 + e_{145} + e_{246} - e_{347} - 3e_{12456} - e_{13457} - e_{23467} + \omega)$	$(t-i)^5(t+i)^3$
0	$\frac{1}{2}(e_4 + e_{12456} + e_{13457} + e_{23467})$	$(t-i)^4(t+i)^4$
−1	$\frac{1}{4}(-e_4 - e_{145} - e_{246} + e_{347} - 3e_{12456} - e_{13457} - e_{23467} - \omega)$	$(t-i)^3(t+i)^5$
−2	$\frac{1}{2}(-e_{145} - e_{12456} - e_{13457} - \omega)$	$(t-i)^2(t+i)^6$
−3	$\frac{1}{4}(e_4 - e_{145} - e_{246} - e_{347} - e_{12456} - e_{13457} - e_{23467} - 3\omega)$	$(t-i)(t+i)^7$
−4	$-\omega = -e_{1234567}$	$(t+i)^8$

TABLE 8. Square roots of $-\mathbf{1}$ in $C\ell_{5,2} \cong \mathcal{M}(8,\mathbb{C})$, $d=4$

k	f_k	$\Delta_k(t)$
4	$\omega = e_{1234567}$	$(t-i)^8$
3	$\frac{1}{4}(-e_{23}+e_{123}+e_{2346}+e_{2357}-e_{12346}$ $-e_{12357}+e_{234567}+3\omega)$	$(t-i)^7(t+i)$
2	$\frac{1}{2}(e_{123}-e_{12346}-e_{12357}+\omega)$	$(t-i)^6(t+i)^2$
1	$\frac{1}{4}(-e_{23}+e_{123}-e_{2346}+e_{2357}-3e_{12346}$ $-e_{12357}-e_{234567}+\omega)$	$(t-i)^5(t+i)^3$
0	$\frac{1}{2}(e_{23}+e_{12346}+e_{12357}+e_{234567})$	$(t-i)^4(t+i)^4$
-1	$\frac{1}{4}(-e_{23}-e_{123}+e_{2346}-e_{2357}-3e_{12346}$ $-e_{12357}-e_{234567}-\omega)$	$(t-i)^3(t+i)^5$
-2	$\frac{1}{2}(-e_{123}-e_{12346}-e_{12357}-\omega)$	$(t-i)^2(t+i)^6$
-3	$\frac{1}{4}(-e_{23}-e_{123}-e_{2346}-e_{2357}-e_{12346}$ $-e_{12357}+e_{234567}-3\omega)$	$(t-i)(t+i)^7$
-4	$-\omega = -e_{1234567}$	$(t+i)^8$

Appendix B. A Sample Maple Worksheet

In this appendix we show a computation of roots of $-\mathbf{1}$ in $C\ell_{3,0}$ in CLIFFORD. Although these computations certainly can be performed by hand, as shown in Section 7, they illustrate how CLIFFORD can be used instead especially when extending these computations to higher dimensions.[23] To see the actual Maple worksheets where these computations have been performed, see [21].

```
> restart:with(Clifford):with(linalg):with(asvd):
> p,q:=3,0; ##<<-- selecting signature
> B:=diag(1$p,-1$q): ##<<-- defining diagonal bilinear form
> eval(makealiases(p+q)): ##<<-- defining aliases
> clibas:=cbasis(p+q); ##assigning basis for Cl(3,0)
```

$$p,\ q := 3,\ 0$$

$$clibas := [Id,\ e1,\ e2,\ e3,\ e12,\ e13,\ e23,\ e123]$$

```
> data:=clidata(); ##<<-- displaying information about Cl(3,0)
```

$$data := \left[complex,\ 2,\ simple,\ \frac{Id}{2}+\frac{e1}{2},\ [Id,\ e2,\ e3,\ e23],\ [Id,\ e23],\ [Id,\ e2]\right]$$

```
> MM:=matKrepr(); ##<<-- displaying default matrices to generators

Cliplus has been loaded. Definitions for type/climon and
type/clipolynom now include &C and &C[K]. Type ?cliprod for
help.
```

[23] In showing Maple display we have edited Maple output to save space. Package asvd is a supplementary package written by the third author and built into CLIFFORD. The primary purpose of asvd is to compute Singular Value Decomposition in Clifford algebras [1].

$$MM := [e1 = \begin{bmatrix} 1 & 0 \\ 0 & -1 \end{bmatrix}, e2 = \begin{bmatrix} 0 & 1 \\ 1 & 0 \end{bmatrix}, e3 = \begin{bmatrix} 0 & -e23 \\ e23 & 0 \end{bmatrix}]$$

Pauli algebra representation displayed in (7.1):

```
> sigma[1]:=evalm(rhs(MM[1]));
> sigma[2]:=evalm(rhs(MM[2]));
> sigma[3]:=evalm(rhs(MM[3]));
```

$$\sigma_1, \sigma_2, \sigma_3 := \begin{bmatrix} 0 & 1 \\ 1 & 0 \end{bmatrix}, \begin{bmatrix} 0 & -e23 \\ e23 & 0 \end{bmatrix}, \begin{bmatrix} 1 & 0 \\ 0 & -1 \end{bmatrix}$$

We show how we represent the imaginary unit i in the field $\mathbb{C}$ and the diagonal matrix $\operatorname{diag}(i,i)$:

```
> ii:=e23; ##<<-- complex imaginary unit
> II:=diag(ii,ii); ##<<-- diagonal matrix diag(i,i)
```

$$ii := e23$$

$$II := \begin{bmatrix} e23 & 0 \\ 0 & e23 \end{bmatrix}$$

We compute matrices $m_1, m_2, \ldots, m_8$ representing each basis element in $C\ell_{3,0}$ isomorphic with $\mathbb{C}(2)$. Note that in our representation element e_{23} in $C\ell_{3,0}$ is used to represent the imaginary unit i.

```
> for i from 1 to nops(clibas) do
  lprint(`The basis element`,clibas[i],`is represented by the following
  matrix:`);
  M[i]:=subs(Id=1,matKrepr(clibas[i])) od;
```

```
`The basis element`, Id, `is represented by the following matrix:`
```

$$M_1 := \begin{bmatrix} 1 & 0 \\ 0 & 1 \end{bmatrix}$$

```
`The basis element`, e1, `is represented by the following matrix:`
```

$$M_2 := \begin{bmatrix} 1 & 0 \\ 0 & -1 \end{bmatrix}$$

```
`The basis element`, e2, `is represented by the following matrix:`
```

$$M_3 := \begin{bmatrix} 0 & 1 \\ 1 & 0 \end{bmatrix}$$

```
`The basis element`, e3, `is represented by the following matrix:`
```

$$M_4 := \begin{bmatrix} 0 & -e23 \\ e23 & 0 \end{bmatrix}$$

```
`The basis element`, e12, `is represented by the following matrix:`
```

$$M_5 := \begin{bmatrix} 0 & 1 \\ -1 & 0 \end{bmatrix}$$

```
`The basis element`, e13, `is represented by the following matrix:`
```

$$M_6 := \begin{bmatrix} 0 & -e23 \\ -e23 & 0 \end{bmatrix}$$

```
`The basis element`, e23, `is represented by the following matrix:`
```

$$M_7 := \begin{bmatrix} e23 & 0 \\ 0 & -e23 \end{bmatrix}$$

```
`The basis element`, e123, `is represented by the following matrix:`
```

$$M_8 := \begin{bmatrix} e23 & 0 \\ 0 & e23 \end{bmatrix}$$

We will use the procedure `phi` from the `asvd` package which gives an isomorphism from $\mathbb{C}(2)$ to $C\ell_{3,0}$. This way we can find the image in $C\ell_{3,0}$ of any complex 2×2 complex matrix A. Knowing the image of each matrix $m_1, m_2, \ldots, m_8$ in terms of the Clifford polynomials in $C\ell_{3,0}$, we can easily find the image of A in our default spinor representation of $C\ell_{3,0}$ which is built into CLIFFORD.

Procedure `Centralizer` computes a centralizer of f with respect to the Clifford basis L:

```
> Centralizer:=proc(f,L) local c,LL,m,vars,i,eq,sol;
  m:=add(c[i]*L[i],i=1..nops(L));
  vars:=[seq(c[i],i=1..nops(L))];
  eq:=clicollect(cmul(f,m)-cmul(m,f));
  if eq=0 then return L end if:
  sol:=op(clisolve(eq,vars));
  m:=subs(sol,m);
  m:=collect(m,vars);
  return sort([coeffs(m,vars)],bygrade);
  end proc:
```

Procedures `Scal` and `Spec` compute the scalar and the pseudoscalar parts of f.

```
> Scal:=proc(f) local p: return scalarpart(f); end proc:
> Spec:=proc(f) local N; global p,q;
  N:=p+q:
  return coeff(vectorpart(f,N),op(cbasis(N,N)));
  end proc:
```

The matrix idempotents in $\mathbb{C}(2)$ displayed in (7.2) are as follows:

```
> d:=1:Eps[1]:=sigma[1] &cm sigma[1];
```

```
> Eps[0]:=evalm(1/2*(1+sigma[3]));
> Eps[-1]:=diag(0,0);
```

$$Eps_1,\ Eps_0,\ Eps_{-1} := \begin{bmatrix} 1 & 0 \\ 0 & 1 \end{bmatrix},\ \begin{bmatrix} 1 & 0 \\ 0 & 0 \end{bmatrix},\ \begin{bmatrix} 0 & 0 \\ 0 & 0 \end{bmatrix}$$

This function `ff` computes matrix square root of -1 corresponding to the matrix idempotent *eps*:

```
> ff:=eps->evalm(II &cm (2*eps-1));
```

$$ff := eps \to \mathrm{evalm}(II\ \&\mathrm{cm}\ (2\ eps - 1))$$

We compute matrix square roots of -1 which correspond to the idempotents Eps_1, Eps_0, Eps_{-1}, and their characteristic and minimal polynomials. Note that in Maple the default imaginary unit is denoted by I.

```
> F[1]:=ff(Eps[1]); ##<<-- this square root of -1 corresponds to Eps[1]
  Delta[1]:=charpoly(subs(e23=I,evalm(F[1])),t);
  Mu[1]:=minpoly(subs(e23=I,evalm(F[1])),t);
```

$$F_1 := \begin{bmatrix} e23 & 0 \\ 0 & e23 \end{bmatrix},\quad \Delta_1 := (t-I)^2,\quad M_1 := t - I$$

```
> F[0]:=ff(Eps[0]); ##<<-- this square root of -1 corresponds to Eps[0]
  Delta[0]:=charpoly(subs(e23=I,evalm(F[0])),t);
  Mu[0]:=minpoly(subs(e23=I,evalm(F[0])),t);
```

$$F_0 := \begin{bmatrix} e23 & 0 \\ 0 & -e23 \end{bmatrix},\quad \Delta_0 := (t-I)\,(t+I),\quad M_0 := 1 + t^2$$

```
> F[-1]:=ff(Eps[-1]); ##<<-- this square root of -1 corresponds to Eps[-1]
  Delta[-1]:=charpoly(subs(e23=I,evalm(F[-1])),t);
  Mu[-1]:=minpoly(subs(e23=I,evalm(F[-1])),t);
```

$$F_{-1} := \begin{bmatrix} -e23 & 0 \\ 0 & -e23 \end{bmatrix},\quad \Delta_{-1} := (t+I)^2,\quad M_{-1} := t + I$$

Now, we can find square roots of -1 in $C\ell_{3,0}$ which correspond to the matrix square roots F_{-1}, F_0, F_1 *via* the isomorphism $\phi : C\ell_{3,0} \to \mathbb{C}(2)$ realized with the procedure `phi`.

First, we let `reprI` denote element in $C\ell_{3,0}$ which represents the diagonal $(2d) \times (2d)$ with $I = i$ on the diagonal where $i^2 = -1$. This element will replace the imaginary unit I in the minimal polynomials.

```
> reprI:=phi(diag(I$(2*d)),M);
```

$$reprI := e123$$

Now, we compute the corresponding square roots f_1, f_0, f_{-1} in $C\ell_{3,0}$.

```
> f[1]:=phi(F[1],M); ##<<-- element in Cl(3,0) corresponding to F[1]
  cmul(f[1],f[1]); ##<<-- checking that this element is a root of -1
  Mu[1]; ##<<-- recalling minpoly of matrix F[1]
  subs(e23=I,evalm(subs(t=evalm(F[1]),Mu[1]))); ##<<-- F[1] in Mu[1]
  mu[1]:=subs(I=reprI,Mu[1]); ##<<-- defining minpoly of f[1]
  cmul(f[1]-reprI,Id); ##<<-- verifying that f[1] satisfies mu[1]
```

$$f_1 := e123$$

$$-Id, \quad t - I, \quad \begin{bmatrix} 0 & 0 \\ 0 & 0 \end{bmatrix}$$

$$\mu_1 := t - e123, \quad 0$$

```
> f[0]:=phi(F[0],M); ##<<-- element in Cl(3,0) corresponding to F[0]
  cmul(f[0],f[0]); ##<<-- checking that this element is a root of -1
  Mu[0]; ##<<-- recalling minpoly of matrix F[0]
  subs(e23=I,evalm(subs(t=evalm(F[0]),Mu[0]))); ##<<-- F[0] in Mu[0]
  mu[0]:=subs(I=reprI,Mu[0]); ##<<-- defining minpoly of f[0]
  cmul(f[0]-reprI,f[0]+reprI); ##<<-- f[0] satisfies mu[0]
```

$$f_0 := e23$$

$$-Id, \quad 1 + t^2, \quad \begin{bmatrix} 0 & 0 \\ 0 & 0 \end{bmatrix}$$

$$\mu_0 := 1 + t^2, \quad 0$$

```
> f[-1]:=phi(F[-1],M); ##<<-- element in Cl(3,0) corresponding to F[-1]
  cmul(f[-1],f[-1]); ##<<-- checking that this element is a root of -1
  Mu[-1]; ##<<-- recalling minpoly of matrix F[-1]
  subs(e23=I,evalm(subs(t=evalm(F[-1]),Mu[-1]))); ##<<-- F[-1] in Mu[-1]
  mu[-1]:=subs(I=reprI,Mu[-1]); ##<<-- defining minpoly of f[-1]
  cmul(f[-1]+reprI,Id); ##<<-- f[-1] satisfies mu[-1]
```

$$f_{-1} := -e123$$

$$-Id, \quad t + I, \quad \begin{bmatrix} 0 & 0 \\ 0 & 0 \end{bmatrix}$$

$$\mu_{-1} := t + e123, \quad 0$$

Functions RdimCentralizer and RdimConjugClass of d and k compute the real dimension of the centralizer $\mathrm{Cent}(f)$ and the conjugacy class of f (see (7.4)).

```
> RdimCentralizer:=(d,k)->2*((d+k)^2+(d-k)^2); ##<<-- from the theory
> RdimConjugClass:=(d,k)->4*(d^2-k^2); ##<<-- from the theory
```

$$RdimCentralizer := (d,\, k) \to 2\,(d+k)^2 + 2\,(d-k)^2$$

$$RdimConjugClass := (d,\, k) \to 4\,d^2 - 4\,k^2$$

Now, we compute the centralizers of the roots and use notation d, k, n_1, n_2 displayed in Examples.

Case $k = 1$:

```
> d:=1:k:=1:n1:=d+k;n2:=d-k;
  A1:=diag(I$n1,-I$n2); ##<<-- this is the first matrix root of -1
```

$$n1 := 2, \quad n2 := 0, \quad A1 := \begin{bmatrix} I & 0 \\ 0 & I \end{bmatrix}$$

```
> f[1]:=phi(A1,M); cmul(f[1],f[1]); Scal(f[1]), Spec(f[1]);
```

$$f_1 := e123, \quad -Id, \quad 0, \quad 1$$

```
> LL1:=Centralizer(f[1],clibas); ##<<-- centralizer of f[1]
  dimCentralizer:=nops(LL1); ##<<-- real dimension of centralizer of f[1]
  RdimCentralizer(d,k); ##<<-- dimension of centralizer of f[1] from theory
  evalb(dimCentralizer=RdimCentralizer(d,k)); ##<<-- checking
  equality
```

$$LL1 := [Id,\ e1,\ e2,\ e3,\ e12,\ e13,\ e23,\ e123]$$
$$dimCentralizer := 8, \quad 8, \quad true$$

Case $k = 0$:

```
> d:=1:k:=0:n1:=d+k;n2:=d-k;
  A0:=diag(I$n1,-I$n2); ##<<-- this is the second matrix root of -1
```

$$n1 := 1, \quad n2 := 1, \quad A0 := \begin{bmatrix} I & 0 \\ 0 & -I \end{bmatrix}$$

```
> f[0]:=phi(A0,M); cmul(f[0],f[0]); Scal(f[0]), Spec(f[0]);
```

$$f_0 := e23, \quad -Id, \quad 0, \quad 0$$

```
> LL0:=Centralizer(f[0],clibas); ##<<-- centralizer of f[0]
  dimCentralizer:=nops(LL0); ##<<-- real dimension of centralizer of f[0]
  RdimCentralizer(d,k); ##<<-- dimension of centralizer of f[0] from theory
  evalb(dimCentralizer=RdimCentralizer(d,k)); ##<<-- checking equality
```

$$LL0 := [Id,\ e1,\ e23,\ e123]$$
$$dimCentralizer := 4, \quad 4, \quad true$$

Case $k = -1$:

```
> d:=1:k:=-1:n1:=d+k;n2:=d-k;
  Am1:=diag(I$n1,-I$n2); ##<<-- this is the third matrix root of -1
```

$$n1 := 0, \quad n2 := 2, \quad Am1 := \begin{bmatrix} -I & 0 \\ 0 & -I \end{bmatrix}$$

```
> f[-1]:=phi(Am1,M); cmul(f[-1],f[-1]); Scal(f[-1]), Spec(f[-1]);
```

$$f_{-1} := -e123, \quad -Id, \quad 0, \quad -1$$

```
> LLm1:=Centralizer(f[-1],clibas); ##<<-- centralizer of f[-1]
  dimCentralizer:=nops(LLm1); ##<<-- real dimension of centralizer of f[-1]
  RdimCentralizer(d,k); ##<<--dimension of centralizer of f[-1] from theory
  evalb(dimCentralizer=RdimCentralizer(d,k)); ##<<-- checking equality
```

$$LLm1 := [Id,\ e1,\ e2,\ e3,\ e12,\ e13,\ e23,\ e123]$$

$$dimCentralizer := 8, \quad 8, \quad true$$

We summarize roots of -1 in $C\ell_{3,0}$:

```
> 'F[1]'=evalm(F[1]); ##<<-- square root of -1 in C(2)
  Mu[1]; ##<<-- minpoly of matrix F[1]
  'f[1]'=f[1]; ##<<-- square root of -1 in Cl(3,0)
  mu[1]; ##<<-- minpoly of element f[1]
```

$$F_1 = \begin{bmatrix} e23 & 0 \\ 0 & e23 \end{bmatrix}, \quad t - I$$

$$f_1 = e123, \quad t - e123$$

```
> 'F[0]'=evalm(F[0]); ##<<-- square root of -1 in C(2)
  Mu[0]; ##<<-- minpoly of matrix F[0]
  'f[0]'=f[0]; ##<<-- square root of -1 in Cl(3,0)
  mu[0]; ##<<-- minpoly of element f[0]
```

$$F_0 = \begin{bmatrix} e23 & 0 \\ 0 & -e23 \end{bmatrix}, \quad 1 + t^2$$

$$f_0 = e23, \quad 1 + t^2$$

```
> 'F[-1]'=evalm(F[-1]); ##<<-- square root of -1 in C(2)
  Mu[-1]; ##<<-- minpoly of matrix F[-1]
  'f[-1]'=f[-1]; ##<<-- square root of -1 in Cl(3,0)
  mu[-1]; ##<<-- minpoly of element f[-1]
```

$$F_{-1} = \begin{bmatrix} -e23 & 0 \\ 0 & -e23 \end{bmatrix}, \quad t + I$$

$$f_{-1} = -e123, \quad t + e123$$

Finaly, we verify that roots f_1 and f_{-1} are related *via* the reversion:

```
> reversion(f[1])=f[-1]; evalb(%);
```

$$-e123 = -e123, \quad true$$

References

[1] R. Abłamowicz. Computations with Clifford and Grassmann algebras. *Advances in Applied Clifford Algebras*, 19(3–4):499–545, 2009.

[2] R. Abłamowicz and B. Fauser. CLIFFORD with bigebra – a Maple package for computations with Clifford and Grassmann algebras. Available at `http://math.tntech.edu/rafal/`, ©1996–2012.

[3] R. Abłamowicz, B. Fauser, K. Podlaski, and J. Rembieliński. Idempotents of Clifford algebras. *Czechoslovak Journal of Physics*, 53(11):949–954, 2003.

[4] J. Armstrong. The center of an algebra. Weblog, available at `http://unapologetic.wordpress.com/2010/10/06/the-center-of-an-algebra/`, accessed 22 March 2011.

[5] M. Bahri, E. Hitzer, R. Ashino, and R. Vaillancourt. Windowed Fourier transform of two-dimensional quaternionic signals. *Applied Mathematics and Computation*, 216(8):2366–2379, June 2010.

[6] M. Bahri, E.M.S. Hitzer, and S. Adji. Two-dimensional Clifford windowed Fourier transform. In E.J. Bayro-Corrochano and G. Scheuermann, editors, *Geometric Algebra Computing in Engineering and Computer Science*, pages 93–106. Springer, London, 2010.

[7] F. Brackx, R. Delanghe, and F. Sommen. *Clifford Analysis*, volume 76. Pitman, Boston, 1982.

[8] T. Bülow. *Hypercomplex Spectral Signal Representations for the Processing and Analysis of Images*. PhD thesis, University of Kiel, Germany, Institut für Informatik und Praktische Mathematik, Aug. 1999.

[9] T. Bülow, M. Felsberg, and G. Sommer. Non-commutative hypercomplex Fourier transforms of multidimensional signals. In G. Sommer, editor, *Geometric computing with Clifford Algebras: Theoretical Foundations and Applications in Computer Vision and Robotics*, pages 187–207, Berlin, 2001. Springer.

[10] C. Chevalley. *The Theory of Lie Groups*. Princeton University Press, Princeton, 1957.

[11] W.K. Clifford. Applications of Grassmann's extensive algebra. *American Journal of Mathematics*, 1(4):350–358, 1878.

[12] L. Dorst and J. Lasenby, editors. *Guide to Geometric Algebra in Practice*. Springer, Berlin, 2011.

[13] J. Ebling and G. Scheuermann. Clifford Fourier transform on vector fields. *IEEE Transactions on Visualization and Computer Graphics*, 11(4):469–479, July 2005.

[14] M. Felsberg. *Low-Level Image Processing with the Structure Multivector*. PhD thesis, Christian-Albrechts-Universität, Institut für Informatik und Praktische Mathematik, Kiel, 2002.

[15] D. Hestenes. *Space-Time Algebra*. Gordon and Breach, London, 1966.

[16] D. Hestenes. *New Foundations for Classical Mechanics*. Kluwer, Dordrecht, 1999.

[17] D. Hestenes and G. Sobczyk. *Clifford Algebra to Geometric Calculus*. D. Reidel Publishing Group, Dordrecht, Netherlands, 1984.

[18] E. Hitzer. Quaternion Fourier transform on quaternion fields and generalizations. *Advances in Applied Clifford Algebras*, 17(3):497–517, May 2007.

[19] E. Hitzer. OPS-QFTs: A new type of quaternion Fourier transform based on the orthogonal planes split with one or two general pure quaternions. In *International Conference on Numerical Analysis and Applied Mathematics*, volume 1389 of *AIP*

Conference Proceedings, pages 280–283, Halkidiki, Greece, 19–25 September 2011. American Institute of Physics.

[20] E. Hitzer and R. Abłamowicz. Geometric roots of -1 in Clifford algebras $C\ell_{p,q}$ with $p+q \leq 4$. *Advances in Applied Clifford Algebras*, 21(1):121–144, 2010. Published online 13 July 2010.

[21] E. Hitzer, J. Helmstetter, and R. Abłamowicz. Maple worksheets created with `CLIFFORD` for a verification of results presented in this chapter. Available at: `http://math.tntech.edu/rafal/publications.html`, ©2012.

[22] E. Hitzer and B. Mawardi. Uncertainty principle for Clifford geometric algebras $C\ell_{n,0}, n = 3(\bmod 4)$ based on Clifford Fourier transform. In T. Qian, M.I. Vai, and Y. Xu, editors, *Wavelet Analysis and Applications*, Applied and Numerical Harmonic Analysis, pages 47–56. Birkhäuser Basel, 2007.

[23] E.M.S. Hitzer and B. Mawardi. Clifford Fourier transform on multivector fields and uncertainty principles for dimensions $n = 2(\bmod 4)$ and $n = 3(\bmod 4)$. *Advances in Applied Clifford Algebras*, 18(3-4):715–736, 2008.

[24] R.A. Horn and C.R. Johnson. *Matrix Analysis.* Cambridge University Press, Cambridge, 1985.

[25] W.M. Incorporated. Maple, a general purpose computer algebra system. `http://www.maplesoft.com`, © 2012.

[26] C. Li, A. McIntosh, and T. Qian. Clifford algebras, Fourier transform and singular convolution operators on Lipschitz surfaces. *Revista Matematica Iberoamericana*, 10(3):665–695, 1994.

[27] H. Li. *Invariant Algebras and Geometric Reasoning.* World Scientific, Singapore, 2009.

[28] P. Lounesto. *Clifford Algebras and Spinors*, volume 286 of *London Mathematical Society Lecture Notes.* Cambridge University Press, 1997.

[29] B. Mawardi and E.M.S. Hitzer. Clifford Fourier transformation and uncertainty principle for the Clifford algebra $C\ell_{3,0}$. *Advances in Applied Clifford Algebras*, 16(1):41–61, 2006.

[30] A. McIntosh. Clifford algebras, Fourier theory, singular integrals, and harmonic functions on Lipschitz domains. In J. Ryan, editor, *Clifford Algebras in Analysis and Related Topics*, chapter 1. CRC Press, Boca Raton, 1996.

[31] T. Qian. Paley-Wiener theorems and Shannon sampling in the Clifford analysis setting. In R. Abłamowicz, editor, *Clifford Algebras – Applications to Mathematics, Physics, and Engineering*, pages 115–124. Birkäuser, Basel, 2004.

[32] S. Said, N. Le Bihan, and S.J. Sangwine. Fast complexified quaternion Fourier transform. *IEEE Transactions on Signal Processing*, 56(4):1522–1531, Apr. 2008.

[33] S.J. Sangwine. Biquaternion (complexified quaternion) roots of -1. *Advances in Applied Clifford Algebras*, 16(1):63–68, June 2006.

[34] Wikipedia article. Conjugacy class. Available at `http://en.wikipedia.org/wiki/Conjugacy_class`, accessed 19 March 2011.

[35] Wikipedia article. Inner automorphism. Available at `http://en.wikipedia.org/wiki/Inner_automorphism`, accessed 19 March 2011.

Eckhard Hitzer
College of Liberal Arts, Department of Material Science
International Christian University
181-8585 Tokyo, Japan
e-mail: hitzer@icu.ac.jp

Jacques Helmstetter
Univesité Grenoble I
Institut Fourier (Mathématiques)
B.P. 74
F-38402 Saint-Martin d'Hères, France
e-mail: Jacques.Helmstetter@ujf-grenoble.fr

Rafał Abłamowicz
Department of Mathematics, Box 5054
Tennessee Technological University
Cookeville, TN 38505, USA
e-mail: rablamowicz@tntech.edu

Quaternion and Clifford–Fourier Transforms and Wavelets
Trends in Mathematics, 155–176

8 A General Geometric Fourier Transform

Roxana Bujack, Gerik Scheuermann and Eckhard Hitzer

Abstract. The increasing demand for Fourier transforms on geometric algebras has resulted in a large variety. Here we introduce one single straightforward definition of a general geometric Fourier transform covering most versions in the literature. We show which constraints are additionally necessary to obtain certain features such as linearity or a shift theorem. As a result, we provide guidelines for the target-oriented design of yet unconsidered transforms that fulfill requirements in a specific application context. Furthermore, the standard theorems do not need to be shown in a slightly different form every time a new geometric Fourier transform is developed since they are proved here once and for all.

Mathematics Subject Classification (2010). Primary 15A66, 11E88; secondary 42A38, 30G35.

Keywords. Fourier transform, geometric algebra, Clifford algebra, image processing, linearity, scaling, shift.

1. Introduction

The Fourier transform by Jean Baptiste Joseph Fourier is an indispensable tool in many fields of mathematics, physics, computer science and engineering, especially for the analysis and solution of differential equations, or in signal and image processing, fields which cannot be imagined without it. The kernel of the Fourier transform consists of the complex exponential function. With the square root of minus one, the imaginary unit i, as part of the argument it is periodic and therefore suitable for the analysis of oscillating systems.

William Kingdon Clifford created the geometric algebras in 1878, [8]. They usually contain continuous submanifolds of geometric square roots of -1 [16, 17]. Each multivector has a natural geometric interpretation so the generalization of the Fourier transform to multivector-valued functions in the geometric algebras is very reasonable. It helps to interpret the transform, apply it in a target-oriented way to the specific underlying problem and it allows a new point of view on fluid mechanics.

Many different application-oriented definitions of Fourier transforms in geometric algebras have been developed. For example the Clifford–Fourier transform introduced by Jancewicz [19] and expanded by Ebling and Scheuermann [10] and Hitzer and Mawardi [18] or the one established by Sommen in [21] and re-established by Bülow [7]. Further we have the quaternionic Fourier transform by Ell [11] and later by Bülow [7], the spacetime Fourier transform by Hitzer [15], the Clifford–Fourier transform for colour images by Batard *et al.* [1], the Cylindrical Fourier transform by Brackx *et al.* [6], the transforms by Felsberg [13] or Ell and Sangwine [20, 12]. All these transforms have different interesting properties and deserve to be studied independently from one another. But the analysis of their similarities reveals a lot about their qualities, too. We concentrate on this matter and summarize all of them in one general definition.

Recently there have been very successful approaches by De Bie, Brackx, De Schepper and Sommen to construct Clifford–Fourier transforms from operator exponentials and differential equations [3, 4, 9, 5]. The definition presented in this chapter does not cover all of them, partly because their closed integral form is not always known or is highly complicated, and partly because they can be produced by combinations and functions of our transforms.

We focus on continuous geometric Fourier transforms over flat spaces $\mathbb{R}^{p,q}$ in their integral representation. That way their finite, regular discrete equivalents as used in computational signal and image processing can be intuitively constructed and direct applicability to the existing practical issues and easy numerical manageability are ensured.

2. Definition of the GFT

We examine geometric algebras $C\ell_{p,q}, p+q=n\in\mathbb{N}$ over $\mathbb{R}^{p+q}$ [14] generated by the associative, bilinear geometric product with neutral element 1 satisfying

$$\boldsymbol{e}_j\boldsymbol{e}_k+\boldsymbol{e}_k\boldsymbol{e}_j=\epsilon_j\delta_{jk}, \tag{2.1}$$

for all $j,k\in\{1,\dots,n\}$ with the Kronecker symbol δ and

$$\epsilon_j=\begin{cases}1 & \forall j=1,\dots,p,\\ -1 & \forall j=p+1,\dots,n.\end{cases} \tag{2.2}$$

For the sake of brevity we want to refer to arbitrary multivectors

$$\boldsymbol{A}=\sum_{k=0}^{n}\sum_{1\le j_1<\dots<j_k\le n} a_{j_1\dots j_k}\boldsymbol{e}_{j_1}\dots\boldsymbol{e}_{j_k}\in C\ell_{p,q}, \tag{2.3}$$

where $a_{j_1\dots j_k}\in\mathbb{R}$, as

$$\boldsymbol{A}=\sum_j a_j\boldsymbol{e}_j. \tag{2.4}$$

where each of the 2^n multi-indices $\boldsymbol{j} \subseteq \{1, \ldots, n\}$ indicates a basis vector of $C\ell_{p,q}$ by $\boldsymbol{e_j} = \boldsymbol{e}_{j_1} \ldots \boldsymbol{e}_{j_k}$, $1 \leq j_1 < \cdots < j_k \leq n, \boldsymbol{e}_\emptyset = \boldsymbol{e}_0 = 1$ and its associated coefficient $a_{\boldsymbol{j}} = a_{j_1 \ldots j_k} \in \mathbb{R}$.

Definition 2.1. The **exponential function** of a multivector $\boldsymbol{A} \in C\ell_{p,q}$ is defined by the power series

$$e^{\boldsymbol{A}} := \sum_{j=0}^{\infty} \frac{\boldsymbol{A}^j}{j!}. \tag{2.5}$$

Lemma 2.2. *For two multivectors $\boldsymbol{AB} = \boldsymbol{BA}$ that commute we have*

$$e^{\boldsymbol{A}+\boldsymbol{B}} = e^{\boldsymbol{A}} e^{\boldsymbol{B}}. \tag{2.6}$$

Proof. Analogous to the exponent rule of real matrices. □

Notation 2.3. For each geometric algebra $C\ell_{p,q}$ we will write $\mathscr{I}^{p,q} = \{i \in C\ell_{p,q}, i^2 \in \mathbb{R}^-\}$ to denote the real multiples of all geometric square roots of -1, compare [16] and [17]. We choose the symbol $\mathscr{I}$ to be reminiscent of the imaginary numbers.

Definition 2.4. Let $C\ell_{p,q}$ be a geometric algebra, $\boldsymbol{A} : \mathbb{R}^m \to C\ell_{p,q}$ be a multivector field and $\boldsymbol{x}, \boldsymbol{u} \in \mathbb{R}^m$ vectors. A **Geometric Fourier Transform** (GFT) $\mathcal{F}_{F_1,F_2}(\boldsymbol{A})$ is defined by two ordered finite sets $F_1 = \{f_1(\boldsymbol{x}, \boldsymbol{u}), \ldots, f_\mu(\boldsymbol{x}, \boldsymbol{u})\}$, $F_2 = \{f_{\mu+1}(\boldsymbol{x}, \boldsymbol{u}), \ldots, f_\nu(\boldsymbol{x}, \boldsymbol{u})\}$ of mappings $f_k(\boldsymbol{x}, \boldsymbol{u}) : \mathbb{R}^m \times \mathbb{R}^m \to \mathscr{I}^{p,q}, \forall k = 1, \ldots, \nu$ and the calculation rule

$$\mathcal{F}_{F_1,F_2}(\boldsymbol{A})(\boldsymbol{u}) := \int_{\mathbb{R}^m} \prod_{f \in F_1} e^{-f(\boldsymbol{x},\boldsymbol{u})} \boldsymbol{A}(\boldsymbol{x}) \prod_{f \in F_2} e^{-f(\boldsymbol{x},\boldsymbol{u})} \, \mathrm{d}^m \boldsymbol{x}. \tag{2.7}$$

This definition combines many Fourier transforms into a single general one. It enables us to prove the well-known theorems which depend only on the properties of the chosen mappings.

Example. Depending on the choice of F_1 and F_2 we obtain previously published transforms.

1. In the case of $\boldsymbol{A} : \mathbb{R}^n \to \mathcal{G}^{n,0}, n = 2 \pmod 4$ or $n = 3 \pmod 4$, we can reproduce the Clifford–Fourier transform introduced by Jancewicz [19] for $n = 3$ and expanded by Ebling and Scheuermann [10] for $n = 2$ and Hitzer and Mawardi [18] for $n = 2 \pmod 4$ or $n = 3 \pmod 4$ using the configuration

$$\begin{aligned} F_1 &= \emptyset, \\ F_2 &= \{f_1\}, \\ f_1(\boldsymbol{x}, \boldsymbol{u}) &= 2\pi i_n \boldsymbol{x} \cdot \boldsymbol{u}, \end{aligned} \tag{2.8}$$

   with i_n being the pseudoscalar of $G^{n,0}$.
2. Choosing multivector fields $\mathbb{R}^n \to \mathcal{G}^{0,n}$,

$$\begin{aligned} F_1 &= \emptyset, \\ F_2 &= \{f_1, \ldots, f_n\}, \\ f_k(\boldsymbol{x}, \boldsymbol{u}) &= 2\pi \boldsymbol{e}_k x_k u_k, \quad \forall k = 1, \ldots, n \end{aligned} \tag{2.9}$$

   we have the Sommen–Bülow–Clifford–Fourier transform from [21, 7].

3. For $\boldsymbol{A} : \mathbb{R}^2 \to \mathcal{G}^{0,2} \approx \mathbb{H}$ the quaternionic Fourier transform [11, 7] is generated by

$$\begin{aligned} F_1 &= \{f_1\}, \\ F_2 &= \{f_2\}, \\ f_1(\boldsymbol{x}, \boldsymbol{u}) &= 2\pi i x_1 u_1, \\ f_2(\boldsymbol{x}, \boldsymbol{u}) &= 2\pi j x_2 u_2. \end{aligned} \tag{2.10}$$

4. Using $\mathcal{G}^{3,1}$ we can build the spacetime, respectively the volume-time, Fourier transform from [15][1] with the $\mathcal{G}^{3,1}$-pseudoscalar i_4 as follows

$$\begin{aligned} F_1 &= \{f_1\}, \\ F_2 &= \{f_2\}, \\ f_1(\boldsymbol{x}, \boldsymbol{u}) &= \boldsymbol{e}_4 x_4 u_4, \\ f_2(\boldsymbol{x}, \boldsymbol{u}) &= \epsilon_4 \boldsymbol{e}_4 i_4 (x_1 u_1 + x_2 u_2 + x_3 u_3). \end{aligned} \tag{2.11}$$

5. The Clifford–Fourier transform for colour images by Batard, Berthier and Saint-Jean [1] for $m = 2, n = 4$, $\boldsymbol{A} : \mathbb{R}^2 \to \mathcal{G}^{4,0}$, a fixed bivector $\boldsymbol{B}$, and the pseudoscalar i can intuitively be written as

$$\begin{aligned} F_1 &= \{f_1\}, \\ F_2 &= \{f_2\}, \\ f_1(\boldsymbol{x}, \boldsymbol{u}) &= \frac{1}{2}(x_1 u_1 + x_2 u_2)(\boldsymbol{B} + i\boldsymbol{B}), \\ f_2(\boldsymbol{x}, \boldsymbol{u}) &= -\frac{1}{2}(x_1 u_1 + x_2 u_2)(\boldsymbol{B} + i\boldsymbol{B}), \end{aligned} \tag{2.12}$$

but $(\boldsymbol{B} + i\boldsymbol{B})$ does not square to a negative real number, see [16]. The special property that $\boldsymbol{B}$ and $i\boldsymbol{B}$ commute allows us to express the formula using

$$\begin{aligned} F_1 &= \{f_1, f_2\}, \\ F_2 &= \{f_3, f_4\}, \\ f_1(\boldsymbol{x}, \boldsymbol{u}) &= \frac{1}{2}(x_1 u_1 + x_2 u_2)\boldsymbol{B}, \\ f_2(\boldsymbol{x}, \boldsymbol{u}) &= \frac{1}{2}(x_1 u_1 + x_2 u_2)i\boldsymbol{B}, \\ f_3(\boldsymbol{x}, \boldsymbol{u}) &= -\frac{1}{2}(x_1 u_1 + x_2 u_2)\boldsymbol{B}, \\ f_4(\boldsymbol{x}, \boldsymbol{u}) &= -\frac{1}{2}(x_1 u_1 + x_2 u_2)i\boldsymbol{B}, \end{aligned} \tag{2.13}$$

which fulfills the conditions of Definition 2.4.

[1] Please note that Hitzer uses a different notation in [15]. His $\boldsymbol{x} = t\boldsymbol{e}_0 + x_1\boldsymbol{e}_1 + x_2\boldsymbol{e}_2 + x_3\boldsymbol{e}_3$ corresponds to our $\boldsymbol{x} = x_1\boldsymbol{e}_1 + x_2\boldsymbol{e}_2 + x_3\boldsymbol{e}_3 + x_4\boldsymbol{e}_4$, with $\boldsymbol{e}_0\boldsymbol{e}_0 = \epsilon_0 = -1$ being equivalent to our $\boldsymbol{e}_4\boldsymbol{e}_4 = \epsilon_4 = -1$.

6. Using $\mathcal{G}^{0,n}$ and

$$\begin{aligned} F_1 &= \{f_1\}, \\ F_2 &= \emptyset, \\ f_1(\boldsymbol{x}, \boldsymbol{u}) &= -\boldsymbol{x} \wedge \boldsymbol{u} \end{aligned} \tag{2.14}$$

produces the cylindrical Fourier transform as introduced by Brackx, de Schepper and Sommen in [6].

3. General Properties

First we prove general properties valid for arbitrary sets F_1, F_2.

Theorem 3.1 (Existence). *The geometric Fourier transform exists for all integrable multivector fields $\boldsymbol{A} \in L_1(\mathbb{R}^n)$.*

Proof. The property

$$f_k^2(\boldsymbol{x}, \boldsymbol{u}) \in \mathbb{R}^- \tag{3.1}$$

of the mappings f_k for $k = 1, \ldots, \nu$ leads to

$$\frac{f_k^2(\boldsymbol{x}, \boldsymbol{u})}{|f_k^2(\boldsymbol{x}, \boldsymbol{u})|} = -1 \tag{3.2}$$

for all $f_k(\boldsymbol{x}, \boldsymbol{u}) \neq 0$. So using the decomposition

$$f_k(\boldsymbol{x}, \boldsymbol{u}) = \frac{f_k(\boldsymbol{x}, \boldsymbol{u})}{|f_k(\boldsymbol{x}, \boldsymbol{u})|} |f_k(\boldsymbol{x}, \boldsymbol{u})| \tag{3.3}$$

we can write $\forall j \in \mathbb{N}$

$$f_k^j(\boldsymbol{x}, \boldsymbol{u}) = \begin{cases} (-1)^l |f_k(\boldsymbol{x}, \boldsymbol{u})|^j & \text{for } j = 2l, l \in \mathbb{N}_0 \\ (-1)^l \dfrac{f_k(\boldsymbol{x}, \boldsymbol{u})}{|f_k(\boldsymbol{x}, \boldsymbol{u})|} |f_k(\boldsymbol{x}, \boldsymbol{u})|^j & \text{for } j = 2l+1, l \in \mathbb{N}_0 \end{cases} \tag{3.4}$$

which results in

$$\begin{aligned} e^{-f_k(\boldsymbol{x}, \boldsymbol{u})} &= \sum_{j=0}^{\infty} \frac{(-f_k(\boldsymbol{x}, \boldsymbol{u}))^j}{j!} \\ &= \sum_{j=0}^{\infty} \frac{(-1)^j |f_k(\boldsymbol{x}, \boldsymbol{u})|^{2j}}{(2j)!} \\ &\quad - \frac{f_k(\boldsymbol{x}, \boldsymbol{u})}{|f_k(\boldsymbol{x}, \boldsymbol{u})|} \sum_{j=0}^{\infty} \frac{(-1)^j |f_k(\boldsymbol{x}, \boldsymbol{u})|^{2j+1}}{(2j+1)!} \\ &= \cos\left(|f_k(\boldsymbol{x}, \boldsymbol{u})|\right) - \frac{f_k(\boldsymbol{x}, \boldsymbol{u})}{|f_k(\boldsymbol{x}, \boldsymbol{u})|} \sin\left(|f_k(\boldsymbol{x}, \boldsymbol{u})|\right). \end{aligned} \tag{3.5}$$

Because of

$$\begin{aligned}|e^{-f_k(\boldsymbol{x},\boldsymbol{u})}| &= \left|\cos(|f_k(\boldsymbol{x},\boldsymbol{u})|) - \frac{f_k(\boldsymbol{x},\boldsymbol{u})}{|f_k(\boldsymbol{x},\boldsymbol{u})|}\sin(|f_k(\boldsymbol{x},\boldsymbol{u})|)\right| \\ &\leq |\cos(|f_k(\boldsymbol{x},\boldsymbol{u})|)| + \left|\frac{f_k(\boldsymbol{x},\boldsymbol{u})}{|f_k(\boldsymbol{x},\boldsymbol{u})|}\right| |\sin(|f_k(\boldsymbol{x},\boldsymbol{u})|)| \\ &\leq 2\end{aligned} \tag{3.6}$$

the magnitude of the improper integral

$$\begin{aligned}|\mathcal{F}_{F_1,F_2}(\boldsymbol{A})(\boldsymbol{u})| &= \left|\int_{\mathbb{R}^m} \prod_{f\in F_1} e^{-f(\boldsymbol{x},\boldsymbol{u})} \boldsymbol{A}(\boldsymbol{x}) \prod_{f\in F_2} e^{-f(\boldsymbol{x},\boldsymbol{u})} \, \mathrm{d}^m\boldsymbol{x}\right| \\ &\leq \int_{\mathbb{R}^m} \prod_{f\in F_1} \left|e^{-f(\boldsymbol{x},\boldsymbol{u})}\right| |\boldsymbol{A}(\boldsymbol{x})| \prod_{f\in F_2} \left|e^{-f(\boldsymbol{x},\boldsymbol{u})}\right| \mathrm{d}^m\boldsymbol{x} \\ &\leq \int_{\mathbb{R}^m} \prod_{f\in F_1} 2\,|\boldsymbol{A}(\boldsymbol{x})| \prod_{f\in F_2} 2\,\mathrm{d}^m\boldsymbol{x} \\ &= 2^\nu \int_{\mathbb{R}^m} |\boldsymbol{A}(\boldsymbol{x})| \, \mathrm{d}^m\boldsymbol{x}\end{aligned} \tag{3.7}$$

is finite and therefore the geometric Fourier transform exists. □

Theorem 3.2 (Scalar linearity). *The geometric Fourier transform is linear with respect to scalar factors. Let $b, c \in \mathbb{R}$ and $\boldsymbol{A}, \boldsymbol{B}, \boldsymbol{C} : \mathbb{R}^m \to C\ell_{p,q}$ be three multivector fields that satisfy $\boldsymbol{A}(\boldsymbol{x}) = b\boldsymbol{B}(\boldsymbol{x}) + c\boldsymbol{C}(\boldsymbol{x})$, then*

$$\mathcal{F}_{F_1,F_2}(\boldsymbol{A})(\boldsymbol{u}) = b\mathcal{F}_{F_1,F_2}(\boldsymbol{B})(\boldsymbol{u}) + c\mathcal{F}_{F_1,F_2}(\boldsymbol{C})(\boldsymbol{u}). \tag{3.8}$$

Proof. The assertion is an easy consequence of the distributivity of the geometric product over addition, the commutativity of scalars and the linearity of the integral. □

4. Bilinearity

All geometric Fourier transforms from the introductory example can also be expressed in terms of a stronger claim. The mappings $f_1, \ldots, f_\nu$, with the first μ terms to the left of the argument function and the $\nu - \mu$ others on the right of it, are all bilinear and therefore take the form

$$\begin{aligned}f_k(\boldsymbol{x},\boldsymbol{u}) &= f_k\left(\sum_{j=1}^m x_j \boldsymbol{e}_j, \sum_{l=1}^m u_l \boldsymbol{e}_l\right) \\ &= \sum_{j,l=1}^m x_j f_k(\boldsymbol{e}_j, \boldsymbol{e}_l) u_l = \boldsymbol{x}^T M_k \boldsymbol{u},\end{aligned} \tag{4.1}$$

$\forall k = 1, \ldots, \nu$, where $M_k \in (\mathscr{I}^{p,q})^{m\times m}, (M_k)_{jl} = f_k(\boldsymbol{e}_j, \boldsymbol{e}_l)$ according to Notation 2.3.

Example. Ordered in the same way as in the previous example, the geometric Fourier transforms expressed in the way of (4.1) take the following shapes:

1. In the Clifford–Fourier transform f_1 can be written with
$$M_1 = 2\pi i_n \operatorname{Id}. \tag{4.2}$$
2. The $\nu = m = n$ mappings $f_k, k = 1, \dots, n$ of the Bülow–Clifford–Fourier transform can be expressed using
$$(M_k)_{lj} = \begin{cases} 2\pi \boldsymbol{e}_k & \text{for } k = l = j, \\ 0 & \text{otherwise.} \end{cases} \tag{4.3}$$
3. Similarly the quaternionic Fourier transform is generated using
$$\begin{aligned} (M_1)_{l\iota} &= \begin{cases} 2\pi i & \text{for } l = \iota = 1, \\ 0 & \text{otherwise,} \end{cases} \\ (M_2)_{l\iota} &= \begin{cases} 2\pi j & \text{for } l = \iota = 2, \\ 0 & \text{otherwise.} \end{cases} \end{aligned} \tag{4.4}$$
4. We can build the spacetime Fourier transform with
$$\begin{aligned} (M_1)_{lj} &= \begin{cases} \boldsymbol{e}_4 & \text{for } l = j = 1, \\ 0 & \text{otherwise,} \end{cases} \\ (M_2)_{lj} &= \begin{cases} \epsilon_4 \boldsymbol{e}_4 i_4 & \text{for } l = j \in \{2,3,4\}, \\ 0 & \text{otherwise.} \end{cases} \end{aligned} \tag{4.5}$$
5. The Clifford–Fourier transform for colour images can be described by
$$\begin{aligned} M_1 &= \frac{1}{2}\boldsymbol{B} \operatorname{Id}, \\ M_2 &= \frac{1}{2} i\boldsymbol{B} \operatorname{Id}, \\ M_3 &= -\frac{1}{2}\boldsymbol{B} \operatorname{Id}, \\ M_4 &= -\frac{1}{2} i\boldsymbol{B} \operatorname{Id}. \end{aligned} \tag{4.6}$$
6. The cylindrical Fourier transform can also be reproduced with mappings satisfying (4.1) because we can write
$$\begin{aligned} \boldsymbol{x} \wedge \boldsymbol{u} = {} & \boldsymbol{e}_1\boldsymbol{e}_2 x_1 u_2 - \boldsymbol{e}_1\boldsymbol{e}_2 x_2 u_1 \\ & + \cdots \\ & + \boldsymbol{e}_{m-1}\boldsymbol{e}_m x_{m-1} u_m - \boldsymbol{e}_{m-1}\boldsymbol{e}_m x_m u_{m-1} \end{aligned} \tag{4.7}$$
and set
$$(M_1)_{lj} = \begin{cases} 0 & \text{for } l = j, \\ \boldsymbol{e}_l \boldsymbol{e}_j & \text{otherwise.} \end{cases} \tag{4.8}$$

Theorem 4.1 (Scaling). *Let $0 \neq a \in \mathbb{R}$ be a real number, $\boldsymbol{A}(\boldsymbol{x}) = \boldsymbol{B}(a\boldsymbol{x})$ two multivector fields and all F_1, F_2 be bilinear mappings then the geometric Fourier transform satisfies*

$$\mathcal{F}_{F_1,F_2}(\boldsymbol{A})(\boldsymbol{u}) = |a|^{-m}\mathcal{F}_{F_1,F_2}(\boldsymbol{B})\left(\frac{\boldsymbol{u}}{a}\right). \tag{4.9}$$

Proof. A change of coordinates together with the bilinearity proves the assertion by

$$\begin{aligned}\mathcal{F}_{F_1,F_2}(\boldsymbol{A})(\boldsymbol{u}) &= \int_{\mathbb{R}^m} \prod_{f\in F} e^{-f(\boldsymbol{x},\boldsymbol{u})}\boldsymbol{B}(a\boldsymbol{x}) \prod_{f\in B} e^{-f(\boldsymbol{x},\boldsymbol{u})}\,\mathrm{d}^m\boldsymbol{x}\\ &\overset{a\boldsymbol{x}=\boldsymbol{y}}{=} \int_{\mathbb{R}^m} \prod_{f\in F} e^{-f(\frac{\boldsymbol{y}}{a},\boldsymbol{u})}\boldsymbol{B}(\boldsymbol{y}) \prod_{f\in B} e^{-f(\frac{\boldsymbol{y}}{a},\boldsymbol{u})}|a|^{-m}\,\mathrm{d}^m\boldsymbol{y}\\ &\overset{f\text{ bilin.}}{=} |a|^{-m}\int_{\mathbb{R}^m} \prod_{f\in F} e^{-f(\boldsymbol{y},\frac{\boldsymbol{u}}{a})}\boldsymbol{B}(\boldsymbol{y}) \prod_{f\in B} e^{-f(\boldsymbol{y},\frac{\boldsymbol{u}}{a})}\,\mathrm{d}^m\boldsymbol{y} \qquad (4.10)\\ &= |a|^{-m}\mathcal{F}_{F_1,F_2}(\boldsymbol{B})\left(\frac{\boldsymbol{u}}{a}\right).\end{aligned}$$

□

5. Products with Invertible Factors

To obtain properties of the GFT like linearity with respect to arbitrary multivectors or a shift theorem we will have to change the order of multivectors and products of exponentials. Since the geometric product usually is neither commutative nor anticommutative this is not trivial. In this section we provide useful Lemmata that allow a swap if at least one of the factors is invertible. For more information see [14] and [17].

Remark 5.1. Every multiple of a square root of -1, $i \in \mathscr{I}^{p,q}$ is invertible, since from $i^2 = -r, r \in \mathbb{R}\setminus\{0\}$ follows $i^{-1} = -i/r$. Because of that, for all $\boldsymbol{u}, \boldsymbol{x} \in \mathbb{R}^m$ a function $f_k(\boldsymbol{x},\boldsymbol{u}) : \mathbb{R}^m \times \mathbb{R}^m \to \mathscr{I}^{p,q}$ is pointwise invertible.

Definition 5.2. For an invertible multivector $\boldsymbol{B} \in C\ell_{p,q}$ and an arbitrary multivector $\boldsymbol{A} \in C\ell_{p,q}$ we define

$$\begin{aligned}\boldsymbol{A}_{\boldsymbol{c}^0(\boldsymbol{B})} &= \frac{1}{2}(\boldsymbol{A} + \boldsymbol{B}^{-1}\boldsymbol{A}\boldsymbol{B}),\\ \boldsymbol{A}_{\boldsymbol{c}^1(\boldsymbol{B})} &= \frac{1}{2}(\boldsymbol{A} - \boldsymbol{B}^{-1}\boldsymbol{A}\boldsymbol{B}).\end{aligned} \tag{5.1}$$

Lemma 5.3. *Let $\boldsymbol{B} \in C\ell_{p,q}$ be invertible with the unique inverse $\boldsymbol{B}^{-1} = \bar{\boldsymbol{B}}/\boldsymbol{B}^2$, $\boldsymbol{B}^2 \in \mathbb{R}\setminus\{0\}$. Every multivector $\boldsymbol{A} \in C\ell_{p,q}$ can be expressed unambiguously by the sum of $\boldsymbol{A}_{\boldsymbol{c}^0(\boldsymbol{B})} \in C\ell_{p,q}$ that commutes and $\boldsymbol{A}_{\boldsymbol{c}^1(\boldsymbol{B})} \in C\ell_{p,q}$ that anticommutes with respect to $\boldsymbol{B}$. That means*

$$\begin{aligned}\boldsymbol{A} &= \boldsymbol{A}_{\boldsymbol{c}^0(\boldsymbol{B})} + \boldsymbol{A}_{\boldsymbol{c}^1(\boldsymbol{B})},\\ \boldsymbol{A}_{\boldsymbol{c}^0(\boldsymbol{B})}\boldsymbol{B} &= \boldsymbol{B}\boldsymbol{A}_{\boldsymbol{c}^0(\boldsymbol{B})},\\ \boldsymbol{A}_{\boldsymbol{c}^1(\boldsymbol{B})}\boldsymbol{B} &= -\boldsymbol{B}\boldsymbol{A}_{\boldsymbol{c}^1(\boldsymbol{B})}.\end{aligned} \tag{5.2}$$

Proof. We will only prove the assertion for $\boldsymbol{A}_{\boldsymbol{c}^0(\boldsymbol{B})}$.

Existence: With Definition 5.2 we get

$$\begin{aligned}\boldsymbol{A}_{\boldsymbol{c}^0(\boldsymbol{B})} + \boldsymbol{A}_{\boldsymbol{c}^1(\boldsymbol{B})} &= \frac{1}{2}(\boldsymbol{A} + \boldsymbol{B}^{-1}\boldsymbol{A}\boldsymbol{B} + \boldsymbol{A} - \boldsymbol{B}^{-1}\boldsymbol{A}\boldsymbol{B}) \\ &= \boldsymbol{A}\end{aligned} \tag{5.3}$$

and considering

$$\boldsymbol{B}^{-1}\boldsymbol{A}\boldsymbol{B} = \frac{\bar{\boldsymbol{B}}\boldsymbol{A}\boldsymbol{B}}{\boldsymbol{B}^2} = \boldsymbol{B}\boldsymbol{A}\boldsymbol{B}^{-1} \tag{5.4}$$

we also get

$$\begin{aligned}\boldsymbol{A}_{\boldsymbol{c}^0(\boldsymbol{B})}\boldsymbol{B} &= \frac{1}{2}(\boldsymbol{A} + \boldsymbol{B}^{-1}\boldsymbol{A}\boldsymbol{B})\boldsymbol{B} \\ &= \frac{1}{2}(\boldsymbol{A} + \boldsymbol{B}\boldsymbol{A}\boldsymbol{B}^{-1})\boldsymbol{B} \\ &= \frac{1}{2}(\boldsymbol{A}\boldsymbol{B} + \boldsymbol{B}\boldsymbol{A}) \\ &= \boldsymbol{B}\frac{1}{2}(\boldsymbol{B}^{-1}\boldsymbol{A}\boldsymbol{B} + \boldsymbol{A}) \\ &= \boldsymbol{B}\boldsymbol{A}_{\boldsymbol{c}^0(\boldsymbol{B})}\end{aligned} \tag{5.5}$$

Uniqueness: From the first claim in (5.2) we get

$$\boldsymbol{A}_{\boldsymbol{c}^1(\boldsymbol{B})} = \boldsymbol{A} - \boldsymbol{A}_{\boldsymbol{c}^0(\boldsymbol{B})}, \tag{5.6}$$

together with the third one this leads to

$$\begin{aligned}(\boldsymbol{A} - \boldsymbol{A}_{\boldsymbol{c}^0(\boldsymbol{B})})\boldsymbol{B} &= -\boldsymbol{B}(\boldsymbol{A} - \boldsymbol{A}_{\boldsymbol{c}^0(\boldsymbol{B})}) \\ \boldsymbol{A}\boldsymbol{B} - \boldsymbol{A}_{\boldsymbol{c}^0(\boldsymbol{B})}\boldsymbol{B} &= -\boldsymbol{B}\boldsymbol{A} + \boldsymbol{B}\boldsymbol{A}_{\boldsymbol{c}^0(\boldsymbol{B})} \\ \boldsymbol{A}\boldsymbol{B} + \boldsymbol{B}\boldsymbol{A} &= \boldsymbol{A}_{\boldsymbol{c}^0(\boldsymbol{B})}\boldsymbol{B} + \boldsymbol{B}\boldsymbol{A}_{\boldsymbol{c}^0(\boldsymbol{B})}\end{aligned} \tag{5.7}$$

and from the second claim finally follows

$$\begin{aligned}\boldsymbol{A}\boldsymbol{B} + \boldsymbol{B}\boldsymbol{A} &= 2\boldsymbol{B}\boldsymbol{A}_{\boldsymbol{c}^0(\boldsymbol{B})} \\ \frac{1}{2}(\boldsymbol{B}^{-1}\boldsymbol{A}\boldsymbol{B} + \boldsymbol{A}) &= \boldsymbol{A}_{\boldsymbol{c}^0(\boldsymbol{B})}.\end{aligned} \tag{5.8}$$

The derivation of the expression for $\boldsymbol{A}_{\boldsymbol{c}^1(\boldsymbol{B})}$ works analogously. □

Corollary 5.4 (Decomposition w.r.t. commutativity). *Let $\boldsymbol{B} \in C\ell_{p,q}$ be invertible, then $\forall \boldsymbol{A} \in C\ell_{p,q}$*

$$\boldsymbol{B}\boldsymbol{A} = (\boldsymbol{A}_{\boldsymbol{c}^0(\boldsymbol{B})} - \boldsymbol{A}_{\boldsymbol{c}^1(\boldsymbol{B})})\boldsymbol{B}. \tag{5.9}$$

Definition 5.5. For $d \in \mathbb{N}, \boldsymbol{A} \in C\ell_{p,q}$, the ordered set $B = \{\boldsymbol{B}_1, \dots, \boldsymbol{B}_d\}$ of invertible multivectors and any multi-index $\boldsymbol{j} \in \{0,1\}^d$ we define

$$\begin{aligned}\boldsymbol{A}_{\boldsymbol{c}^{\boldsymbol{j}}(\overrightarrow{B})} &:= ((\boldsymbol{A}_{\boldsymbol{c}^{j_1}(\boldsymbol{B}_1)})_{\boldsymbol{c}^{j_2}(\boldsymbol{B}_2)} \cdots)_{\boldsymbol{c}^{j_d}(\boldsymbol{B}_d)}, \\ \boldsymbol{A}_{\boldsymbol{c}^{\boldsymbol{j}}(\overleftarrow{B})} &:= ((\boldsymbol{A}_{\boldsymbol{c}^{j_d}(\boldsymbol{B}_d)})_{\boldsymbol{c}^{j_{d-1}}(\boldsymbol{B}_{d-1})} \cdots)_{\boldsymbol{c}^{j_1}(\boldsymbol{B}_1)}\end{aligned} \tag{5.10}$$

recursively with $\boldsymbol{c}^0, \boldsymbol{c}^1$ as in Definition 5.2.

Example. Let $\boldsymbol{A} = a_0 + a_1\boldsymbol{e}_1 + a_2\boldsymbol{e}_2 + a_{12}\boldsymbol{e}_{12} \in \mathcal{G}^{2,0}$ then, for example

$$\begin{aligned}\boldsymbol{A}_{\boldsymbol{c}^0(\boldsymbol{e}_1)} &= \frac{1}{2}(\boldsymbol{A} + \boldsymbol{e}_1^{-1}\boldsymbol{A}\boldsymbol{e}_1)\\ &= \frac{1}{2}(\boldsymbol{A} + a_0 + a_1\boldsymbol{e}_1 - a_2\boldsymbol{e}_2 - a_{12}\boldsymbol{e}_{12})\\ &= a_0 + a_1\boldsymbol{e}_1\end{aligned} \tag{5.11}$$

and further

$$\begin{aligned}\boldsymbol{A}_{\boldsymbol{c}^{0,0}(\overrightarrow{\boldsymbol{e}_1,\boldsymbol{e}_2})} &= (\boldsymbol{A}_{\boldsymbol{c}^0(\boldsymbol{e}_1)})_{\boldsymbol{c}^0(\boldsymbol{e}_2)}\\ &= (a_0 + a_1\boldsymbol{e}_1)_{\boldsymbol{c}^0(\boldsymbol{e}_2)} = a_0.\end{aligned} \tag{5.12}$$

The computation of the other multi-indices with $d = 2$ works analogously and therefore

$$\begin{aligned}\boldsymbol{A} &= \sum_{\boldsymbol{j}\in\{0,1\}^d} \boldsymbol{A}_{\boldsymbol{c}^{\boldsymbol{j}}(\boldsymbol{e}_1,\boldsymbol{e}_2)}\\ &= \boldsymbol{A}_{\boldsymbol{c}^{00}(\overrightarrow{\boldsymbol{e}_1,\boldsymbol{e}_2})} + \boldsymbol{A}_{\boldsymbol{c}^{01}(\overrightarrow{\boldsymbol{e}_1,\boldsymbol{e}_2})} + \boldsymbol{A}_{\boldsymbol{c}^{10}(\overrightarrow{\boldsymbol{e}_1,\boldsymbol{e}_2})} + \boldsymbol{A}_{\boldsymbol{c}^{11}(\overrightarrow{\boldsymbol{e}_1,\boldsymbol{e}_2})}\\ &= a_0 + a_1\boldsymbol{e}_1 + a_2\boldsymbol{e}_2 + a_{12}\boldsymbol{e}_{12}.\end{aligned} \tag{5.13}$$

Lemma 5.6. *Let $d \in \mathbb{N}, B = \{\boldsymbol{B}_1, \dots, \boldsymbol{B}_d\}$ be invertible multivectors and for $\boldsymbol{j} \in \{0,1\}^d$ let $|\boldsymbol{j}| := \sum_{k=1}^d j_k$, then $\forall \boldsymbol{A} \in C\ell_{p,q}$*

$$\begin{aligned}\boldsymbol{A} &= \sum_{\boldsymbol{j}\in\{0,1\}^d} \boldsymbol{A}_{\boldsymbol{c}^{\boldsymbol{j}}(\overrightarrow{B})},\\ \boldsymbol{A}\boldsymbol{B}_1\dots\boldsymbol{B}_d &= \boldsymbol{B}_1\dots\boldsymbol{B}_d \sum_{\boldsymbol{j}\in\{0,1\}^d} (-1)^{|\boldsymbol{j}|}\boldsymbol{A}_{\boldsymbol{c}^{\boldsymbol{j}}(\overrightarrow{B})},\\ \boldsymbol{B}_1\dots\boldsymbol{B}_d\boldsymbol{A} &= \sum_{\boldsymbol{j}\in\{0,1\}^d} (-1)^{|\boldsymbol{j}|}\boldsymbol{A}_{\boldsymbol{c}^{\boldsymbol{j}}(\overleftarrow{B})}\boldsymbol{B}_1\dots\boldsymbol{B}_d.\end{aligned} \tag{5.14}$$

Proof. Apply Lemma 5.3 repeatedly. □

Remark 5.7. The distinction of the two directions can be omitted using the equality

$$\boldsymbol{A}_{\boldsymbol{c}^{\boldsymbol{j}}(\overrightarrow{\boldsymbol{B}_1,\dots,\boldsymbol{B}_d})} = \boldsymbol{A}_{\boldsymbol{c}^{\boldsymbol{j}}(\overleftarrow{\boldsymbol{B}_d,\dots,\boldsymbol{B}_1})}. \tag{5.15}$$

We established it for the sake of notational brevity and will not formulate nor prove every assertion for both directions.

Lemma 5.8. *Let $F = \{f_1(\boldsymbol{x},\boldsymbol{u}), \dots, f_d(\boldsymbol{x},\boldsymbol{u})\}$ be a set of pointwise invertible functions then the ordered product of their exponentials and an arbitrary multivector $\boldsymbol{A} \in C\ell_{p,q}$ satisfies*

$$\prod_{k=1}^d e^{-f_k(\boldsymbol{x},\boldsymbol{u})}\boldsymbol{A} = \sum_{\boldsymbol{j}\in\{0,1\}^d} \boldsymbol{A}_{\boldsymbol{c}^{\boldsymbol{j}}(\overleftarrow{F})}(\boldsymbol{x},\boldsymbol{u}) \prod_{k=1}^d e^{-(-1)^{j_k} f_k(\boldsymbol{x},\boldsymbol{u})}, \tag{5.16}$$

where $A_{\boldsymbol{c}^{\boldsymbol{j}}(\overleftarrow{F})}(\boldsymbol{x},\boldsymbol{u}) := A_{\boldsymbol{c}^{\boldsymbol{j}}(\overleftarrow{F(\boldsymbol{x},\boldsymbol{u})})}$ is a multivector-valued function $\mathbb{R}^m \times \mathbb{R}^m \to C\ell_{p,q}$.

Proof. For all $\boldsymbol{x}, \boldsymbol{u} \in \mathbb{R}^m$ the commutation properties of $f_k(\boldsymbol{x}, \boldsymbol{u})$ dictate the properties of $e^{-f_k(\boldsymbol{x},\boldsymbol{u})}$ by

$$\begin{aligned} e^{-f_k(\boldsymbol{x},\boldsymbol{u})}\boldsymbol{A} &\overset{\text{Def. 2.1}}{=} \sum_{l=0}^{\infty} \frac{(-f_k(\boldsymbol{x},\boldsymbol{u}))^l}{l!}\boldsymbol{A} \\ &\overset{\text{Lem. 5.3}}{=} \sum_{l=0}^{\infty} \frac{(-f_k(\boldsymbol{x},\boldsymbol{u}))^l}{l!}\left(\boldsymbol{A}_{\boldsymbol{c}^0(f_k(\boldsymbol{x},\boldsymbol{u}))} + \boldsymbol{A}_{\boldsymbol{c}^1(f_k(\boldsymbol{x},\boldsymbol{u}))}\right). \end{aligned} \tag{5.17}$$

The shape of this decomposition of $\boldsymbol{A}$ may depend on $\boldsymbol{x}$ and $\boldsymbol{u}$. To stress this fact we will interpret $\boldsymbol{A}_{\boldsymbol{c}^0(f_k(\boldsymbol{x},\boldsymbol{u}))}$ as a multivector function and write $\boldsymbol{A}_{\boldsymbol{c}^0(f_k)}(\boldsymbol{x},\boldsymbol{u})$. According to Lemma 5.3 we can move $\boldsymbol{A}_{\boldsymbol{c}^0(f_k)}(\boldsymbol{x},\boldsymbol{u})$ through all factors, because it commutes. Analogously swapping $\boldsymbol{A}_{\boldsymbol{c}^1(f_k)}(\boldsymbol{x},\boldsymbol{u})$ will change the sign of each factor because it anticommutes. Hence we get

$$\begin{aligned} &= \boldsymbol{A}_{\boldsymbol{c}^0(f_k)}(\boldsymbol{x},\boldsymbol{u})\sum_{l=0}^{\infty}\frac{(-f_k(\boldsymbol{x},\boldsymbol{u}))^l}{l!} + \boldsymbol{A}_{\boldsymbol{c}^1(f_k)}(\boldsymbol{x},\boldsymbol{u})\sum_{l=0}^{\infty}\frac{(f_k(\boldsymbol{x},\boldsymbol{u}))^l}{l!} \\ &= \boldsymbol{A}_{\boldsymbol{c}^0(f_k)}(\boldsymbol{x},\boldsymbol{u})e^{-f_k(\boldsymbol{x},\boldsymbol{u})} + \boldsymbol{A}_{\boldsymbol{c}^1(f_k)}(\boldsymbol{x},\boldsymbol{u})e^{f_k(\boldsymbol{x},\boldsymbol{u})}. \end{aligned} \tag{5.18}$$

Applying this repeatedly to the product we can deduce

$$\begin{aligned} \prod_{k=1}^{d} e^{-f_k(\boldsymbol{x},\boldsymbol{u})}\boldsymbol{A} &= \prod_{k=1}^{d-1} e^{-f_k(\boldsymbol{x},\boldsymbol{u})}\begin{pmatrix} \boldsymbol{A}_{\boldsymbol{c}^0(f_d)}(\boldsymbol{x},\boldsymbol{u})e^{-f_d(\boldsymbol{x},\boldsymbol{u})} \\ +\boldsymbol{A}_{\boldsymbol{c}^1(f_d)}(\boldsymbol{x},\boldsymbol{u})e^{f_d(\boldsymbol{x},\boldsymbol{u})} \end{pmatrix} \\ &= \prod_{k=1}^{d-2} e^{-f_k(\boldsymbol{x},\boldsymbol{u})}\begin{pmatrix} \boldsymbol{A}_{\boldsymbol{c}^{0,0}(\overleftarrow{f_{d-1},f_d})}(\boldsymbol{x},\boldsymbol{u})e^{-f_{d-1}(\boldsymbol{x},\boldsymbol{u})}e^{-f_d(\boldsymbol{x},\boldsymbol{u})} \\ +\boldsymbol{A}_{\boldsymbol{c}^{1,0}(\overleftarrow{f_{d-1},f_d})}(\boldsymbol{x},\boldsymbol{u})e^{f_{d-1}(\boldsymbol{x},\boldsymbol{u})}e^{-f_d(\boldsymbol{x},\boldsymbol{u})} \\ +\boldsymbol{A}_{\boldsymbol{c}^{0,1}(\overleftarrow{f_{d-1},f_d})}(\boldsymbol{x},\boldsymbol{u})e^{-f_{d-1}(\boldsymbol{x},\boldsymbol{u})}e^{f_d(\boldsymbol{x},\boldsymbol{u})} \\ +\boldsymbol{A}_{\boldsymbol{c}^{1,1}(\overleftarrow{f_{d-1},f_d})}(\boldsymbol{x},\boldsymbol{u})e^{f_{d-1}(\boldsymbol{x},\boldsymbol{u})}e^{f_d(\boldsymbol{x},\boldsymbol{u})} \end{pmatrix} \\ &\quad\vdots \qquad\qquad \vdots \qquad\qquad \vdots \\ &= \sum_{\boldsymbol{j}\in\{0,1\}^d} \boldsymbol{A}_{\boldsymbol{c}^{\boldsymbol{j}}(\overleftarrow{F})}(\boldsymbol{x},\boldsymbol{u})\prod_{k=1}^{d} e^{-(-1)^{j_k}f_k(\boldsymbol{x},\boldsymbol{u})}. \end{aligned} \tag{5.19}$$

□

6. Separable GFT

From now on we want to restrict ourselves to an important group of geometric Fourier transforms whose square roots of -1 are independent from the first argument.

Definition 6.1. We call a GFT **left (right) separable**, if

$$f_l = |f_l(\boldsymbol{x},\boldsymbol{u})|\, i_l(\boldsymbol{u}), \tag{6.1}$$

$\forall l = 1,\dots,\mu$, $(l = \mu+1,\dots,\nu)$, where $|f_l(\boldsymbol{x},\boldsymbol{u})| : \mathbb{R}^m \times \mathbb{R}^m \to \mathbb{R}$ is a real function and $i_l : \mathbb{R}^m \to \mathscr{I}^{p,q}$ a function that does not depend on $\boldsymbol{x}$.

Example. The first five transforms from the introductory example are separable, while the cylindrical transform (vi) can not be expressed as in (6.1) except for the two-dimensional case.

We have seen in the proof of Lemma 5.8 that the decomposition of a constant multivector $\boldsymbol{A}$ with respect to a product of exponentials generally results in multivector-valued functions $\boldsymbol{A}_{\boldsymbol{c}^j(F)}(\boldsymbol{x},\boldsymbol{u})$ of $\boldsymbol{x}$ and $\boldsymbol{u}$. Separability guarantees independence from $\boldsymbol{x}$ and therefore allows separation from the integral.

Corollary 6.2 (Decomposition independent from $\boldsymbol{x}$). *Consider a set of functions $F = \{f_1(\boldsymbol{x},\boldsymbol{u}),\dots,f_d(\boldsymbol{x},\boldsymbol{u})\}$ satisfying condition (6.1) then the ordered product of their exponentials and an arbitrary multivector $\boldsymbol{A} \in C\ell_{p,q}$ satisfies*

$$\prod_{k=1}^{d} e^{-f_k(\boldsymbol{x},\boldsymbol{u})}\boldsymbol{A} = \sum_{\boldsymbol{j}\in\{0,1\}^d} \boldsymbol{A}_{\boldsymbol{c}^j(\overleftarrow{F})}(\boldsymbol{u}) \prod_{k=1}^{d} e^{-(-1)^{j_k} f_k(\boldsymbol{x},\boldsymbol{u})}. \tag{6.2}$$

Remark 6.3. If a GFT can be expressed as in 6.1 but with multiples of square roots of -1, $i_k \in \mathscr{I}^{p,q}$, which are independent from $\boldsymbol{x}$ and $\boldsymbol{u}$, the parts $\boldsymbol{A}_{\boldsymbol{c}^j(\overleftarrow{F})}$ of $\boldsymbol{A}$ will be constants. Note that the first five GFTs from the reference example satisfy this stronger condition, too.

Definition 6.4. For a set of functions $F = \{f_1(\boldsymbol{x},\boldsymbol{u}),\dots,f_d(\boldsymbol{x},\boldsymbol{u})\}$ and a multi-index $\boldsymbol{j} \in \{0,1\}^d$, we define the set of functions $F(\boldsymbol{j})$ by

$$F(\boldsymbol{j}) := \left\{(-1)^{j_1} f_1(\boldsymbol{x},\boldsymbol{u}),\dots,(-1)^{j_d} f_d(\boldsymbol{x},\boldsymbol{u})\right\}. \tag{6.3}$$

Theorem 6.5 (Left and right products). *Let $\boldsymbol{C} \in C\ell_{p,q}$ and $\boldsymbol{A},\boldsymbol{B} : \mathbb{R}^m \to C\ell_{p,q}$ be two multivector fields with $\boldsymbol{A}(\boldsymbol{x}) = \boldsymbol{C}\boldsymbol{B}(\boldsymbol{x})$ then a left separable geometric Fourier transform obeys*

$$\mathcal{F}_{F_1,F_2}(\boldsymbol{A})(\boldsymbol{u}) = \sum_{\boldsymbol{j}\in\{0,1\}^\mu} \boldsymbol{C}_{\boldsymbol{c}^j(\overleftarrow{F_1})}(\boldsymbol{u})\mathcal{F}_{F_1(\boldsymbol{j}),F_2}(\boldsymbol{B})(\boldsymbol{u}). \tag{6.4}$$

If $\boldsymbol{A}(\boldsymbol{x}) = \boldsymbol{B}(\boldsymbol{x})\boldsymbol{C}$ we analogously get

$$\mathcal{F}_{F_1,F_2}(\boldsymbol{A})(\boldsymbol{u}) = \sum_{\boldsymbol{k}\in\{0,1\}^{(\nu-\mu)}} \mathcal{F}_{F_1,F_2(\boldsymbol{k})}(\boldsymbol{B})(\boldsymbol{u})\boldsymbol{C}_{\boldsymbol{c}^k(\overrightarrow{F_2})}(\boldsymbol{u}) \tag{6.5}$$

for a right separable GFT.

Proof. We restrict ourselves to the proof of the first assertion.

$$\begin{aligned}\mathcal{F}_{F_1,F_2}(\boldsymbol{A})(\boldsymbol{u}) &= \int_{\mathbb{R}^m} \prod_{f\in F_1} e^{-f(\boldsymbol{x},\boldsymbol{u})}\boldsymbol{C}\boldsymbol{B}(\boldsymbol{x}) \prod_{f\in F_2} e^{-f(\boldsymbol{x},\boldsymbol{u})}\,\mathrm{d}^m\boldsymbol{x}\\ &\overset{\text{Lem. 5.8}}{=} \int_{\mathbb{R}^m} \left(\sum_{\boldsymbol{j}\in\{0,1\}^\mu} \boldsymbol{C}_{\boldsymbol{c}^j(\overleftarrow{F_1})}(\boldsymbol{u}) \prod_{l=1}^{\mu} e^{-(-1)^{j_l} f_l(\boldsymbol{x},\boldsymbol{u})}\right)\\ &\qquad\qquad \boldsymbol{B}(\boldsymbol{x}) \prod_{f\in F_2} e^{-f(\boldsymbol{x},\boldsymbol{u})}\,\mathrm{d}^m\boldsymbol{x}\end{aligned}$$

$$= \sum_{j\in\{0,1\}^\mu} \boldsymbol{C}_{\boldsymbol{c}^j(\overleftarrow{F_1})}(\boldsymbol{u}) \int_{\mathbb{R}^m} \prod_{l=1}^{\mu} e^{-(-1)^{j_l} f_l(\boldsymbol{x},\boldsymbol{u})}$$

$$\boldsymbol{B}(\boldsymbol{x}) \prod_{f\in F_2} e^{-f(\boldsymbol{x},\boldsymbol{u})}\,\mathrm{d}^m\boldsymbol{x}$$

$$= \sum_{j\in\{0,1\}^\mu} \boldsymbol{C}_{\boldsymbol{c}^j(\overleftarrow{F_1})}(\boldsymbol{u}) \mathcal{F}_{F_1(\boldsymbol{j}),F_2}(\boldsymbol{B})(\boldsymbol{u}).$$

The second one follows in the same way. □

Corollary 6.6 (Uniform constants). *Let the claims from Theorem 6.5 hold. If the constant $\boldsymbol{C}$ satisfies $\boldsymbol{C} = \boldsymbol{C}_{\boldsymbol{c}^j(\overleftarrow{F_1})}(\boldsymbol{u})$ for a multi-index $\boldsymbol{j} \in \{0,1\}^\mu$ then the theorem simplifies to*

$$\mathcal{F}_{F_1,F_2}(\boldsymbol{A})(\boldsymbol{u}) = \boldsymbol{C}\mathcal{F}_{F_1(\boldsymbol{j}),F_2}(\boldsymbol{B})(\boldsymbol{u}) \tag{6.6}$$

for $\boldsymbol{A}(\boldsymbol{x}) = \boldsymbol{C}\boldsymbol{B}(\boldsymbol{x})$ respectively

$$\mathcal{F}_{F_1,F_2}(\boldsymbol{A})(\boldsymbol{u}) = \mathcal{F}_{F_1,F_2(\boldsymbol{k})}(\boldsymbol{B})(\boldsymbol{u})\boldsymbol{C} \tag{6.7}$$

for $\boldsymbol{A}(\boldsymbol{x}) = \boldsymbol{B}(\boldsymbol{x})\boldsymbol{C}$ and $\boldsymbol{C} = \boldsymbol{C}_{\boldsymbol{c}^k(\overrightarrow{F_2})}(\boldsymbol{u})$ for a multi-index $\boldsymbol{k} \in \{0,1\}^{(\nu-\mu)}$.[2]

Corollary 6.7 (Left and right linearity). *The geometric Fourier transform is left (respectively right) linear if F_1 (respectively F_2) consists only of functions f_k with values in the center of $C\ell_{p,q}$, that means $\forall \boldsymbol{x},\boldsymbol{u} \in \mathbb{R}^m, \forall \boldsymbol{A} \in C\ell_{p,q} : \boldsymbol{A}f_k(\boldsymbol{x},\boldsymbol{u}) = f_k(\boldsymbol{x},\boldsymbol{u})\boldsymbol{A}$.*

Remark 6.8. Note that for empty sets F_1 (or F_2) necessarily all elements satisfy commutativity and therefore the condition in Corollary 6.7.

The different appearances of Theorem 6.5 are summarized in Table 1 and Table 2.

We have seen how to change the order of a multivector and a product of exponentials in the previous section. To get a shift theorem we will have to separate sums appearing in the exponent and sort the resulting exponentials with respect to the summands. Note that Corollary 6.2 can be applied in two ways here, because exponentials appear on both sides.

Not every factor will need to be swapped with every other. So, to keep things short, we will make use of the notation $\boldsymbol{c}^{(J)_l}(f_1,\ldots,f_l,0,\ldots,0)$ for $l \in \{1,\ldots,d\}$ instead of distinguishing between differently sized multi-indices for every l that appears. The zeros at the end substitutionally indicate real numbers. They commute with every multivector. That implies, that for the last $d-l$ factors no swap and therefore no separation needs to be made. It would also be possible to use the notation $\boldsymbol{c}^{(J)_l}(f_1,\ldots,f_{l-1},0,\ldots,0)$ for $l \in \{1,\ldots,d\}$, because every function commutes with itself. The choice we have made means that no exceptional treatment

[2]Corollary 6.6 follows directly from $(\boldsymbol{C}_{\boldsymbol{c}^j(\overleftarrow{F_1})})_{\boldsymbol{c}^k(\overleftarrow{F_1})} = 0$ for all $\boldsymbol{k} \neq \boldsymbol{j}$ because no non-zero component of $\boldsymbol{C}$ can commute and anticommute with respect to a function in F_1.

Table 1. Theorem 6.5 (Left products) applied to the GFTs of the first example enumerated in the same order.
Notations: on the LHS $\mathcal{F}_{F_1,F_2} = \mathcal{F}_{F_1,F_2}(\boldsymbol{A})(\boldsymbol{u})$, on the RHS $\mathcal{F}_{F_1',F_2'} = \mathcal{F}_{F_1',F_2'}(\boldsymbol{B})(\boldsymbol{u})$

	GFT	$\boldsymbol{A}(\boldsymbol{x}) = \boldsymbol{C}\boldsymbol{B}(\boldsymbol{x})$
1.	Clifford	$\mathcal{F}_{f_1} = \boldsymbol{C}\mathcal{F}_{f_1}$
2.	Bülow	$\mathcal{F}_{f_1,\dots,f_n} = \boldsymbol{C}\mathcal{F}_{f_1,\dots,f_n}$
3.	Quaternionic	$\mathcal{F}_{f_1,f_2} = \boldsymbol{C}_{\boldsymbol{c}^0(i)}\mathcal{F}_{f_1,f_2} + \boldsymbol{C}_{\boldsymbol{c}^1(i)}\mathcal{F}_{-f_1,f_2}$
4.	Spacetime	$\mathcal{F}_{f_1,f_2} = \boldsymbol{C}_{\boldsymbol{c}^0(\boldsymbol{e}_4)}\mathcal{F}_{f_1,f_2} + \boldsymbol{C}_{\boldsymbol{c}^1(\boldsymbol{e}_4)}\mathcal{F}_{-f_1,f_2}$
5.	Colour Image	$\mathcal{F}_{f_1,f_2,f_3,f_4} = \boldsymbol{C}_{\boldsymbol{c}^{00}(\overleftarrow{\boldsymbol{B},i\boldsymbol{B}})}\mathcal{F}_{f_1,f_2,f_3,f_4}$
		$+\boldsymbol{C}_{\boldsymbol{c}^{10}(\overleftarrow{\boldsymbol{B},i\boldsymbol{B}})}\mathcal{F}_{-f_1,f_2,f_3,f_4}$
		$+\boldsymbol{C}_{\boldsymbol{c}^{01}(\overleftarrow{\boldsymbol{B},i\boldsymbol{B}})}\mathcal{F}_{f_1,-f_2,f_3,f_4}$
		$+\boldsymbol{C}_{\boldsymbol{c}^{11}(\overleftarrow{\boldsymbol{B},i\boldsymbol{B}})}\mathcal{F}_{-f_1,-f_2,f_3,f_4}$
6.	Cylindrical $n = 2$	$\mathcal{F}_{f_1} = \boldsymbol{C}_{\boldsymbol{c}^0(\boldsymbol{e}_{12})}\mathcal{F}_{f_1} + \boldsymbol{C}_{\boldsymbol{c}^1(\boldsymbol{e}_{12})}\mathcal{F}_{-f_1}$
	Cylindrical $n \neq 2$	-

of f_1 is necessary. But please note that the multivectors $(J)_l$ indicating the commutative and anticommutative parts will all have zeros from l to d and therefore form a strictly triangular matrix.

Lemma 6.9. *Let a set of functions $F = \{f_1(\boldsymbol{x},\boldsymbol{u}),\dots,f_d(\boldsymbol{x},\boldsymbol{u})\}$ fulfil* (6.1) *and be linear with respect to $\boldsymbol{x}$. Further let $J \in \{0,1\}^{d\times d}$ be a strictly lower triangular matrix, that is associated column by column with a multi-index $\boldsymbol{j} \in \{0,1\}^d$ by $\forall k = 1,\dots,d : (\sum_{l=1}^d J_{l,k}) \bmod 2 = j_k$, with $(J)_l$ being its lth row, then*

$$\prod_{l=1}^{d} e^{-f_l(\boldsymbol{x}+\boldsymbol{y},\boldsymbol{u})} = \sum_{\boldsymbol{j}\in\{0,1\}^d} \sum_{\substack{J\in\{0,1\}^{d\times d},\\ \sum_{l=1}^d (J)_l \bmod 2 = \boldsymbol{j}}} \prod_{l=1}^{d} e^{-f_l(\boldsymbol{x},\boldsymbol{u})}_{\boldsymbol{c}^{(J)_l}(\overleftarrow{f_1,\dots,f_l,0,\dots,0})} \prod_{l=1}^{d} e^{-(-1)^{j_l} f_l(\boldsymbol{y},\boldsymbol{u})} \tag{6.8}$$

or alternatively with strictly upper triangular matrices J:

$$\prod_{l=1}^{d} e^{-f_l(\boldsymbol{x}+\boldsymbol{y},\boldsymbol{u})} = \sum_{\boldsymbol{j}\in\{0,1\}^d} \sum_{\substack{J\in\{0,1\}^{d\times d},\\ \sum_{l=1}^d (J)_l \bmod 2 = \boldsymbol{j}}} \prod_{l=1}^{d} e^{-(-1)^{j_l} f_l(\boldsymbol{x},\boldsymbol{u})} \prod_{l=1}^{d} e^{-f_l(\boldsymbol{y},\boldsymbol{u})}_{\boldsymbol{c}^{(J)_l}(\overrightarrow{0,\dots,0,f_l,\dots,f_d})}. \tag{6.9}$$

We do not explicitly indicate the dependence of the partition on $\boldsymbol{u}$ as in Corollary 6.2, because the functions in the exponents already contain this dependence. Please note that the decomposition is pointwise.

TABLE 2. Theorem 6.5 (Right products) applied to the GFTs of the first example, enumerated in the same order.
Notations: on the LHS $\mathcal{F}_{F_1,F_2} = \mathcal{F}_{F_1,F_2}(\boldsymbol{A})(\boldsymbol{u})$, on the RHS $\mathcal{F}_{F_1',F_2'} = \mathcal{F}_{F_1',F_2'}(\boldsymbol{B})(\boldsymbol{u})$

	GFT	$\boldsymbol{A}(\boldsymbol{x}) = \boldsymbol{B}(\boldsymbol{x})\boldsymbol{C}$
1.	Clif. $n = 2 \pmod 4$	$\mathcal{F}_{f_1} = \mathcal{F}_{f_1}\boldsymbol{C}_{\boldsymbol{c}^0(i)} + \mathcal{F}_{-f_1}\boldsymbol{C}_{\boldsymbol{c}^1(i)}$
	Clif. $n = 3 \pmod 4$	$\mathcal{F}_{f_1} = \mathcal{F}_{f_1}\boldsymbol{C}$
2.	Bülow	$\mathcal{F}_{f_1,\dots,f_n} = \sum_{\boldsymbol{k}\in\{0,1\}^n} \mathcal{F}_{(-1)^{k_1}f_1,\dots,(-1)^{k_n}f_n}\boldsymbol{C}_{\boldsymbol{c}^{\boldsymbol{k}}(\overrightarrow{f_1,\dots,f_n})}$
3.	Quaternionic	$\mathcal{F}_{f_1,f_2} = \mathcal{F}_{f_1,f_2}\boldsymbol{C}_{\boldsymbol{c}^0(j)} + \mathcal{F}_{f_1,-f_2}\boldsymbol{C}_{\boldsymbol{c}^1(j)}$
4.	Spacetime	$\mathcal{F}_{f_1,f_2} = \mathcal{F}_{f_1,f_2}\boldsymbol{C}_{\boldsymbol{c}^0(\boldsymbol{e}_4 i_4)} + \mathcal{F}_{f_1,-f_2}\boldsymbol{C}_{\boldsymbol{c}^1(\boldsymbol{e}_4 i_4)}$
5.	Colour Image	$\mathcal{F}_{f_1,f_2,f_3,f_4} = \mathcal{F}_{f_1,f_2,f_3,f_4}\boldsymbol{C}_{\boldsymbol{c}^{00}(\overrightarrow{\boldsymbol{B},i\boldsymbol{B}})} + \mathcal{F}_{f_1,f_2,-f_3,f_4}\boldsymbol{C}_{\boldsymbol{c}^{10}(\overrightarrow{\boldsymbol{B},i\boldsymbol{B}})} + \mathcal{F}_{f_1,f_2,f_3,-f_4}\boldsymbol{C}_{\boldsymbol{c}^{01}(\overrightarrow{\boldsymbol{B},i\boldsymbol{B}})} + \mathcal{F}_{f_1,f_2,-f_3,-f_4}\boldsymbol{C}_{\boldsymbol{c}^{11}(\overrightarrow{\boldsymbol{B},i\boldsymbol{B}})}$
6.	Cylindrical	$\mathcal{F}_{f_1} = \mathcal{F}_{f_1}\boldsymbol{C}$

Proof. We will only prove the first assertion. The second one follows analogously by applying Corollary 6.2 the other way around.

$$\prod_{l=1}^{d} e^{-f_l(\boldsymbol{x}+\boldsymbol{y},\boldsymbol{u})} \overset{F\text{ lin.}}{=} \prod_{l=1}^{d} e^{-f_l(\boldsymbol{x},\boldsymbol{u})-f_l(\boldsymbol{y},\boldsymbol{u})} \tag{6.10}$$

$$\overset{\text{Lem. 2.2}}{=} \prod_{l=1}^{d} e^{-f_l(\boldsymbol{x},\boldsymbol{u})} e^{-f_l(\boldsymbol{y},\boldsymbol{u})} \tag{6.11}$$

$$= e^{-f_1(\boldsymbol{x},\boldsymbol{u})} e^{-f_1(\boldsymbol{y},\boldsymbol{u})} \prod_{l=2}^{d} e^{-f_l(\boldsymbol{x},\boldsymbol{u})} e^{-f_l(\boldsymbol{y},\boldsymbol{u})} \tag{6.12}$$

$$\overset{\text{Cor. 6.2}}{=} e^{-f_1(\boldsymbol{x},\boldsymbol{u})} \big(e^{-f_2(\boldsymbol{x},\boldsymbol{u})}_{\boldsymbol{c}^0(f_1)} e^{-f_1(\boldsymbol{y},\boldsymbol{u})} e^{-f_2(\boldsymbol{y},\boldsymbol{u})} + e^{-f_2(\boldsymbol{x},\boldsymbol{u})}_{\boldsymbol{c}^1(f_1)} e^{f_1(\boldsymbol{y},\boldsymbol{u})} e^{-f_2(\boldsymbol{y},\boldsymbol{u})} \big) \prod_{l=3}^{d} e^{-f_l(\boldsymbol{x},\boldsymbol{u})} e^{-f_l(\boldsymbol{y},\boldsymbol{u})}. \tag{6.13}$$

Now we use Corollary 6.2 to step by step rearrange the order of the product.

$$\overset{\text{Cor. 6.2}}{=} e^{-f_1(\boldsymbol{x},\boldsymbol{u})} \Big(e^{-f_2(\boldsymbol{x},\boldsymbol{u})}_{\boldsymbol{c}^0(f_1)} e^{-f_3(\boldsymbol{x},\boldsymbol{u})}_{\boldsymbol{c}^{00}(\overleftarrow{f_1,f_2})} e^{-f_1(\boldsymbol{y},\boldsymbol{u})} e^{-f_2(\boldsymbol{y},\boldsymbol{u})} e^{-f_3(\boldsymbol{y},\boldsymbol{u})}$$

$$
\begin{aligned}
&+ e^{-f_2(\boldsymbol{x},\boldsymbol{u})}_{\boldsymbol{c}^0(f_1)} e^{-f_3(\boldsymbol{x},\boldsymbol{u})}_{\boldsymbol{c}^{01}(\overleftarrow{f_1,f_2})} e^{-f_1(\boldsymbol{y},\boldsymbol{u})} e^{f_2(\boldsymbol{y},\boldsymbol{u})} e^{-f_3(\boldsymbol{y},\boldsymbol{u})} \\
&+ e^{-f_2(\boldsymbol{x},\boldsymbol{u})}_{\boldsymbol{c}^0(f_1)} e^{-f_3(\boldsymbol{x},\boldsymbol{u})}_{\boldsymbol{c}^{10}(\overleftarrow{f_1,f_2})} e^{f_1(\boldsymbol{y},\boldsymbol{u})} e^{-f_2(\boldsymbol{y},\boldsymbol{u})} e^{-f_3(\boldsymbol{y},\boldsymbol{u})} \\
&+ e^{-f_2(\boldsymbol{x},\boldsymbol{u})}_{\boldsymbol{c}^0(f_1)} e^{-f_3(\boldsymbol{x},\boldsymbol{u})}_{\boldsymbol{c}^{11}(\overleftarrow{f_1,f_2})} e^{f_1(\boldsymbol{y},\boldsymbol{u})} e^{f_2(\boldsymbol{y},\boldsymbol{u})} e^{-f_3(\boldsymbol{y},\boldsymbol{u})} \\
&+ e^{-f_2(\boldsymbol{x},\boldsymbol{u})}_{\boldsymbol{c}^1(f_1)} e^{-f_3(\boldsymbol{x},\boldsymbol{u})}_{\boldsymbol{c}^{00}(\overleftarrow{f_1,f_2})} e^{f_1(\boldsymbol{y},\boldsymbol{u})} e^{-f_2(\boldsymbol{y},\boldsymbol{u})} e^{-f_3(\boldsymbol{y},\boldsymbol{u})} \\
&+ e^{-f_2(\boldsymbol{x},\boldsymbol{u})}_{\boldsymbol{c}^1(f_1)} e^{-f_3(\boldsymbol{x},\boldsymbol{u})}_{\boldsymbol{c}^{01}(\overleftarrow{f_1,f_2})} e^{f_1(\boldsymbol{y},\boldsymbol{u})} e^{f_2(\boldsymbol{y},\boldsymbol{u})} e^{-f_3(\boldsymbol{y},\boldsymbol{u})} \\
&+ e^{-f_2(\boldsymbol{x},\boldsymbol{u})}_{\boldsymbol{c}^1(f_1)} e^{-f_3(\boldsymbol{x},\boldsymbol{u})}_{\boldsymbol{c}^{10}(\overleftarrow{f_1,f_2})} e^{-f_1(\boldsymbol{y},\boldsymbol{u})} e^{-f_2(\boldsymbol{y},\boldsymbol{u})} e^{-f_3(\boldsymbol{y},\boldsymbol{u})} \\
&+ e^{-f_2(\boldsymbol{x},\boldsymbol{u})}_{\boldsymbol{c}^1(f_1)} e^{-f_3(\boldsymbol{x},\boldsymbol{u})}_{\boldsymbol{c}^{11}(\overleftarrow{f_1,f_2})} e^{-f_1(\boldsymbol{y},\boldsymbol{u})} e^{f_2(\boldsymbol{y},\boldsymbol{u})} e^{-f_3(\boldsymbol{y},\boldsymbol{u})} \Bigg) \\
&\prod_{l=4}^{d} e^{-f_l(\boldsymbol{x},\boldsymbol{u})} e^{-f_l(\boldsymbol{y},\boldsymbol{u})}. \qquad (6.14)
\end{aligned}
$$

There are only 2^δ ways of distributing the signs of δ exponents, so some of the summands can be combined.

$$
\begin{aligned}
= e^{-f_1(\boldsymbol{x},\boldsymbol{u})} \Bigg(&\left(e^{-f_2(\boldsymbol{x},\boldsymbol{u})}_{\boldsymbol{c}^0(f_1)} e^{-f_3(\boldsymbol{x},\boldsymbol{u})}_{\boldsymbol{c}^{00}(\overleftarrow{f_1,f_2})} + e^{-f_2(\boldsymbol{x},\boldsymbol{u})}_{\boldsymbol{c}^1(f_1)} e^{-f_3(\boldsymbol{x},\boldsymbol{u})}_{\boldsymbol{c}^{10}(\overleftarrow{f_1,f_2})} \right) \\
&e^{-f_1(\boldsymbol{y},\boldsymbol{u})} e^{-f_2(\boldsymbol{y},\boldsymbol{u})} e^{-f_3(\boldsymbol{y},\boldsymbol{u})} \\
+ &\left(e^{-f_2(\boldsymbol{x},\boldsymbol{u})}_{\boldsymbol{c}^0(f_1)} e^{-f_3(\boldsymbol{x},\boldsymbol{u})}_{\boldsymbol{c}^{01}(\overleftarrow{f_1,f_2})} + e^{-f_2(\boldsymbol{x},\boldsymbol{u})}_{\boldsymbol{c}^1(f_1)} e^{-f_3(\boldsymbol{x},\boldsymbol{u})}_{\boldsymbol{c}^{11}(\overleftarrow{f_1,f_2})} \right) \\
&e^{-f_1(\boldsymbol{y},\boldsymbol{u})} e^{f_2(\boldsymbol{y},\boldsymbol{u})} e^{-f_3(\boldsymbol{y},\boldsymbol{u})} \\
+ &\left(e^{-f_2(\boldsymbol{x},\boldsymbol{u})}_{\boldsymbol{c}^0(f_1)} e^{-f_3(\boldsymbol{x},\boldsymbol{u})}_{\boldsymbol{c}^{10}(\overleftarrow{f_1,f_2})} + e^{-f_2(\boldsymbol{x},\boldsymbol{u})}_{\boldsymbol{c}^1(f_1)} e^{-f_3(\boldsymbol{x},\boldsymbol{u})}_{\boldsymbol{c}^{00}(\overleftarrow{f_1,f_2})} \right) \\
&e^{f_1(\boldsymbol{y},\boldsymbol{u})} e^{-f_2(\boldsymbol{y},\boldsymbol{u})} e^{-f_3(\boldsymbol{y},\boldsymbol{u})} \\
+ &\left(e^{-f_2(\boldsymbol{x},\boldsymbol{u})}_{\boldsymbol{c}^0(f_1)} e^{-f_3(\boldsymbol{x},\boldsymbol{u})}_{\boldsymbol{c}^{11}(\overleftarrow{f_1,f_2})} + e^{-f_2(\boldsymbol{x},\boldsymbol{u})}_{\boldsymbol{c}^1(f_1)} e^{-f_3(\boldsymbol{x},\boldsymbol{u})}_{\boldsymbol{c}^{01}(\overleftarrow{f_1,f_2})} \right) e^{f_1(\boldsymbol{y},\boldsymbol{u})} \\
&e^{f_2(\boldsymbol{y},\boldsymbol{u})} e^{-f_3(\boldsymbol{y},\boldsymbol{u})} \Bigg) \prod_{l=4}^{d} e^{-f_l(\boldsymbol{x},\boldsymbol{u})} e^{-f_l(\boldsymbol{y},\boldsymbol{u})}. \qquad (6.15)
\end{aligned}
$$

To get a compact notation we expand all multi-indices by adding zeros until they have the same length. Note that the last non-zero argument in terms like $\boldsymbol{c}^{000}(\overleftarrow{f_1,0,0})$ always coincides with the exponent of the corresponding factor. Because of that it will always commute and could also be replaced by a zero.

$$
\begin{aligned}
= &e^{-f_1(\boldsymbol{x},\boldsymbol{u})}_{\boldsymbol{c}^{000}(\overleftarrow{f_1,0,0})} \\
&\Bigg(\left(e^{-f_2(\boldsymbol{x},\boldsymbol{u})}_{\boldsymbol{c}^{000}(\overleftarrow{f_1,f_2,0})} e^{-f_3(\boldsymbol{x},\boldsymbol{u})}_{\boldsymbol{c}^{000}(\overleftarrow{f_1,f_2,f_3})} + e^{-f_2(\boldsymbol{x},\boldsymbol{u})}_{\boldsymbol{c}^{100}(\overleftarrow{f_1,f_2,0})} e^{-f_3(\boldsymbol{x},\boldsymbol{u})}_{\boldsymbol{c}^{100}(\overleftarrow{f_1,f_2,f_3})} \right) \\
&e^{-f_1(\boldsymbol{y},\boldsymbol{u})} e^{-f_2(\boldsymbol{y},\boldsymbol{u})} e^{-f_3(\boldsymbol{y},\boldsymbol{u})}
\end{aligned}
$$

$$+\left(e^{-f_2(\boldsymbol{x},\boldsymbol{u})}_{\boldsymbol{c}^{000}(\overleftarrow{f_1,f_2,0})}e^{-f_3(\boldsymbol{x},\boldsymbol{u})}_{\boldsymbol{c}^{010}(\overleftarrow{f_1,f_2,f_3})}+e^{-f_2(\boldsymbol{x},\boldsymbol{u})}_{\boldsymbol{c}^{100}(\overleftarrow{f_1,f_2,0})}e^{-f_3(\boldsymbol{x},\boldsymbol{u})}_{\boldsymbol{c}^{110}(\overleftarrow{f_1,f_2,f_3})}\right)$$
$$e^{-f_1(\boldsymbol{y},\boldsymbol{u})}e^{f_2(\boldsymbol{y},\boldsymbol{u})}e^{-f_3(\boldsymbol{y},\boldsymbol{u})}$$
$$+\left(e^{-f_2(\boldsymbol{x},\boldsymbol{u})}_{\boldsymbol{c}^{000}(\overleftarrow{f_1,f_2,0})}e^{-f_3(\boldsymbol{x},\boldsymbol{u})}_{\boldsymbol{c}^{100}(\overleftarrow{f_1,f_2,f_3})}+e^{-f_2(\boldsymbol{x},\boldsymbol{u})}_{\boldsymbol{c}^{100}(\overleftarrow{f_1,f_2,0})}e^{-f_3(\boldsymbol{x},\boldsymbol{u})}_{\boldsymbol{c}^{000}(\overleftarrow{f_1,f_2,f_3})}\right)$$
$$e^{f_1(\boldsymbol{y},\boldsymbol{u})}e^{-f_2(\boldsymbol{y},\boldsymbol{u})}e^{-f_3(\boldsymbol{y},\boldsymbol{u})}$$
$$+\left(e^{-f_2(\boldsymbol{x},\boldsymbol{u})}_{\boldsymbol{c}^{000}(\overleftarrow{f_1,f_2,0})}e^{-f_3(\boldsymbol{x},\boldsymbol{u})}_{\boldsymbol{c}^{110}(\overleftarrow{f_1,f_2,f_3})}+e^{-f_2(\boldsymbol{x},\boldsymbol{u})}_{\boldsymbol{c}^{100}(\overleftarrow{f_1,f_2,0})}e^{f_3(\boldsymbol{x},\boldsymbol{u})}_{\boldsymbol{c}^{010}(\overleftarrow{f_1,f_2,f_3})}\right)$$
$$\left.e^{f_1(\boldsymbol{y},\boldsymbol{u})}e^{f_2(\boldsymbol{y},\boldsymbol{u})}e^{-f_3(\boldsymbol{y},\boldsymbol{u})}\right)\prod_{l=4}^{d}e^{-f_l(\boldsymbol{x},\boldsymbol{u})}e^{-f_l(\boldsymbol{y},\boldsymbol{u})} \tag{6.16}$$

For $\delta = 3$ we look at all strictly lower triangular matrices $J \in \{0,1\}^{\delta\times\delta}$ with the property

$$\forall k = 1,\ldots,\delta : \left(\sum_{l=1}^{\delta}(J)_{l,k}\right) \bmod 2 = j_k. \tag{6.17}$$

That means the lth row $(J)_l$ of J contains a multi-index $(J)_l \in \{0,1\}^\delta$, with the last $\delta - l - 1$ entries being zero and the kth column sum being even when $j_k = 0$ and odd when $j_k = 1$. For example, the first multi-index is $\boldsymbol{j} = (0,0,0)$. There are only two different strictly lower triangular matrices that have columns summing up to even numbers:

$$J = \begin{pmatrix} 0 & 0 & 0 \\ 0 & 0 & 0 \\ 0 & 0 & 0 \end{pmatrix} \text{ and } J = \begin{pmatrix} 0 & 0 & 0 \\ 1 & 0 & 0 \\ 1 & 0 & 0 \end{pmatrix}. \tag{6.18}$$

The first row of each contains the multi-index that belongs to $e^{-f_1(\boldsymbol{x},\boldsymbol{u})}$, the second one belongs to $e^{-f_2(\boldsymbol{x},\boldsymbol{u})}$ and so on. So the summands with exactly these multi-indices are the ones assigned to the product of exponentials whose signs are invariant during the reordering. With this notation and all $J \in \{0,1\}^{3\times 3}$ that satisfy the property (6.17) we can write

$$\prod_{l=1}^{d}e^{-f_l(\boldsymbol{x}+\boldsymbol{y},\boldsymbol{u})} = \sum_{\boldsymbol{j}\in\{0,1\}^3}\sum_J\prod_{l=1}^{3}e^{-f_l(\boldsymbol{x},\boldsymbol{u})}_{\boldsymbol{c}^{(J)_l}(\overleftarrow{f_1,\ldots,f_l,0,\ldots,0})}\prod_{l=1}^{3}e^{-(-1)^{j_l}f_l(\boldsymbol{y},\boldsymbol{u})}$$
$$\prod_{l=4}^{d}e^{-f_l(\boldsymbol{x},\boldsymbol{u})}e^{-f_l(\boldsymbol{y},\boldsymbol{u})}. \tag{6.19}$$

Using mathematical induction with matrices $J \in \{0,1\}^{\delta\times\delta}$ as introduced above for growing δ and Corollary 6.2 repeatedly until we reach $\delta = d$ we get

$$= \sum_{\boldsymbol{j}\in\{0,1\}^d}\sum_J\prod_{l=1}^{d}e^{-f_l(\boldsymbol{x},\boldsymbol{u})}_{\boldsymbol{c}^{(J)_l}(\overleftarrow{f_1,\ldots,f_l,0,\ldots,0})}\prod_{l=1}^{d}e^{-(-1)^{j_l}f_l(\boldsymbol{y},\boldsymbol{u})}. \tag{6.20}$$

□

Remark 6.10. The number of summands actually appearing is usually much smaller than in Theorem 6.11. It is determined by the number of distinct strictly lower (upper) triangular matrices J with entries being either zero or one, namely:

$$2^{\frac{d(d-1)}{2}}. \tag{6.21}$$

Theorem 6.11 (Shift). *Let $\boldsymbol{A}(\boldsymbol{x}) = \boldsymbol{B}(\boldsymbol{x}-\boldsymbol{x}_0)$ be multivector fields, F_1, F_2, linear with respect to $\boldsymbol{x}$, and let $\boldsymbol{j} \in \{0,1\}^{\mu}, \boldsymbol{k} \in \{0,1\}^{(\nu-\mu)}$ be multi-indices, and $F_1(\boldsymbol{j}), F_2(\boldsymbol{k})$ be as introduced in Definition* 6.4*, then a separable GFT suffices*

$$\begin{aligned}\mathcal{F}_{F_1,F_2}(\boldsymbol{A})(\boldsymbol{u}) = &\sum_{\boldsymbol{j},\boldsymbol{k}}\sum_{J,K}\prod_{l=1}^{\mu} e^{-f_l(\boldsymbol{x}_0,\boldsymbol{u})}_{\boldsymbol{c}^{(J)_l}(\overleftarrow{f_1,\dots,f_l,0,\dots,0})}\mathcal{F}_{F_1(\boldsymbol{j}),F_2(\boldsymbol{k})}(\boldsymbol{B})(\boldsymbol{u})\\ &\prod_{l=\mu+1}^{\nu} e^{-f_l(\boldsymbol{x}_0,\boldsymbol{u})}_{\boldsymbol{c}^{(K)_{l-\mu}}(\overrightarrow{0,\dots,0,f_l,\dots,f_\nu})},\end{aligned} \tag{6.22}$$

where $J \in \{0,1\}^{\mu\times\mu}$ and $K \in \{0,1\}^{(\nu-\mu)\times(\nu-\mu)}$ are the strictly lower, respectively upper, triangular matrices with rows $(J)_l, (K)_{l-\mu}$ summing up to $(\sum_{l=1}^{\mu}(J)_l)$ mod $2 = \boldsymbol{j}$ respectively $\left(\sum_{l=\mu+1}^{\nu}(K)_{l-\mu}\right)$ mod $2 = \boldsymbol{k}$ as in Lemma 6.9.

Proof. First we rewrite the transformed function in terms of $\boldsymbol{B}(\boldsymbol{y})$ using a change of coordinates.

$$\begin{aligned}\mathcal{F}_{F_1,F_2}(\boldsymbol{A})(\boldsymbol{u}) &= \int_{\mathbb{R}^m}\prod_{l=1}^{\mu} e^{-f_l(\boldsymbol{x},\boldsymbol{u})}\boldsymbol{A}(\boldsymbol{x})\prod_{l=\mu+1}^{\nu} e^{-f_l(\boldsymbol{x},\boldsymbol{u})}\,\mathrm{d}^m\boldsymbol{x}\\ &= \int_{\mathbb{R}^m}\prod_{l=1}^{\mu} e^{-f_l(\boldsymbol{x},\boldsymbol{u})}\boldsymbol{B}(\boldsymbol{x}-\boldsymbol{x}_0)\prod_{l=\mu+1}^{\nu} e^{-f_l(\boldsymbol{x},\boldsymbol{u})}\,\mathrm{d}^m\boldsymbol{x}\\ &\overset{\boldsymbol{y}=\boldsymbol{x}-\boldsymbol{x}_0}{=} \int_{\mathbb{R}^m}\prod_{l=1}^{\mu} e^{-f_l(\boldsymbol{y}+\boldsymbol{x}_0,\boldsymbol{u})}\boldsymbol{B}(\boldsymbol{y})\prod_{l=\mu+1}^{\nu} e^{-f_l(\boldsymbol{y}+\boldsymbol{x}_0,\boldsymbol{u})}\,\mathrm{d}^m\boldsymbol{y}\end{aligned} \tag{6.23}$$

Now we separate and sort the factors using Lemma 6.9.

$$\begin{aligned}\overset{\text{Lem. 6.9}}{=} \int_{\mathbb{R}^m}&\sum_{\boldsymbol{j}\in\{0,1\}^{\mu}}\ \sum_{\substack{J\in\{0,1\}^{\mu\times\mu}\\ \sum(J)_l \bmod 2=\boldsymbol{j}}}\\ &\prod_{l=1}^{\mu} e^{-f_l(\boldsymbol{x}_0,\boldsymbol{u})}_{\boldsymbol{c}^{(J)_l}(\overleftarrow{f_1,\dots,f_l,0,\dots,0})}\prod_{l=1}^{\mu} e^{-(-1)^{j_l}f_l(\boldsymbol{y},\boldsymbol{u})}\boldsymbol{B}(\boldsymbol{y})\\ &\sum_{\boldsymbol{k}\in\{0,1\}^{(\nu-\mu)}}\ \sum_{\substack{K\in\{0,1\}^{(\nu-\mu)\times(\nu-\mu)}\\ \sum(K)_l \bmod 2=\boldsymbol{k}}}\\ &\prod_{l=\mu+1}^{\nu} e^{-(-1)^{k_{l-\mu}}f_l(\boldsymbol{y},\boldsymbol{u})}\prod_{l=\mu+1}^{\nu} e^{-f_l(\boldsymbol{x}_0,\boldsymbol{u})}_{\boldsymbol{c}^{(K)_{l-\mu}}(\overrightarrow{0,\dots,0,f_l,\dots,f_\nu})}\,\mathrm{d}^m\boldsymbol{y}\end{aligned}$$

$$
\begin{aligned}
&= \sum_{\boldsymbol{j},\boldsymbol{k}} \sum_{J,K} \prod_{l=1}^{\mu} e^{-f_l(\boldsymbol{x}_0,\boldsymbol{u})}_{\mathbf{c}^{(J)_l}(\overleftarrow{f_1,\ldots,f_l,0,\ldots,0})} \qquad (6.24)\\
&\quad \mathcal{F}_{F_1(\boldsymbol{j}),F_2(\boldsymbol{k})}(\boldsymbol{B})(\boldsymbol{u}) \prod_{l=\mu+1}^{\nu} e^{-f_l(\boldsymbol{x}_0,\boldsymbol{u})}_{\mathbf{c}^{(K)_{l-\mu}}(\overrightarrow{0,\ldots,0,f_l,\ldots,f_\nu})} \qquad \square
\end{aligned}
$$

Corollary 6.12 (Shift). *Let $\boldsymbol{A}(\boldsymbol{x}) = \boldsymbol{B}(\boldsymbol{x} - \boldsymbol{x}_0)$ be multivector fields, F_1 and F_2 each consisting of mutually commutative functions*[3] *being linear with respect to $\boldsymbol{x}$, then the GFT obeys*

$$
\mathcal{F}_{F_1,F_2}(\boldsymbol{A})(\boldsymbol{u}) = \prod_{l=1}^{\mu} e^{-f_l(\boldsymbol{x}_0,\boldsymbol{u})} \mathcal{F}_{F_1,F_2}(\boldsymbol{B})(\boldsymbol{u}) \prod_{l=\mu+1}^{\nu} e^{-f_l(\boldsymbol{x}_0,\boldsymbol{u})}. \qquad (6.25)
$$

Remark 6.13. For sets F_1, F_2 that each consist of less than two functions the condition of Corollary 6.12 is necessarily satisfied, compare, *e.g.*, the Clifford–Fourier transform, the quaternionic transform or the spacetime Fourier transform listed in the preceeding examples.

The specific forms taken by our standard examples are summarized in Table 3. As expected they are often shorter than what could be expected from Remark 6.10.

7. Conclusions and Outlook

For multivector fields over $\mathbb{R}^{p,q}$ with values in any geometric algebra $G^{p,q}$ we have successfully defined a general geometric Fourier transform. It covers all popular Fourier transforms from current literature in the introductory example. Its existence, independent of the specific choice of functions F_1, F_2, can be proved for all integrable multivector fields, see Theorem 3.1. Theorem 3.2 shows that our geometric Fourier transform is generally linear over the field of real numbers. All transforms from the reference example consist of bilinear F_1 and F_2. We proved that this property is sufficient to ensure the scaling property of Theorem 4.1.

If a general geometric Fourier transform is separable as introduced in Definition 6.1, then Theorem 6.5 (Left and right products) guarantees that constant factors can be separated from the vector field to be transformed. As a consequence general linearity is achieved by choosing F_1, F_2 with values in the centre of the geometric algebra $C\ell_{p,q}$, compare Corollary 6.7. All examples except for the cylindrical Fourier transform [6] satisfy this claim.

Under the condition of linearity with respect to the first argument of the functions of the sets F_1 and F_2 additionally to the separability property just mentioned, we have also proved a shift property (Theorem 6.11).

In future publications we are going to state the necessary constraints for a generalized convolution theorem, invertibility, derivation theorem and we will examine how simplifications can be achieved based on symmetry properties of the

[3]Cross commutativity between F_1 and F_2 is not necessary.

TABLE 3. Theorem 6.11 (Shift) applied to the GFTs of the first example, enumerated in the same order.
Notations: on the LHS $\mathcal{F}_{F_1,F_2} = \mathcal{F}_{F_1,F_2}(\boldsymbol{A})(\boldsymbol{u})$, on the RHS $\mathcal{F}_{F_1',F_2'} = \mathcal{F}_{F_1',F_2'}(\boldsymbol{B})(\boldsymbol{u})$. In the second row K represents all strictly upper triangular matrices $\in \{0,1\}^{n\times n}$ with rows $(K)_{l-\mu}$ summing up to $\left(\sum_{l=\mu+1}^{\nu}(K)_{l-\mu}\right) \bmod 2 = \boldsymbol{k}$. The simplified shape of the colour image FT results from the commutativity of $\boldsymbol{B}$ and $i\boldsymbol{B}$ and application of Lemma 2.2.

	GFT	$\boldsymbol{A}(\boldsymbol{x}) = \boldsymbol{B}(\boldsymbol{x} - \boldsymbol{x}_0)$
1.	Clifford	$\mathcal{F}_{f_1} = \mathcal{F}_{f_1} e^{-2\pi i \boldsymbol{x}_0 \cdot \boldsymbol{u}}$
2.	Bülow	$\mathcal{F}_{f_1,\dots,f_n} = \sum_{\boldsymbol{k}\in\{0,1\}^n} \sum_K \mathcal{F}_{(-1)^{k_1} f_1,\dots,(-1)^{k_n} f_n} \prod_{l=1}^{n} e^{-2\pi x_{0k} u_k}_{\overrightarrow{\boldsymbol{c}^{(K)_l}(0,\dots,0,f_l,\dots,f_n)}}$
3.	Quaternionic	$\mathcal{F}_{f_1,f_2} = e^{-2\pi i x_{01} u_1} \mathcal{F}_{f_1,f_2} e^{-2\pi j x_{02} u_2}$
4.	Spacetime	$\mathcal{F}_{f_1,f_2} = e^{-\boldsymbol{e}_4 x_{04} u_4} \mathcal{F}_{f_1,f_2} e^{-\epsilon_4 \boldsymbol{e}_4 i_4 (x_1 u_1 + x_2 u_2 + x_3 u_3)}$
5.	Colour Image	$\mathcal{F}_{f_1,f_2,f_3,f_4} = e^{-\frac{1}{2}(x_{01}u_1 + x_{02}u_2)(\boldsymbol{B}+i\boldsymbol{B})} \mathcal{F}_{f_1,f_2,f_3,f_4} e^{\frac{1}{2}(x_{01}u_1 + x_{02}u_2)(\boldsymbol{B}+i\boldsymbol{B})}$
6.	Cyl. $n = 2$	$\mathcal{F}_{f_1} = e^{\boldsymbol{x}_0 \wedge \boldsymbol{u}} \mathcal{F}_{f_1}$
	Cyl. $n \neq 2$	-

multivector fields to be transformed. We will also construct generalized geometric Fourier transforms in a broad sense from combinations of the ones introduced in this chapter and from decomposition into their sine and cosine parts which will also cover the vector and bivector Fourier transforms of [9]. It would further be of interest to extend our approach to Fourier transforms defined on spheres or other non-Euclidean manifolds, to functions in the Schwartz space and to square-integrable functions.

References

[1] T. Batard, M. Berthier, and C. Saint-Jean. Clifford Fourier transform for color image processing. In Bayro-Corrochano and Scheuermann [2], pages 135–162.

[2] E.J. Bayro-Corrochano and G. Scheuermann, editors. *Geometric Algebra Computing in Engineering and Computer Science.* Springer, London, 2010.

[3] F. Brackx, N. De Schepper, and F. Sommen. The Clifford–Fourier transform. *Journal of Fourier Analysis and Applications*, 11(6):669–681, 2005.

[4] F. Brackx, N. De Schepper, and F. Sommen. The two-dimensional Clifford–Fourier transform. *Journal of Mathematical Imaging and Vision*, 26(1):5–18, 2006.

[5] F. Brackx, N. De Schepper, and F. Sommen. The Clifford–Fourier integral kernel in even dimensional Euclidean space. *Journal of Mathematical Analysis and Applications*, 365(2):718–728, 2010.

[6] F. Brackx, N. De Schepper, and F. Sommen. The Cylindrical Fourier Transform. In Bayro-Corrochano and Scheuermann [2], pages 107–119.

[7] T. Bülow. *Hypercomplex Spectral Signal Representations for the Processing and Analysis of Images.* PhD thesis, University of Kiel, Germany, Institut für Informatik und Praktische Mathematik, Aug. 1999.

[8] W.K. Clifford. Applications of Grassmann's extensive algebra. *American Journal of Mathematics*, 1(4):350–358, 1878.

[9] H. De Bie and F. Sommen. Vector and bivector Fourier transforms in Clifford analysis. In K. Guerlebeck and C. Koenke, editors, *18th International Conference on the Application of Computer Science and Mathematics in Architecture and Civil Engineering*, page 11, 2009.

[10] J. Ebling. *Visualization and Analysis of Flow Fields using Clifford Convolution.* PhD thesis, University of Leipzig, Germany, 2006.

[11] T.A. Ell. Quaternion-Fourier transforms for analysis of 2-dimensional linear time-invariant partial-differential systems. In *Proceedings of the 32nd Conference on Decision and Control*, pages 1830–1841, San Antonio, Texas, USA, 15–17 December 1993. IEEE Control Systems Society.

[12] T.A. Ell and S.J. Sangwine. Hypercomplex Fourier transforms of color images. *IEEE Transactions on Image Processing*, 16(1):22–35, Jan. 2007.

[13] M. Felsberg. *Low-Level Image Processing with the Structure Multivector.* PhD thesis, Christian-Albrechts-Universität, Institut für Informatik und Praktische Mathematik, Kiel, 2002.

[14] D. Hestenes and G. Sobczyk. *Clifford Algebra to Geometric Calculus.* D. Reidel Publishing Group, Dordrecht, Netherlands, 1984.

[15] E. Hitzer. Quaternion Fourier transform on quaternion fields and generalizations. *Advances in Applied Clifford Algebras*, 17(3):497–517, May 2007.

[16] E. Hitzer and R. Abłamowicz. Geometric roots of -1 in Clifford algebras $C\ell_{p,q}$ with $p+q \leq 4$. *Advances in Applied Clifford Algebras*, 21(1):121–144, 2010. Published online 13 July 2010.

[17] E. Hitzer, J. Helmstetter, and R. Abłamowicz. Square roots of -1 in real Clifford algebras. In K. Gürlebeck, editor, *9th International Conference on Clifford Algebras and their Applications*, Weimar, Germany, 15–20 July 2011. 12 pp.

[18] E.M.S. Hitzer and B. Mawardi. Clifford Fourier transform on multivector fields and uncertainty principles for dimensions $n = 2(\text{mod}\,4)$ and $n = 3(\text{mod}\,4)$. *Advances in Applied Clifford Algebras*, 18(3-4):715–736, 2008.

[19] B. Jancewicz. Trivector Fourier transformation and electromagnetic field. *Journal of Mathematical Physics*, 31(8):1847–1852, 1990.

[20] S.J. Sangwine and T.A. Ell. The discrete Fourier transform of a colour image. In J.M. Blackledge and M.J. Turner, editors, *Image Processing II Mathematical Methods, Algorithms and Applications*, pages 430–441, Chichester, 2000. Horwood Publishing for Institute of Mathematics and its Applications. Proceedings Second IMA Conference on Image Processing, De Montfort University, Leicester, UK, September 1998.

[21] F. Sommen. Hypercomplex Fourier and Laplace transforms I. *Illinois Journal of Mathematics*, 26(2):332–352, 1982.

Roxana Bujack
Universität Leipzig
Institut für Informatik
Johannisgasse 26
D-04103 Leipzig, Germany
e-mail: bujack@informatik.uni-leipzig.de

Gerik Scheuermann
Universität Leipzig
Institut für Informatik
Johannisgasse 26
D-04103 Leipzig, Germany
e-mail: scheuermann@informatik.uni-leipzig.de

Eckhard Hitzer
College of Liberal Arts, Department of Material Science,
International Christian University,
181-8585 Tokyo, Japan
e-mail: hitzer@icu.ac.jp

Quaternion and Clifford–Fourier Transforms and Wavelets
Trends in Mathematics, 177–195

9 Clifford–Fourier Transform and Spinor Representation of Images

Thomas Batard and Michel Berthier

Abstract. We propose in this chapter to introduce a spinor representation for images based on the work of T. Friedrich. This spinor representation generalizes the usual Weierstrass representation of minimal surfaces (*i.e.*, surfaces with constant mean curvature equal to zero) to arbitrary surfaces (immersed in $\mathbb{R}^3$). We investigate applications to image processing focusing on segmentation and Clifford–Fourier analysis. All these applications involve sections of the spinor bundle of image graphs, that is spinor fields, satisfying the so-called Dirac equation.

Mathematics Subject Classification (2010). Primary 68U10, 53C27; secondary 53A05, 43A32.

Keywords. Image processing, spin geometry, Clifford–Fourier transform.

1. Introduction

The idea of this chapter is to perform grey-level image processing using the geometric information given by the Gauss map variations of image graphs. While it is well known that one can parameterize the Gauss map of a minimal surface by a meromorphic function (see below), it is a much more recent result (see [5]) that such a parametrization can be extended to arbitrary surfaces of $\mathbb{R}^3$ when dealing with spin geometry.

Let us first recall that a minimal surface Σ immersed in $\mathbb{R}^3$, that is a surface with constant mean curvature equal to zero, can be described with one holomorphic function φ and one meromorphic function ψ such that the product $\varphi\psi^2$ is holomorphic. This is the so-called Weierstrass representation of Σ (see [6] or [8] for details). The function ψ is nothing else but the composition of the Gauss map of Σ with the stereographic projection from the unit sphere to the complex plane.

This work was partially supported by ONR Grant N00014-09-1-0493.

The main result of T. Friedrich in [5] states that there is a one-to-one correspondance between spinor fields φ^* of constant length on a Riemannian surface (Σ, g) and satisfying

$$D\varphi^* = H\varphi^* \tag{1.1}$$

where D is a Dirac operator in one hand, and isometric immersions of Σ in $\mathbb{R}^3$ with mean curvature equal to H, on the other hand. The Weierstrass representation appears to be the particular case corresponding to $H \equiv 0$.

Let us describe now the method introduced in the following. Let

$$\begin{aligned} \chi : \Omega \subset \mathbb{R}^2 &\longrightarrow \mathbb{R}^3 \\ (x, y) &\longmapsto (x, y, I(x, y)) \end{aligned} \tag{1.2}$$

be the immersion in the three-dimensional Euclidean space of a grey-level image I defined on a domain Ω of $\mathbb{R}^2$. The first step (see § 2) consists in computing the spinor field φ^* that describes the image surface Σ. We follow here the paper of T. Friedrich [5]: φ^* is obtained from the restriction to the surface Σ of a parallel spinor ϕ on $\mathbb{R}^3$. The computation of φ^* requires us to deal with irreducible representations of the complex Clifford algebra $C\ell_{3,0} \otimes \mathbb{C}$ and with the generalized Weierstrass representation of Σ based on period forms. In practice, φ^* is given by a field of elements of $\mathbb{C}^2$.

As said before, the spinor field φ^* characterizes the geometry of the surface Σ immersed in $\mathbb{R}^3$ by the parametrization (1.2). In the same way that the normal of a minimal surface is parameterized by the meromorphic function ψ, the normal of the surface Σ is parameterized by the spinor field φ^*. The latter explains how the tangent plane to Σ varies in the ambient space.

There are many reasons to believe that such a generalized Weierstrass parametrization may reveal itself to be an efficient tool in the context of image processing:

1. The field φ^* of elements of $\mathbb{C}^2$ (see (2.26)) encodes the Riemannian structure of the surface Σ in a very tractable way (although the definition of φ^* may appear quite complicated).
2. The geometrical methods based on the study of the so-called structure tensor involve only the eigenvalues of the structure tensor, that means in some sense the values of the first fundamental form of the surface. The spinor field φ^* contains both intrinsic and extrinsic information. Studying the variations of φ^* allows us to get not only information about the variations (derivative) of the first fundamental form, but also about the geometric embedding of the surface Σ and in particular about the mean curvature.
3. We are dealing here with first-order instead of zero-order geometric variations of Σ. As shown later, this appears to be more relevant by taking into account both edges and textures.
4. As will be detailed in the sequel, the spinor field φ^* can be decomposed as a series of basic spinor fields using a suitable Clifford–Fourier transform. This series corresponds to a harmonic decomposition of the surface Σ adapted to

the Riemannian geometry. This is in fact the main novelty of this chapter since the usual techniques of Fourier analysis do not involve geometric data.
5. One can envisage the possibility of performing diffusion in this context. The usual Laplace Beltrami operator can be replaced by the squared Atiyah Singer Dirac operator [7] (the Atiyah Singer Dirac operator acting as an elliptic operator of order one on spinor fields).

To illustrate some of these ideas, we investigate rapidly in § 3 applications to segmentation and more precisely to edge and texture detection. As stated before, the basic idea is to replace the usual order-one structure tensor by an order-two structure tensor called the spinor tensor obtained from the derivative of the spinor field φ^*. This spinor tensor measures the variations of the unit normal of the image surface. Experiments show that this approach is particularly well adapted to texture detection.

We define in § 4 the Clifford–Fourier transform of a spinor field. For this, we follow the approach of [3] that relies on a spin generalization of the usual notion of group character. We are led to compute the group morphisms from $\mathbb{Z}/M\mathbb{Z} \times \mathbb{Z}/N\mathbb{Z}$ to Spin(3). Since this last group acts on the sections of the spinor bundle, a Clifford–Fourier transform can be defined by averaging this action. One of the key ideas here is to split the spinor bundle of the surface according to the Clifford multiplication by the bivector coding the tangent plane to the surface. This has two advantages: the first one is to involve the geometry in the process, the second one is to reduce the computation of the Clifford–Fourier transform to two usual complex Fourier transforms. It is important to notice that although the Fourier transform we propose is, as usual, a global transformation on the image, the way it is computed takes into account local geometric data. We finally introduce the harmonic decomposition mentioned above and show some results of filtering on standard images.

The reader will find in Appendix A the mathematical definitions and results used throughout the text.

2. Spinor Representation of Images

This section is devoted to the explicit computation of the spinor field φ^* of a given surface immersed in Euclidean space. It is obtained as the restriction of a constant spinor field of $\mathbb{R}^3$ the components of which are determined using period forms.

2.1. Spinors and Graphs

Let $I : \Omega \longrightarrow \mathbb{R}$ be a differentiable function defined on a domain Ω of $\mathbb{R}^2$. We consider the surface Σ immersed in $\mathbb{R}^3$ by the parametrization:

$$\chi(x, y) = (x, y, I(x, y)). \tag{2.1}$$

Also, let g be the metric on Σ induced by the Euclidean metric of $\mathbb{R}^3$. The couple (Σ, g) is a Riemannian surface of global chart (Ω, χ). We denote by M the Riemannian manifold $(\mathbb{R}^3, \| \ \|_2)$ and by (z_1, z_2, ν) an orthonormal frame field of

M with (z_1, z_2) an orthonormal frame field on Σ, and by ν the global unit field normal to Σ. One can choose (z_1, z_2, ν) with the following matrix representation

$$\begin{pmatrix} \dfrac{I_x}{\sqrt{(I_x^2+I_y^2)(I_x^2+I_y^2+1)}} & \dfrac{-I_y}{\sqrt{I_x^2+I_y^2}} & \dfrac{-I_x}{\sqrt{I_x^2+I_y^2+1}} \\ \dfrac{I_y}{\sqrt{(I_x^2+I_y^2)(I_x^2+I_y^2+1)}} & \dfrac{I_x}{\sqrt{I_x^2+I_y^2}} & \dfrac{-I_y}{\sqrt{I_x^2+I_y^2+1}} \\ \dfrac{I_x^2+I_y^2}{\sqrt{(I_x^2+I_y^2)(I_x^2+I_y^2+1)}} & 0 & \dfrac{1}{\sqrt{I_x^2+I_y^2+1}} \end{pmatrix}. \tag{2.2}$$

Note that z_1 and z_2 are not defined when $I_x = I_y = 0$. This has no consequence in the sequel since we deal only with the normal ν.

Following [5] the surface Σ can be represented by a spinor field φ^* with constant length satisfying the Dirac equation:

$$D\varphi^* = H\varphi^* \tag{2.3}$$

where H denotes the mean curvature of Σ. We recall here the basic idea (see Appendix A for notations and definitions). Let ϕ be a parallel spinor field of M, *i.e.*, satisfying

$$\nabla_X^M \phi = 0 \tag{2.4}$$

for all vector fields X on M. Let also φ be the restriction $\phi_{|\Sigma}$ of ϕ to Σ. The spinor field φ decomposes into

$$\varphi = \varphi^+ + \varphi^- \tag{2.5}$$

with

$$\varphi^+ = \frac{1}{2}(\varphi + i\nu\cdot\varphi) \qquad \varphi^- = \frac{1}{2}(\varphi - i\nu\cdot\varphi) \tag{2.6}$$

and satisfies

$$D\varphi = -H\cdot\nu\cdot\varphi. \tag{2.7}$$

This last equation reads

$$D(\varphi^+ + \varphi^-) = -H\cdot\nu\cdot(\varphi^+ + \varphi^-) \tag{2.8}$$

and implies

$$D\varphi^+ = -iH\varphi^- \qquad D\varphi^- = iH\varphi^+. \tag{2.9}$$

If we set $\varphi^* = \varphi^+ - i\varphi^-$ then $D\varphi^* = H\varphi^*$ and φ^* is of constant length.

Proposition 2.1. *The spinor fields φ^+, φ^- and φ^* are given by*

$$\varphi^+ = \frac{1}{2}\begin{pmatrix} \left(1 - \frac{I_y}{\sqrt{1+I_x^2+I_y^2}}\right) u + \left(\frac{I_x - i}{\sqrt{1+I_x^2+I_y^2}}\right) v \\ \left(1 + \frac{I_y}{\sqrt{1+I_x^2+I_y^2}}\right) v + \left(\frac{I_x + i}{\sqrt{1+I_x^2+I_y^2}}\right) u \end{pmatrix} \tag{2.10}$$

$$\varphi^- = \frac{1}{2}\begin{pmatrix} \left(1 + \frac{I_y}{\sqrt{1+I_x^2+I_y^2}}\right) u - \left(\frac{I_x - i}{\sqrt{1+I_x^2+I_y^2}}\right) v \\ \left(1 - \frac{I_y}{\sqrt{1+I_x^2+I_y^2}}\right) v - \left(\frac{I_x + i}{\sqrt{1+I_x^2+I_y^2}}\right) u \end{pmatrix} \tag{2.11}$$

and

$$\varphi^* = \frac{1}{2}(1-i)\begin{pmatrix} \left(1 - \frac{iI_y}{\sqrt{1+I_x^2+I_y^2}}\right) u + \left(\frac{1 + iI_x}{\sqrt{1+I_x^2+I_y^2}}\right) v \\ \left(1 + \frac{iI_y}{\sqrt{1+I_x^2+I_y^2}}\right) v + \left(\frac{iI_x - 1}{\sqrt{1+I_x^2+I_y^2}}\right) u \end{pmatrix} \tag{2.12}$$

where u and v are (constant) complex numbers.

Proof. Since ϕ is a parallel spinor field on M, $\phi = (u, v)$ where u and v are two (constant) complex numbers. Let ρ_2 be the irreducible complex representation of $\mathbb{C}l(3)$ described in Appendix A.1. Recall that

$$\nu = \frac{1}{\Delta}(-I_x e_1 - I_y e_2 + e_3) \tag{2.13}$$

where $\Delta = \sqrt{I_x^2 + I_y^2 + 1}$, so that

$$\rho_2(\nu) = -\frac{I_x}{\Delta}\begin{pmatrix} 0 & i \\ i & 0 \end{pmatrix} - \frac{I_y}{\Delta}\begin{pmatrix} -i & 0 \\ 0 & i \end{pmatrix} + \frac{1}{\Delta}\begin{pmatrix} 0 & -1 \\ 1 & 0 \end{pmatrix}. \tag{2.14}$$

By definition:

$$\nu \cdot \varphi = \rho_2(\nu)\begin{pmatrix} u \\ v \end{pmatrix}. \tag{2.15}$$

Simple computations lead now to the result. □

The next step consists in computing the components (u, v) of the constant field ϕ. This is done by considering a quaternionic structure on the spinor bundle $S(\Sigma)$ of the surface Σ and period forms.

2.2. Quaternionic Structure and Period Forms

Let I be the complex structure on $S(\Sigma)$ given by the multiplication by i. A quaternionic structure on $S(\Sigma)$ is a linear map J that satisfies $J^2 = -Id$ and $IJ = -JI$. In the sequel J is given by

$$J\begin{pmatrix} \varphi_1 \\ \varphi_2 \end{pmatrix} = \begin{pmatrix} -\overline{\varphi_2} \\ \overline{\varphi_1} \end{pmatrix}. \tag{2.16}$$

If we write $\varphi_1 = \alpha_1 + i\beta_1$ and $\varphi_2 = \alpha_2 + i\beta_2$, the corresponding quaternion is given by

$$\varphi_1 + \varphi_2 j = (\alpha_1 + i\beta_1) + (\alpha_2 + i\beta_2)j = \alpha_1 + i\beta_1 + \alpha_2 j + \beta_2 k \tag{2.17}$$

and

$$j(\varphi_1 + \varphi_2 j) = -\overline{\varphi_2} + \overline{\varphi_1} j, \tag{2.18}$$

i.e., J is the left multiplication by j. Since

$$S^+(\Sigma) = \left\{ \begin{pmatrix} \varphi_1 \\ \varphi_2 \end{pmatrix}, \ \varphi_1 = \frac{I_x - i}{I_y + \Delta}\varphi_2 \right\} \tag{2.19}$$

and

$$S^+(\Sigma) = \left\{ \begin{pmatrix} \varphi_1 \\ \varphi_2 \end{pmatrix}, \ \varphi_1 = \frac{I_x - i}{I_y - \Delta}\varphi_2 \right\} \tag{2.20}$$

then $JS^+(\Sigma) \subset S^-(\Sigma)$ and $JS^-(\Sigma) \subset S^+(\Sigma)$. We also denote by J the quaternionic structure (obtained in the same way) on $S(M)$.

Let us consider $\phi = (u, v)$ a constant spinor field on M and φ^* its restriction on Σ. Let also $f : \mathbb{R}^3 \longrightarrow \mathbb{R}$ and $g : \mathbb{R}^3 \longrightarrow \mathbb{C}$ be the functions defined by

$$f(m) = -\Im(m \cdot \phi, \phi) \tag{2.21}$$

and

$$g(m) = i(m \cdot \phi, J(\phi)) \tag{2.22}$$

where (,) denotes the Hermitian product. Using the representation ρ_2, one can check that

$$m \cdot \phi = \begin{pmatrix} -im_2 u + (im_1 - m_3)v \\ (im_1 + m_3)u + im_2 v \end{pmatrix} \tag{2.23}$$

for $m = (m_1, m_2, m_3)$. The equations $f(m) = m_1$ and $g(m) = m_2 + im_3$ are equivalent to:

$$|u|^2 = |v|^2, \quad u\overline{v} = -\frac{1}{2} \tag{2.24}$$

and

$$uv = -\frac{1}{2}, \quad u^2 + v^2 = 1, \quad u^2 = v^2. \tag{2.25}$$

This implies $u = \pm 1/\sqrt{2}$ and $v = -u$.

Definition 2.2. The spinor representation of the image given by the parametrization (2.1) is defined by

$$\varphi^* = \frac{1}{2\sqrt{2}}(1-i)\begin{pmatrix} \left(1 - \dfrac{1+i(\ I_x + I_y)}{\sqrt{1+I_x^2+I_y^2}}\right) \\ -\left(1 + \dfrac{1+i(-I_x + I_y)}{\sqrt{1+I_x^2+I_y^2}}\right) \end{pmatrix}. \tag{2.26}$$

This means that $u = 1/\sqrt{2}$ and $v = -1/\sqrt{2}$ in the expression (2.12).

The two 1-forms

$$\eta_f(X) = 2\,\Re(X \cdot (\varphi^*)^+, (\varphi^*)^-) = -\Im(X \cdot \varphi, \varphi) \tag{2.27}$$

$$\begin{aligned}\eta_g(X) &= i(X \cdot (\varphi^*)^+, J((\varphi^*)^+)) + i(X \cdot (\varphi^*)^-, J((\varphi^*)^-)) \\ &= i(X \cdot \varphi, J(\varphi))\end{aligned} \tag{2.28}$$

are exact and verify $d(f_{|\Sigma}) = \eta_f$, $d(g_{|\Sigma}) = \eta_g$. The generalized Weierstrass parametrization is actually given by the isometric immersion:

$$\int (\eta_f, \eta_g) : \Sigma \longrightarrow M. \tag{2.29}$$

2.3. Dirac Equation and Mean Curvature

We only mention here some results that can be used when dealing with diffusion. We do not go into further details since we will not treat this problem in the present chapter. Let (Σ, g) be an oriented two-dimensional Riemannian manifold and φ a spinor field without zeros solution of the Dirac equation $D\varphi = \lambda\varphi$. Then φ defines an isometric immersion

$$(\widetilde{\Sigma}, |\varphi|^4 g) \longrightarrow \mathbb{R}^3 \tag{2.30}$$

with mean curvature $H = \lambda/|\varphi|^2$ (see [5]).

3. Spinors and Segmentation

The aim of this section is to introduce the spinor tensor corresponding to the variations of the unit normal and to show its capability to detect both edges and textures.

3.1. The Spinor Tensor

We propose here to deal with a second-order version of the classical approach of edge detection based on the so-called structure tensor (see [10]). Instead of measuring edges from eigenvalues of the Riemannian metric, we focus here on the eigenvalues of the tensor obtained from the derivative of the spinor field φ^*. More precisely let

$$\varphi = \begin{pmatrix} \varphi_1 \\ \varphi_2 \end{pmatrix} \tag{3.1}$$

be a section of the spinor bundle $S(\Sigma)$ given in an orthonormal frame, *i.e.*, $|\varphi|^2 = |\varphi_1|^2 + |\varphi_2|^2$ and let $X = (X_1, X_2)$ be a section of the tangent bundle $T(\Sigma)$. We consider the connection ∇ on $S(\Sigma)$ given by the connection 1-form $\omega = 0$. Thus

$$\nabla_X \varphi = \begin{pmatrix} X_1 \dfrac{\partial \varphi_1}{\partial x} + X_2 \dfrac{\partial \varphi_1}{\partial y} \\ X_1 \dfrac{\partial \varphi_2}{\partial x} + X_2 \dfrac{\partial \varphi_2}{\partial y} \end{pmatrix} \tag{3.2}$$

and

$$\begin{aligned} |\nabla_X \varphi|^2 = X_1^2 \left| \frac{\partial \varphi_1}{\partial x} \right|^2 + 2X_1X_2 \Re \left(\frac{\partial \varphi_1}{\partial x} \overline{\frac{\partial \varphi_1}{\partial y}} \right) + X_2^2 \left| \frac{\partial \varphi_1}{\partial y} \right|^2 \\ + X_1^2 \left| \frac{\partial \varphi_2}{\partial x} \right|^2 + 2X_1X_2 \Re \left(\frac{\partial \varphi_2}{\partial x} \overline{\frac{\partial \varphi_2}{\partial y}} \right) + X_2^2 \left| \frac{\partial \varphi_2}{\partial y} \right|^2 . \end{aligned} \tag{3.3}$$

If we denote

$$G_\varphi = \begin{pmatrix} \left| \dfrac{\partial \varphi_1}{\partial x} \right|^2 + \left| \dfrac{\partial \varphi_2}{\partial x} \right|^2 & \Re \left(\dfrac{\partial \varphi_1}{\partial x} \overline{\dfrac{\partial \varphi_1}{\partial y}} + \dfrac{\partial \varphi_2}{\partial x} \overline{\dfrac{\partial \varphi_2}{\partial y}} \right) \\ \Re \left(\dfrac{\partial \varphi_1}{\partial x} \overline{\dfrac{\partial \varphi_1}{\partial y}} + \dfrac{\partial \varphi_2}{\partial x} \overline{\dfrac{\partial \varphi_2}{\partial y}} \right) & \left| \dfrac{\partial \varphi_1}{\partial y} \right|^2 + \left| \dfrac{\partial \varphi_2}{\partial y} \right|^2 \end{pmatrix} \tag{3.4}$$

then

$$(X_1 \; X_2) G_\varphi (X_1 \; X_2)^T = |\nabla_X \varphi|^2 . \tag{3.5}$$

G_φ is a field of real symmetric matrices.

As in the case of the usual structure tensor (*i.e.*, Di Zenzo tensor, see [10]) the optima of $|\nabla_X \varphi|^2$ under the constraint $\|X\| = 1$ (for the Euclidean norm) are given by the field of eigenvalues of G_φ. Applying the above formula to the spinor φ^* of Definition 2.2 leads to

$$G_{\varphi^*} = \frac{1}{2(1 + I_x^2 + I_y^2)^2} \begin{pmatrix} G_{\varphi^*}^{11} & G_{\varphi^*}^{12} \\ G_{\varphi^*}^{21} & G_{\varphi^*}^{22} \end{pmatrix} \tag{3.6}$$

with

$$\begin{aligned} G_{\varphi^*}^{11} &= I_{xx}^2 + I_{xy}^2 + I_{xx}^2 I_y^2 + I_{xy}^2 I_x^2 - 2I_{xx}I_{xy}I_xI_y \\ G_{\varphi^*}^{22} &= I_{yy}^2 + I_{xy}^2 + I_{yy}^2 I_x^2 + I_{xy}^2 I_y^2 - 2I_{yy}I_{xy}I_xI_y \\ G_{\varphi^*}^{12} &= I_{xx}I_{xy} + I_{xy}I_{yy} + I_{xx}I_{xy}I_y^2 + I_{xy}I_{yy}I_x^2 - I_{xy}^2 I_xI_y - I_{xx}I_{yy}I_xI_y \\ G_{\varphi^*}^{21} &= G_{\varphi^*}^{12} . \end{aligned} \tag{3.7}$$

Definition 3.1. The tensor G_{φ^*} is called the spinor tensor of the surface Σ.

Note that as already mentioned this last tensor corresponds to the tensor involved in the measure of the variations of the unit normal ν introduced in § 2.1. Indeed, we have

$$(X_1\ X_2)G_{\varphi^*}(X_1\ X_2)^T = \|d_X\nu\|^2. \tag{3.8}$$

3.2. Experiments

We compare in Figure 1 the edge and texture detection methods based on the usual structure tensor (Figure 1(b) and 1(d)) and on the spinor tensor (Figure 1(e) and 1(f)).

The structure tensor only takes into account the first-order derivatives of the function I. The subsequent segmentation method detects the strongest grey-level variations of the image. As a consequence, this method provides thick edges, as can be observed.

The spinor tensor takes into account the second-order derivatives of the function I too. By definition, it measures the strongest variations of the unit normal to the surface parametrized by the graph of I. We observe that this new approach provides thinner edges than the first one. It appears also to be more relevant to detect textures.

4. Spinors and Clifford–Fourier Transform

We first define a Clifford–Fourier transform using spin characters that is group morphisms from $\mathbb{R}^2$ to Spin(3). Then, we introduce a harmonic decomposition of spinor fields and show some results of filtering applied to images.

4.1. Clifford–Fourier Transform with Spin Characters

Let us recall the idea of the construction of the Clifford–Fourier transform for colour image processing introduced in [3]. From the mathematical viewpoint, a Fourier transform is defined through group actions and more precisely through irreducible and unitary representations of the involved group. This is closely related to the well-known shift theorem stating that:

$$\mathcal{F}f_\alpha(u) = e^{i\alpha u}\mathcal{F}f(u) \tag{4.1}$$

where $f_\alpha(u) = f(\alpha + u)$. The group morphism

$$\alpha \longmapsto e^{i\alpha u} \tag{4.2}$$

is a so-called *character* of the additive group $(\mathbb{R}, +)$, that is an irreducible unitary representation of dimension 1.

The definition proposed in [3] relies on a Clifford generalization of this notion by introducing spin characters. It can be shown that the group morphisms from $\mathbb{Z}/M\mathbb{Z} \times \mathbb{Z}/N\mathbb{Z}$ to Spin(3) are given by

$$\rho_{u,v,B} : (m, n) \longmapsto e^{2\pi(um/M+vn/N)B} \tag{4.3}$$

(a) (b)

(c) Segmentation of (a) *via* structure tensor (d) Segmentation of (b) *via* structure tensor

(e) Segmentation of (a) *via* spinor tensor (e) Segmentation of (b) *via* spinor tensor

FIGURE 1. Segmentation: Structure tensor vs. spinor tensor

where

$$e^{2\pi\left(\frac{um}{M}+\frac{vn}{N}\right)B} = \cos 2\pi\left(\frac{um}{M}+\frac{vn}{N}\right) + \sin 2\pi\left(\frac{um}{M}+\frac{vn}{N}\right) B \tag{4.4}$$

$(u, v) \in \mathbb{Z}/M\mathbb{Z} \times \mathbb{Z}/N\mathbb{Z}$, and

$$B = \gamma_1 e_1 e_2 + \gamma_2 e_1 e_3 + \gamma_3 e_2 e_3 \tag{4.5}$$

is a unit bivector, *i.e.*, $\gamma_1^2 + \gamma_2^2 + \gamma_3^2 = 1$. The map $\rho_{u,v,B}$ is called a spin character of the group $\mathbb{Z}/M\mathbb{Z} \times \mathbb{Z}/N\mathbb{Z}$. Recalling that Spin(3) acts on the sections of the spinor bundle, we are led to propose the following definition.

Definition 4.1. The Clifford–Fourier transform of a spinor φ of $S(\Sigma)$ is given by

$$\mathcal{F}(\varphi)(u, v) = \sum_{\substack{n\in\mathbb{Z}/N\mathbb{Z}\\ m\in\mathbb{Z}/M\mathbb{Z}}} \rho_{u,v,z_1\wedge z_2\,(m,n)}(-m, -n) \cdot \varphi(m, n) \tag{4.6}$$

where (z_1, z_2) is an orthonormal frame of $T(\Sigma)$.

Since the spinor bundle of Σ splits into

$$S(\Sigma) = S^+_{z_1\wedge z_2}(\Sigma) \oplus S^-_{z_1\wedge z_2}(\Sigma) \tag{4.7}$$

we have

$$\begin{aligned}\rho_{u,v,z_1\wedge z_2(m,n)}(-m, -n) \cdot \varphi(m, n) &= e^{2\pi i\left(\frac{um}{M}+\frac{vn}{N}\right)} \varphi^+(m, n)\, v_{-i}(m, n)\\ &\quad + e^{-2\pi i\left(\frac{um}{M}+\frac{vn}{N}\right)} \varphi^-(m, n)\, v_i(m, n)\end{aligned} \tag{4.8}$$

where v_{-i}, respectively v_i, is the unit eigenspinor field of eigenvalue $-i$, respectively i, relatively to the operator $z_1 \wedge z_2\,\cdot$ (here $\cdot$ denotes the Clifford multiplication). Consequently

$$\mathcal{F}(\varphi)(u, v) = \left(\widehat{\varphi^+}^{-1}(u, v),\ \widehat{\varphi^-}(u, v)\right) \tag{4.9}$$

in the frame (v_{-i}, v_i), where $\widehat{\ }$ and $\widehat{\ }^{-1}$ denote the Fourier transform on

$$L^2(\mathbb{Z}/M\mathbb{Z} \times \mathbb{Z}/N\mathbb{Z}, \mathbb{C}),$$

also called discrete Fourier transform, and its inverse.

4.2. Spinor Field Decomposition

The inverse Clifford–Fourier transform of φ is

$$\mathcal{F}^{-1}(\varphi)(u, v) = \sum_{\substack{n\in\mathbb{Z}/N\mathbb{Z}\\ m\in\mathbb{Z}/M\mathbb{Z}}} \rho_{u,v,z_1\wedge z_2(m,n)}(m, n) \cdot \varphi(m, n) \tag{4.10}$$

This means that every spinor field φ may be written as a superposition of basic spinor fields, *i.e.*,

$$\varphi = \sum \varphi_{m,n} \tag{4.11}$$

where

$$\varphi_{m,n} : (u, v) \longmapsto \rho_{u,v,z_1\wedge z_2(m,n)}(m, n) \cdot \mathcal{F}(\varphi)(m, n) \tag{4.12}$$

Following the splitting $S(\Sigma) = S^+_{z_1\wedge z_2}(\Sigma) \oplus S^-_{z_1\wedge z_2}(\Sigma)$, we have

$$\varphi_{m,n} = \left(\varphi^+_{m,n}, \varphi^-_{m,n}\right)$$

in the frame (v_{-i}, v_i), with

$$\varphi^+_{m,n} \colon (u,v) \longmapsto e^{-2\pi i(um/M+vn/N)} \widehat{\varphi^+}^{-1}(m,n)$$

and

$$\varphi^-_{m,n} \colon (u,v) \longmapsto e^{2\pi i(um/M+vn/N)} \widehat{\varphi^-}(m,n)$$

Moreover,

$$|\varphi_{m,n}|^2 = |\varphi^+_{m,n}|^2 + |\varphi^-_{m,n}|^2$$

since $S^+_{z_1\wedge z_2}(\Sigma)$ and $S^-_{z_1\wedge z_2}(\Sigma)$ are orthogonal.

4.3. Experiments

Let us now give an example of applications of the Clifford–Fourier transform on spinor fields to image processing. In order to perform filtering with the decomposition (4.11), we proceed as follows. Let I be a grey-level image, and φ^* be the corresponding spinor representation given in Definition 2.2. We apply a Gaussian mask T_σ of variance σ in the spectrum $\mathcal{F}\varphi^*$ of φ^*. Then, we consider the norm of its inverse Fourier transform, *i.e.*, $|\mathcal{F}^{-1}T_\sigma\mathcal{F}\varphi^*|$ and the function $|\mathcal{F}^{-1}T_\sigma\mathcal{F}\varphi^*|\,I$.

Figures 2 and 3 show results of this process for different values of σ (left column $|\mathcal{F}^{-1}T_\sigma\mathcal{F}\varphi^*|$ and right column $|\mathcal{F}^{-1}T_\sigma\mathcal{F}\varphi^*|\,I$). It is clear that for σ sufficiently high, we have $|\mathcal{F}^{-1}T_\sigma\mathcal{F}\varphi^*|\,I \simeq I$ and $|\mathcal{F}^{-1}T_\sigma\mathcal{F}\varphi^*| \simeq 1$ since $|\varphi^*| = 1$. This explains why the two left lower images are almost white and the two right lower images are almost the same as the originals.

We can see in the left columns of Figures 2 and 3 that the filtering acts through φ^* as a smoothing of the geometry of the image. More precisely, when σ is small, the modulus $|\mathcal{F}^{-1}T_\sigma\mathcal{F}\varphi^*|$ is small at points corresponding to nearly all the geometric variations of the image. When σ increases the modulus is affected only at points corresponding to the strongest geometric variations, *i.e.*, to both edges and textures (and also where the noise is high).

The right columns of Figures 2 and 3 show that the filtering acts through $|\mathcal{F}^{-1}T_\sigma\mathcal{F}\varphi^*|\,I$ as a diffusion that leaves the geometric data untouched (the higher the value of σ, the more important is the diffusion). This appears clearly in Figure 4 (compare the plumes of the hat) or in Figure 5 (compare the hair).

These experiments show that our approach is relevant to deal with harmonic analysis together with Riemannian geometry.

Conclusion

Spin geometry is a powerful mathematical tool to deal with many theoretical and applied geometric problems. In this chapter we have shown how to take advantage of the generalized Weierstrass representation to perform grey-level image processing, in particular edge and texture detection. Our main contribution is the

FIGURE 2. Left: $|\mathcal{F}^{-1}(T_\sigma\, \mathcal{F}\varphi^*)|$ for $\sigma = 100, 1000, 10000, 100000$ (from top to bottom). Right: $|\mathcal{F}^{-1}(T_\sigma\, \mathcal{F}\varphi^*)|I$

definition of a Clifford–Fourier transform for spinor fields that relies on a generalization of the usual notion of character (the spin character). One important fact is that this new transform takes into account the Riemannian geometry of the

FIGURE 3. Left: $|\mathcal{F}^{-1}(T_\sigma\, \mathcal{F}\varphi^*)|$ for $\sigma = 100, 1000, 10000, 100000$ (from top to bottom). Right: $|\mathcal{F}^{-1}(T_\sigma\, \mathcal{F}\varphi^*)|I$

image surface by involving the spinor field that parameterizes the normal and the bivector field coding the tangent plane. We have also introduced what appears to

FIGURE 4. Left: original. Right: $|\mathcal{F}^{-1}T_\sigma\mathcal{F}\varphi^*|\,I$ with $\sigma = 100$

FIGURE 5. Left: original. Right: $|\mathcal{F}^{-1}T_\sigma\mathcal{F}\varphi^*|\,I$ with $\sigma = 100$

be a harmonic decomposition of the parametrization and investigated applications to filtering.

Note that there are only two cases where the Grassmannian $G_{n,2}$ of 2-planes in $\mathbb{R}^n$ admits a rational parametrization. In fact, one can show that $G_{3,2} \simeq \mathbb{C}P^1$ and $G_{4,2} \simeq \mathbb{C}P^1 \times \mathbb{C}P^1$ (see [9]). The case treated here corresponds to $G_{3,2}$. As a consequence the generalization to colour images is not straightforward. Nevertheless, a quite different approach is possible to tackle this problem and will be the subject of a forthcoming paper.

Let us also mention that one may envisage performing diffusion on grey-level images through the heat equation given by the Dirac operator. The latter is well known be a square root of the Laplacian. Preliminary results are discussed in [2] that show that this diffusion better preserves edges and textures than the usual Riemannian approaches.

Appendix A. Mathematical Background

We recall here some definitions and results concerning spin geometry. The reader may refer to [7] for details and conventions. We focus on the particular case of an oriented surface immersed in $\mathbb{R}^3$.

A.1. Complex Representations of $C\ell_{3,0} \otimes \mathbb{C}$

Let (e_1, e_2, e_3) be an orthonormal basis of $\mathbb{R}^3$. The Clifford algebra $C\ell_{3,0}$ is the quotient of the tensor algebra of the vectorial space $\mathbb{R}^3$ by the ideal generated by the elements $u \otimes u + Q(u)$ where Q is the Euclidean quadratic form. It can be shown that $C\ell_{3,0}$ is isomorphic to the product $\mathbb{H} \times \mathbb{H}$ of two copies of the quaternion algebra. The complex Clifford algebra $C\ell_{3,0} \otimes \mathbb{C}$ is isomorphic to $\mathbb{C}(2) \oplus \mathbb{C}(2)$ where $\mathbb{C}(2)$ denotes the algebra of 2×2-matrices with complex entries. This decomposition is given by

$$C\ell_{3,0} \otimes \mathbb{C} \simeq (C\ell_{3,0} \otimes \mathbb{C})^+ \oplus (C\ell_{3,0} \otimes \mathbb{C})^- \tag{A.1}$$

where

$$(C\ell_{3,0} \otimes \mathbb{C})^{\pm} = (1 \pm \omega_3) C\ell_{3,0} \otimes \mathbb{C} \tag{A.2}$$

and ω_3 is the pseudoscalar $e_1 e_2 e_3$. More precisely, the subalgebra $(C\ell_{3,0} \otimes \mathbb{C})^+$ is generated by the elements

$$\alpha_1 = \frac{1 + e_1 e_2 e_3}{2}, \ \alpha_2 = \frac{e_2 e_3 - e_1}{2}, \ \alpha_3 = \frac{e_2 + e_1 e_3}{2}, \ \alpha_4 = \frac{e_3 - e_1 e_2}{2} \tag{A.3}$$

and an isomorphism with $\mathbb{C}(2)$ is given by sending these elements to the matrices

$$A_1 = \begin{pmatrix} 1 & 0 \\ 0 & 1 \end{pmatrix}, A_2 = \begin{pmatrix} 0 & i \\ i & 0 \end{pmatrix}, A_3 = \begin{pmatrix} i & 0 \\ 0 & -i \end{pmatrix}, A_4 = \begin{pmatrix} 0 & 1 \\ -1 & 0 \end{pmatrix}. \tag{A.4}$$

In the same way, $(C\ell_{3,0} \otimes \mathbb{C})^-$ is generated by

$$\beta_1 = \frac{1 - e_1 e_2 e_3}{2}, \beta_2 = \frac{e_2 e_3 + e_1}{2}, \beta_3 = \frac{e_1 e_3 - e_2}{2}, \beta_4 = \frac{-e_3 - e_1 e_2}{2} \tag{A.5}$$

and an isomorphism is given by sending these elements to the above matrices A_1, A_2, A_3 and A_4.

Let us denote by ρ the natural representation of $\mathbb{C}(2)$ on $\mathbb{C}^2$. The two equivalent classes ρ_1 and ρ_2 of irreducible complex representations of $C\ell_{3,0} \otimes \mathbb{C}$ are given by

$$\rho_1(\varphi_1 + \varphi_2) = \rho(\varphi_1) \quad \rho_2(\varphi_1 + \varphi_2) = \rho(\varphi_2). \tag{A.6}$$

They are characterized by

$$\rho_1(\omega_3) = Id \ \text{ and } \ \rho_2(\omega_3) = -Id \tag{A.7}$$

For the sake of completeness, let us list these representations explicitly:

$$\begin{aligned} \rho_1(1) = \rho(\alpha_1) &= A_1, & \rho_1(e_1) = \rho(-\alpha_2) &= -A_2 \\ \rho_1(e_2) = \rho(\alpha_3) &= A_3, & \rho_1(e_3) = \rho(\alpha_4) &= A_4 \\ \rho_1(e_1 e_2) = \rho(-\alpha_4) &= -A_4, & \rho_1(e_1 e_3) = \rho(\alpha_3) &= A_3 \\ \rho_1(e_2 e_3) = \rho(\alpha_2) &= A_2, & \rho_1(\omega_3) = \rho(\alpha_1) &= A_1 \end{aligned} \tag{A.8}$$

and

$$\begin{aligned}
\rho_2(1) &= \rho(\beta_1) = A_1, & \rho_2(e_1) &= \rho(\beta_2) = A_2\\
\rho_2(e_2) &= \rho(-\beta_3) = -A_3, & \rho_2(e_3) &= \rho(-\beta_4) = -A_4\\
\rho_2(e_1e_2) &= \rho(-\beta_4) = -A_4, & \rho_2(e_1e_3) &= \rho(\beta_3) = A_3\\
\rho_2(e_2e_3) &= \rho(\beta_2) = A_2, & \rho_2(\omega_3) &= \rho(-\beta_1) = -A_1.
\end{aligned} \tag{A.9}$$

The complex spin representation of Spin(3) is the homomorphism

$$\Delta_3 : \mathrm{Spin}(3) \longrightarrow \mathbb{C}(2) \tag{A.10}$$

given by restricting an irreducible complex representation of $C\ell_{3,0}\otimes\mathbb{C}$ to the spinor group $\mathrm{Spin}(3) \subset (C\ell_{3,0}\otimes\mathbb{C})^0$ (see for example [4] for the definition of the Spin group). Note that Δ_3 is independant of the chosen representation.

A.2. Spin Structures and Spinor Bundles

Let us denote by M the Riemannian manifold $\mathbb{R}^3$ and $P_{SO}(M)$ the principal $SO(3)$-bundle of oriented orthonormal frames of M. A spin structure on M is a principal Spin(3)-bundle $P_{\mathrm{Spin}}(M)$ together with a 2-sheeted covering

$$P_{\mathrm{Spin}}(M) \longrightarrow P_{SO}(M) \tag{A.11}$$

that is compatible with $SO(3)$ and Spin(3) actions. The Spinor bundle $S(M)$ is the bundle associated to the spin structure $P_{\mathrm{Spin}}(M)$ and the complex spin representation Δ_3. More precisely, it is the quotient of the product $P_{\mathrm{Spin}}(M)\times\mathbb{C}^2$ by the action

$$\mathrm{Spin}(3)\times P_{\mathrm{Spin}}(M)\times\mathbb{C}^2 \longrightarrow P_{\mathrm{Spin}}(M)\times\mathbb{C}^2 \tag{A.12}$$

that sends (τ, p, z) to $(p\tau^{-1}, \Delta_3(\tau)z)$. We will write

$$S(M) = P_{\mathrm{Spin}}(M)\times_{\Delta_3}\mathbb{C}^2. \tag{A.13}$$

It appears that the fiber bundle $S(M)$ is a bundle of complex left modules over the Clifford bundle $Cl(M) = P_{\mathrm{Spin}}(M)\times_{Ad}\mathbb{C}l(3)$ of M. In the sequel

$$(u,\phi) \longmapsto u\cdot\phi \tag{A.14}$$

denotes the corresponding multiplication for $u \in T(M)$ and ϕ a section of $S(M)$.

We consider now an oriented surface Σ embedded in M. Let us denote by (z_1, z_2) an orthonormal frame of $T(\Sigma)$ and ν the global unit field normal to Σ. Using the map

$$(z_1, z_2) \longmapsto (z_1, z_2, \nu) \tag{A.15}$$

it is possible to pull back the bundle $P_{\mathrm{Spin}}(M)_{|\Sigma}$ to obtain a spin structure $P_{\mathrm{Spin}}(\Sigma)$ on Σ. Since $C\ell_{2,0}\otimes\mathbb{C}$ is isomorphic to $(C\ell_{3,0}\otimes\mathbb{C})^0$ under the map α defined by

$$\alpha(\eta^0+\eta^1) = \eta^0+\eta^1\nu \tag{A.16}$$

the algebra $C\ell_{2,0}\otimes\mathbb{C}$ acts on $\mathbb{C}^2$ *via* ρ_2. This representation leads to the complex spinor representation Δ_2 of Spin(2). It can be shown that the induced bundle

$$S(\Sigma) = P_{\mathrm{Spin}}(\Sigma)\times_{\Delta_3\circ\alpha}\mathbb{C}^2 \tag{A.17}$$

coincides with the spinor bundle of the induced spin structure on Σ. Once again $S(\Sigma)$ is a bundle of complex left modules over the Clifford bundle $Cl(\Sigma)$ of Σ: the Clifford multiplication is given by the map

$$(v,\varphi) \longmapsto v \cdot \nu \cdot \varphi \tag{A.18}$$

for $v \in T(\Sigma)$ and φ a section of $T(\Sigma)$.

The Spinor bundle $S(\Sigma)$ decomposes into

$$S(\Sigma) = S^+(\Sigma) \oplus S^-(\Sigma) \tag{A.19}$$

where

$$S^\pm(\Sigma) = \{\varphi \in S(\Sigma),\ i \cdot z_1 \cdot z_2 \cdot \varphi = \pm\varphi\} \tag{A.20}$$

(compare [5]). Since $\rho_2(z_1 z_2 \nu)$ is minus the identity, this is equivalent to

$$S^\pm(\Sigma) = \{\varphi \in S(\Sigma),\ i\nu \cdot \varphi = \pm\varphi\}. \tag{A.21}$$

A.3. Spinor Connections and Dirac Operators

Let ∇^M and ∇^Σ be the Levi–Civita connections on the tangent bundles $T(M)$ and $T(\Sigma)$ respectively. The classical Gauss formula asserts that

$$\nabla^M_X Y = \nabla^\Sigma_X Y - \langle \nabla^M_X \nu, Y\rangle \nu \tag{A.22}$$

where X and Y are vector fields on Σ. A similar formula exists when dealing with spinor fields. Let us first recall that one may construct on $S(M)$ and $S(\Sigma)$ some spinor Levi–Civita connections compatible with the Clifford multiplication, that is connections which we continue to denote by ∇^M and ∇^Σ verifying

$$\nabla^M_X (Y \cdot \varphi) = (\nabla^M_X Y) \cdot \varphi + Y \cdot \nabla^M_X \varphi \tag{A.23}$$

when X and Y are vector fields on M and φ is a section of $S(M)$ and a similar formula for ∇^Σ. The analog of the Gauss formula reads

$$\nabla^M_X \varphi = \nabla^\Sigma_X \varphi - \frac{1}{2}(\nabla^M_X \nu) \cdot \nu \cdot \varphi \tag{A.24}$$

for φ a section of $S(\Sigma)$ and X a vector field on Σ (see [1] for a proof). If (z_1, z_2) is an orthonormal frame of $T(\Sigma)$, following [5], the Dirac operator on $S(\Sigma)$ is defined by

$$D = z_1 \cdot \nabla^\Sigma_{z_1} + z_2 \cdot \nabla^\Sigma_{z_2} \tag{A.25}$$

and it can be verified that $DS^\pm(\Sigma) \subset S^\mp(\Sigma)$.

Let now ϕ and φ be respectively a section of $S(M)$ and the section of $S(\Sigma)$ given by the restriction $\phi_{|\Sigma}$. We obtain from the Gauss spinor formula

$$z_1 \cdot \nabla^M_{z_1}\phi + z_2 \cdot \nabla^M_{z_2}\phi = D\varphi - \frac{1}{2}(z_1 \cdot (\nabla^M_{z_1}\nu) \cdot \nu \cdot \varphi + z_2 \cdot (\nabla^M_{z_2}\nu) \cdot \nu \cdot \varphi). \tag{A.26}$$

Since

$$z_1 \cdot (\nabla^M_{z_1}\nu) + z_2 \cdot (\nabla^M_{z_2}\nu) = -2H \tag{A.27}$$

where H is the mean curvature of Σ, it follows that

$$D\varphi = z_1 \cdot \nabla^M_{z_1}\phi + z_2 \cdot \nabla^M_{z_2}\phi - H \cdot \nu \cdot \varphi. \tag{A.28}$$

References

[1] C. Bar. Metrics with harmonic spinors. *Geometric and Functional Analysis*, 6(6):899–942, 1996.

[2] T. Batard and M. Berthier. The spinor representation of images. In K. Gürlebeck, editor, *9th International Conference on Clifford Algebras and their Applications*, Weimar, Germany, 15–20 July 2011.

[3] T. Batard, M. Berthier, and C. Saint-Jean. Clifford Fourier transform for color image processing. In E.J. Bayro-Corrochano and G. Scheuermann, editors, *Geometric Algebra Computing in Engineering and Computer Science*, pages 135–162. Springer, London, 2010.

[4] T. Batard, C.S. Jean, and M. Berthier. A metric approach to nD images edge detection with Clifford algebras. *Journal of Mathematical Imaging and Vision*, 33:296–312, 2009.

[5] T. Friedrich. On the spinor representation of surfaces in Euclidean 3-space. *Journal of Geometry and Physics*, 28:143–157, 1998.

[6] B. Lawson. *Lectures on minimal manifolds, volume I*, volume 9 of *Mathematics Lecture Series*. Publish or Perish Inc., Wilmington, Del., second edition, 1980.

[7] B. Lawson and M.-L. Michelson. *Spin Geometry*. Princeton University Press, Princeton, New Jersey, 1989.

[8] R. Osserman. *A survey of minimal surfaces*. Dover Publications, Inc., New York, second edition, 1986.

[9] I.A. Taimanov. Two-dimensional Dirac operator and surface theory. *Russian Mathematical Surveys*, 61(1):79–159, 2006.

[10] S.D. Zenzo. A note on the gradient of a multi-image. *Computer Vision, Graphics, and Image Processing*, 33:116–125, 1986.

Thomas Batard
Department of Applied Mathematics
Tel Aviv University
Ramat Aviv
Tel Aviv 69978, Israel
e-mail: thomas.batard@gmail.com

Michel Berthier
MIA Laboratory
La Rochelle University
Avenue Michel Crépeau
F-17042 La Rochelle, France
e-mail: michel.berthier@univ-lr.fr

Quaternion and Clifford–Fourier Transforms and Wavelets
Trends in Mathematics, 197–219

10 Analytic Video (2D + t) Signals Using Clifford–Fourier Transforms in Multiquaternion Grassmann–Hamilton–Clifford Algebras

P.R. Girard, R. Pujol, P. Clarysse, A. Marion,
R. Goutte and P. Delachartre

Abstract. We present an algebraic framework for (2D + t) video analytic signals and a numerical implementation thereof using Clifford biquaternions and Clifford–Fourier transforms. Though the basic concepts of Clifford–Fourier transforms are well known, an implementation of analytic video sequences using multiquaternion algebras does not seem to have been realized so far. After a short presentation of multiquaternion Clifford algebras and Clifford–Fourier transforms, a brief pedagogical review of 1D and 2D quaternion analytic signals using right quaternion Fourier transforms is given. Then, the biquaternion algebraic framework is developed to express Clifford–Fourier transforms and (2D + t) video analytic signals in standard and polar form constituted by a scalar, a pseudoscalar and six phases. The phase extraction procedure is fully detailed. Finally, a numerical implementation using discrete fast Fourier transforms of an analytic multiquaternion video signal is provided.

Mathematics Subject Classification (2010). Primary 15A66; secondary 42A38.

Keywords. Multiquaternion Clifford algebras, Clifford biquaternions, Clifford–Fourier transforms, 2D + t analytic video signals.

1. Introduction

In recent decades, new algebraic structures based on Clifford algebras have been developed [19,31]. Many of these developments use a geometric approach whereas we propose an algebraic multiquaternion approach [10–12]. This chapter focuses on a (2D + t) Clifford biquaternion analytic signal using Clifford–Fourier transforms [1,29]. After a short pedagogical review of 1D complex and 2D quaternion analytic signals using right quaternion Fourier transforms, we shall provide a concrete algebraic Clifford biquaternion framework for the Clifford–Fourier transform, the analytic Clifford–Fourier transform and the analytic signal both in standard and polar form. Finally, we shall give a numerical implementation, using discrete fast Fourier transforms, of a (2D + t) video analytic signal.

2. Multiquaternion Grassmann–Hamilton–Clifford Algebras

In 1844, in his *Ausdehnungslehre*, Hermann Günther Grassmann (1809–1877) laid the foundations of an n-dimensional associative multivector calculus [14–16, 27]. A year before, in 1843, William Rowan Hamilton (1805–1865) discovered the quaternions and later on, was to introduce biquaternions [6, 22–24]. In 1878, William Kingdon Clifford (1845–1879) was to give a concise definition of Clifford algebras and to demonstrate a theorem relating Grassmann's system to Hamilton's quaternions [4], [5, pp. 266–276]. Due to this close connection, we shall refer to Clifford algebras also as Grassmann–Hamilton–Clifford algebras or multiquaternion algebras [13, 17, 18].

Definition 2.1. A Clifford algebra C_n is an algebra composed of n generators $\boldsymbol{e}_1, \boldsymbol{e}_2, \dots, \boldsymbol{e}_n$ multiplying according to the rule $\boldsymbol{e}_i\boldsymbol{e}_j = -\boldsymbol{e}_j\boldsymbol{e}_i$ $(i \neq j)$ and such that $\boldsymbol{e}_i^2 = \pm 1$. The algebra C_n contains 2^n elements constituted by the n generators, the various products $\boldsymbol{e}_i\boldsymbol{e}_j, \boldsymbol{e}_i\boldsymbol{e}_j\boldsymbol{e}_k, \dots$ and the unit element 1.

Theorem 2.2. *If $n = 2m$ (m : integer), the Clifford algebra C_{2m} is the tensor product of m quaternion algebras. If $n = 2m - 1$, the Clifford algebra C_{2m-1} is the tensor product of $m - 1$ quaternion algebras and the algebra $(1, \epsilon)$ where ϵ is the product of the $2m$ generators ($\epsilon = \boldsymbol{e}_1\boldsymbol{e}_2 \dots \boldsymbol{e}_{2m}$) of the algebra C_{2m}.*

Examples of Clifford algebras are the complex numbers $\mathbb{C}$ (with $\boldsymbol{e}_1 = i$), quaternions $\mathbb{H}$ ($\boldsymbol{e}_1 = \boldsymbol{i}, \boldsymbol{e}_2 = \boldsymbol{j}$), biquaternions ($\boldsymbol{e}_1 = I\boldsymbol{i}, \boldsymbol{e}_2 = I\boldsymbol{j}, \boldsymbol{e}_3 = I\boldsymbol{k}, I^2 = -1$, I commuting with $\boldsymbol{i}, \boldsymbol{j}, \boldsymbol{k}, \boldsymbol{e}_i^2 = 1$) and tetraquaternions $\mathbb{H} \otimes \mathbb{H}$ ($\boldsymbol{e}_0 = \boldsymbol{j}, \boldsymbol{e}_1 = \boldsymbol{k}\boldsymbol{I}, \boldsymbol{e}_2 = \boldsymbol{k}\boldsymbol{J}, \boldsymbol{e}_3 = \boldsymbol{k}\boldsymbol{K}$, where the small $\boldsymbol{i}, \boldsymbol{j}, \boldsymbol{k}$ commute with the capital $\boldsymbol{I}, \boldsymbol{J}, \boldsymbol{K}$). These examples prove the Clifford theorem up to dimension $n = 4$.

3. Analytic Signal in N Dimensions

Consider a Grassmann–Hamilton–Clifford algebra having n generators $\boldsymbol{e}_n$ (with $\boldsymbol{e}_n^2 = -1$) and let A be a general element of this algebra [7, 21]. Call $K[A]$ the conjugate of A such that

$$K(AB) = K(B)K(A), K(\boldsymbol{e}_n) = -\boldsymbol{e}_n. \tag{3.1}$$

Given a function $f(\boldsymbol{x})$ having its value in the Clifford algebra with $\boldsymbol{x} = (x_1, x_2, \dots, x_n)$, let $F(\boldsymbol{u})$ with $\boldsymbol{u} = (u_1, u_2, \dots, u_n)$ denote the Clifford–Fourier transform

$$F(\boldsymbol{u}) = \int_{\mathbb{R}^n} f(\boldsymbol{x}) \prod_{k=1}^{n} e^{-\boldsymbol{e}_k 2\pi u_k x_k} d^n\boldsymbol{x}. \tag{3.2}$$

The inverse Clifford–Fourier transform is given by

$$f(\boldsymbol{x}) = \int_{\mathbb{R}^n} F(\boldsymbol{u}) \prod_{k=0}^{n-1} e^{e_{n-k} 2\pi u_{n-k} x_{n-k}} d^n \boldsymbol{u}. \tag{3.3}$$

Proof. Inserting $F(\boldsymbol{u})$ in the equation above, one obtains after integration respectively over $d^n\boldsymbol{u}$ and $d^n\boldsymbol{x}'$ [29]

$$\int_{\mathbb{R}^{2n}} f(\boldsymbol{x}') \prod_{j=1}^{n} e^{-e_j 2\pi u_j x'_j} d^n \boldsymbol{x}' \prod_{k=0}^{n-1} e^{e_{n-k} 2\pi u_{n-k} x_{n-k}} d^n \boldsymbol{u} \tag{3.4}$$

$$= \int_{\mathbb{R}^n} f(\boldsymbol{x}') \delta^n (\boldsymbol{x} - \boldsymbol{x}') \, d^n \boldsymbol{x}' = f(\boldsymbol{x}).$$

□

Given two functions $f(\boldsymbol{x}), g(\boldsymbol{x})$ taking their values in the Clifford algebra, define the following product

$$\langle f(\boldsymbol{x}), g(\boldsymbol{x}) \rangle = \frac{1}{2} \int_{\mathbb{R}^n} [fK(g) + gK(f)] \, d^n \boldsymbol{x} \tag{3.5}$$

and similarly $\langle F(\boldsymbol{u}), G(\boldsymbol{u}) \rangle$. These products satisfy Plancherel's theorem

$$\langle f(\boldsymbol{x}), g(\boldsymbol{x}) \rangle = \langle F(\boldsymbol{u}), G(\boldsymbol{u}) \rangle \tag{3.6}$$

Proof.

$$f(\boldsymbol{x}) = \int_{\mathbb{R}^n} F(\boldsymbol{u}) e^{e_n 2\pi u_n x_n} \dots e^{e_1 2\pi u_1 x_1} d^n \boldsymbol{u} \tag{3.7}$$

$$Kg(\boldsymbol{x}) = \int_{\mathbb{R}^n} e^{-e_1 2\pi u'_1 x_1} \dots e^{-e_n 2\pi u'_n x_n} K[G(\boldsymbol{u}')] \, d^n \boldsymbol{u}'. \tag{3.8}$$

After integration over $d^n\boldsymbol{x}$, one has

$$\frac{1}{2} \int_{\mathbb{R}^n} f(\boldsymbol{x}) K[g(\boldsymbol{x})] \, d^n \boldsymbol{x} = \frac{1}{2} \int_{\mathbb{R}^{2n}} F(\boldsymbol{u}) \, \delta^n (\boldsymbol{u} - \boldsymbol{u}') \, K[G(\boldsymbol{u}')] \, d^n \boldsymbol{u} d^n \boldsymbol{u}'$$

$$= \frac{1}{2} \int_{\mathbb{R}^n} F(\boldsymbol{u}) \, K[G(\boldsymbol{u})] \, d^n \boldsymbol{u}; \tag{3.9}$$

similarly,

$$\frac{1}{2} \int_{\mathbb{R}^n} g(\boldsymbol{x}) K[f(\boldsymbol{x})] \, d^n \boldsymbol{x} = \frac{1}{2} \int_{\mathbb{R}^{2n}} G(\boldsymbol{u}) \, \delta^n (\boldsymbol{u} - \boldsymbol{u}') \, K[F(\boldsymbol{u}')] \, d^n \boldsymbol{u} d^n \boldsymbol{u}'$$

$$= \frac{1}{2} \int_{\mathbb{R}^n} G(\boldsymbol{u}) \, K[F(\boldsymbol{u})] \, d^n \boldsymbol{u}. \tag{3.10}$$

Hence, adding the last two equations, one obtains equation (3.5). □

Next consider a real scalar function $f(\boldsymbol{x})$ and its Clifford–Fourier transform $F(\boldsymbol{u})$, the analytic Clifford–Fourier transform $F_A(\boldsymbol{u})$ and the analytic signal $f_A(\boldsymbol{x})$ are respectively defined by

$$F_A(\boldsymbol{u}) = \prod_{k=1}^{n} [1 + \mathrm{sign}\,(u_k)]\, F(\boldsymbol{u}) \tag{3.11}$$

$$f_A(\boldsymbol{x}) = \int_{\mathbb{R}^n} F_A(\boldsymbol{u}) \prod_{k=0}^{n-1} e^{\boldsymbol{e}_{n-k} 2\pi u_{n-k} x_{n-k}} d^n \boldsymbol{u} \tag{3.12}$$

with $\mathrm{sign}(u_k)$ yielding $-1, 0$, or 1 depending on whether u_k is negative, zero or positive. In particular, the analytic signal satisfies the Parseval relation

$$\langle f_A(\boldsymbol{x}), f_A(\boldsymbol{x})\rangle = \langle F_A(\boldsymbol{u}), F_A(\boldsymbol{u})\rangle . \tag{3.13}$$

4. Analytic Signals in 1 and 2 Dimensions

4.1. Complex 1D Analytic Signal

The one-dimensional analytic signal notion was introduced by D. Gabor [9] and J. Ville [30] in 1948. Within the Clifford algebra framework it can be presented as follows. Using the complex numbers as Clifford algebra ($\boldsymbol{e}_1 = i$) for modeling 1D physics, the Clifford–Fourier transform of a scalar function $f(x_1)$ is simply the standard Fourier transform

$$F(u_1) = \int_{-\infty}^{\infty} f(x_1) e^{-i2\pi u_1 x_1} dx_1 \tag{4.1}$$

with the inverse transformation

$$f(x_1) = \int_{-\infty}^{\infty} F(u_1) e^{i2\pi u_1 x_1} du_1. \tag{4.2}$$

Furthermore, one defines the scalar product of two functions $f(x_1)$, $g(x_1)$ as

$$\langle f(x_1), g(x_1)\rangle = \frac{1}{2}\int_{-\infty}^{\infty} (fg^* + gf^*)\, dx_1 \tag{4.3}$$

where $f^*(x_1), g^*(x_1)$ are the complex conjugate functions. A few properties of the Fourier transform are recalled in Table 1.

Under an orthogonal symmetry transforming the basis vector $\boldsymbol{e}_1$ into $-\boldsymbol{e}_1$ one has $F(-u_1) = F(u_1)^*$ where $F(u_1)^*$ is the complex conjugate of $F(u_1)$. Hence, only one orthant ($u_1 \geq 0$) of the Fourier domain is necessary to obtain the signal. The analytic Fourier transform is defined by

$$F_A(u_1) = [1 + \mathrm{sign}(u_1)]\, F(u_1) \tag{4.4}$$

and the analytic signal by

$$f_A(x_1) = \int_{-\infty}^{\infty} F_A(u_1) e^{i2\pi u_1 x_1} du_1. \tag{4.5}$$

TABLE 1. Properties of the complex Fourier transform ($a, b \in \mathbb{R}$)

Property	Complex function	Fourier transform
Linearity	$af(x_1) + bg(x_1)$	$aF(u_1) + bG(u_1)$
Translation	$f(x_1 - x_0)$	$e^{-i2\pi u_1 x_0} F(u_1)$
Scaling	$f(ax_1)$	$\frac{1}{\lvert a \rvert} F\left(\frac{u_1}{a}\right)$
Partial derivative	$\frac{\partial^r f(x_1)}{\partial x_1^r}$	$(i2\pi u_1)^r F(u_1)$
Plancherel	$\langle f, g \rangle =$	$\langle F, G \rangle$
Parseval	$\langle f, f \rangle =$	$\langle F, F \rangle$

The analytic signal satisfies Plancherel's theorem

$$\int_{-\infty}^{\infty} |f_A(x_1)|^2 \, dx_1 = \int_{-\infty}^{\infty} |F_A(u_1)|^2 \, du_1. \tag{4.6}$$

As example, consider the function $f(x) = a \cos \omega t$ which yields the analytic signal $f_A = ae^{i\omega t}$. The instantaneous amplitude and the instantaneous phase of a real signal f at a given position x can be defined as the modulus and the angular argument of the complex-valued analytic signal. Therefore, the analytic signal plays a key role in one-dimensional signal processing. It is also important to note that the analytic signal concept is global, *i.e.*, $f_A(x)$ depends on the whole of the original signal and not only on values at positions near x. This concept can be summarized into three main properties:

1. The analytic signal has a one-sided spectrum;
2. The original signal can be reconstructed from its analytic signal (in particular, the real part of the analytic signal is equal to the original signal);
3. The local amplitude (envelope) and the local phase of the original signal can be derived as the modulus and angular argument of the analytic signal respectively.

4.2. Quaternion 2D Analytic Signal

4.2.1. Clifford Algebra. The analytic signal notion can be extended to the 2D case in several ways [2, 3, 8, 20, 25]. Taking the quaternions $\mathbb{H}$ as a Clifford algebra, it contains the elements

$$[1, \boldsymbol{e}_1 = \boldsymbol{i}, \boldsymbol{e}_2 = \boldsymbol{j}, \boldsymbol{e}_1\boldsymbol{e}_2 = \boldsymbol{k}]. \tag{4.7}$$

Hence, a general element q can be expressed as

$$\begin{aligned} q &= q_0 + \boldsymbol{i}q_1 + \boldsymbol{j}q_2 + \boldsymbol{k}q_3 \\ &= [q_0, q_1, q_2, q_3] \end{aligned} \tag{4.8}$$

the conjugate q_c being

$$q_c = [q_0, -q_1, -q_2, -q_3]. \tag{4.9}$$

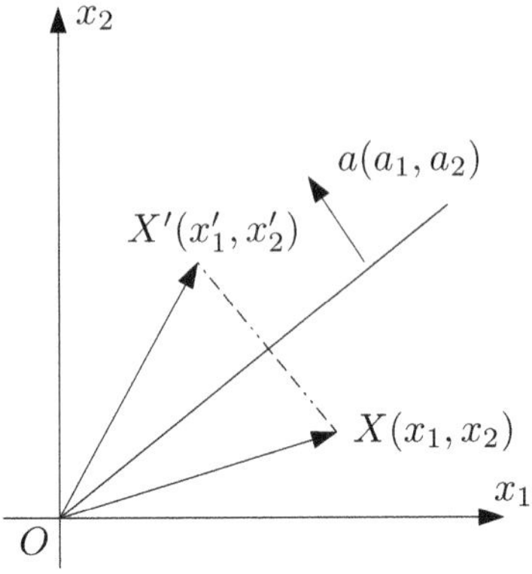

FIGURE 1. Orthogonal symmetry with respect to a straight line.

A 2D vector is expressed by $X = \boldsymbol{i}q_1 + \boldsymbol{j}q_2$ with $XX_c = -X^2 = (q_1)^2 + (q_2)^2$, a bivector by $B = \boldsymbol{k}q_3$ and a scalar by q_0.

Under an orthogonal symmetry with respect to a straight line passing through the origin and perpendicular to the unit vector $a = \boldsymbol{i}a_1 + \boldsymbol{j}a_2$, $(aa_c = 1)$, a vector is transformed into the vector $X' = \frac{aXa}{aa_c} = aXa$ [10, p. 76], a bivector $B = X \wedge Y = -\frac{1}{2}(XY - YX)$ into $B' = -aBa$, whereas a scalar remains invariant (Figure 1). Hence, the entire quaternion transforms as

$$q' = a^* q a_c^* \tag{4.10}$$

where $a^* = \boldsymbol{k}a$ is the dual of a ($a_c^* = a_c\boldsymbol{k}_c = a\boldsymbol{k}$). One verifies the conservation of the square of the norm of the quaternion $q'q'_c = qq_c$ under an orthogonal symmetry. Equation (4.10) allows us to deduce the involution formulas developed below which are useful for the comprehension of the properties of the analytic Clifford quaternion Fourier transform and the extraction of phases. In particular, if $a = \boldsymbol{i}$ (orthogonal symmetry with respect to the y-axis), one has

$$q' = K_1(q) = \boldsymbol{i}^* q \boldsymbol{i}_c^* = \boldsymbol{kiqik} = -\boldsymbol{j}q\boldsymbol{j} = q_0 - \boldsymbol{i}q_1 + \boldsymbol{j}q_2 - \boldsymbol{k}q_3 \tag{4.11}$$

where $K_1(q)$ is an involution of q. Similarly, if $a = \boldsymbol{j}$ (orthogonal symmetry with respect to the x-axis)

$$q' = K_2(q) = \boldsymbol{j}^* q \boldsymbol{j}_c^* = \boldsymbol{kjqjk} = -\boldsymbol{i}q\boldsymbol{i} = q_0 + \boldsymbol{i}q_1 - \boldsymbol{j}q_2 - \boldsymbol{k}q_3. \tag{4.12}$$

A combination of the two above orthogonal symmetries ($\boldsymbol{i} \to -\boldsymbol{i}, \boldsymbol{j} \to -\boldsymbol{j}$) yields a rotation of π around the origin

$$q' = K_{12}(q) = \boldsymbol{ijqji} = -\boldsymbol{k}q\boldsymbol{k} = q_0 - \boldsymbol{i}q_1 - \boldsymbol{j}q_2 + \boldsymbol{k}q_3. \tag{4.13}$$

TABLE 2. Properties of the right quaternion Fourier transform ($a, b \in \mathbb{R}$)

Property	scalar functions	RQFT
Linearity	$af(x_1, x_2) + bg(x_1, x_2)$	$aF(u_1, u_2) + bG(u_1, u_2)$
Translation	$f(x_1 - a_1, x_2 - a_2)$	$e^{-\boldsymbol{i}2\pi u_1 a_1} F(u_1, u_2) e^{-\boldsymbol{j}2\pi u_2 a_2}$
Scaling	$f(a_1 x_1, a_2 x_2)$	$\frac{1}{\lvert a_1\rvert}\frac{1}{\lvert a_2\rvert} F\left(\frac{u_1}{a_1}, \frac{u_2}{a_2}\right)$
Partial derivative	$\frac{\partial^r f}{\partial x_1^r}(x_1, x_2)$	$F(u_1, u_2)(\boldsymbol{i}2\pi u_1)^r$
$(q = 4p)$	$\frac{\partial^q f}{\partial x_2^q}(x_1, x_2)$	$(2\pi u_2)^q F(u_1, u_2)$
$(q = 4p + 1)$		$(2\pi u_2)^q jK_1[F(u_1, u_2)]$
$(q = 4p + 2)$		$-(2\pi u_2)^q F(u_1, u_2)$
$(q = 4p + 3)$		$-(2\pi u_2)^q jK_1[F(u_1, u_2)]$
Plancherel	$\langle f, g\rangle =$	$\langle F, G\rangle$
Parseval	$\langle f, f\rangle =$	$\langle F, F\rangle$

4.2.2. Right Quaternion Fourier Transform (RQFT). As Clifford–Fourier transform, we shall take the right quaternion Fourier transform [1,26,28]

$$F(u_1, u_2) = \int_{\mathbb{R}^2} f(x_1, x_2) e^{-\boldsymbol{i}2\pi u_1 x_1} e^{-\boldsymbol{j}2\pi u_2 x_2} dx_1 dx_2 \tag{4.14}$$

and its inverse transform

$$f(x_1, x_2) = \int_{\mathbb{R}^2} F(u_1, u_2) e^{\boldsymbol{j}2\pi u_2 x_2} e^{\boldsymbol{i}2\pi u_1 x_1} du_1 du_2. \tag{4.15}$$

The scalar product of two quaternion functions $f(\boldsymbol{x})$, $g(\boldsymbol{x})$ with $(\boldsymbol{x}) = (x_1, x_2)$ is defined by

$$\langle f(\boldsymbol{x}), g(\boldsymbol{x})\rangle = \frac{1}{2}\int_{\mathbb{R}^2} (f g_c + g f_c)\, dx_1 dx_2 \tag{4.16}$$

where g_c, f_c are the quaternion conjugates. Table 2 lists a few properties of the right Fourier transform.

From the definition of the QFT and the above orthogonal symmetry properties, it follows that with $(\boldsymbol{u}) = (u_1, u_2)$

$$F(-u_1, u_2) = K_1[F(\boldsymbol{u})] = -\boldsymbol{j}F(\boldsymbol{u})\boldsymbol{j} \tag{4.17}$$

$$F(u_1, -u_2) = K_2[F(\boldsymbol{u})] = -\boldsymbol{i}F(\boldsymbol{u})\boldsymbol{i} \tag{4.18}$$

$$F(-u_1, -u_2) = K_{12}[F(\boldsymbol{u})] = -\boldsymbol{k}F(\boldsymbol{u})\boldsymbol{k} \tag{4.19}$$

Hence, only one orthant of the Fourier space is necessary to represent the entire Fourier space (Figure 2).

4.2.3. Analytic Signal. The analytic quaternion Fourier transform (Figure 3) is defined as

$$F_A(\boldsymbol{u}) = [1 + \operatorname{sign}(u_1)]\,[1 + \operatorname{sign}(u_2)]\,F(\boldsymbol{u}) \tag{4.20}$$

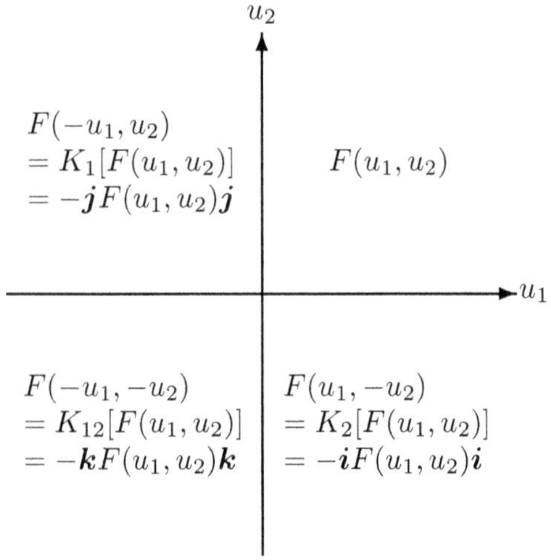

FIGURE 2. The quaternion Fourier spectrum of a real signal can be reconstructed from only one quadrant of the Fourier plane.

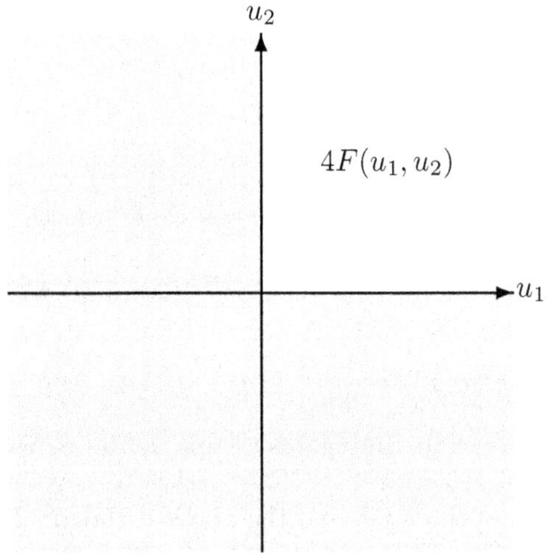

FIGURE 3. The spectrum of the analytic quaternion Fourier transform is obtained from only one quadrant.

and the analytic signal by

$$f_A(\boldsymbol{x}) = \int_{\mathbb{R}^2} F_A(\boldsymbol{u}) e^{\boldsymbol{j}2\pi u_2 x_2} e^{\boldsymbol{i}2\pi u_1 x_1} du_1 du_2. \tag{4.21}$$

The analytic signal satisfies Plancherel's theorem

$$\int_{\mathbb{R}^2} f_A(\boldsymbol{x}) \left[f_A(\boldsymbol{x})\right]_c dx_1 dx_2 = \int_{\mathbb{R}^2} F_A(\boldsymbol{u}) \left[F_A(\boldsymbol{u})\right]_c du_1 du_2. \tag{4.22}$$

TABLE 3. Examples of analytic signals.

Function	Analytic signal
$\cos\omega_1 x_1 \cos\omega_2 x_2$	$e^{\boldsymbol{j}\omega_2 x_2} e^{\boldsymbol{i}\omega_1 x_1}$
$\sin\omega_1 x_1 \sin\omega_2 x_2$	$\boldsymbol{k} e^{\boldsymbol{j}\omega_2 x_2} e^{\boldsymbol{i}\omega_1 x_1}$
$\sin\omega_1 x_1 \cos\omega_2 x_2$	$-\boldsymbol{i} e^{\boldsymbol{j}\omega_2 x_2} e^{\boldsymbol{i}\omega_1 x_1}$
$\cos\omega_1 x_1 \sin\omega_2 x_2$	$-\boldsymbol{j} e^{\boldsymbol{j}\omega_2 x_2} e^{\boldsymbol{i}\omega_1 x_1}$
$\cos(\omega_1 x_1 \pm \omega_2 x_2 + \varphi)$	$(1 \mp \boldsymbol{k})\, e^{\boldsymbol{j}\varphi} e^{\boldsymbol{j}\omega_2 x_2} e^{\boldsymbol{i}\omega_1 x_1}$

Examples of analytic signals are given in Table 3. The analytic signal being a quaternion, it can be represented as

$$q = |q|\, e^{\boldsymbol{k}\theta_3} e^{\boldsymbol{j}\theta_2} e^{\boldsymbol{i}\theta_1} = |q|\, r \tag{4.23}$$

with $|q| = \sqrt{qq_c}$ and where $r = e^{\boldsymbol{k}\theta_3} e^{\boldsymbol{j}\theta_2} e^{\boldsymbol{i}\theta_1}$ is a unit quaternion ($rr_c = 1$). The procedure of extracting the triplet of phases $(\theta_1, \theta_2, \theta_3)$ is recalled below [29]. Using the involutions K_1, K_2, K_{12} defined above one has (with $*$ representing the quaternion multiplication)

$$\begin{aligned} d_2 &= r * [K_2(r)]_c \\ &= [\cos 2\theta_2 \cos 2\theta_3, 0, \sin 2\theta_2, \cos 2\theta_2 \sin 2\theta_3] && (4.24) \\ d_{12} &= [K_{12}(r)]_c * r \\ &= [\cos 2\theta_1 \cos 2\theta_2, \cos 2\theta_2 \sin 2\theta_1, \sin 2\theta_2, 0]\,. && (4.25) \end{aligned}$$

From d_2 one obtains $\theta_2 = \frac{1}{2}\arcsin d_2(3)$ where $d_2(3)$ means the third component of d_2. If $\cos 2\theta_2 \neq 0$, one has

$$\theta_3 = \frac{\operatorname{Arg}[d_2(1) + \boldsymbol{i} d_2(4)]}{2}, \quad \theta_1 = \frac{\operatorname{Arg}[d_{12}(1) + \boldsymbol{i} d_{12}(2)]}{2}. \tag{4.26}$$

If $\cos 2\theta_2 = 0$ ($\theta_2 = \pm\pi/4$), one has an indeterminacy and only $(\theta_1 \pm \theta_3)$ can be determined; adopting the choice $\theta_3 = 0$, one has

$$d_1 = r * [K_1(r)]_c = [\cos 2\theta_1, 0, 0, \mp \sin 2\theta_1] \tag{4.27}$$

and thus

$$\theta_1 = \mp \frac{Arg[d_1(1) + \boldsymbol{i} d_1(4)]}{2}. \tag{4.28}$$

Finally, having determined the phases $(\theta_1, \theta_2, \theta_3)$ one computes $e^{\boldsymbol{k}\theta_3} e^{\boldsymbol{j}\theta_2} e^{\boldsymbol{i}\theta_1}$; if it is equal to $-r$ and $\theta_3 \geq 0$, one takes $\theta_3 - \pi$; if one has $-r$ and $\theta_3 < 0$ then one takes $\theta_3 + \pi$. Hence, the domain of the phases is

$$(\theta_1, \theta_2, \theta_3) \in \left[-\frac{\pi}{2}, \frac{\pi}{2}\right] \cup \left[-\frac{\pi}{2}, \frac{\pi}{2}\right] \cup [-\pi, \pi[\,. \tag{4.29}$$

5. Clifford Biquaternion 2D + t Analytic Signal: Implementation

5.1. Clifford Biquaternion Algebra

In 2D + t dimensions, we shall take as Clifford algebra, Clifford biquaternions having as generators ($\boldsymbol{e}_1 = \epsilon i, \boldsymbol{e}_2 = \epsilon j, \boldsymbol{e}_3 = \epsilon k, \epsilon = i'I, \epsilon^2 = 1, \boldsymbol{e}_i^2 = -1$) with i' designating the usual complex imaginary ($i'^2 = -1$) and where the tensor product $I = 1 \otimes i$ ($I^2 = -1$) commutes with $i = i \otimes 1, j = j \otimes 1, k = k \otimes 1$; $\boldsymbol{e}_1$ corresponds to the time axis x_1, $\boldsymbol{e}_2$ and $\boldsymbol{e}_3$ correspond respectively to the x_2 and x_3 axes. The full algebra contains the elements

$$\begin{bmatrix} 1 & i = \boldsymbol{e}_2\boldsymbol{e}_3 & j = \boldsymbol{e}_3\boldsymbol{e}_1 & k = \boldsymbol{e}_1\boldsymbol{e}_2 \\ \epsilon = i'I = -\boldsymbol{e}_1\boldsymbol{e}_2\boldsymbol{e}_3 & \boldsymbol{e}_1 = \epsilon i & \boldsymbol{e}_2 = \epsilon j & \boldsymbol{e}_3 = \epsilon k \end{bmatrix}. \tag{5.1}$$

Hence, a general element of the algebra can be expressed as a Clifford biquaternion (a clifbquat for short)

$$\begin{aligned} A &= p + \epsilon q \\ &= (p_0 + ip_1 + jp_2 + kp_3) + \epsilon(q_0 + iq_1 + jq_2 + kq_3) \\ &= [p_0, p_1, p_2, p_3] + \epsilon\,[q_0, q_1, q_2, q_3] \end{aligned} \tag{5.2}$$

where p, q are quaternions. The product of two clifbquats A and $B = p' + \epsilon q'$ is given by

$$\begin{aligned} AB &= (p + \epsilon q)\,(p' + \epsilon q') \\ &= (pp' + qq') + \epsilon\,(pq' + qp')\,. \end{aligned} \tag{5.3}$$

The conjugate of A is

$$\begin{aligned} A_c &= p_c + \epsilon q_c \\ &= [p_0, -p_1, -p_2, -p_3] + \epsilon\,[q_0, -q_1, -q_2, -q_3] \end{aligned} \tag{5.4}$$

where p_c, q_c are respectively the quaternion conjugates of p and q. The complex conjugate of A is

$$\overline{A} = p - \epsilon q. \tag{5.5}$$

A vector is expressed by

$$a = \epsilon(ia_1 + ja_2 + ka_3) \tag{5.6}$$

with $aa_c = a_1^2 + a_2^2 + a_3^2$, a bivector by $B = a \wedge b = -\frac{1}{2}(ab - ba)$, a trivector by $T = c \wedge B = \frac{1}{2}(cB + Bc)$. A unit vector a is defined by $aa_c = 1$.

Under an orthogonal symmetry with respect to a hyperplane (space of dimension $n - 1$) perpendicular to a unit vector a, a vector X is transformed into the vector $X' = aXa$, a bivector B into $B' = -aBa$ and a trivector T into $T' = aTa$, whereas a scalar remains invariant. These formulas allow us to derive the transformation of a clifbquat under an arbitrary orthogonal symmetry (Figure 4). In particular, if $a = \epsilon i$ (orthogonal symmetry with respect to the hyperplane Ox_2x_3), one has the involution

$$A' = K_1(A) = [p_0, p_1, -p_2, -p_3] + \epsilon\,[-q_0, -q_1, q_2, q_3]\,. \tag{5.7}$$

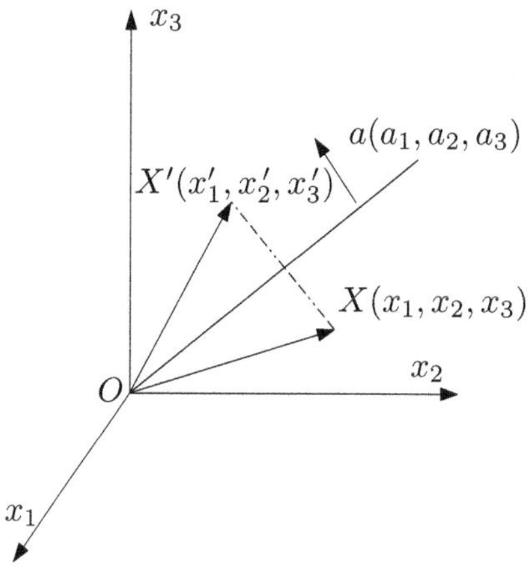

FIGURE 4. Orthogonal symmetry with respect to a hyperplane.

Similarly, if $a = \epsilon j$ (orthogonal symmetry with respect to the hyperplane Ox_1x_3), one has

$$A' = K_2(A) = [p_0, -p_1, p_2, -p_3] + \epsilon[-q_0, q_1, -q_2, q_3]. \tag{5.8}$$

If $a = \epsilon k$, one has

$$A' = K_3(A) = [p_0, -p_1, -p_2, p_3] + \epsilon[-q_0, q_1, q_2, -q_3]. \tag{5.9}$$

A combination of K_1 followed by K_2 leads to a rotation of π around the origin

$$A' = K_{12}(A) = rAr_c = [p_0, -p_1, -p_2, p_3] + \epsilon[q_0, -q_1, -q_2, q_3] \tag{5.10}$$

with $r = ji = -k$. Similarly,

$$A' = K_{13}(A) = [p_0, -p_1, p_2, -p_3] + \epsilon[q_0, -q_1, q_2, -q_3] \tag{5.11}$$

$$A' = K_{23}(A) = [p_0, p_1, -p_2, -p_3] + \epsilon[q_0, q_1, -q_2, -q_3]. \tag{5.12}$$

A combination of three symmetries leads to

$$A' = K_{123}(A) = [p_0, p_1, p_2, p_3] + \epsilon[-q_0, -q_1, -q_2, -q_3]. \tag{5.13}$$

5.2. Clifford–Fourier Transform

The Clifford–Fourier 2D + t transform and its inverse are respectively defined with $\boldsymbol{x} = (x_1, x_2, x_3)$, and $\boldsymbol{u} = (u_1, u_2, u_3)$ by

$$F(\boldsymbol{u}) = \int_{R^3} f(\boldsymbol{x}) e^{-\epsilon i 2\pi u_1 x_1} e^{-\epsilon j 2\pi u_2 x_2} e^{-\epsilon k 2\pi u_3 x_3} dx_1 dx_2 dx_3 \tag{5.14}$$

$$f(\boldsymbol{x}) = \int_{R^3} F(\boldsymbol{u}) e^{\epsilon k 2\pi u_3 x_3} e^{\epsilon j 2\pi u_2 x_2} e^{\epsilon 2\pi u_1 x_1} du_1 du_2 du_3. \tag{5.15}$$

TABLE 4. Computation principle of the Clifford–Fourier transform (CFT) with a clifbquat entry (for the inverse CFT, the formulas are the same except that the FFT is replaced by an IFFT and the integration order is reversed: x_3, x_2, x_1).

CFT of $f(\mathbf{x}) = [p_1, p_2, p_3, p_4] + \epsilon\,[p_5, p_6, p_7, p_8]$
real component: $p_\mu(\mathbf{x}) = a(\mathbf{x})$
FFT on x_1 : $a_1 + \epsilon i a_2$
FFT on x_2 : $(a_{1\alpha} + \epsilon j a_{1\beta}) + \epsilon i\,(a_{2\alpha} + \epsilon j a_{2\beta})$
FFT on x_3 : $(a_{1\alpha\gamma} + \epsilon k a_{1\alpha\delta}) + \epsilon j\,(a_{1\beta\gamma} + \epsilon k a_{1\beta\delta})$ $+\epsilon i\,[(a_{2\alpha\gamma} + \epsilon k a_{2\alpha\delta}) + \epsilon j\,(a_{2\beta\gamma} + \epsilon k a_{2\beta\delta})]$
$F_\mu = CFT(p_\mu) = [a_{1\alpha\gamma}, a_{1\beta\delta}, -a_{2\alpha\delta}, a_{2\beta\gamma}] + \epsilon\,[-a_{2\beta\delta}, a_{2\alpha\gamma}, a_{1\beta\gamma}, a_{1\alpha\delta}]$
CFT(f)=$(F_1 + iF_2 + jF_3 + kF_4) + \epsilon\,(F_5 + iF_6 + jF_7 + kF_8)$

Considering two clifbquat functions $f(\boldsymbol{x})$ and $g(\boldsymbol{x})$, one defines the product

$$\langle f, g\rangle = \frac{1}{2}\int_{\mathbb{R}^3} (f g_c + g f_c)\, dx_1 dx_2 dx_3 \tag{5.16}$$

which satisfies Plancherel's theorem

$$\langle f(\boldsymbol{x}), g(\boldsymbol{x})\rangle = \langle F(\boldsymbol{u}), G(\boldsymbol{u})\rangle\,. \tag{5.17}$$

To compute the direct Clifford–Fourier transform (CFT), one proceeds in cascade integrating first with respect to x_1 using a standard FFT. A second FFT (integration with respect to x_2) is then applied to each real component of the previous complex number. Then, a third FFT (integration on x_3) is applied on each of the resulting real components. Finally, all the components are properly displayed as a clifbquat. For the inverse Clifford–Fourier transform, one proceeds in the same way on each real component of the clifbquat using an IFFT and reversing the order of integration (see Table 4).

The above involutions lead to the following symmetries of the Clifford–Fourier transform of a scalar function $f(x_1, x_2, x_3)$

$$F(-u_1, u_2, u_3) = K_1\,[F(\boldsymbol{u})]\,, \qquad F(u_1, -u_2, u_3) = K_2\,[F(\boldsymbol{u})] \tag{5.18}$$

$$F(u_1, u_2, -u_3) = K_3\,[F(\boldsymbol{u})]\,, \qquad F(-u_1, -u_2, u_3) = K_{12}\,[F(\boldsymbol{u})] \tag{5.19}$$

$$F(-u_1, u_2, -u_3) = K_{13}\,[F(\boldsymbol{u})]\,, \qquad F(u_1, -u_2, -u_3) = K_{23}\,[F(\boldsymbol{u})] \tag{5.20}$$

$$F(-u_1, -u_2, -u_3) = K_{123}\,[F(\boldsymbol{u})]\,. \tag{5.21}$$

Hence, only one orthant of the Fourier space is necessary to represent the entire Fourier space. A few properties of the Clifford–Fourier transform of a scalar function are given in Table 5.

5.3. Analytic Signal

The analytic Clifford–Fourier transform is defined by

$$F_A(\boldsymbol{u}) = [1 + \operatorname{sign}(u_1)]\,[1 + \operatorname{sign}(u_2)]\,[1 + \operatorname{sign}(u_3)]\,F(\boldsymbol{u}) \tag{5.22}$$

TABLE 5. Properties of the Clifford–Fourier transform of a scalar function ($a, b \in \mathbb{R}$).

Property	Functions	CFT
Linearity	$af(\boldsymbol{x}) + bg(\boldsymbol{x})$	$aF(\boldsymbol{u}) + bG(\boldsymbol{u})$
Translation	$f(x_1 - a_1, x_2, x_3 - a_3)$	$e^{-\epsilon i 2\pi u_1 a_1} F(\boldsymbol{u}) e^{-\epsilon k 2\pi u_3 a_3}$
Scaling	$f(a_1 x_1, a_2 x_2, a_3 x_3)$	$\frac{1}{\lvert a_1\rvert}\frac{1}{\lvert a_2\rvert}\frac{1}{\lvert a_3\rvert} F\left(\frac{u_1}{a_1}, \frac{u_2}{a_2}, \frac{u_3}{a_3}\right)$
Partial derivative	$\frac{\partial^r f}{\partial x_1^r}(x_1, x_2, x_3)$	$F(\boldsymbol{u})(\epsilon i 2\pi u_1)^r$
$(q = 4p)$	$\frac{\partial^q f}{\partial x_2^q}(x_1, x_2, x_3)$	$(2\pi u_2)^q F(\boldsymbol{u})$
$(q = 4p + 1)$		$(2\pi u_2)^q \epsilon j K_1[F(\boldsymbol{u})]$
$(q = 4p + 2)$		$-(2\pi u_2)^q F(\boldsymbol{u})$
$(q = 4p + 3)$		$-(2\pi u_2)^q \epsilon j K_1[F(\boldsymbol{u})]$
	$\frac{\partial^r f}{\partial x_3^r}(x_1, x_2, x_3)$	$F(\boldsymbol{u})(\epsilon k 2\pi u_3)^r$
Plancherel	$\langle f, g\rangle =$	$\langle F, G\rangle$
Parseval	$\langle f, f\rangle =$	$\langle F, F\rangle$

TABLE 6. Examples of analytic signals.

Function: $f(\boldsymbol{x})$	Analytic signal: $f_A(\boldsymbol{x})$
$\cos\omega_1 x_1 \cos\omega_2 x_2 \cos\omega_3 x_3$	$e^{\epsilon k\omega_3 x_3} e^{\epsilon j\omega_2 x_2} e^{\epsilon i\omega_1 x_1}$
$\cos\omega_1 x_1 \sin\omega_2 x_2 \sin\omega_3 x_3$	$i e^{\epsilon k\omega_3 x_3} e^{\epsilon j\omega_2 x_2} e^{\epsilon i\omega_1 x_1}$
$\sin\omega_1 x_1 \sin\omega_2 x_2 \sin\omega_3 x_3$	$\epsilon e^{\epsilon k\omega_3 x_3} e^{\epsilon j\omega_2 x_2} e^{\epsilon i\omega_1 x_1}$
$\cos\omega_1 x_1 \cos\omega_2 x_2 \sin\omega_3 x_3$	$-\epsilon k e^{\epsilon k\omega_3 x_3} e^{\epsilon j\omega_2 x_2} e^{\epsilon i\omega_1 x_1}$
$\cos(\omega_1 x_1 + \omega_2 x_2 + \omega_3 x_3)$	$(1 - i + j - k)\, e^{\epsilon k\omega_3 x_3} e^{\epsilon j\omega_2 x_2} e^{\epsilon i\omega_1 x_1}$ $= 2e^{j\pi/4} e^{-k\pi/4} e^{\epsilon k\omega_3 x_3} e^{\epsilon j\omega_2 x_2} e^{\epsilon i\omega_1 x_1}$

and the analytic signal by

$$f_A(\boldsymbol{x}) = \int_{R^3} F_A(\boldsymbol{u}) e^{\epsilon k 2\pi u_3 x_3} e^{\epsilon j 2\pi u_2 x_2} e^{\epsilon i 2\pi u_1 x_1} du_1 du_2 du_3. \tag{5.23}$$

Examples of analytic signals are given in Table 6. The analytic signal being a Clifford biquaternion, it can be represented as $A = \lambda a$ where $\lambda = \alpha + \epsilon\beta$ consists of a scalar and a pseudo-scalar and where a is a unit Clifford biquaternion ($aa_c = 1$). Writing, $AA_c = \lambda^2 = g_1 + \epsilon g_2$, one finds

$$\alpha = \sqrt{\frac{g_1 + \sqrt{g_1^2 - g_2^2}}{2}}, \quad \beta = \sqrt{\frac{g_1 - \sqrt{g_1^2 - g_2^2}}{2}} \tag{5.24}$$

(if $\alpha = \beta = 0$, one adopts the choice $a = 1$, in order to have a unit Clifford biquaternion a in all cases). The unit Clifford biquaternion a is obtained *via* the relation

$$a = \lambda^{-1} A = (\alpha + \epsilon\beta)^{-1} A = \left(\frac{\alpha - \epsilon\beta}{\alpha^2 - \beta^2} \right) A, \tag{5.25}$$

and is decomposed according to $a = br$, with $b = b_1 + \epsilon(\boldsymbol{i}b_2 + \boldsymbol{j}b_3 + \boldsymbol{k}b_4)$ being a unit Clifford biquaternion such that $\overline{b_c} = b$ and where $r = r_1 + \boldsymbol{i}r_2 + \boldsymbol{j}r_3 + \boldsymbol{k}r_4$ with $rr_c = 1$. Using a procedure similar to that used in special relativity [11, p. 82], one has

$$b = \frac{1 + d}{\sqrt{2 + d + d_c}} \tag{5.26}$$

with $d = a(\overline{a_c})$. If $d = -1$, one has a particular case which has to be treated so as to yield a unit Clifford biquaternion b in all cases. For example, if $a = \pm\epsilon$, (with $d = -1$) one takes $b = \epsilon\boldsymbol{i}$; if $a = \epsilon(\boldsymbol{i}a_1 + \boldsymbol{j}a_2 + \boldsymbol{k}a_3)$, one adopts

$$b = \epsilon \frac{(\boldsymbol{i}a_1 + \boldsymbol{j}a_2 + \boldsymbol{k}a_3)}{\sqrt{a_1^2 + a_2^2 + a_3^2}}. \tag{5.27}$$

The unit Clifford biquaternion r is then obtained as $r = b_c a$. Both, b and r can be put into a polar form, respectively

$$b = e^{\epsilon j \varphi_2} \left[e^{\epsilon k \varphi_3} \left(e^{\epsilon i \varphi_1} \right) e^{\epsilon k \varphi_3} \right] e^{\epsilon j \varphi_2} \tag{5.28}$$

$$= \cos\varphi_1 \cos 2\varphi_2 \cos 2\varphi_3 + \epsilon(\imath \sin\varphi_1 + \jmath \cos\varphi_1 \cos 2\varphi_3 \sin 2\varphi_2 + k \cos\varphi_1 \sin 2\varphi_3) \tag{5.29}$$

with $\overline{b_c} = b$ and $r = e^{i\theta_1} e^{k\theta_3} e^{j\theta_2}$. The phases of b are extracted according to the rules:

$$\varphi_1 = \arcsin b(2,2) \tag{5.30}$$

where $b(2,2)$ means the second component of the second quaternion of b. If $\cos\varphi_1 \neq 0$,

$$\varphi_3 = \frac{1}{2} \arcsin \left(\frac{b(2,4)}{\cos\varphi_1} \right); \tag{5.31}$$

if $\cos\varphi_1 \neq 0$ and $\cos 2\varphi_3 \neq 0$

$$\varphi_2 = \frac{1}{2} \arcsin \left(\frac{b(2,3)}{\cos\varphi_1 \cos 2\varphi_3} \right). \tag{5.32}$$

The particular cases are treated as follows. If $\varphi_1 = \pm\frac{\pi}{2}$, the choice $\varphi_2 = \varphi_3 = 0$ is adopted; if $\varphi_1 \neq \pm\frac{\pi}{2}$ and $\varphi_3 = \pm\frac{\pi}{4}$, one chooses $\varphi_2 = 0$. The sign of the reconstructed b is then compared to that of the initial b; if it is opposed, φ_1 is replaced by $[\varphi_1 - \pi \operatorname{sign}(\varphi_1)]$. The phases of $r = e^{i\theta_1} e^{k\theta_3} e^{j\theta_2}$ are extracted according to the procedure presented for the 2D analytic signal by Sommer [29, p. 194] (except that eventual phase shifts are reported on θ_3 rather than on θ_1).

Within the 2D + t Clifford biquaternion framework, the procedure of extracting the triplet of phases $(\theta_1, \theta_2, \theta_3)$ is as follows. From the relations

$$\begin{aligned} d_1 &= r * [K_2(r)]_c \\ &= [\cos 2\theta_1 \cos 2\theta_3, \sin 2\theta_1 \cos 2\theta_3, 0, \sin 2\theta_3] \end{aligned} \tag{5.33}$$

$$\begin{aligned} d_2 &= [K_1(r)]_c * r \\ &= [\cos 2\theta_2 \cos 2\theta_3, 0, \cos 2\theta_3 \sin 2\theta_2, \sin 2\theta_3] \end{aligned} \tag{5.34}$$

one obtains $\theta_3 = \frac{1}{2} \arcsin d_1(4)$ where $d_1(4)$ means the fourth component of d_1. If $\cos 2\theta_3 \neq 0$, one has

$$\theta_1 = \frac{1}{2} \operatorname{Arg}[d_1(1) + id_1(2)], \quad \theta_2 = \frac{1}{2} \operatorname{Arg}[d_2(1) + id_2(3)]. \tag{5.35}$$

If $\cos 2\theta_3 = 0$, one has an indeterminacy and only $(\theta_1 \pm \theta_2)$ can be determined from the relation

$$d_3 = [K_{12}(r)]_c * r = [\cos 2(\theta_1 \mp \theta_2), 0, \mp \sin 2(\theta_1 \mp \theta_2), 0]; \tag{5.36}$$

adopting the choice $\theta_1 = 0$, one has

$$\theta_2 = \frac{1}{2} \operatorname{Arg}[d_3(1) + id_3(3)]. \tag{5.37}$$

Finally, having determined the phases $(\theta_1, \theta_2, \theta_3)$ one computes $e^{i\theta_1} e^{k\theta_3} e^{j\theta_2}$; if it is equal to $-r$ and $\theta_3 \geq 0$, one takes $\theta_3 - \pi$; if one has $-r$ and $\theta_3 < 0$ then one takes $\theta_3 + \pi$. Hence, the phase domains for r and for b are respectively

$$(\theta_1, \theta_2, \theta_3) \in \left[-\frac{\pi}{2}, \frac{\pi}{2}\right] \cup \left[-\frac{\pi}{2}, \frac{\pi}{2}\right] \cup [-\pi, \pi[\tag{5.38}$$

$$(\varphi_1, \varphi_2, \varphi_3) \in [-\pi, \pi[\cup \left[-\frac{\pi}{2}, \frac{\pi}{2}\right] \cup \left[-\frac{\pi}{2}, \frac{\pi}{2}\right]. \tag{5.39}$$

Finally, the analytic signal is characterized by a scalar, a pseudo-scalar and the six phases above.

6. Results

6.1. Introductory Example

One considers the function

$$f(t, x, y) = \cos(\omega_1 t + k_1 x + p_1 y) + \cos(\omega_2 t + k_2 x + p_2 y) \tag{6.1}$$

with $\omega_i = \Omega \pm \frac{\omega}{2}$, $k_i = K \pm \frac{k}{2}$, $P_i = P \pm \frac{p}{2}$ and $\omega_2 > \omega_1$, $k_2 > k_1$, $p_2 > p_1$.

The analytic signal is given by

$$f_A(t,x,y) = \begin{bmatrix} 2\cos(\omega t + kx + py)\cos(\Omega t + Kx + Py), \\ -2\cos(\omega t - kx + py)\cos(\Omega t - Kx + Py), \\ 2\cos(\omega t + kx - py)\cos(\Omega t - Kx - Py), \\ -2\cos(\omega t - kx - py)\cos(\Omega t - Kx - Py) \end{bmatrix} + \epsilon \begin{bmatrix} 2\cos(\omega t - kx + py)\sin(\Omega t - Kx + Py), \\ 2\cos(\omega t + kx + py)\sin(\Omega t + Kx + Py), \\ -2\cos(\omega t - kx - py)\sin(\Omega t - Kx - Py), \\ -2\cos(\omega t + kx - py)\sin(\Omega t + Kx - Py) \end{bmatrix}. \tag{6.2}$$

One then computes

$$d_4 = [f_A(1)]^2 + [f_A(6)]^2 = 4\cos^2(\omega t + kx + py) \tag{6.3}$$

$$d_5 = f_A f_{Ac} = \begin{matrix} [8(1 + \cos 2\omega t \cos 2kx \cos 2py), 0, 0, 0] \\ +\epsilon\,[8 \sin 2\omega t \sin 2Kx \cos 2py, 0, 0, 0] \end{matrix}. \tag{6.4}$$

The phase and group velocities are obtained as follows. Write

$$\Phi_1 = \Omega t + Kx + Py, \Phi_2 = \omega t + kx + py; \tag{6.5}$$

one has

$$\Phi_1 = \arctan \frac{f_A(6)}{f_A(1)} \tag{6.6}$$

$$\Phi_2 = \frac{1}{2} \arccos\left(\frac{d_4}{2} - 1\right). \tag{6.7}$$

Hence the phase and group velocities are

$$v_{\varphi x} = \frac{\frac{\partial \Phi_1}{\partial t}}{\frac{\partial \Phi_1}{\partial x}}, \quad v_{\varphi y} = \frac{\frac{\partial \Phi_1}{\partial t}}{\frac{\partial \Phi_1}{\partial y}} \tag{6.8}$$

$$v_{gx} = \frac{\frac{\partial \Phi_2}{\partial t}}{\frac{\partial \Phi_2}{\partial x}}, \quad v_{gy} = \frac{\frac{\partial \Phi_2}{\partial t}}{\frac{\partial \Phi_2}{\partial y}}. \tag{6.9}$$

Finally, one might notice that the scalar part of d_5 contains only the modulation of the signal.

6.2. Analytic Video Signal 2D + t Implementation

The implementation considers the scalar function

$$f(x_1, x_2, x_3) = \cos x_1 \cos x_2 \cos x_3 \tag{6.10}$$

(x_1 being the time and x_2, x_3 respectively x, y) where (x_1, x_2, x_3) vary between $[0, 2\pi]$. The explicit form of the analytic video signal is

$$A = \left[\begin{array}{c} \left(\begin{array}{c} \cos x_1 \cos x_2 \cos x_3, -\cos x_1 \sin x_2 \sin x_3, \\ \sin x_1 \cos x_2 \sin x_3, -\sin x_1 \sin x_2 \cos x_3 \end{array} \right) \\ +\epsilon \left(\begin{array}{c} \sin x_1 \sin x_2 \sin x_3, \sin x_1 \cos x_2 \cos x_3, \\ \cos x_1 \sin x_2 \cos x_3, \cos x_1 \cos x_2 \sin x_3 \end{array} \right) \end{array} \right] \tag{6.11}$$

with $AA_c = 1$, hence $\lambda = 1$. The eight components of A are represented at the time $x_1 = \frac{2\pi}{8}$ s in Figure 5

$$A = (a, b, c, d) + \epsilon\, (e, f, g, h) . \tag{6.12}$$

The eight components in polar form $(\alpha, \theta_1, \theta_2, \theta_3, \beta, \varphi_1, \varphi_2, \varphi_3)$ are represented in Figure 6.

7. Conclusion

This chapter has presented a concrete algebraic framework, *i.e.*, Clifford's biquaternions for the expression of Clifford–Fourier transforms and 2D + t analytic signals. Then, we have shown how to put the analytic signal into a polar form constituted by a scalar, a pseudoscalar and six phases. Finally, using discrete fast Fourier transforms we have implemented numerically the Clifford biquaternion Fourier transform, the analytic Fourier transform and the analytic signal both in standard and polar form. Our next objective will be to extract physical information from the phases in medical images.

Appendix A. Examples of (2D + t) Clifford–Fourier Transforms

A.1. Example 1

Signal

$$f(x_1, x_2, x_3) = \cos 2\pi f_1 x_1 \cos 2\pi f_2 x_2 \cos 2\pi f_3 x_3 .$$

Clifford–Fourier transform

$$\begin{aligned} F(u_1, u_2, u_3) = {} & \frac{1}{8} \left[\delta\,(f_1 - u_1) + \delta\,(f_1 + u_1) \right] \left[\delta\,(f_2 - u_2) + \delta\,(f_2 + u_2) \right] \\ & \times \left[\delta\,(f_3 - u_3) + \delta\,(f_3 + u_3) \right] . \end{aligned}$$

Analytic Clifford–Fourier transform

$$F_A(u_1, u_2, u_3) - \delta\,(f_1 - u_1)\, \delta\,(f_2 - u_2)\, \delta\,(f_3 - u_3) .$$

Analytic signal

$$f_A(x_1, x_2, x_3) = e^{ck 2\pi f_3 x_3} e^{cj 2\pi f_2 x_2} e^{\epsilon i 2\pi f_1 x_1} .$$

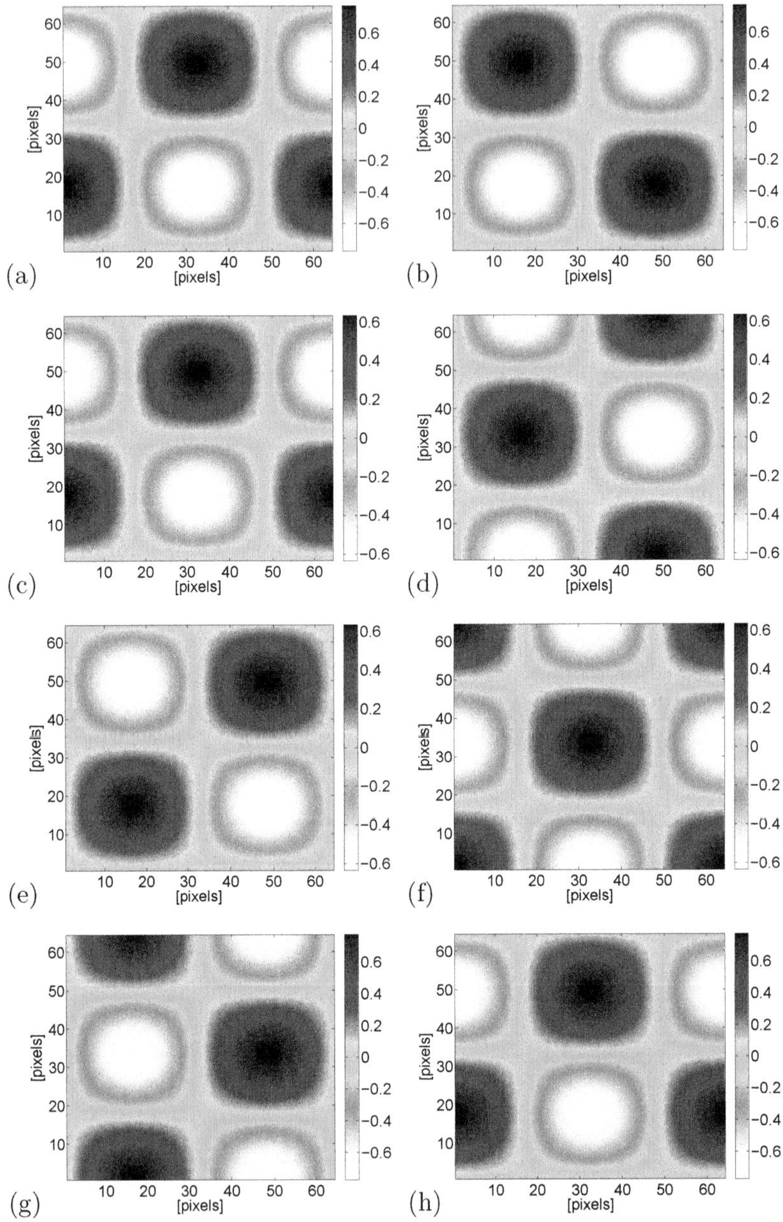

FIGURE 5. The eight components of the analytic video signal $A = (a, b, c, d) + \epsilon(e, f, g, h)$ at the time $x_1 = \frac{2\pi}{8}$ s.

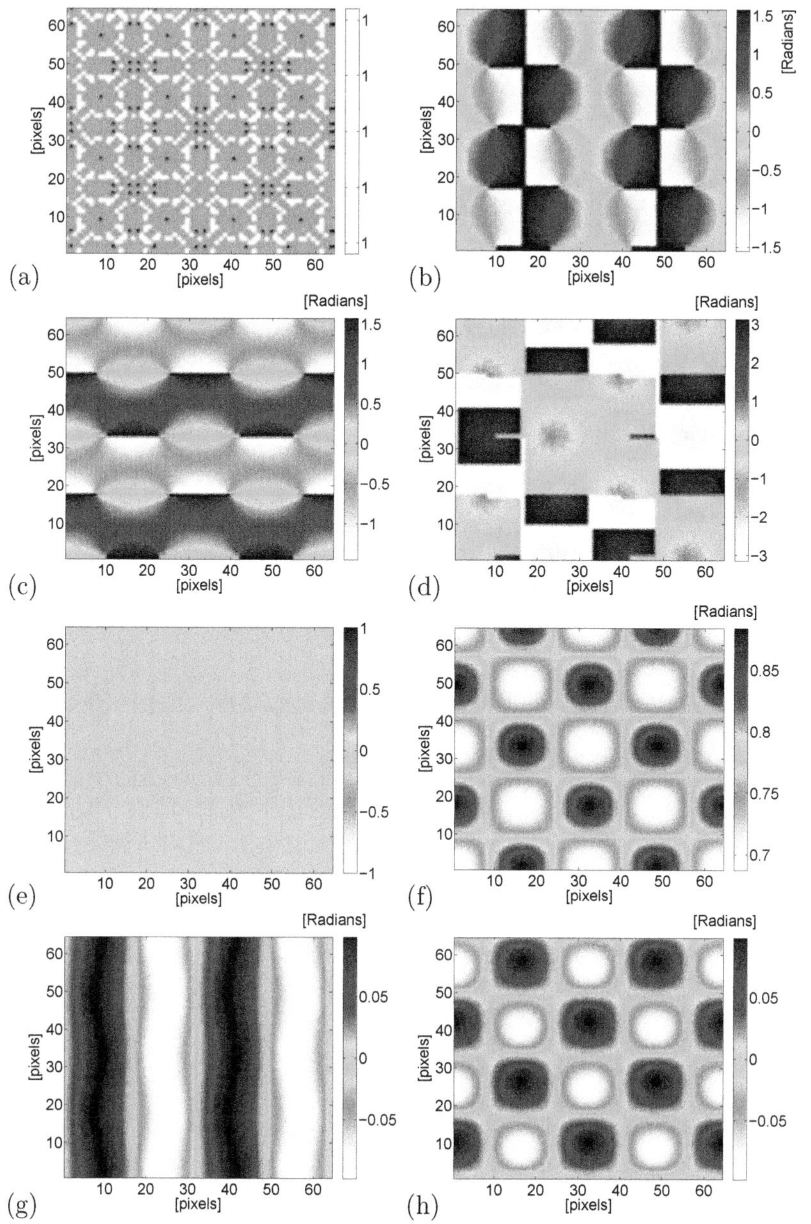

FIGURE 6. Polar form of the analytic video signal at the time $x_1 = \frac{2\pi}{8}$ s, (a) is the scalar, (b, c, d) are the three phases $(\theta_1, \theta_2, \theta_3)$; (e) is the pseudo-scalar and (f, g, h) are the three phases $(\varphi_1, \varphi_2, \varphi_3)$.

A.2. Example 2

Signal

$$f(x_1, x_2, x_3) = \cos(2\pi f_1 x_1 + 2\pi f_2 x_2 + 2\pi f_3 x_3) = f_1 + f_2 + f_3 + f_4$$

with

$$\begin{aligned}
f_1 &= \cos 2\pi f_1 x_1 \cos 2\pi f_2 \pi x_2 \cos 2\pi f_3 x_3 \\
f_2 &= -\sin 2\pi f_1 x_1 \sin 2\pi f_2 \pi x_2 \cos 2\pi f_3 x_3 \\
f_3 &= -\sin 2\pi f_1 x_1 \cos 2\pi f_2 \pi x_2 \sin 2\pi f_3 x_3 \\
f_4 &= -\cos 2\pi f_1 x_1 \sin 2\pi f_2 \pi x_2 \sin 2\pi f_3 x_3.
\end{aligned}$$

Clifford–Fourier transform

$$F(u_1, u_2, u_3) = F_1 + F_2 + F_3 + F_4, \quad F_i = CFT(f_i)$$

where

$$\begin{aligned}
F_1 = \frac{1}{8} & [\delta(f_1 - u_1) + \delta(f_1 + u_1)] [\delta(f_2 - u_2) + \delta(f_2 + u_2)] \\
& \times [\delta(f_3 - u_3) + \delta(f_3 + u_3)]
\end{aligned}$$

and

$$\begin{aligned}
F_2 &= \frac{-1}{8} [-\epsilon i\delta(f_1 - u_1) + \epsilon i\delta(f_1 + u_1)] [-\epsilon j\delta(f_2 - u_2) + \epsilon j\delta(f_2 + u_2)] \\
& \quad \times [-\delta(f_3 - u_3) + \delta(f_3 + u_3)] \\
&= \frac{-k}{8} [-\delta(f_1 - u_1) + \delta(f_1 + u_1)] [-\delta(f_2 - u_2) + \delta(f_2 + u_2)] \\
& \quad \times [-\delta(f_3 - u_3) + \delta(f_3 + u_3)]
\end{aligned}$$

and

$$\begin{aligned}
F_3 &= \frac{-1}{8} [-\epsilon i\delta(f_1 - u_1) + \epsilon i\delta(f_1 + u_1)] [\delta(f_2 - u_2) + \delta(f_2 + u_2)] \\
& \quad \times [-\epsilon k\delta(f_3 - u_3) + \epsilon k\delta(f_3 + u_3)] \\
&= \frac{j}{8} [-\delta(f_1 - u_1) + \delta(f_1 + u_1)] [\delta(f_2 - u_2) + \delta(f_2 + u_2)] \\
& \quad \times [-\delta(f_3 - u_3) + \delta(f_3 + u_3)]
\end{aligned}$$

and

$$\begin{aligned}
F_4 &= \frac{-1}{8} [\delta(f_1 - u_1) + \delta(f_1 + u_1)] [-\epsilon j\delta(f_2 - u_2) + \epsilon j\delta(f_2 + u_2)] \\
& \quad \times [-\epsilon k\delta(f_3 - u_3) + \epsilon k\delta(f_3 + u_3)] \\
&= \frac{-i}{8} [\delta(f_1 - u_1) + \delta(f_1 + u_1)] [-\delta(f_2 - u_2) + \delta(f_2 + u_2)] \\
& \quad \times [-\delta(f_3 - u_3) + \delta(f_3 + u_3)].
\end{aligned}$$

Analytic Clifford–Fourier transform

$$F_A(u_1, u_2, u_3) = (1 - i + j - k)\, \delta(f_1 - u_1)\, \delta(f_2 - u_2)\, \delta(f_3 - u_3)$$

Analytic signal

$$f_A(x_1, x_2, x_3) = (1 - i + j - k) e^{\epsilon k 2\pi f_3 x_3} e^{\epsilon j 2\pi f_2 x_2} e^{\epsilon i 2\pi f_1 x_1} = p + \epsilon q$$

where

$$p = \begin{bmatrix} \cos(2\pi f_1 x_1 + 2\pi f_2 x_2 + 2\pi f_3 x_3), -\cos(2\pi f_1 x_1 - 2\pi f_2 x_2 + 2\pi f_3 x_3), \\ \cos(2\pi f_1 x_1 + 2\pi f_2 x_2 - 2\pi f_3 x_3), -\cos(2\pi f_1 x_1 - 2\pi f_2 x_2 - 2\pi f_3 x_3) \end{bmatrix}$$

$$q = \begin{bmatrix} \sin(2\pi f_1 x_1 - 2\pi f_2 x_2 + 2\pi f_3 x_3), \quad \sin(2\pi f_1 x_1 + 2\pi f_2 x_2 + 2\pi f_3 x_3), \\ -\sin(2\pi f_1 x_1 - 2\pi f_2 x_2 - 2\pi f_3 x_3), -\sin(2\pi f_1 x_1 + 2\pi f_2 x_2 - 2\pi f_3 x_3) \end{bmatrix}$$

References

[1] F. Brackx, N. De Schepper, and F. Sommen. The two-dimensional Clifford–Fourier transform. *Journal of Mathematical Imaging and Vision*, 26(1):5–18, 2006.

[2] T. Bülow. *Hypercomplex Spectral Signal Representations for the Processing and Analysis of Images.* PhD thesis, University of Kiel, Germany, Institut für Informatik und Praktische Mathematik, Aug. 1999.

[3] T. Bülow and G. Sommer. A novel approach to the 2D analytic signal. In F. Solina and A. Leonardis, editors, *Computer Analysis of Images and Patterns*, volume 1689 of *Lecture Notes in Computer Science*, pages 25–32. Springer, Berlin/Heidelberg, 1999.

[4] W.K. Clifford. Applications of Grassmann's extensive algebra. *American Journal of Mathematics*, 1(4):350–358, 1878.

[5] W.K. Clifford. *Mathematical Papers.* Chelsea Publishing Company, New York, 1968. First published 1882, edited by R. Tucker.

[6] M.J. Crowe. *A History of Vector analysis: The Evolution of the Idea of a Vectorial System.* University of Notre Dame, Notre Dame, London, 1967.

[7] P. Delachartre, P. Clarysse, R. Goutte, and P.R. Girard. Mise en oeuvre du signal analytique dans les algèbres de Clifford. In *GRETSI*, page 4, Bordeaux, France, 2011.

[8] T.A. Ell. *Hypercomplex Spectral Transformations.* PhD thesis, University of Minnesota, June 1992.

[9] D. Gabor. Theory of communication. *Journal of the Institution of Electrical Engineers*, 93(26):429–457, 1946. Part III.

[10] P.R. Girard. *Quaternions, Algèbre de Clifford et Physique Relativiste.* PPUR, Lausanne, 2004.

[11] P.R. Girard. *Quaternions, Clifford Algebras and Relativistic Physics.* Birkhäuser, Basel, 2007. Translation of [10].

[12] P.R. Girard. Quaternion Grassmann-Hamilton-Clifford-algebras: new mathematical tools for classical and relativistic modeling. In O. Dössel and W.C. Schlegel, editors, *World Congress on Medical Physics and Biomedical Engineering, September* 7–12, 2009, volume 25/IV of *IFMBE Proceedings*, pages 65–68. Springer, 2010.

[13] P.R. Girard. Multiquaternion Grassmann-Hamilton-Clifford algebras in physics and engineering: a short historical perspective. In K. Gürlebeck, editor, *9th International Conference on Clifford Algebras and their Applications*, page 9. Weimar, Germany, 15–20 July 2011.

[14] H. Grassmann. *Die lineale Ausdehungslehre: ein neuer Zweig der Mathematik, dargestellt und durch Anwendungen auf die übrigen Zweige der Mathematik, wie auch die Statik, Mechanik, die Lehre von Magnetismus und der Krystallonomie erläutert.* Wigand, Leipzig, second, 1878 edition, 1844.

[15] H. Grassmann. Der Ort der Hamilton'schen Quaternionen in der Ausdehnungslehre. *Mathematische Annalen*, 12:375–386, 1877.

[16] H. Grassmann. *Gesammelte mathematische und physikalische Werke.* B.G. Teubner, Leipzig, 1894–1911. 3 volumes in 6 parts.

[17] A. Gsponer and J.P. Hurni. Quaternions in mathematical physics (1): Alphabetical bibliography. Preprint, available at: `http://www.arxiv.org/abs/arXiv:mathph/0510059`, 2008. 1430 references.

[18] A. Gsponer and J.P. Hurni. Quaternions in mathematical physics (2): Analytical bibliography. Preprint, available at: `http://www.arxiv.org/abs/arXiv:mathphy/0511092`, 2008. 1100 references.

[19] K. Gürlebeck and W. Sprößig. *Quaternionic and Clifford Calculus for Physicists and Engineers.* Wiley, Aug. 1997.

[20] S.L. Hahn. Multidimensional complex signals with single-orthant spectra. *Proceedings of the IEEE*, 80(8):1287–1300, Aug. 1992.

[21] S.L. Hahn and K.M. Snopek. The unified theory of n-dimensional complex and hypercomplex analytic signals. *Bulletin of the Polish Academy of Sciences Technical Sciences*, 59(2):167–181, 2011.

[22] H. Halberstam and R.E. Ingram, editors. *The Mathematical Papers of Sir William Rowan Hamilton*, volume III Algebra. Cambridge University Press, Cambridge, 1967.

[23] W.R. Hamilton. *Elements of Quaternions.* Chelsea Publishing Company, New York, reprinted 1969 edition, 1969. 2 volumes (1899–1901).

[24] T.L. Hankins. *Sir William Rowan Hamilton.* Johns Hopkins University Press, Baltimore, London, 1980.

[25] J.P. Havlicek, J.W. Havlicek, and A.C. Bovik. The analytic image. In *Proceedings* 1997 *International Conference on Image Processing* (*ICIP* '97), volume 2, pages 446–449, Washington, DC, USA, October 26–29 1997.

[26] E. Hitzer. Quaternion Fourier transform on quaternion fields and generalizations. *Advances in Applied Clifford Algebras*, 17(3):497–517, May 2007.

[27] H.-J. Petsche. *Grassmann.* Birkhäuser, Basel, 2006.

[28] S.J. Sangwine and T.A. Ell. The discrete Fourier transform of a colour image. In J.M. Blackledge and M.J. Turner, editors, *Image Processing II Mathematical Methods, Algorithms and Applications*, pages 430–441, Chichester, 2000. Horwood Publishing for Institute of Mathematics and its Applications. Proceedings Second IMA Conference on Image Processing, De Montfort University, Leicester, UK, September 1998.

[29] G. Sommer, editor. *Geometric computing with Clifford Algebras: Theoretical Foundations and Applications in Computer Vision and Robotics.* Springer, Berlin, 2001.

[30] J. Ville. Théorie et applications de la notion de signal analytique. *Câbles et Transmission*, 2A:61–74, 1948.

[31] J. Vince. *Geometric Algebra for Computer Graphics.* Springer, London, 2008.

P.R. Girard, P. Clarysse, A. Marion, R. Goutte and P. Delachartre
Université de Lyon, CREATIS; CNRS UMR 5220
Inserm U1044; INSA-Lyon;
Université Lyon 1, France
Bât. Blaise Pascal
7 avenue Jean Capelle
F-69621 Villeurbanne, France
e-mail: patrick.girard@creatis.insa-lyon.fr
patrick.clarysse@creatis.insa-lyon.fr
adrien.marion@creatis.insa-lyon.fr
robert.goutte@creatis.insa-lyon.fr
philippe.delachartre@creatis.insa-lyon.fr

R. Pujol
Université de Lyon
Pôle de Mathématiques, INSA-Lyon
Bât. Léonard de Vinci
21 avenue Jean Capelle
F-69621 Villeurbanne, France
e-mail: romaric.pujol@insa-lyon.fr

Quaternion and Clifford–Fourier Transforms and Wavelets
Trends in Mathematics, 221–246

11 Generalized Analytic Signals in Image Processing: Comparison, Theory and Applications

Swanhild Bernstein, Jean-Luc Bouchot, Martin Reinhardt and Bettina Heise

Abstract. This article is intended as a mathematical overview of the generalizations of analytic signals to higher-dimensional problems, as well as of their applications to and of their comparison on artificial and real-world image samples.

We first start by reviewing the basic concepts behind analytic signal theory and derive its mathematical background based on boundary value problems of one-dimensional analytic functions. Following that, two generalizations are motivated by means of higher-dimensional complex analysis or Clifford analysis. Both approaches are proven to be valid generalizations of the known analytic signal concept.

In the last part we experimentally motivate the choice of such higher-dimensional analytic or monogenic signal representations in the context of image analysis. We see how one can take advantage of one or the other representation depending on the application.

Mathematics Subject Classification (2010). Primary 94A12; secondary 44A12, 30G35.

Keywords. Monogenic signal, monogenic functional theory, image processing, texture, Riesz transform.

1. Introduction

In the past years and since the pioneer work of Gabor [11], the analytic signal has attracted much interest in signal processing and information theory. Due to an orthogonal decomposition of oscillating signals into envelope and instantaneous phase or respectively into energetic and structural components, this concept has become very suitable for analyzing signals. In this context such a property is called a split of identity and allows to separate the different characteristics of a signal into useful components.

While this approach has given rise to many one-dimensional signal processing methods, other developments have been directed towards higher-dimensional generalizations. Of particular interest is the two-dimensional case, *i.e.*, how to deal with images in an analytic way. As it will be demonstrated in our chapter, two main directions have been taken, one based on multidimensional complex analysis and another one based on Clifford analysis.

This article is intended as an overview of the mathematical concepts behind analytic signals based on the Hilbert transform (Section 2). Then, the mathematical generalizations are detailed in Section 3. The end of that section is dedicated to illustrative examples of the detailed differences between the two generalizations. Section 4 describes the use of spinors for image analysis tasks. The last section of this article (Section 5) illustrates their applications like demodulation of two-dimensional AM-FM signals as provided, *e.g.*, in interferometry and some applications to the processing of natural images.

2. Analytic Signal Theory and Signal Decomposition

Analytic signals were introduced for signal processing in the context of communication theory in the late 40s [11]. Since then, there has been a growing interest in the analytic signal as a useful tool for representing real-valued signals [25]. We start here by first reviewing the basics of analytic signal theory and the Hilbert transform and see why the so-called *split of identity* is an interesting property. In the last part we review the mathematical basics and see how we can derive the analytic signal from a boundary value problem in complex analysis.

2.1. Basic Analytic Signal Theory and the Hilbert Transform

Definition 2.1 (One-dimensional Fourier Transform). In the following, we use as Fourier transform $\mathcal{F}$:

$$\mathcal{F}(f)(u) = \widehat{f}(u) = \frac{1}{\sqrt{2\pi}} \int_{\mathbb{R}} f(t) e^{-itu} \mathrm{d}t \tag{2.1}$$

for $t \in \mathbb{R}, u \in \mathbb{R}$ and $f \in L^2(\mathbb{R})$

Definition 2.2 (Hilbert Transform). The Hilbert transform of a signal $f \in L^2(\mathbb{R})$ (or more generally $f \in L^p(\mathbb{R}), 1 < p < \infty$) is defined, either in the spatial domain as a convolution with the Hilbert kernel (2.2), or as a Fourier multiplier (2.3):

$$\mathcal{H}f = h * f \tag{2.2}$$

$$\mathcal{F}(\mathcal{H}f)(u) = -i\,\mathrm{sign}(u)\mathcal{F}(f)(u) \tag{2.3}$$

where we have made use of two functions:

- The Hilbert kernel $h(t) = \frac{1}{\pi t}$.
- The operator $\mathrm{sign}(u) = \begin{cases} 1 & u > 0 \\ 0 & u = 0 \\ -1 & u < 0. \end{cases}$

Following its definition, we notice that the Hilbert transform acts as an asymmetric phase shift: if we write $\pm i = e^{\pm i\pi/2}$, the phase of the Fourier spectrum of the Hilbert transform is obtained by a rotation of $\pm 90°$.

Proposition 2.3 (Properties of the Hilbert Transform). *Given a signal f the following hold true:*

- $\forall u \neq 0,\ |\mathcal{H}f(u)| = |\mathcal{F}(f)(u)|$,
- $\mathcal{H}\mathcal{H}f = -f \Rightarrow \mathcal{H}^{-1} = -\mathcal{H}$.

Note that a constant function being not in L^2 can not be reconstructed in this way.

The analytic signal is computed as a complex combination of both the original signal and its Hilbert transform:

Definition 2.4 (Analytic Signal).

$$f_A = f + i\mathcal{H}f. \tag{2.4}$$

Due to its definition, an analytic signal has a one-sided Fourier spectrum. Moreover, its values are doubled on the positive side. We also remark that it is possible to recover the original signal from its analytic description by taking the real part.

The following proposition holds:

Proposition 2.5.

$$\langle f, \mathcal{H}f\rangle_{L_2} = 0 \qquad \textit{Orthogonality}, \tag{2.5}$$

$$\|f\|_2^2 = \|\mathcal{H}f\|_2^2 \qquad \textit{Energy conservation}. \tag{2.6}$$

The energy equality is valid only if the DC, or zero-frequency component of the signal is neglected [10].

Note that it is possible to write the complex analytic signal in polar coordinates. In this case we have: $\forall t \in \mathbb{R}, f_A(t) = A(t)e^{i\phi(t)}$. A is called the local amplitude and ϕ is called the local phase. These local features are defined as follows [11]:

Definition 2.6 (Local features).

$$A(t) = \sqrt{f(t)^2 + \mathcal{H}f(t)^2} \tag{2.7}$$

$$\phi(t) = \arctan\left(\frac{\mathcal{H}f(t)}{f(t)}\right) = \arctan\left(\frac{\Im(f_A(t))}{\Re(f_A(t))}\right). \tag{2.8}$$

Proposition 2.7 (Invariance–Equivariance, Split of identity [10]). *The local phase, together with the local amplitude, fulfil the properties of invariance and equivariance:*

- *The local phase depends only on the local structure.*
- *The local amplitude depends only on the local energy.*

If moreover these features constitute a complete description of the signal, they are said to perform a split of identity.

However as stated in [10], a split of identity is strictly valid only for band-limited signals with a local zero mean property.

If these conditions are fulfilled the analytic signal representation relies on an orthogonal decomposition of the *structural information* (the local phase), and the *energetic information* (the local amplitude).

This split of identity is illustrated in Figure 1. The first plot represents three signals. They are sine waves generated from a mother sine wave (the red one). The blue curve corresponds to a modification in terms of amplitude of the red one, while the green curve has half the frequency of the red one. Figures 1(b) and 1(c) are respectively the local amplitudes and phases of these three signals. Note that a small phase shift has been added to the blue curve for better readability. We can clearly see that due to the split of identity, modifying one local characteristic of the signal does not affect the other one and vice versa.

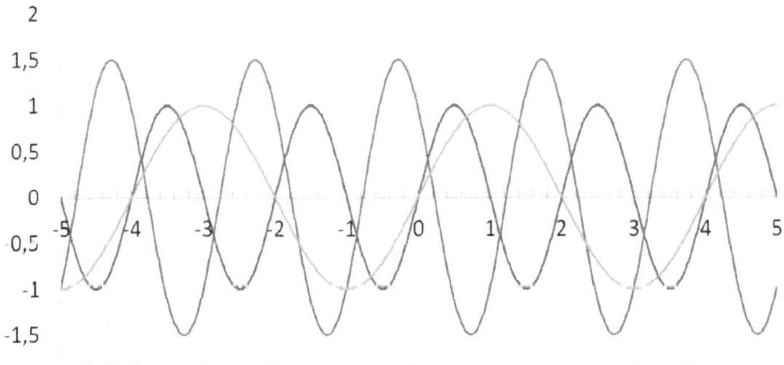

(a) Signals with varying frequency or amplitude

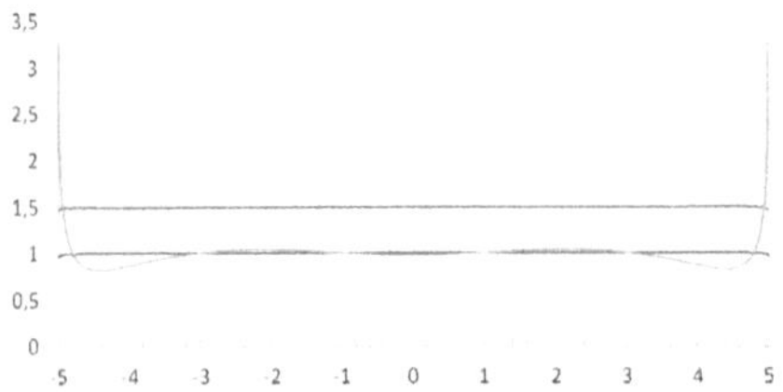

(b) Local amplitudes

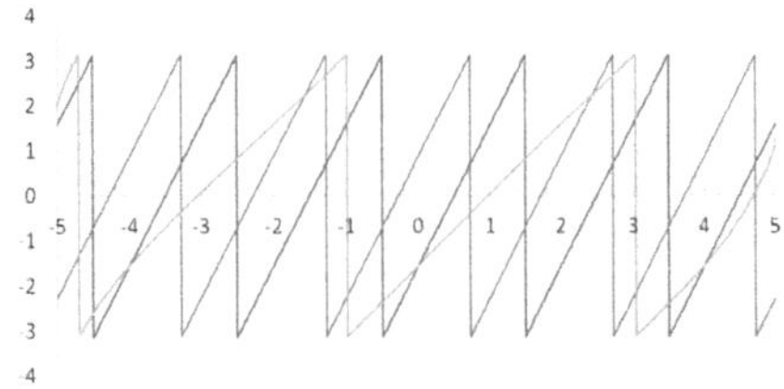

(c) Local phases

FIGURE 1. Illustration of the split of identity. (Explanation in the text.)

2.2. From Analytic Function to Analytic Signal

While the analytic signal is a very common concept in the field of signal theory, its basic mathematics can be derived from the theory of analytic functions. The close connection can be understood when considering the following Riemann–Hilbert problem with respect to the complex parameter $z = x + iy$:

$$\frac{\partial F}{\partial \bar{z}} = 0, \qquad z \in \mathbb{C}, y \geq 0, \tag{2.9}$$

$$\Re(F(x)) = f(x), \qquad x \in \mathbb{R}. \tag{2.10}$$

One solution of this problem is given by the Cauchy integral

$$F(z) = F_\Gamma f(z) := \frac{1}{2\pi i} \int_{\mathbb{R}} \frac{1}{\tau - z} f(\tau) \mathrm{d}\tau. \tag{2.11}$$

Of course this solution is unique only up to a constant. Normally, this constant will be fixed by the condition $\Im(F(z_0)) = c$, *i.e.*, the imaginary part of F given at an interior point.

When we now consider the trace of F_Γ, *i.e.*, the boundary value, we arrive at the so-called Plemelj–Sokhotzki formula:

$$\mathrm{tr} F_\Gamma f = \frac{1}{2}(I + i\mathcal{H})f = \frac{1}{2}f + \frac{1}{2}i\mathcal{H}f =: P_\Gamma f. \tag{2.12}$$

Up to the factor $1/2$ this corresponds to our above definition of an analytic signal.

In this way an analytic signal represents the boundary values of an analytic function in the upper half-plane (or for periodic functions in the unit disc). Starting from this concept we are now going to take a look at higher-dimensional generalizations.

3. Higher-dimensional Generalizations

Different approaches have been studied in past years to extend the definition of an analytic signal to higher-dimensional spaces. Two of them have gained the greatest interest based respectively on multidimensional complex analysis and Clifford analysis.

3.1. Using Multiple Complex Variables

3.1.1. Mathematics. In 1998 Bülow proposed a definition of a hypercomplex signal based on the so-called partial and total Hilbert transform [6]. To adapt this to our point of view, that analytic signals are functions in a Hardy space, we consider the following Riemann–Hilbert problem in $\mathbb{C}^2$:

$$\frac{\partial F}{\partial \bar{z}_1} = 0, \qquad (z_1, z_2) \in \mathbb{C}^2, y_1, y_2 \geq 0, \tag{3.1}$$

$$\frac{\partial F}{\partial \bar{z}_2} = 0, \qquad (z_1, z_2) \in \mathbb{C}^2, y_1, y_2 \geq 0, \tag{3.2}$$

$$\Re(F(x_1, x_2)) = f(x_1, x_2), \qquad x_1, x_2 \in \mathbb{R}^2. \tag{3.3}$$

For the solution, (see, *e.g.*, [8] or [21]), we need to point out that the domain is a poly-domain in the sense of $\mathbb{C}^n$, so that we can give it in the form of a Cauchy integral:

$$F(z_1, z_2) = \frac{1}{4\pi^2} \int_{\mathbb{R}^2} \frac{1}{(\boldsymbol{\xi}_1 - z_1)(\boldsymbol{\xi}_2 - z_2)} f(\boldsymbol{\xi}_1, \boldsymbol{\xi}_2) \mathrm{d}\boldsymbol{\xi}_1 \mathrm{d}\boldsymbol{\xi}_2. \tag{3.4}$$

Now again looking at the corresponding Plemelj–Sokhotzki formula we get

$$\begin{aligned} \mathrm{tr}F(x_1, x_2) = &\frac{1}{4} f(x_1, x_2) - \frac{1}{4} \int_{\mathbb{R}^2} \frac{1}{(\boldsymbol{\xi}_1 - x_1)(\boldsymbol{\xi}_2 - x_2)} f(\boldsymbol{\xi}_1, \boldsymbol{\xi}_2) \mathrm{d}\boldsymbol{\xi}_1 \mathrm{d}\boldsymbol{\xi}_2 \\ &+ i\frac{1}{4} \left(\int_{\mathbb{R}} \frac{1}{\boldsymbol{\xi}_1 - x_1} f(\boldsymbol{\xi}_1, x_2) \mathrm{d}\boldsymbol{\xi}_1 + \int_{\mathbb{R}} \frac{1}{\boldsymbol{\xi}_2 - x_2} f(x_1, \boldsymbol{\xi}_2) \mathrm{d}\boldsymbol{\xi}_2 \right), \end{aligned} \tag{3.5}$$

which up to the factor 1/4 corresponds to the definition of an analytic signal given by Hahn [13]. Here

$$\mathcal{H}_1 f(x_1, x_2) = \int_{\mathbb{R}} \frac{1}{\boldsymbol{\xi}_1 - x_1} f(\boldsymbol{\xi}_1, x_2) \mathrm{d}\boldsymbol{\xi}_1 \tag{3.6}$$

$$\mathcal{H}_2 f(x_1, x_2) = \int_{\mathbb{R}} \frac{1}{\boldsymbol{\xi}_2 - x_2} f(x_1, \boldsymbol{\xi}_2) \mathrm{d}\boldsymbol{\xi}_2 \tag{3.7}$$

is called a partial Hilbert transform, and

$$\mathcal{H}_T f = \frac{1}{4} \int_{\mathbb{R}^2} \frac{1}{(\boldsymbol{\xi}_1 - x_1)(\boldsymbol{\xi}_2 - x_2)} f(\boldsymbol{\xi}_1, \boldsymbol{\xi}_2) \mathrm{d}\boldsymbol{\xi}_1 \mathrm{d}\boldsymbol{\xi}_2 \tag{3.8}$$

a total Hilbert transform. On the level of Fourier symbols we get

$$\mathcal{F}(\mathrm{tr}F)(u_1, u_2) = (1 + \mathrm{sign} u_1)(1 + \mathrm{sign} u_2) \mathcal{F} f(u_1, u_2). \tag{3.9}$$

Let us now take a look at the definition of Bülow. To this end we consider F to be a function of two variables z_1 and $\mathfrak{z}_2$ with two different imaginary units $\boldsymbol{i}$ and $\boldsymbol{j}$ (with $\boldsymbol{i}^2 = \boldsymbol{j}^2 = -1$), *i.e.*, $z_1 = x_1 + \boldsymbol{i} y_1$ and $\mathfrak{z}_2 = x_2 + \boldsymbol{j} y_2$. We remark that both imaginary units can be understood as elements of the quaternionic basis with multiplication rules $\boldsymbol{ij} = -\boldsymbol{ji} = \boldsymbol{k}$. In this way the above Riemann–Hilbert problem can be rewritten as

$$\frac{\partial}{\partial \bar{z}_1} F = 0, \qquad (z_1, \mathfrak{z}_2) \in \mathbb{C}^2, y_1, y_2 \geq 0, \tag{3.10}$$

$$F \frac{\partial}{\partial \bar{\mathfrak{z}}_2} = 0, \qquad (z_1, \mathfrak{z}_2) \in \mathbb{C}^2, y_1, y_2 \geq 0, \tag{3.11}$$

$$\Re(F(x_1, x_2)) = f(x_1, x_2), \qquad x_1, x_2 \in \mathbb{R}^2, \tag{3.12}$$

where the second equation should be understood as $\partial_{\bar{\mathfrak{z}}_2}$ being applied from the right due to the non-commutativity of the complex units $\boldsymbol{i}$ and $\boldsymbol{j}$.

The solution is given by

$$F(z_1, \mathfrak{z}_2) = \frac{1}{4\pi^2} \int_{\mathbb{R}^2} \frac{1}{(\boldsymbol{\xi}_1 - z_1)(\boldsymbol{\xi}_2 - \mathfrak{z}_2)} f(\boldsymbol{\xi}_1, \boldsymbol{\xi}_2) \mathrm{d}\boldsymbol{\xi}_1 \mathrm{d}\boldsymbol{\xi}_2, \tag{3.13}$$

so that we get from the Plemelj–Sokhotzki formulae

$$\mathrm{tr}F(x_1, x_2) = \frac{1}{4}(I + \boldsymbol{i}H_1)(I + \boldsymbol{j}H_2)f(x_1, x_2) \tag{3.14}$$

$$= \frac{1}{4}(f + \boldsymbol{i}\mathcal{H}_1 f + \boldsymbol{j}\mathcal{H}_2 f + \boldsymbol{k}\mathcal{H}_T f)(x_1, x_2). \tag{3.15}$$

While (3.15) is a quaternionic-valued function, it still corresponds to a boundary value of a function holomorphic in two variables. For the representation in the Fourier domain one has to keep in mind that one has to apply one Fourier transform with respect to the complex plane in $\boldsymbol{i}$, and one Fourier transform with respect to the complex plane generated by $\boldsymbol{j}$. Taking into account that $\boldsymbol{ij} = -\boldsymbol{ji}$ one arrives at the so-called quaternionic Fourier transform [16, 6]:

$$\mathcal{QF}f = \int_{\mathbb{R}^2} e^{\boldsymbol{i}x_1\xi_1} f(x_1, x_2) e^{\boldsymbol{j}x_2\xi_2} \mathrm{d}x_1 \mathrm{d}x_2, \tag{3.16}$$

and the following representation in Fourier symbols

$$\mathcal{QF}(\mathrm{tr}F)(u_1, u_2) = (1 + \mathrm{sign}u_1)(1 + \mathrm{sign}u_2)\mathcal{QF}f(u_1, u_2). \tag{3.17}$$

3.1.2. Image Analysis. In image analysis problems, we can introduce the following features according to [13]

Amplitude. The local amplitude of a multidimensional analytic signal is defined in a similar way as in the one-dimensional case:

$$A_A(x, y) = \sqrt{|f(x, y)|^2 + |\mathcal{H}_1 f(x, y)|^2 + |\mathcal{H}_2 f(x, y)|^2 + |\mathcal{H}_T f(x, y)|^2} \tag{3.18}$$

This is also called *energetic information.*

Phase. The phase is a feature describing how much a vector or quaternion number diverge from the real axis. It is defined in a manner similar to the classical complex plane.

$$\phi_A = \arctan\left(\frac{\sqrt{\mathcal{H}_1 f^2 + \mathcal{H}_2 f^2 + \mathcal{H}_T f^2}}{f}\right). \tag{3.19}$$

This angle ϕ_A is what is called phase or *structural information.*

Orientation. Because we are currently considering 2D signals (that is, images), we can also describe orientation information, as the principal direction carrying the phase information. The imaginary plane, spanned by $\{\boldsymbol{i}, \boldsymbol{j}\}$, is two dimensional and therefore we can also define an angle θ_A in this plane:

$$\theta_A = \arctan\left(\frac{\mathcal{H}_2 f}{\mathcal{H}_1 f}\right). \tag{3.20}$$

This new angle is called the orientation of the signal or *geometric information.*

3.2. Using Clifford Analysis

Another approach to higher dimensions is Clifford analysis.

3.2.1. Mathematics. Here we use Clifford algebra $C\ell_{0,n}$ [4]. This is the free algebra constructed over $\mathbb{R}^n$ generated modulo the relation

$$x^2 = -|x|^2 \boldsymbol{e}_0, \quad x \in \mathbb{R}^n \tag{3.21}$$

where $\boldsymbol{e}_0$ is the identity of $C\ell_{0,n}$.

For the algebra $C\ell_{0,n}$ we have the anti-commutation relationship

$$\boldsymbol{e}_i\boldsymbol{e}_j + \boldsymbol{e}_j\boldsymbol{e}_i = -2\delta_{ij}\boldsymbol{e}_0, \tag{3.22}$$

where δ_{ij} is the Kronecker symbol. Each element x of $\mathbb{R}^n$ may be represented by

$$x = \sum_{i=1}^{n} x_i \boldsymbol{e}_i. \tag{3.23}$$

A first-order differential operator which factorizes the Laplacian is given by the Dirac operator

$$Df(x) = \sum_{j=1}^{n} \frac{\partial f}{\partial x_j}. \tag{3.24}$$

The Riemann–Hilbert problem for the Dirac operator in $\mathbb{R}^3$ can be stated in the form

$$DF(x) = 0, \qquad x \in \mathbb{R}^3, x_3 > 0, \tag{3.25}$$

$$\Re\left(F(x_1, x_2)\right) = f(x_1, x_2), \qquad x_1, x_2 \in \mathbb{R}^2. \tag{3.26}$$

To solve this problem we follow the same idea as above.

$$F_\Gamma f = \int_{\mathbb{R}^2} \frac{x-y}{|x-y|^2} \boldsymbol{e}_3 f(x_1, x_2) \mathrm{d}x_1 \mathrm{d}x_2 \tag{3.27}$$

$$\begin{aligned} \mathrm{tr} F_\Gamma f &= \frac{1}{2}(I + S_\Gamma) f \\ &= \frac{1}{2} f(\tilde{y}_1, \tilde{y}_2) + \frac{1}{2} \int_{\mathbb{R}^2} \frac{\boldsymbol{e}_1(x_1 - \tilde{y}_1) + \boldsymbol{e}_2(x_2 - \tilde{y}_2)}{|x-y|^2} \boldsymbol{e}_3 f(x_1, x_2) \mathrm{d}x_1 \mathrm{d}x_2. \end{aligned} \tag{3.28}$$

Because quaternions $\mathbb{H}$ are isomorphic to the even subalgebra $C\ell_{0,3}^+$, *i.e.*, all elements of the form

$$c_0 + c_1\boldsymbol{e}_1\boldsymbol{e}_2 + c_2\boldsymbol{e}_1\boldsymbol{e}_3 + c_3\boldsymbol{e}_2\boldsymbol{e}_3, \quad c_0, c_1, c_2, c_3 \in \mathbb{R} \tag{3.29}$$

we can set $\boldsymbol{i} = \boldsymbol{e}_1\boldsymbol{e}_2$ and $\boldsymbol{j} = \boldsymbol{e}_2\boldsymbol{e}_3$ so that

$$\mathrm{tr} F_\Gamma f = \frac{1}{2}(I + S_\Gamma) f \tag{3.30}$$

$$= \frac{1}{2} f(\tilde{y}_1, \tilde{y}_2) + \frac{1}{2} \int_{\mathbb{R}^2} \frac{\boldsymbol{i}(x_1 - \tilde{y}_1) + \boldsymbol{j}(x_2 - \tilde{y}_2)}{|x-y|^2} f(x_1, x_2) \mathrm{d}x_1 \mathrm{d}x_2. \tag{3.31}$$

Up to the factor $1/2$ this is the monogenic signal $f_M = f + \boldsymbol{i}\mathcal{R}_1 f + \boldsymbol{j}\mathcal{R}_2 f := f + (\boldsymbol{i}, \boldsymbol{j})\mathcal{R} f$ of Sommer and Felsberg [10]. Here $\mathcal{R}_1$, $\mathcal{R}_2$ and $\mathcal{R}$ denote respectively the first and second component of the Riesz transform, and the Riesz transform itself [23]. Defined as Fourier multipliers, it holds:

$$\widehat{\mathcal{R}f}(u_1,u_2) = \frac{i(u_1,u_2)}{\|(u_1,u_2)\|_2}\widehat{f}(u_1,u_2), \tag{3.32}$$

$$\widehat{\mathcal{R}_1 f}(u_1,u_2) = \frac{iu_1}{\|(u_1,u_2)\|_2}\widehat{f}(u_1,u_2), \tag{3.33}$$

$$\widehat{\mathcal{R}_2 f}(u_1,u_2) = \frac{iu_2}{\|(u_1,u_2)\|_2}\widehat{f}(u_1,u_2), \tag{3.34}$$

where $\|(u_1,u_2)\|_2 = \sqrt{u_1^2+u_2^2}$. An equivalent definition in the spatial domain can be obtained by convolution with the two-dimensional Riesz kernel, *i.e.*, for $m = 1, 2$

$$\mathcal{R}_i f = c\frac{x_i}{\|x\|_2^3} * f, \tag{3.35}$$

with c being a constant.

3.2.2. Image Analysis. Following [10], three features can be computed, and will also be denoted as *energetic, structural and geometrical information*, as already introduced for the multidimensional analytic signal.

Amplitude. The local amplitude of a monogenic signal is defined in a similar manner as for the analytic signal:

$$A_M(x,y) = \sqrt{|f(x,y)|^2 + |\mathcal{R}f(x,y)|^2} = \sqrt{f_M(x,y)\overline{f_M(x,y)}}, \tag{3.36}$$

where the overbar denotes the conjugation of a quaternion.

Phase.

$$\phi_M(x,y) = \arctan\frac{|\mathcal{R}f(x,y)|}{f(x,y)}, \tag{3.37}$$

and we also have that ϕ_M denotes the angle between $A(x,y)$ and f_M (in the plane spanned by the two complex vectors). This yields values $\phi_M \in [-\pi/2;\pi/2]$.

An alternative but equivalent definition is using the arccosine:

$$\phi_M = \arccos\frac{f}{|f_M|}. \tag{3.38}$$

In (3.38), we have $\phi_M \in [0;\pi]$.

Orientation. Once again, we can derive an orientation $\theta_M \in [-\pi,\pi]$ based on the monogenic signal which represents the direction of the phase information.

$$\theta_M = \arctan\frac{\mathcal{R}_2 f}{\mathcal{R}_1 f}. \tag{3.39}$$

We note that this definition actually only provides an orientation modulo π. To determine the orientation respectively the direction modulo 2π, a further orientation unwrapping step or sign estimation is needed [18, 5].

Representation. In the rest of this chapter the different features of images are represented by using different colourmaps: a grey colour map is used for the amplitude representation, which is normalized between $[0, 1]$, a jet colormap for the phase (running between blue and red and mapping the interval $[0, \pi)$) [24], and a HSV colormap for the orientation (running between red, blue and red in a periodic way and mapping the interval $[-\pi/2, \pi/2)$), as depicted in Figure 2 (left, resp. from top to bottom) [24]. As a last representation, we use a colormap according to the Middlebury representation [1] where the colour encodes the orientation and the intensity is computed according to the phase (or structure) information. This representation can be seen on the right-hand side of Figure 2.

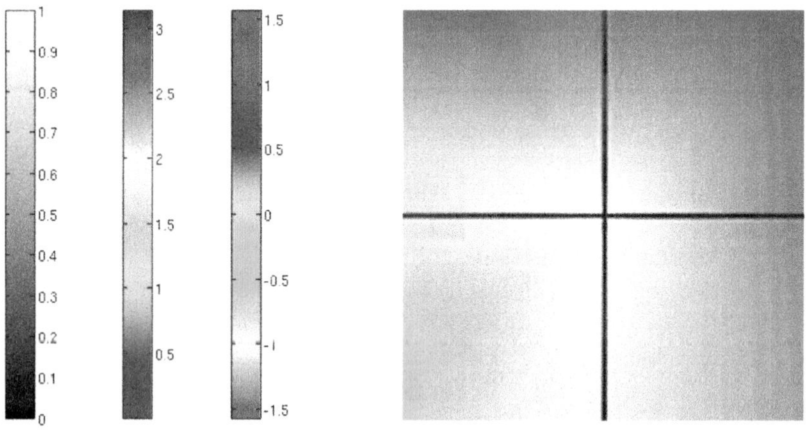

(a) Grey, jet, HSV [24] (b) Middlebury [1]

FIGURE 2. Scales used for the different colour coding. (See explanation in text.)

3.3. Illustrations

We want here to illustrate the differences between the generalizations proposed. We will visually assess the characteristics of both approaches first applied to a Siemens star[1] then to a checkerboard image. Both examples are interesting for their regularity (point symmetry for the star and many horizontal and vertical line symmetries for the checkerboard).

An example of such star is depicted in Figure 3(a). The two other images in the first row of Figure 3 illustrate the two components of the Riesz transform. As we can see, and we will come back to this property later, the partial Riesz transforms show in some directions behaviour similar to steered derivatives. The

[1]The Siemens star is a test image used to characterize the resolution of different optical and graphical devices such as printers or computer projectors. The image is interesting as it shows much regularity, as well as many intrinsic one-dimensional and two-dimensional parts.

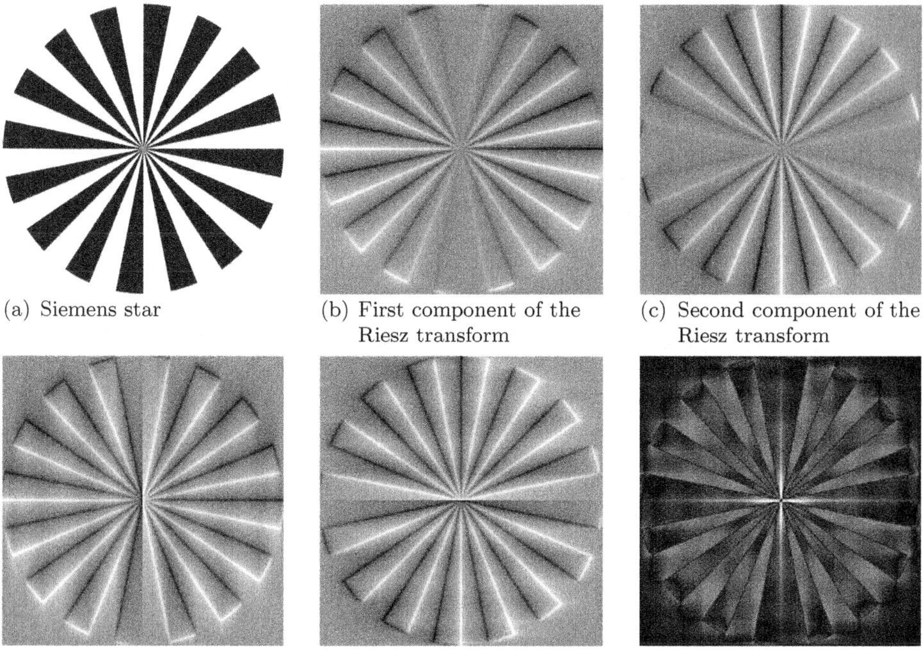

(a) Siemens star

(b) First component of the Riesz transform

(c) Second component of the Riesz transform

(d) First Hilbert transform

(e) Second Hilbert transform

(f) Total Hilbert transform

FIGURE 3. The Siemens star together with the different Riesz and Hilbert transforms presented in this section. (Additional explanation in the text.)

first component tends to emphasize horizontal edges while the second one tends to respond more to vertical ones.

The second row shows the results of applying the different Hilbert transforms to the Siemens star. The two first images represent the results of the two partial Hilbert transforms and the last one depicts the result of the total Hilbert transform. We can notice the high anisotropy of these transforms at, for instance, the strong vertical respectively horizontal delineation through the centres of the images. We can also notice the patchy response of the total Hilbert transform.

As the Riesz kernel in polar coordinate $[r, \alpha]$ of the spatial domain reads

$$R(r, \alpha) \sim \frac{1}{r^2} e^{i\alpha}, \tag{3.40}$$

it exhibits an isotropic behavior with respect to its magnitude. In comparison, the partial and the total Hilbert transforms both induce a strict relationship to the orthogonal coordinate system, and therefore also the two-dimensional analytic signal inherits this characteristic.

Next we consider local features computed according to the formulas introduced above. The results are depicted in Figure 4. The first row corresponds to

(a) Monogenic amplitude (b) Monogenic phase (c) Monogenic orientation (d) Phase-weighted monogenic orientation

(e) Analytic amplitude (f) Analytic phase (g) Analytic orientation (h) Phase-weighted analytic orientation

FIGURE 4. Local features computed with the monogenic signal representation (first row), and the multidimensional analytic signal (second row). The images are depicted in a pseudo-colour representation with amplitude: grey, phase: jet, orientation: HSV-Middlebury.

monogenic features, while the second one corresponds to analytic features. The phase is displayed in a jet colormap, the orientation in a hue-saturation-value HSV colormap. The last column shows the orientations with intensities weighted proportionally to the cosine of the phase. It is shown according to the Middlebury representation[2]: strength (cosine of the phase) is encoded as an intensity value of the colour and the colour itself corresponds to the orientation. The main differences between these two sets of features lie in the shape and in the boundaries. While monogenic features yield rather smooth boundaries, the analytic representation creates abrupt changes due to its anisotropy. We remark that the phase gives reasonable insights into the structure in the images.

In comparison to the Siemens star, the checkerboard example (see Figure 5(a)) shows many orthogonal features. In this case, we see that the partial Hilbert transforms give some good insights into the closeness of an edge and preserve the checkerboard structure (Figure 5(d) and 5(e)), while the Riesz transform gives

[2]The Middlebury benchmark for optical flow is a web resource for comparing results on optical flow computations. The colour error representation is well suited for encoding our orientation. More information can be found in [1].

(a) Checkerboard image

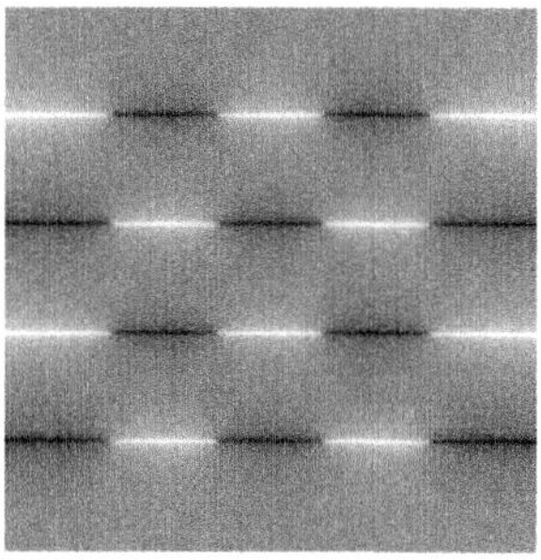
(b) First component of the Riesz transform

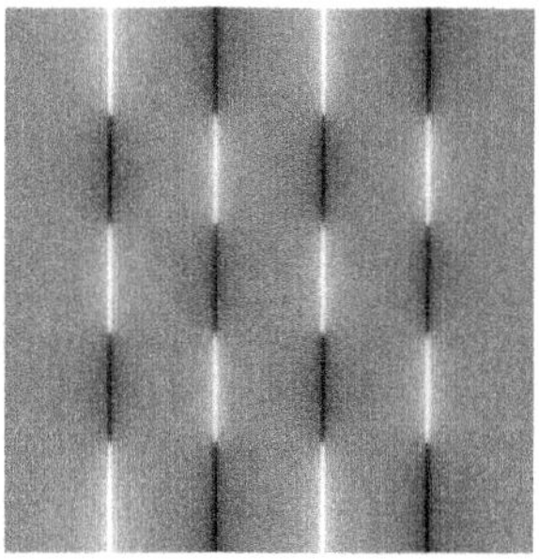
(c) Second component of the Riesz transform

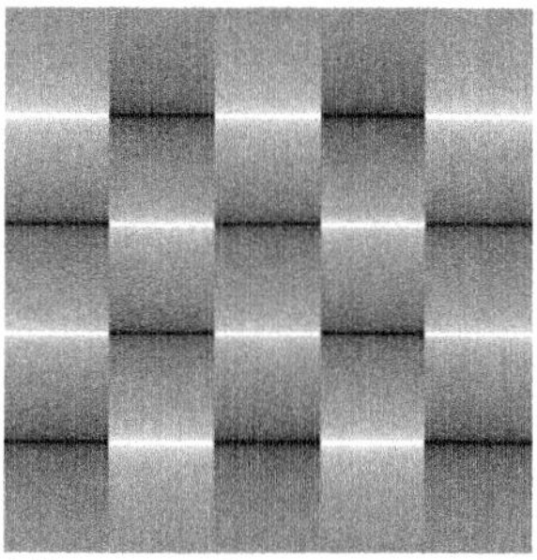
(d) First Hilbert transform

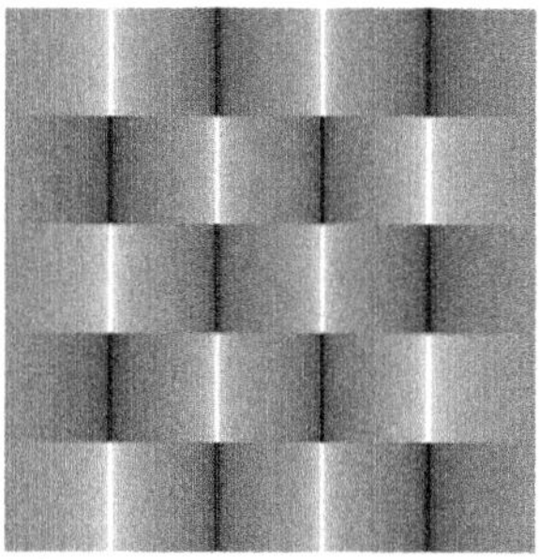
(e) Second Hilbert transform

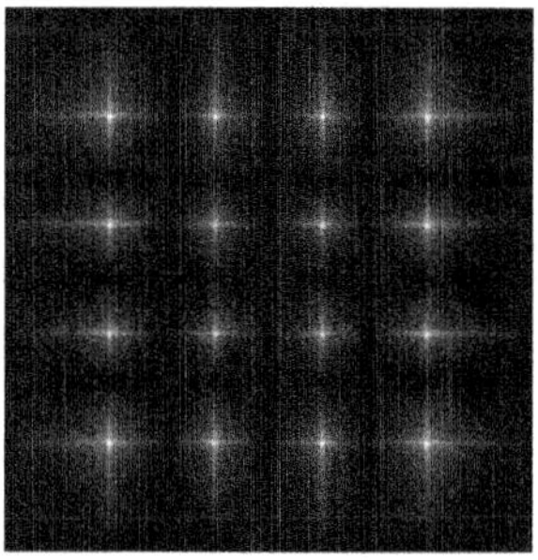
(f) Total Hilbert transform

FIGURE 5. The checkerboard together with the different Riesz and Hilbert transforms presented in this section.

more local responses. The total Hilbert transform acts as an accurate corner detector, as can be seen from its response in Figure 5(f).

When discussing the analytic and monogenic features (Figure 6) we remark that this effect is preserved. The Riesz transform, being well localized at the edges, does not yield many differences inside any of the squares and seems to jump from one extreme to another across the edges. See in particular Figure 6(b) for an illustrative example of the phase. On the other hand, the Hilbert transform contains more neighbourhood information and yields a smoother transition in the phase from one square to the next. These features have to be considered carefully based on the application problem one wishes to solve.

4. The Geometric Approach

For a better understanding of signals a geometric interpretation can help. The following considerations about complex numbers, quaternions, rotations, the unitary group, special unitary and special orthogonal groups, as well as the spin group, are well known and can be found in numerous papers. We would like to suggest the book by Lounesto [19], which provides a comprehensive knowledge of these topics.

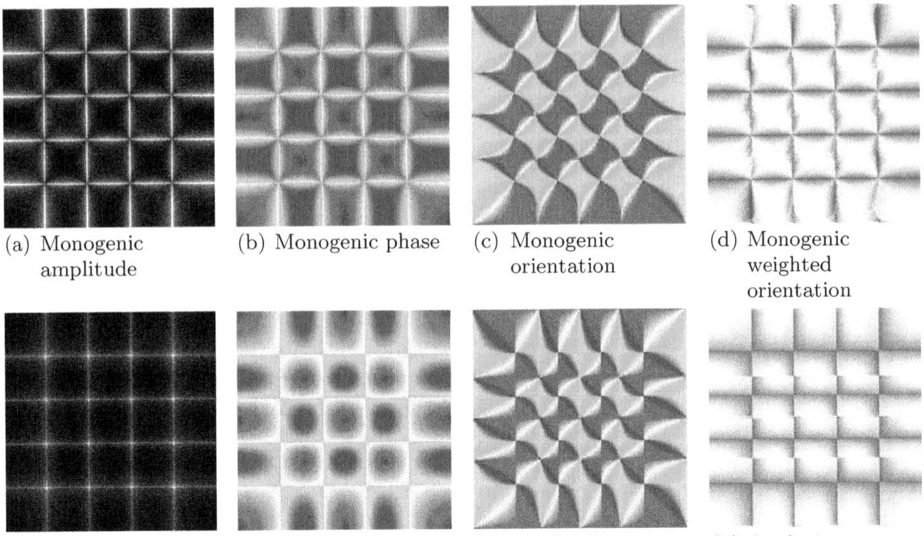

(a) Monogenic amplitude (b) Monogenic phase (c) Monogenic orientation (d) Monogenic weighted orientation

(e) Analytic amplitude (f) Analytic phase (g) Analytic orientation (h) Analytic weighted orientation

FIGURE 6. Local features computed with the monogenic signal representation (first row), and the multidimensional analytic signal (second row). The images are depicted in a pseudo-colour representation with amplitude: grey, phase: jet, orientation: HSV-Middlebury.

The analytic signal $f_A(t) = A(t)e^{i\phi(t)}$ consists of boundary values of an analytic function, but the analytic signal can also be seen as a complex-valued function, where $e^{i\phi(t)} = \cos\phi(t) + i\sin\phi(t)$ has unit modulus and hence can be identified with the unit circle S^1. But there is even more. The set of unit complex numbers is a group with the complex multiplication as group operation, which is the unitary group $U(1) = \{z \in \mathbb{C} : z\overline{z} = 1\}$. On the other hand a unit complex number can also be seen as a rotation in $\mathbb{R}^2$ if we identify the unit complex number with the matrix

$$R_\phi = \begin{pmatrix} \cos\phi & -\sin\phi \\ \sin\phi & \cos\phi \end{pmatrix} \in SO(2), \tag{4.1}$$

i.e., the group of all counter-clockwise rotations in $\mathbb{R}^2$. Now all this can also be described inside Clifford algebras. Let us consider the Clifford algebra $C\ell_{0,2}$ with generators e_1, e_2. The complex numbers can be identified with all elements $x + ye_{12}$, $x, y \in \mathbb{R}$, *i.e.*, the even subalgebra $C\ell_{0,2}^+$ of the Clifford algebra $C\ell_{0,2}$. The rotation (4.1) can also be described by a Clifford multiplication. To see that, we identify $(x,\, y) \in \mathbb{R}^2$ with $xe_1 + ye_2 \in C\ell_{0,2}$, and set

$$R_\phi(x,\, y)^T = (\cos\tfrac{\phi}{2} + e_{12}\sin\tfrac{\phi}{2})^{-1}(xe_1 + ye_2)(\cos\tfrac{\phi}{2} + e_{12}\sin\tfrac{\phi}{2}), \tag{4.2}$$

where $\cos\frac{\phi}{2} + e_{12}\sin\frac{\phi}{2} \in \mathrm{Spin}(2) = \{s \in C\ell^+_{0,2} : s\bar{s} = 1\}$, the spin group of even products of Clifford vectors. It is easily seen that s and $-s$ in Spin(2) represent the same rotation, which means that Spin(2) is a two-fold cover of $SO(2)$. The basis for all these interpretations is the description of complex numbers in a trigonometric way, which is possible by using a logarithm function, which is well known for complex numbers. All of that can be generalized into higher dimensions and has been used for monogenic signals. We will start with quaternions because they are the even subalgebra of the Clifford algebra $C\ell_{0,3}$.

4.1. Quaternions and Rotations

A quaternion $q \in \mathbb{H}$ can be written as

$$q = q_0 + \underline{q} = \mathrm{S}(q) + \mathbf{V}(q) = |q|\,\frac{q}{|q|}, \tag{4.3}$$

where $|q|$ is the absolute value or norm of q in $\mathbb{R}^4$ and $\frac{q}{|q|} \in \mathbb{H}_1$ is a unit quaternion.

Because of

$$\left|\frac{q}{|q|}\right|^2 = \sum_{i=0}^{3} \frac{q_i^2}{|q|^2} = 1, \tag{4.4}$$

the set of unit quaternions $\mathbb{H}_1$ can be identified with S^3, the three-dimensional sphere in $\mathbb{R}^4$.

On the other hand the Clifford algebra $C\ell_{0,3}$ is generated by the elements $\boldsymbol{e}_1, \boldsymbol{e}_2$ and $\boldsymbol{e}_3$ with $\boldsymbol{e}_1^2 = \boldsymbol{e}_2^2 = \boldsymbol{e}_3^2 = -1$ and $\boldsymbol{e}_i\boldsymbol{e}_j + \boldsymbol{e}_j\boldsymbol{e}_i = -2\delta_{i,j}$. Its even subalgebra $C\ell^+_{0,3}$, as defined in (3.29), can be identified with quaternions by $\boldsymbol{e}_1\boldsymbol{e}_2 \sim \boldsymbol{i}$, $\boldsymbol{e}_1\boldsymbol{e}_3 \sim \boldsymbol{j}$ and $\boldsymbol{e}_2\boldsymbol{e}_3 \sim \boldsymbol{k}$.

Furthermore,

$$\mathrm{Spin}(3) = \{u \in C\ell^+_{0,3} : u\bar{u} = 1\} = \mathbb{H}_1. \tag{4.5}$$

That means a unit quaternion can be considered as a spinor. Because Spin(3) is a double cover of the group $SO(3)$, rotations can be described by unit quaternions. The monogenic signal is interpreted as a spinor in [26] and lately in [2].

4.2. Quaternions in Trigonometric Form

In this section we represent the monogenic signal in a similar manner to the analytic signal. The analytic signal is a holomorphic and analytic function and therefore connected to complex numbers. Complex numbers can be written in algebraic or trigonometric form as:

$$z = x + iy = re^{i\phi}.$$

The analytic signal is given by

$$A(t)e^{i\phi(t)},$$

with amplitude $A(t)$ and (local) phase $\phi(t)$. We want to obtain a similar representation of the monogenic signal using quaternions. A simple computation leads to

$$q = |q|\left(\frac{q_0}{|q|} + \frac{\underline{q}}{|\underline{q}|}\frac{|\underline{q}|}{|q|}\right) = |q|\,(\cos\phi + \underline{u}\sin\phi),$$

where $\phi = \arccos \frac{q_0}{|q|}$ and $\underline{u} = \frac{\underline{q}}{|\underline{q}|} \in S^2$. (Alternatively, the argument ϕ can be defined by the arctan.)
We have to mention that this representation is different from all previous ones, specifically the vector $\underline{u}$ is a unit vector in $\mathbb{R}^3$ and not a unit quaternion.
We can represent the quaternion q by its amplitude $|q|$, the phase ϕ and the orientation $\underline{u}$. Moreover,

$$q = |q| \; e^{\underline{u}\phi},$$

where e is the usual exponential function.

With the aid of an appropriate logarithm we can compute $\underline{u}\phi$ from $\frac{q}{|q|} = e^{\underline{u}\phi}$.
Next, we want to explain the orientation $\underline{u}$. We have already obtained that

$$q = |q| \left(\cos\phi + \underline{u}\sin\phi\right),$$

where $\underline{u} = \frac{\underline{q}}{|\underline{q}|} \in S^2$ and $\underline{u}^2 = -1$, *i.e.*, $\underline{u}$ behaves like a complex unit. But because of $\underline{u} \in S^2$ we can express $\underline{u}$ in spherical coordinates. We have

$$\underline{u} = \frac{q_1\boldsymbol{i} + q_2\boldsymbol{j} + q_3\boldsymbol{k}}{|q_1\boldsymbol{i} + q_2\boldsymbol{j} + q_3\boldsymbol{k}|} = \boldsymbol{i}\left(\frac{q_1}{|\underline{q}|} + \frac{(q_2(-\boldsymbol{i}\boldsymbol{j}) + q_3\boldsymbol{j})}{|\underline{q}|}\right), \tag{4.6}$$

and if we set $\cos\theta = \frac{q_1}{|\underline{q}|}$ we get

$$\underline{u} = \boldsymbol{i}(\cos\theta + \underline{\underline{u}}\sin\theta), \quad \underline{\underline{u}} = \frac{\underline{\underline{q}}}{|\underline{\underline{q}}|} \quad \text{and} \quad \underline{\underline{q}} = \boldsymbol{j}q_3 - \boldsymbol{i}\boldsymbol{j}q_2. \tag{4.7}$$

Because of

$$\underline{\underline{q}} = \boldsymbol{j}q_3 - \boldsymbol{i}\boldsymbol{j}q_2 = \boldsymbol{j}(q_3 + \boldsymbol{i}q_2) \tag{4.8}$$

and with $\cos\tau = \frac{q_3}{|\underline{\underline{q}}|}$ we get that

$$\underline{\underline{u}} = \boldsymbol{j}(\cos\tau + \boldsymbol{i}\sin\tau). \tag{4.9}$$

Finally, we put everything together we obtain

$$\begin{aligned}
q &= q_0 + q_1\boldsymbol{i} + q_2\boldsymbol{j} + q_3\boldsymbol{k} && (4.10)\\
&= |q|\left(\cos\phi + \underline{u}\sin\phi\right) = |q|\left(\cos\phi + \boldsymbol{i}\left(\cos\theta + \underline{\underline{u}}\sin\theta\right)\sin\phi\right) && (4.11)\\
&= |q|\left(\cos\phi + \boldsymbol{i}\left(\cos\theta + \boldsymbol{j}\left(\cos\tau + \boldsymbol{i}\sin\tau\right)\sin\theta\right)\sin\phi\right) && (4.12)\\
&= |q|\left(\cos\phi + \boldsymbol{i}\sin\phi\cos\theta + \boldsymbol{j}\sin\phi\sin\theta\sin\tau + \boldsymbol{k}\sin\phi\sin\theta\cos\tau\right), && (4.13)
\end{aligned}$$

where $\phi,\ \theta \in [0,\ \pi]$ and $\tau \in [0,\ 2\pi]$. In case of a reduced quaternion, *i.e.*, $q_3 = 0$, a similar computation leads to

$$\begin{aligned}
q &= q_0 + q_1\boldsymbol{i} + q_2\boldsymbol{j} && (4.14)\\
&= |q|\left(\cos\phi + \underline{u}\sin\phi\right) = |q|\left(\cos\phi + \boldsymbol{i}\left(\cos\theta - \boldsymbol{k}\sin\theta\right)\sin\phi\right) && (4.15)\\
&= |q|\left(\cos\phi + \boldsymbol{i}\sin\phi\cos\theta + \boldsymbol{j}\sin\phi\sin\theta\right), && (4.16)
\end{aligned}$$

where $\phi \in [0,\ \pi]$ and $\theta \in [0,\ 2\pi]$.

It is easily seen that θ can be computed by

$$\tan\theta = \frac{q_2}{q_1} \iff \theta = \arctan\frac{q_2}{q_1}.$$

If we compare that with the monogenic signal

$$f_M(x,y) = f(x,y) + \boldsymbol{i}(\mathcal{R}_1 f)(x,y) + \boldsymbol{j}(\mathcal{R}_2 f)(x,y)$$

we see that (compare with (3.39))

$$\theta = \arctan\frac{(\mathcal{R}_2 f)(x,y)}{(\mathcal{R}_1 f)(x,y)} = \theta_M(x,y). \tag{4.17}$$

Therefore the vector $\underline{u} = \boldsymbol{i}\cos\theta + \boldsymbol{j}\sin\theta$ can also be considered as the orientation.

4.3. Exponential Function and Logarithm for Quaternionic Arguments

The exponential function for quaternions and para-vectors in a Clifford algebra is defined in [12] and many other papers.

Definition 4.1. For $q \in \mathbb{H}$ the exponential function is defined as

$$e^q := \sum_{k=0}^{\infty} \frac{q^k}{k!}. \tag{4.18}$$

Lemma 4.2. *With* $\underline{u} = \frac{\underline{q}}{|\underline{q}|}$ *the exponential function can be written as*

$$e^q = e^{q_0}\left(\cos|\underline{q}| + \underline{u}\sin|\underline{q}|\right) = e^{q_0}e^{\underline{u}|\underline{q}|}. \tag{4.19}$$

Remark 4.3. The formula

$$e^{\underline{u}|\underline{q}|} = \cos|\underline{q}| + \underline{u}\sin|\underline{q}| \tag{4.20}$$

can be considered as a generalized Euler formula.

It is always a challenge to define a logarithm. We will use the following definition.

Definition 4.4. Let $\underline{u} = \frac{\underline{q}}{|\underline{q}|}$. Then the logarithm is defined as

$$\ln q := \begin{cases} \ln|q| + \underline{u}\arccos\frac{q_0}{|q|}, & |\underline{q}| \neq 0, \text{ or } |\underline{q}| = 0 \text{ and } q_0 > 0, \\ \text{undefined}, & |\underline{q}| = 0 \text{ and } q_0 \leq 0. \end{cases} \tag{4.21}$$

Remark 4.5. A logarithm cannot be uniquely defined for -1 because

$$e^{\underline{u}\pi} = \cos\pi + \underline{u}\sin\pi = -1, \tag{4.22}$$

for all $\underline{u} \in S^2$.

Remark 4.6. More precisely, we can define the kth branch, $k \in \mathbb{Z}$, of the logarithm because $\cos t$ is a 2π periodic function.

Theorem 4.7 (Exponential and logarithm function).

1. *For* $|\underline{q}| \neq 0$ *or* $|\underline{q}| = 0$ *and* $q_0 > 0$,
$$e^{\ln q} = q. \tag{4.23}$$
2. *For* $|\underline{q}| \neq k\pi,\ k \in \mathbb{Z}\backslash\{0\}$ *the following holds true*
$$\ln e^{q} = q. \tag{4.24}$$

Lemma 4.8. *For* $q \in \mathbb{H}_1$ *and* $q \neq -1$ *both relations are true:*
$$e^{\ln q} = \ln e^{q} = q. \tag{4.25}$$

5. Applications to Image Analysis

5.1. Motivations

In several imaging applications only intensity-based images (encoded mostly in gray-scale representation) are provided. Apart from monochromatic camera images, we can cite, *e.g.*, computerized tomography images which encodes local absorption inside a body, or optical coherence tomography images which represents the back-scattering at an interface. These kinds of images directly describe natural scenes or physical quantities. In other types of images information is encoded indirectly, *e.g.*, in varying amplitude or frequency of fringe patterns. They are called amplitude modulated (AM) or frequency modulated (FM) signals. Textures can be interpreted as a trade-off between both ideas: they depict natural scenes and can be described as generalized AM-FM signals.

To enrich the information content of a pure intensity image (*i.e.*, images encoded with a single value at each pixel), we test the concept of analytic signals in image processing.

5.2. Application to AM-FM Image Demodulation

Here we study the applicability of the monogenic signal representation to AM-FM signal demodulation, as needed for instance in interferometric imaging [18]. A certain given two-dimensional signal (= an image, Figure 7(a)) exhibits both amplitude modulations (Figure 7(b)) and frequency modulations (Figure 7(c)). The aim is to separate each component of the signal by means of monogenic signal analysis.

The three features described in the previous section are computed and their results are depicted as local orientation in Figure 7(d), local amplitude in Figure 7(e) and local phase in Figure 7(f).

It appears that for such AM-FM signals, the orientation is able to describe the direction of the phase modulation, while the local amplitude gives a good approximation of the amplitude modulation (corresponding to the energy of the two-dimensional signal) and the phase encodes information about the frequency modulation (understood as the structural information).

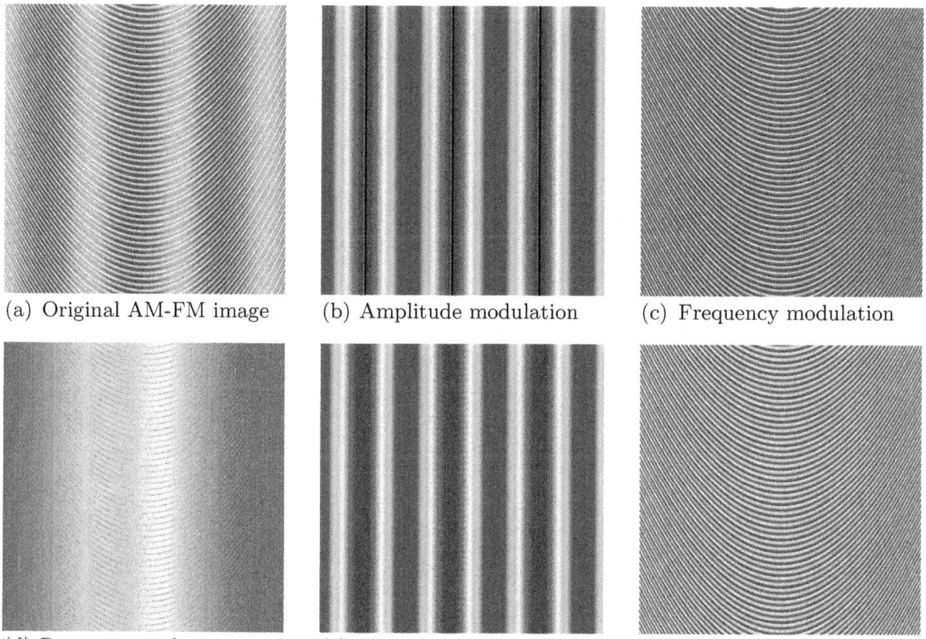

(a) Original AM-FM image (b) Amplitude modulation (c) Frequency modulation

(d) Reconstructed orientation (e) Reconstructed amplitude (f) Reconstructed phase

FIGURE 7. Example of a two-dimensional AM-FM signal. The first row shows the input ground truth image together with its amplitude and frequency modulations. The second row depicts the recovered orientation, amplitude and phases. The images are displayed using the conventional jet colormap.

The next example, in Figure 8(a), shows a fringe pattern as an example of a real-world interferometric AM-FM image. The following images show the monogenic analysis of this image. The local amplitude is depicted beneath the fringe pattern (Figure 8(d)). This image gives us a coarse idea of how much structure is to be found within a given neighbourhood. The second column illustrates the phase calculation either on the whole image (Figure 8(b)) or only where the local amplitude is above a given threshold (Figure 8(e)). The two images in the last column represent the monogenic orientation encoded in HSV without or with the previous mask. As we would expect, illumination changes are appearing in the amplitude, while local structures are contained in both phase and orientation features.

5.3. Application to Texture Analysis

A task of particular interest in artificial vision, is the characterization or description of textures. The problem here is to find interesting features to describe a given texture the best we can in order to classify it for instance [14]. The use of

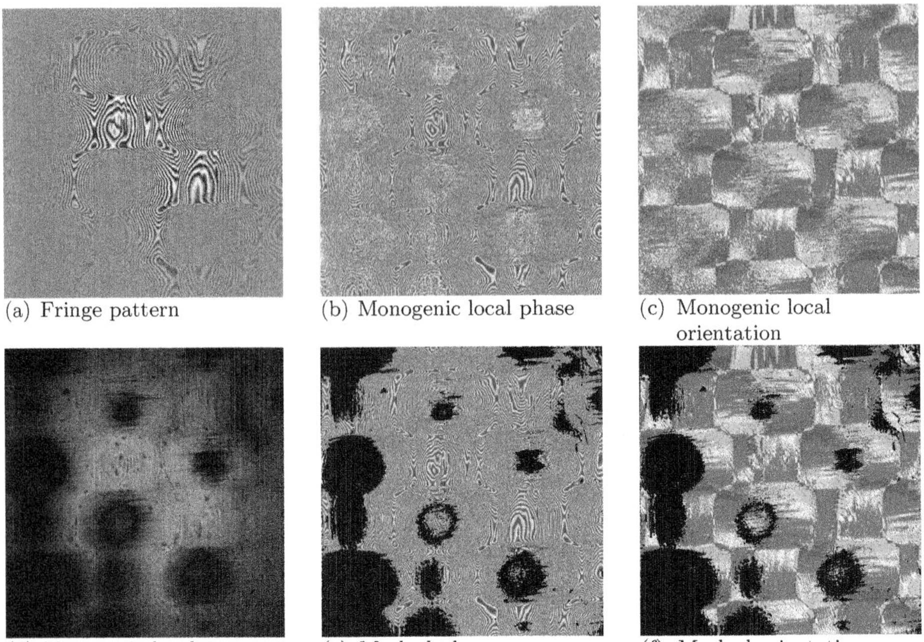

(a) Fringe pattern (b) Monogenic local phase (c) Monogenic local orientation

(d) Monogenic local amplitude (e) Masked phase (f) Masked orientation

FIGURE 8. Example of a fringe pattern and its monogenic decomposition. Phase (second column) is encoded as a jet colormap and orientation as HSV. The two last images show phase and orientation masked with a binary filter set to one when the local amplitude gets over a certain threshold.

steerable filters could both optimize feature computations and affect the classification. In other words, if we can compute good descriptive features, we can better characterize a texture.

Considering textures from a more general viewpoint as approximate AM-FM signals, we examine here the use of monogenic representation for the local characterization of a textured object as depicted in Figure 9(a).

When looking at the local monogenic signal description (amplitude in Figure 9(b), phase in Figure 9(c) and orientation in Figure 9(d)), we indeed see these repetitive features along the textured object. Moreover, these estimated values seem to be robust against small imperfections in the periodicity.

5.4. Applications to Natural Image Scenes

In this part of our chapter, we want to highlight the interest of the monogenic signal for natural images. Such images have completely different characteristics from those introduced above. For instance, natural images are often embedded in a fully cluttered background, encoded with several colour channels, and have

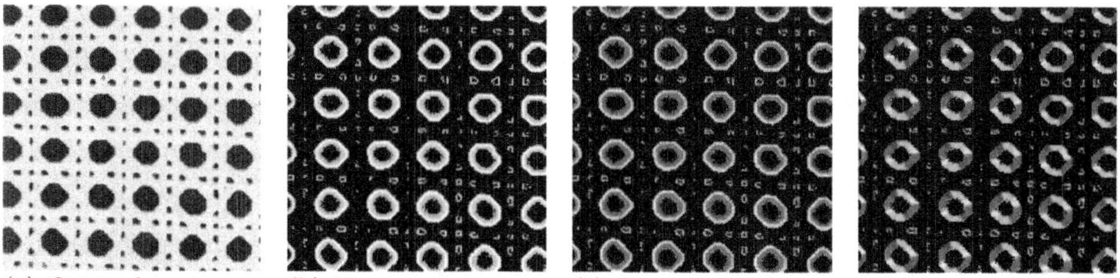

(a) Original texture (b) Monogenic amplitude (c) Monogenic phase (d) Monogenic orientation

FIGURE 9. Example of a textured image superposed with a reliability mask together with its monogenic analysis. Regions with too little amplitude are masked out as unreliable.

information at many different scales, *etc.* In practical applications one needs to apply band-pass filters before analyzing such images [10]. Note that this work considers only grey-scale images, but there is literature dealing with multichannel images [3].

In the following we will describe two tasks useful for image processing. The first part deals with edge detection. We will see how the Riesz transform can be used as an edge detector in images. Then we will see how the orientation estimation is useful, for instance, in computer vision tasks, and how monogenic signal analysis can help with this, as has already been done for structure interpretation [22, 15].

5.4.1. Edge Detection. The Riesz transform acts as an edge detector for several reasons. This becomes clear when one has a closer look at its definition as a Fourier multiplier. Indeed, let us recall the jth Riesz multiplier ($j = 1, 2$, see (3.32)):

$$\widehat{\mathcal{R}_j f} = i\frac{u_j}{|u|}\widehat{f}, \tag{5.1}$$

and we have

$$\widehat{\mathcal{R}_j f} = i\frac{1}{|u|}\widehat{\partial_j f}, \tag{5.2}$$

so that the Riesz transform acts as a normalized derivative operator.

Another (eventually better) way to see this derivative effect is to consider the Fourier multipliers in polar coordinates [17], as given in (3.40). Figure 10 illustrates this behavior. The first column shows examples of natural grey-level images. The second and third columns show the first and second Riesz components respectively. It appears that they act as edge detectors steered in the x and y directions. If we compare the two Riesz components, we can see the response to different kinds of edges.

5.4.2. Orientation Estimation of Edges. An important task in image processing and higher level computer vision is to estimate the orientation of edges. As this is often the first step towards feature description and image interpretation (we

(a) Lena image

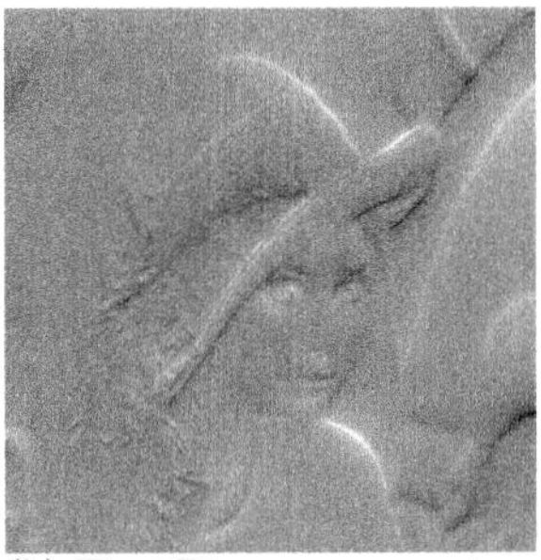

(b) First Riesz component of Lena

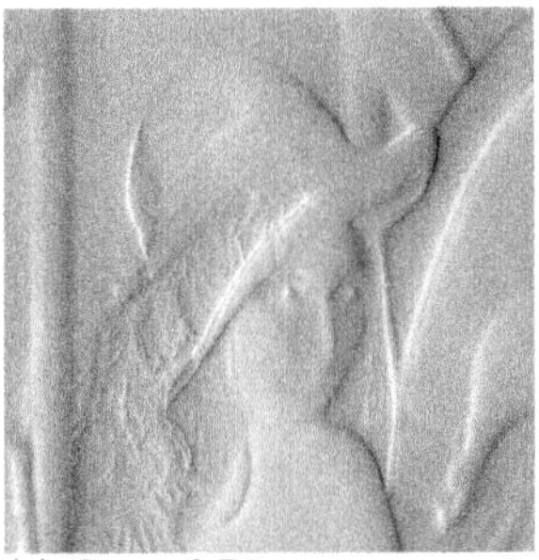

(c) Second Riesz component of Lena

(d) Barbara image

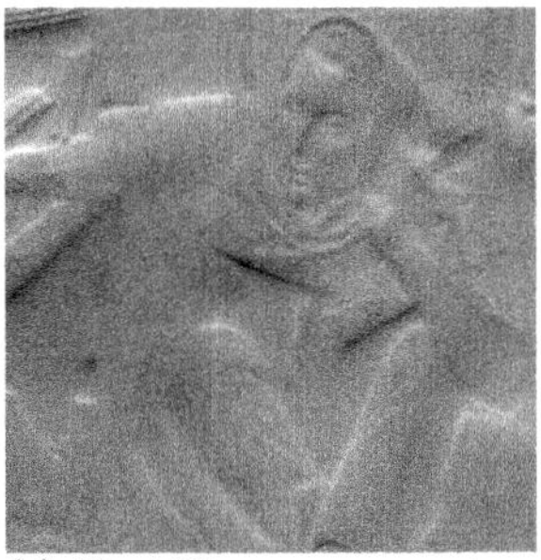

(e) First Riesz component of Barbara

(f) Second Riesz component of Barbara

FIGURE 10. First and second components of the Riesz transform on some natural images. Notice for instance the table leg appearing in Figure 10(f) but not in Figure 10(e), showing the directions of the components.

refer the reader to [7, 20] for some non-exhaustive surveys), one wants to have an orientation estimator which is as reliable as possible.

As stated in earlier sections, an orientation can be computed from an analytic or a monogenic signal analysis. For simplicity, let us consider the case of images, where the input function is defined on a set $D \subset \mathbb{R}^2$. Using polar coordinates in the Fourier domain (ρ, β), it holds that

$$\widehat{\mathcal{R}f} = i(\cos\beta, \sin\beta)^T \widehat{f}, \tag{5.3}$$

but on the other hand, we also have

$$\widehat{\nabla f} = i\rho(\cos\beta, \sin\beta)^T \widehat{f}, \tag{5.4}$$

so that both the gradient and the Riesz operator have similar effects on the angles in the Fourier domain.

It has been shown in [9] that using monogenic orientation estimation increases the robustness compared to the traditional Sobel operator. Moreover in

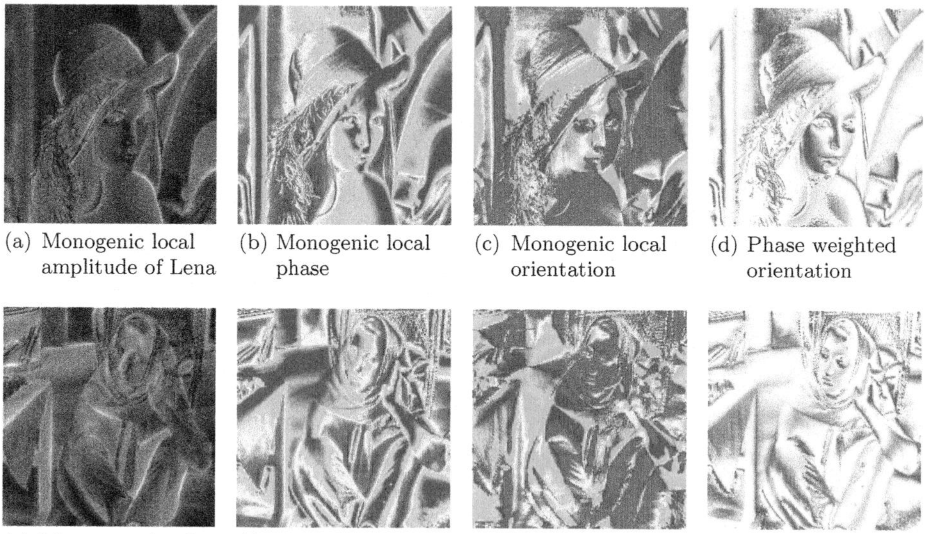

(a) Monogenic local amplitude of Lena (b) Monogenic local phase (c) Monogenic local orientation (d) Phase weighted orientation

(e) Monogenic local amplitude of Barbara (f) Monogenic local phase (g) Monogenic local orientation (h) Phase weighted orientation

FIGURE 11. Local features computed by means of monogenic signal analysis.

their work Felsberg and Sommer introduced an improved version based on local neighbourhood considerations and by using the phase as a confidence value.

Figure 11 illustrates the monogenic analysis of our two test images. The first column represents the local amplitude of the image whereas the second column shows the local monogenic phases. The last two columns illustrate the computation of the monogenic orientation. The colours are encoded on a linear periodic basis according to the Middlebury colour coding. The last column shows exactly the same orientation but with the phase in the important role of intensity information. The basic idea is to keep relevant orientation only where the structural information (*i.e.*, the phase) is high.

Note that we do not discuss here the local-zero mean property in natural image scenes. So, for example, background and illumination effects may influence the procedure and will be discussed elsewhere.

6. Conclusion

In this chapter the specificity and analysis of two generalizations of the analytic signal to higher dimensions have been detailed mathematically, based respectively on multiple complex analysis and Clifford analysis. It is shown that they are both valid extensions of the one-dimensional concept of the analytic signal. The main difference between the two approaches is with regard to rotation invariance due to the point symmetric definition of the sign function in the case of the monogenic approach compared to the single orthant definition of the multidimensional analytic signal.

In a second part we have illustrated the analytic or monogenic analysis of images on both artificial samples and real-world examples in terms of fringe analysis and texture analysis. In the context of AM-FM signal demodulation the monogenic signal analysis yields a robust decomposition into energectic, structural and geometric information. Finally some ideas for the use of generalized analytic signals in higher-level image processing and computer vision tasks were given showing the high potential for further research.

Acknowledgment

We thank the ERASMUS program, and gratefully acknowledge financial support from the Federal Ministry of Economy, Family and Youth, and from the National Foundation for Research, Technology and Development. This work was further supported in part by the Austrian Science Fund under grant number P21496 N23.

References

[1] S. Baker, D. Scharstein, J. Lewis, S. Roth, M. Black, and R. Szeliski. A database and evaluation methodology for optical flow. *International Journal of Computer Vision*, 92:1–31, 2011. See also website: `http://vision.middlebury.edu/flow/`.

[2] T. Batard and M. Berthier. The spinor representation of images. In K. Gürlebeck, editor, *9th International Conference on Clifford Algebras and their Applications*, Weimar, Germany, 15–20 July 2011.

[3] T. Batard, M. Berthier, and C. Saint-Jean. Clifford Fourier transform for color image processing. In E.J. Bayro-Corrochano and G. Scheuermann, editors, *Geometric Algebra Computing in Engineering and Computer Science*, pages 135–162. Springer, London, 2010.

[4] F. Brackx, R. Delanghe, and F. Sommen. *Clifford Analysis*, volume 76. Pitman, Boston, 1982.

[5] T. Bülow, D. Pallek, and G. Sommer. Riesz transform for the isotropic estimation of the local phase of Moiré interferograms. In G. Sommer, N. Krüger, and C. Perwass, editors, *DAGM-Symposium*, Informatik Aktuell, pages 333–340. Springer, 2000.

[6] T. Bülow and G. Sommer. Hypercomplex signals – a novel extension of the analytic signal to the multidimensional case. *IEEE Transactions on Signal Processing*, 49(11):2844–2852, Nov. 2001.

[7] V. Chandrasekhar, D.M. Chen, A. Lin, G. Takacs, S.S. Tsai, N.M. Cheung, Y. Reznik, R. Grzeszczuk, and B. Girod. Comparison of local feature descriptors for mobile visual search. In *Image Processing (ICIP), 2010 17th IEEE International Conference on*, pages 3885–3888. IEEE, 2010.

[8] A. Dzhuraev. On Riemann–Hilbert boundary problem in several complex variables. *Complex Variables and Elliptic Equations*, 29(4):287–303, 1996.

[9] M. Felsberg and G. Sommer. A new extension of linear signal processing for estimating local properties and detecting features. *Proceedings of the DAGM* 2000, pages 195–202, 2000.

[10] M. Felsberg and G. Sommer. The monogenic signal. *IEEE Transactions on Signal Processing*, 49(12):3136–3144, Dec. 2001.

[11] D. Gabor. Theory of communication. *Journal of the Institution of Electrical Engineers*, 93(26):429–457, 1946. Part III.

[12] K. Gürlebeck, K. Habetha, and W. Sprössig. *Holomorphic Functions in the Plane and n-dimensional Space.* Birkhäuser, 2008.

[13] S.L. Hahn. Multidimensional complex signals with single-orthant spectra. *Proceedings of the IEEE*, 80(8):1287–1300, Aug. 1992.

[14] R.M. Haralick, K. Shanmugam, and I.H. Dinstein. Textural features for image classification. *IEEE Transactions on Systems, Man and Cybernetics*, 3(6):610–621, 1973.

[15] B. Heise, S.E. Schausberger, C. Maurer, M. Ritsch-Marte, S. Bernet, and D. Stifter. Enhancing of structures in coherence probe microscopy imaging. In *Proceedings of SPIE*, pages 83350G–83350G–7, 2012.

[16] E. Hitzer. Quaternion Fourier transform on quaternion fields and generalizations. *Advances in Applied Clifford Algebras*, 17(3):497–517, May 2007.

[17] U. Köthe and M. Felsberg. Riesz-transforms versus derivatives: On the relationship between the boundary tensor and the energy tensor. *Scale Space and PDE Methods in Computer Vision*, pages 179–191, 2005.

[18] K.G. Larkin, D.J. Bone, and M.A. Oldfield. Natural demodulation of two-dimensional fringe patterns. I. general background of the spiral phase quadrature transform. *Journal of the Optical Society of America A*, 18(8):1862–1870, 2001.

[19] P. Lounesto. *Clifford Algebras and Spinors*, volume 286 of *London Mathematical Society Lecture Notes.* Cambridge University Press, 1997.

[20] K. Mikolajczyk and C. Schmid. A performance evaluation of local descriptors. *IEEE Transactions on Pattern Analysis and Machine Intelligence*, 27(10):1615–1630, 2005.

[21] W. Rudin. *Function Theory in the Unit Ball of* $\mathbb{C}^n$. Springer, 1980.

[22] V. Schlager, S. Schausberger, D. Stifter, and B. Heise. Coherence probe microscopy imaging and analysis for fiber-reinforced polymers. *Image Analysis*, pages 424–434, 2011.

[23] E.M. Stein. *Singular Integrals and Differentiability Properties of Functions*, volume 30 of *Princeton Mathematical Series.* Princeton University Press, 1970.

[24] The Mathworks, Inc. MATLAB® R2012b documentation: `colormap`. Software documentation available at: `http://www.mathworks.de/help/matlab/ref/colormap.html`, 1994–2012.

[25] J. Ville. Théorie et applications de la notion de signal analytique. *Cables et Transmission*, 2A:61–74, 1948.

[26] D. Zang and G. Sommer. Signal modeling for two-dimensional image structures. *Journal of Visual Communication and Image Representation*, 18(1):81–99, 2007.

Swanhild Bernstein and Martin Reinhardt
Technische Universität Bergakademie Freiberg
Fakultät für Mathematik und Informatik
Institut für Angewandte Analysis
D-09596 Freiberg, Germany
e-mail: swanhild.bernstein@math.tu-freiberg.de
martin.reinhardt.87@googlemail.com

Jean-Luc Bouchot
Johannes Kepler University
Department of Knowledge-Based Mathematical Systems, FLLL
Altenbergerstr., 69
A-4040 Linz, Austria
e-mail: jean-luc.bouchot@jku.at

Bettina Heise
Johannes Kepler University
Department of Knowledge-Based Mathematical Systems, FLLL
and
Christian Doppler Laboratory MS-MACH
Center for Surface- and Nanoanalytics, ZONA
Altenbergerstr., 69
A-4040 Linz, Austria
e-mail: bettina.heise@jku.at

Quaternion and Clifford–Fourier Transforms and Wavelets
Trends in Mathematics, 247–268

12 Colour Extension of Monogenic Wavelets with Geometric Algebra: Application to Color Image Denoising

Raphaël Soulard and Philippe Carré

Abstract. We define a colour monogenic wavelet transform. This is based on recent greyscale monogenic wavelet transforms and an extension to colour signals aimed at defining non-marginal tools. Wavelet based colour image processing schemes have mostly been made by separately using a greyscale tool on every colour channel. This may have some unexpected effects on colours because those marginal schemes are not necessarily justified. Here we propose a definition that considers a colour (vector) image right at the beginning of the mathematical definition so that we can expect to create an actual colour wavelet transform – which has not been done so far to our knowledge. This provides a promising multiresolution colour geometric analysis of images. We show an application of this transform through the definition of a full denoising scheme based on statistical modelling of coefficients.

Mathematics Subject Classification (2010). Primary 68U10; secondary 15A66, 42C40.

Keywords. Colour wavelets, analytic, monogenic, wavelet transforms, image analysis, denoising.

1. Introduction

Wavelets have been widely used for handling images for more than 20 years. It seems that the human visual system sees images through different channels related to particular frequency bands and directions; and wavelets provide such decompositions. Since 2001, the *analytic signal* and its 2D generalizations have brought a great improvement to wavelets [1, 8, 9] by a natural embedding of an AM/FM analysis in the subband coding framework. This yields an efficient representation of

This work is part of the French ANR project VERSO-CAIMAN.

geometric structures in greyscale images thanks to a *local phase* carrying geometric information complementary to an *amplitude envelope* having good invariance properties. So it codes the signal in a more coherent way than standard wavelets. The last and seemingly most appropriate proposition [9] of *analytic wavelets* for image analysis is based on the *monogenic signal* [5] defined with geometric algebra.

In parallel a *colour monogenic signal* was proposed [3] as a mathematical extension of the monogenic signal; paving the way to non-marginal colour tools especially by using geometric algebra and above all by considering a colour signal right at the foundation of the mathematical construction.

We define here a *colour monogenic wavelet transform* that extends the monogenic wavelets of [9] to colour. These *analytic wavelets* are defined for colour 2D signals (images) and avoid the classical pitfall of marginal processing (greyscale tools used separately on colour channels) by relying on a sound mathematical definition. We can therefore expect to handle coherent information of multiresolution colour geometric structure, which would ease wavelet based colour image processing. To our knowledge colour wavelets have not been proposed so far.

We first give a technical study of analytic signals and wavelets with the intent to popularize them since they rely on non-trivial concepts of geometric algebra, complex and harmonic analysis, as well as non-separable wavelet frames. Then we describe our *colour monogenic wavelet transform*, and finally an application to colour denoising will be presented.

Notation:
2D vector coordinates: $\boldsymbol{x} = (x, y)$, $\boldsymbol{\omega} = (\omega_1, \omega_2) \in \mathbb{R}^2$; $\boldsymbol{k} \in \mathbb{Z}^2$
Euclidean norm: $\|\boldsymbol{x}\| = \sqrt{x^2 + y^2}$
Complex imaginary number: $\boldsymbol{j} \in \mathbb{C}$
Argument of a complex number: arg
Convolution symbol: $*$
Fourier transform: $\mathcal{F}$

2. Analytic Signal and 2D Generalization

2.1. Analytic Signal (1D)

An *analytic signal* s_A is a multi-component signal associated to a real signal s to be analyzed. The definition is well known in the 1D case where $s_A(t) = s(t) + \boldsymbol{j}\ (h * s)(t)$ is the complex signal made of s and its Hilbert transform (with $h(t) = 1/\pi t$).

The polar form of the 1D analytic signal provides an AM/FM representation of s with $|s_A|$ being the *amplitude envelope* and $\varphi = \arg(s_A)$ the *instantaneous phase*. This classical tool can be found in many signal processing books and is for example used in communications.

Interestingly, we can also interpret the phase in terms of the signal shape, *i.e.*, there is a direct link between the angle φ and the *local structure* of s. Such a link between a 2D phase and the local geometric structures of an image is very attractive in image processing. That is why there were several attempts to generalize it to 2D signals; and among them the *monogenic signal* introduced by Felsberg [5] seems the most advanced since it is rotation invariant.

2.2. Monogenic Signal (2D)

Without going beyond the strictly necessary details, we review here the key points of the fundamental construction of the monogenic signal, which will be necessary to understand the colour extension.

The definition of the 1D case given above can be interpreted in terms of *signal processing*: the Hilbert transform makes a 'pure $\frac{\pi}{2}$-phase-shift'. But such a phase-shift is not straightforward to define in 2D (similar to many other 1D signal tools) so let us look at the equivalent *complex analysis* definition of the 1D analytic signal. It says that s_A is the *holomorphic extension* of s restricted to the real line. But complex algebra is limited regarding generalizations to higher dimensions. To bypass this limitation we can see a *holomorphic function* as a 2D *harmonic field* that is an equivalent *harmonic analysis* concept involving the 2D Laplace equation $\Delta f = 0$. It then can be generalized within the framework of 3D harmonic fields by using the 3D Laplace operator $\Delta_3 = \left(\frac{\delta^2}{\delta x^2} + \frac{\delta^2}{\delta y^2} + \frac{\delta^2}{\delta z^2}\right)$. The whole generalization relies on this natural choice and the remaining points are analogous to the 1D case (see [5] for more details). Note that in Felsberg's thesis [5] this construction is expressed in terms of *geometric algebra* but here we have avoided this for the sake of simplicity. Finally the 2D monogenic signal s_A associated to s is the 3-vector-valued signal:

$$s_A(\boldsymbol{x}) = \begin{bmatrix} s(\boldsymbol{x}) \\ s_{r1}(\boldsymbol{x}) = \dfrac{x}{2\pi \|\boldsymbol{x}\|^3} * s(\boldsymbol{x}) \\ s_{r2}(\boldsymbol{x}) = \dfrac{y}{2\pi \|\boldsymbol{x}\|^3} * s(\boldsymbol{x}) \end{bmatrix}. \tag{2.1}$$

Where s_{r1} and s_{r2} are analogous to the imaginary part of the complex 1D analytic signal. Interestingly, this construction reveals the two components of a Riesz transform:

$$\mathcal{R}\{s\} = (s_{r1}(\boldsymbol{x}), s_{r2}(\boldsymbol{x})) = \left(\frac{x}{2\pi \|\boldsymbol{x}\|^3} * s(\boldsymbol{x}), \frac{y}{2\pi \|\boldsymbol{x}\|^3} * s(\boldsymbol{x})\right), \tag{2.2}$$

in the same way that the 1D case exhibits a Hilbert transform. Note that we get back to a *signal processing* interpretation since the Riesz transform can also be viewed as a pure 2D phase-shift. In the end, by focusing on the *complex analysis* definition of the analytic signal we end up with a convincing generalization of the Hilbert transform.

Now recall that the motivation to build 2D analytic signals arises from the strong link existing between the phase and the geometric structure. To define the 2D phase related to the Riesz transform the actual monogenic signal must be expressed in spherical coordinates, which yields the following amplitude envelope

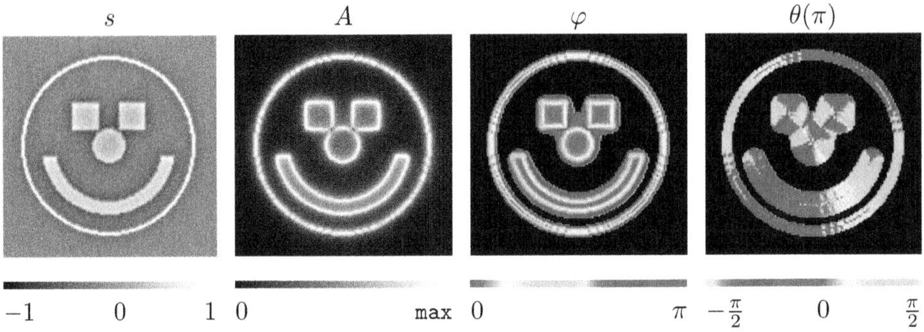

FIGURE 1. Felsberg's monogenic signal associated to a narrow-band signal s. The orientation θ is shown modulo π for visual convenience. Phase values of small coefficients have no meaning so they are replaced by black pixels.

and 2-angle phase:

$$\begin{array}{lll|l} \text{Amplitude:} & A = \sqrt{s^2 + s_{r1}^2 + s_{r2}^2}\,, & & s = A\cos\varphi\,, \\ \text{Orientation:} & \theta = \arg\left(s_{r1} + \boldsymbol{j}s_{r2}\right), & & s_{r1} = A\sin\varphi\cos\theta\,, \\ \text{1D Phase:} & \varphi = \arccos\left(s/A\right), & & s_{r2} = A\sin\varphi\sin\theta\,. \end{array} \qquad (2.3)$$

A monogenic signal analysis is illustrated in Figure 1.

2.3. Physical Interpretation

Felsberg shows a direct link between the angles θ and φ and the geometric local structure of s. The signal is thus expressed as an 'A-strong' (A = amplitude) 1D structure with orientation θ. φ is analogous to the 1D local phase and indicates whether the structure is a line or an edge. A direct drawback is that intrinsically 2D structures are not handled. Yet this tool has found many applications in image analysis, from contour detection to motion estimation (see [9] and references therein p. 1).

From a *signal processing* viewpoint the AM/FM representation provided by an analytic signal is well suited for narrowband signals. That is why it seems natural to embed it within a wavelet transform that performs subband decomposition. We now present the monogenic wavelet analysis proposed by Unser in [9].

3. Monogenic Wavelets

So far there is one proposition of computable monogenic wavelets in the literature [9]. It provides 3D vector-valued monogenic subbands consisting of a rotation-covariant *magnitude* and a new 2D *phase*. This representation – specially defined for 2D signals – is a great theoretical improvement over complex and quaternion wavelets [8, 1], similar to the way that the monogenic signal itself is an improvement over its complex and quaternion counterparts.

The proposition of [9] consists of one real-valued 'primary' wavelet transform in parallel with an associated complex-valued wavelet transform. Both transforms are linked to each other by a Riesz transform so they carry out a monogenic multiresolution analysis. We end up with three vector coefficients forming subbands that are monogenic.

3.1. Primary Transform

The primary transform is real-valued and relies on a dyadic pyramid decomposition tied to a wavelet frame. Only one 2D wavelet is needed and the dyadic downsampling is done only on the low frequency branch; leading to a redundancy of 4 : 3. The scaling function φ_γ and the mother wavelet ψ are defined in the Fourier domain as

$$\varphi_\gamma \overset{\mathcal{F}}{\longleftrightarrow} \frac{\left(4\left(\sin^2 \frac{1}{2}\omega_1 + \sin^2 \frac{1}{2}\omega_2\right) - \frac{8}{3}\sin^2 \frac{1}{2}\omega_1 \sin^2 \frac{1}{2}\omega_2\right)^{\frac{\gamma}{2}}}{\|\boldsymbol{\omega}\|^\gamma}, \tag{3.1}$$

$$\psi(\boldsymbol{x}) = (-\Delta)^{\frac{\gamma}{2}}\, \varphi_{2\gamma}(2\boldsymbol{x})\,. \tag{3.2}$$

Note that φ_γ is a cardinal polyharmonic spline of order γ and spans the space of those splines with its integer shifts. It also generates – as a scaling function – a valid multiresolution analysis.

This particular construction is made by an extension of a wavelet basis (non-redundant) related to a critically-sampled filterbank. This extension to a wavelet frame (redundant) adds some degrees of freedom used by the authors to tune the involved functions. In addition a specific *subband regression* algorithm is used on the synthesis side. The construction is fully described in [10].

3.2. The Monogenic Transform

The second 'Riesz part' transform is a complex-valued extension of the primary one. We define the associated complex-valued wavelet by including the Riesz components

$$\psi' = -\left(\frac{x}{2\pi\,\|\boldsymbol{x}\|^3} * \psi(\boldsymbol{x})\right) + \boldsymbol{j}\left(\frac{y}{2\pi\,\|\boldsymbol{x}\|^3} * \psi(\boldsymbol{x})\right)\,. \tag{3.3}$$

It can be shown that this generates a valid wavelet basis and that it can be extended to the pyramid described above. The joint consideration of both transforms leads to monogenic subbands from which the amplitude and the phase can be extracted with an overall redundancy of 4 : 1. The monogenic wavelet transform by Unser *et al.* is illustrated in Figure 2.

So far no applications of the monogenic wavelets have been proposed. In [9] a demonstration of AM/FM analysis is done with fine orientation estimation and gives very good results in terms of coherency and accuracy. Accordingly this tool may be used for analysis tasks rather than for processing.

Motivated by the powerful analysis provided by the monogenic wavelet transform we propose now to extend it to colour images.

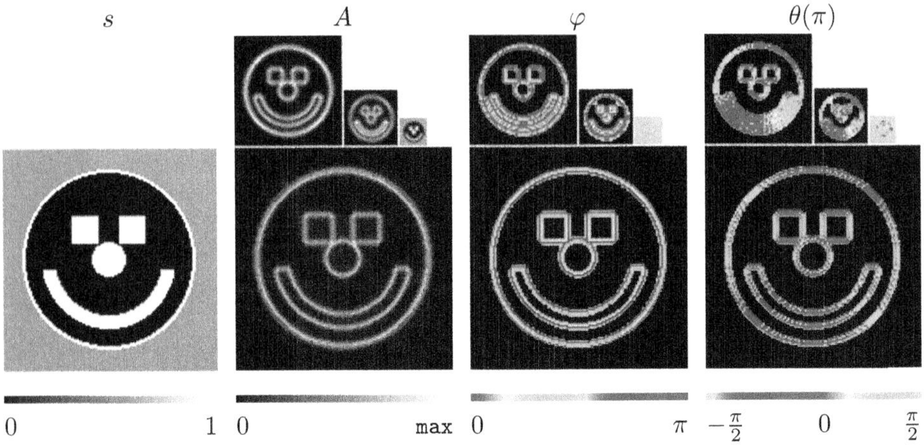

FIGURE 2. Unser's MWT of the image '`face`'. Same graphic chart as Figure 1. We used $\gamma = 3$ and the scales are $i \in \{-1, -2, -3\}$.

4. Colour Monogenic Wavelets

We define here our proposition that combines a fundamental generalization of the monogenic signal for colour signals with the monogenic wavelets described above. The challenge is to avoid the classical *marginal* definition that would apply a *greyscale* monogenic transform to each of the three colour channels of a colour image. We believe that the monogenic signal has a favorable theoretical framework for a colour extension and this is why we propose to start from this particular wavelet transform rather than from a more classical one.

The colour generalization of the monogenic signal can be expressed within the *geometric algebra* framework. This algebra is very general and embeds the complex numbers and quaternions as subalgebras. Its elements are 'multivectors', naturally linked with various geometric entities. The use of this fundamental tool is gaining popularity in the literature because it allows rewriting sophisticated concepts with simpler algebraic expressions and so paves the way to innovative ideas and generalizations in many fields.

For simplicity's sake and since we would not have enough space to present the fundamentals of geometric algebra, we express the construction here in classical terms, as we did above in Section 2.2. We may sometimes point out some necessary specific mechanisms but we refer the reader to [3, 5] for further details.

4.1. The Colour Monogenic Signal

Starting from Felsberg's approach that is originally expressed in the geometric algebra of $\mathbb{R}^3$; the extension proposed in [3] is written in the geometric algebra of $\mathbb{R}^5$ for 3-vector-valued 2D signals of the form (s_R, s_G, s_B). By simply increasing the dimensions we can embed each colour channel along a different axis and the original equation from Felsberg involving a 3D Laplace operator can be generalized in 5D with

$$\Delta_5 = \left(\frac{\delta^2}{\delta x_1^2} + \frac{\delta^2}{\delta x_2^2} + \frac{\delta^2}{\delta x_3^2} + \frac{\delta^2}{\delta x_4^2} + \frac{\delta^2}{\delta x_5^2}\right).$$

Then the system can be simplified by splitting it into three systems each with a 3D Laplace equation, reduced to be able to apply Felsberg's condition to each colour channel. At this stage the importance of *geometric algebra* appears since an algebraic simplification between vectors leads to a 5-vector colour monogenic signal that is non-marginal. Instead of naively applying the Riesz transform to each colour channel, this fundamental generalization leads to the following colour monogenic signal: $s_A = (s_R, s_G, s_B, s_{r1}, s_{r2})$ where s_{r1} and s_{r2} are the Riesz transform applied to $s_R + s_G + s_B$.

Now that the colour extension of Felsberg's monogenic signal has been defined, let us construct the colour extension of the monogenic wavelets.

4.2. The Colour Monogenic Wavelet Transform

We can now define a wavelet transform whose subbands are colour monogenic signals. The goal is to obtain vector coefficients of the form

$$(c_R, c_G, c_B, c_{r1}, c_{r2}) \tag{4.1}$$

such that

$$c_{r1} = \frac{x}{2\pi \|\boldsymbol{x}\|^3} * (c_R + c_G + c_B),$$
$$c_{r2} = \frac{y}{2\pi \|\boldsymbol{x}\|^3} * (c_R + c_G + c_B).$$

It turns out that we can very simply use the transforms presented above by applying the *primary* one on each colour channel and the *Riesz part* on the sum of the three. The five related colour wavelets illustrated in Figure 3 and forming one colour monogenic wavelet ψ_A are:

$$\psi_R = \begin{pmatrix}\psi\\0\\0\end{pmatrix}, \quad \psi_G = \begin{pmatrix}0\\\psi\\0\end{pmatrix}, \quad \psi_B = \begin{pmatrix}0\\0\\\psi\end{pmatrix} \tag{4.2}$$

$$\psi_{r1} = \begin{pmatrix}\frac{x}{2\pi\|\boldsymbol{x}\|^3} * \psi\\ \frac{x}{2\pi\|\boldsymbol{x}\|^3} * \psi\\ \frac{x}{2\pi\|\boldsymbol{x}\|^3} * \psi\end{pmatrix} \quad \psi_{r2} = \begin{pmatrix}\frac{y}{2\pi\|\boldsymbol{x}\|^3} * \psi\\ \frac{y}{2\pi\|\boldsymbol{x}\|^3} * \psi\\ \frac{y}{2\pi\|\boldsymbol{x}\|^3} * \psi\end{pmatrix} \tag{4.3}$$

$$\psi_A = (\psi_R, \psi_G, \psi_B, \psi_{r1}, \psi_{r2}) \tag{4.4}$$

We then get five vector coefficients verifying our conditions, and forming a colour monogenic wavelet transform. The associated decomposition is described by the diagram of Figure 4. This provides a multiresolution colour monogenic analysis made of a 5-vector-valued pyramid transform. The five decompositions of two images are shown in Figure 5 from left to right. Each one consists of four juxtaposed

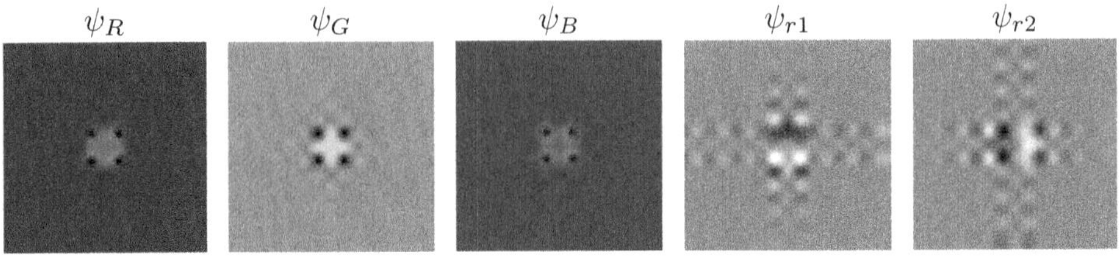

FIGURE 3. Space representation of the 5 colour wavelets.

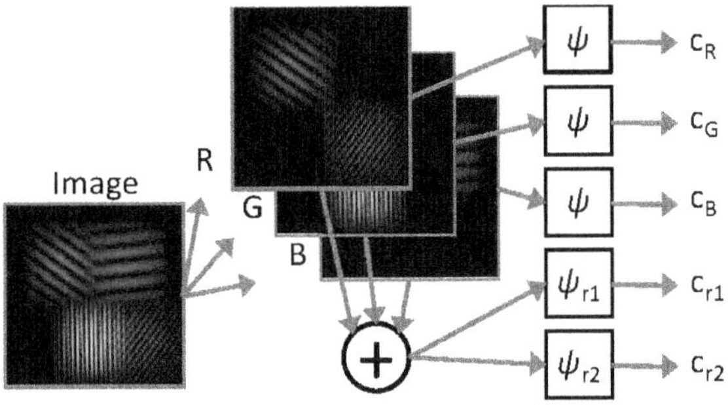

FIGURE 4. Colour MWT scheme. Each colour channel is analyzed with the primary wavelet transform symbolized by a ψ block and the sum '$R + G + B$' is analyzed with the 'Riesz part' wavelet transform (ψ_{r1} and ψ_{r2} blocks).

image-like subbands resulting from a 3-level decomposition. We fixed $\gamma = 3$, because it gave good experimental results.

4.3. Interpretation

Let us look at the first three greymaps. These are the three primary transforms c_R, c_G and c_B where white (respectively black) pixels are high positive (respectively negative) values. Note that our transform is non-separable and so provides at each scale only *one* subband related to *all* orientations. We are not subject to the arbitrarily separated horizontal, vertical and diagonal analyses usual with wavelets. This advantage is even greater in colour. Whereas marginal separable transforms show three arbitrary orientations within each colour channel – which is not easily interpretable – the colour monogenic wavelet transform provides a more compact *energy* representation of the colour image content regardless of the local orientation. The colour information is well separated through c_R, c_G and c_B: see, *e.g.*, that the blue contours of the first image are present only in c_B, and in each of the three decompositions it is clear that every orientation is equally represented all along the round contours. This is different from separable transforms that prefer particular directions. The multiresolution framework causes the horizontal blue

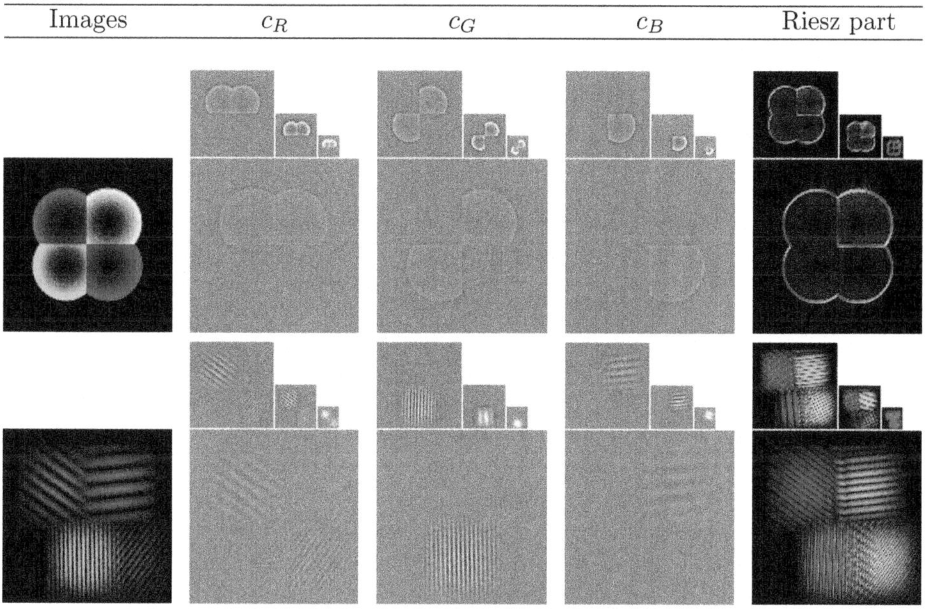

FIGURE 5. Colour MWT of images. The two components of the *Riesz part* are displayed in the same figure part with the magnitude of $c_{r1}+\boldsymbol{j}c_{r2}$ encoded in the intensity, and its argument (local orientation) encoded in the hue.

low-frequency structure of the second image to be coded mainly in the third scale of c_B.

But the directional analysis is not lost thanks to the Riesz part that completes this representation. Now look at the '2-in-1' last decomposition forming the Riesz part. It is displayed in one colour map where the geometric energy $\sqrt{c_{r1}^2 + c_{r2}^2}$ is encoded into the intensity (with respect to the well-known HSV colour space) and the orientation $\arg(c_{r1} + \boldsymbol{j}c_{r2})(\pi)$ is encoded in the hue (*e.g.*, red is for $\{0, \pi\}$ and cyan is for $\pm\frac{\pi}{2}$). This way of displaying the Riesz part reveals well the provided geometric analysis of the image.

The Riesz part gives a precise analysis that is *local* both in space and scale. If there is a local colour geometric structure in the image at a certain scale the Riesz part exhibits a high intensity in the corresponding position and subband. This is completed with an orientation analysis (hue) of the underlying structure. For instance a horizontal (respectively vertical) structure in the image will be coded by an intense cyan (respectively red) point in the corresponding subband. The orientation analysis is strikingly coherent and accurate. See, for example, that colour structures with constant orientation (second image) exhibit a *constant* hue in the Riesz part over the whole structure.

Note that low intensity corresponds to 'no structure', *i.e.*, where the image has no geometric information. It is sensible not to display the orientation (low intensity makes the hue invisible) for these coefficients since the values have no meaning in these cases.

In short, the colour and geometric information of the image are well separated from each other, and the orientation analysis is very accurate. In addition the invariance properties of the primary and Riesz wavelet transforms are kept in the colour extension for a slight overall redundancy of $20 : 9 \approx 2.2$. This transform is non-marginal because the RGB components are considered as well as the intensity $(R + G + B)$ – which involves two different colour spaces.

4.4. Reconstruction Issue

Image processing tasks such as denoising need the synthesis part of filterbanks. In the case of redundant representations, there are often several ways to perfectly reconstruct the transformed image. However, when wavelet coefficients are processed, the reconstruction method affects the way in which the wavelet domain processing will modify the image. In other words, we have to combine the redundant data so that the retrieved graphical elements are consistent with the modifications that have been done to the wavelet coefficients.

This issue occurs in the scalar case, since the pyramids we use have a redundancy of $4 : 3$. The associated reconstruction algorithm has been well defined by the authors in [10] and consists of using the spatial redundancy of each subband at the synthesis stage, by using the so-called *subband regression* algorithm.

In our case, we have to face another kind of redundancy, which stems from the monogenic model. Apart from the wavelet decomposition, the monogenic representation (as well as the analytic representation) is already basically redundant since additional signals are processed (the Riesz part). In our case, the following distinct reconstructions are possible:

- We can reconstruct the whole colour image (R, G, B) solely from the primary part (c_R, c_G, c_B).
- The Riesz part $c_{r1} + \boldsymbol{j} c_{r2}$ can be used to reconstruct $R + G + B$ – which is only a partial reconstruction.
- One can also combine both reconstructions with a specific application driven method.

In every case, the reconstruction is perfect. What is unknown is the meaning of wavelet domain processing with respect to the chosen reconstruction method.

Let us now study the new colour wavelet transform from an experimental point of view.

5. Wavelet Coefficients Study for Denoising

We propose in this section a data restoration algorithm based on the colour monogenic wavelet transform defined above. Classical denoising methods consist in performing non-linear processing in the wavelet domain by thresholding the coefficients. However in our case, information carried by the wavelet coefficients is richer than in the usual orthogonal case. The colour monogenic wavelet decomposition is not orthogonal, so we have to further study modelling of this new kind of coefficient.

In addition, we saw that the colour monogenic decomposition is composed of two kinds of data, *i.e.*:

- The 'primary part': A set of coefficients associated to colorimetric information, forming three pyramids linked to the colour channels.
- The 'Riesz part': A geometric measure composed of a norm and an angle at each point, processed by the Riesz transform and giving some information about shape and structure.

In order to carry out an image restoration algorithm, a coefficient selection process has to be defined.

We first experimentally characterize noise-related coefficients in the different subbands in order to identify their distribution and correlation. Then both the colorimetric information and the geometric information are merged into a single thresholding to perform a colour monogenic wavelet based denoising.

5.1. Modeling of Noise-related Coefficients

Classical wavelet based denoising techniques rely on the assumption that the distribution of noise-related wavelet coefficients can be modelled efficiently by a centred Gaussian law. This usually implies a constant threshold over the whole transform (see [4]). Here we have a non-orthogonal transform so we need to observe the behaviour of the noise term of a noisy image through the colour monogenic wavelet analysis. Despite the singularity of our transform, we retain the classical Gaussian model, a method which will be experimentally validated.

5.1.1. Primary Coefficients. As shown in Figure 6 we observe that the decomposition of the centred Gaussian noise with variance $\sigma^2 = 1$ remains centered and Gaussian even after the decomposition, but with different variances.

Experimental values for the standard deviations are given in Table 1. The continuous theoretical distributions have been plotted above the histograms (left side of Figure 6) so as to confirm the Gaussian model:

$$f = \frac{1}{\sigma_s\sqrt{2\pi}} \exp\left(\frac{-x^2}{2\sigma_s^2}\right) \tag{5.1}$$

where σ_s is the estimated variance of the coefficients. The decomposition is performed through a set of band-pass filters, with very limited correlation between the basis functions, which implies that we retain Gaussian signals with some degree of independence.

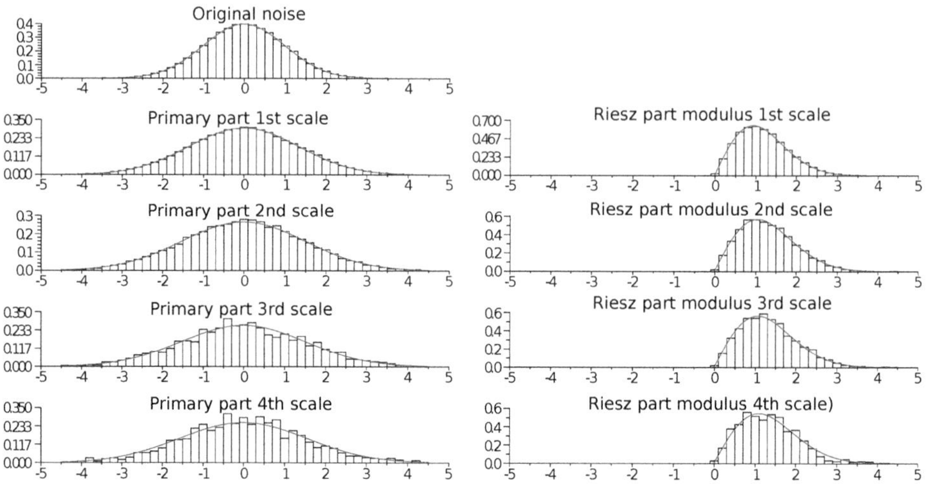

FIGURE 6. Histograms (bars) of primary subbands and modulus of Riesz subbands, and Probability Density Function models (line).

TABLE 1. Standard deviations σ_s of subbands after decomposition of Gaussian noise with variance 1 (with $\gamma = 3$).

Scale s	1(High freq.)	2	3	4
Standard deviation σ_s	1.339	1.502	1.512	1.560

The values of σ_s can also be derived analytically from the definitions of the filters. Recall that linear filtering of a stationary random signal x by a filter H with output y can be studied with power spectral densities $\Psi_x = |\mathcal{F}[x]|^2$ (PSD) and autocorrelations $R_x(\tau) = \mathcal{F}^{-1}\Psi_x$. In particular we have $\Psi_y = \Psi_x |H|^2$. The output variance σ_y^2 is equal to $R_y(0)$ which reduces to

$$\sigma_y^2 = \frac{\sigma_x^2}{4\pi^2} \iint_{(0,0)}^{(2\pi,2\pi)} |H(\boldsymbol{\omega})|^2 \, d\boldsymbol{\omega}.$$

For example, the first scale output of our filterbank is directly linked to the first stage high-pass filter:

$$\sigma_1^2 = \frac{\sigma^2}{4\pi^2} \iint_{(0,0)}^{(2\pi,2\pi)} \left| \frac{\left(4(\sin^2 \frac{1}{2}\omega_1 + \sin^2 \frac{1}{2}\omega_2) - \frac{8}{3}\sin^2 \frac{1}{2}\omega_1 \sin^2 \frac{1}{2}\omega_2\right)^{\gamma}}{2\,\|\boldsymbol{\omega}\|^{\gamma}} \right|^2 d\boldsymbol{\omega} \qquad (5.2)$$

The remaining coefficients are tied to equivalent filters of each filterbank output.

To have more realistic data, we introduce the image `peppers` altered by additive Gaussian white noise (SNR $= 83\,\mathrm{dB}$) in Figure 7.

FIGURE 7. Image **peppers** altered by additive white Gaussian noise.

We illustrate in Figure 8 a histogram of c_R (primary part, red channel) on the first scale. As in classical denoising, we assume that the first scale mainly contains noise related to coefficients since natural images should not have substantial high frequency content. Again, this histogram visually suggests that the Gaussian model is justified.

To finally confirm the Gaussian model, we computed a Kolmogorov-Smirnov test comparing experimental coefficients with a true Gaussian distribution. The test was positive for the image `peppers` with a p-value of 0.9. Let us now study the modelling of the Riesz part coefficients.

5.1.2. Riesz Part Coefficients. Recall that structure information is both carried by an angle and a modulus. Handling and thresholding of circular data such as an angle is a difficult issue. In this exploratory work, and as a natural first step, we will concentrate on the modulus. Moreover, the modulus is tied to certain *amplitude* information related to geometrical structures, and for which thresholding will be relevant. Thresholding an angle is less intuitive.

The Riesz part coefficients can also be viewed as outputs of filtering processes, so we again propose to make a Gaussian assumption for their distribution. Real and imaginary parts of subbands follow centered Gaussian distributions with variance σ_s^2.

Studying the modulus requires us to find the distribution law of the random variable $Z = (X^2 + Y^2)^{\frac{1}{2}}$ where (X, Y) are two independent Gaussian random variables with zero-mean and standard variation σ_s. In this case, it is well known

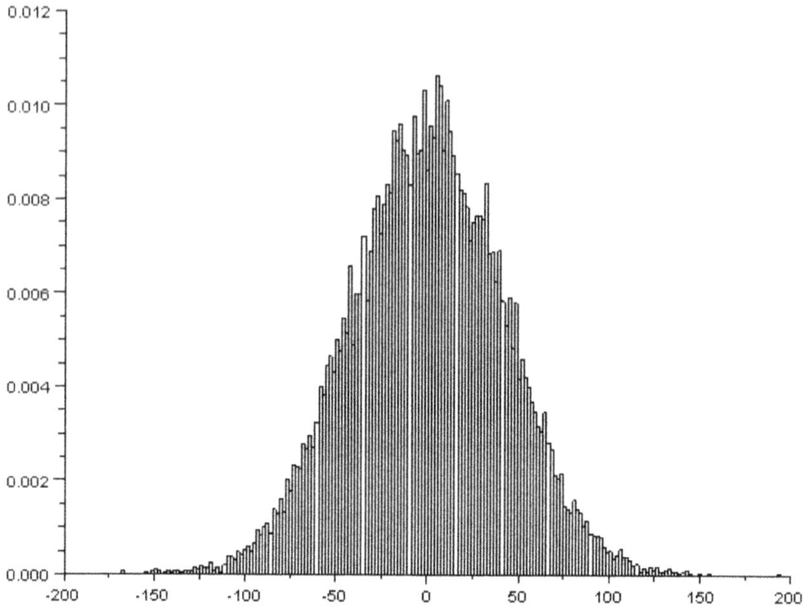

FIGURE 8. Distribution of first scale coefficients of red primary decomposition c_R.

that Z follows a Rayleigh distribution

$$f_Z(m) = \frac{m}{\sigma_s^2} \exp\left(-m^2/2\sigma_s^2\right) \mathbf{1}_{m>0}$$

with $\mathbf{1}_{m>0}$ being the Heaviside step. The moments of Z are then $E(Z) = \sigma_s\sqrt{\pi/2}$ and $V(Z) = \sigma_s^2(2-\pi/2)$. The sequence of operations in the experiment of Figure 6 were:

- Generate a test colour image made of Gaussian noise only,
- analyze it through the colour monogenic wavelet transform,
- process histograms of the modulus of the primary subbands and the Riesz subbands,
- compute standard deviations of the primary subbands to get experimental values for σ_s,
- compute theoretical distributions f and f_Z with the measured σ_s,
- plot histograms and theoretical curves in the same diagram for each scale.

We can see that the modeled Riesz-part curves – on the right side of Figure 6 – correspond well to the measured histograms. Histograms of higher scales are less regular because of the small number of coefficients in those subbands.

Note that the Rayleigh distribution would theoretically be fully justified if statistical independence were guaranteed between X and Y. In our case, the goal

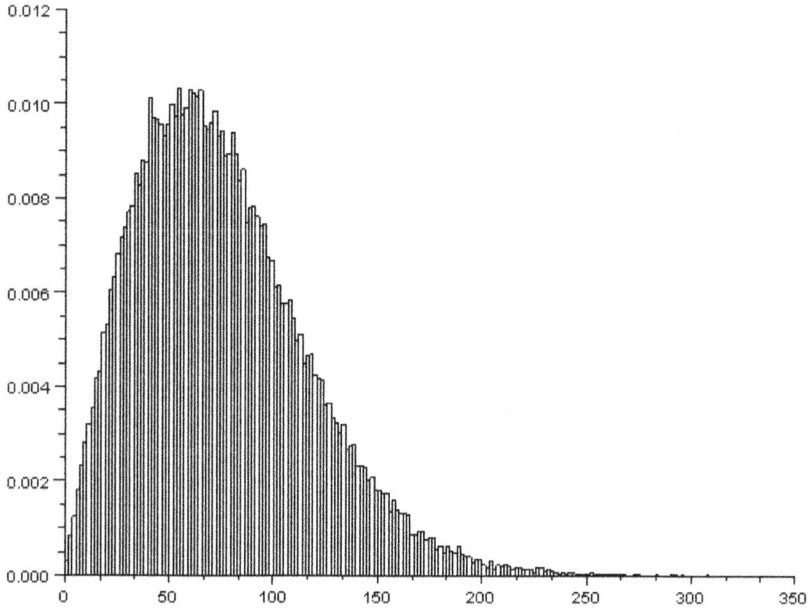

FIGURE 9. Experimental distribution of the Riesz-part modulus on the first scale.

of the modelling is simply to define a threshold so that we can keep the Rayleigh model for the modulus of the Riesz part.

In the case of the noisy natural image `peppers` introduced above, we observe the first scale histogram of the Riesz part modulus in Figure 9. According to the hypothesis that the first scale contains mainly noise, the related histogram is still Rayleigh-shaped.

To further illustrate this choice, Figure 10 shows the Riesz part modulus histograms for several scales of the decomposed noisy image `peppers`. We can see that the dispersion increases when useful information becomes – with scale – more important than noise-related coefficients. Therefore, it is a good idea to estimate the threshold on the first scale, which will discard all noise coefficients while preserving structural information within other scales.

Finally, a Gaussian model according to the primary subbands and a Rayleigh model for the modulus of Riesz subbands will be chosen. We can now apply automatic thresholding to perform colour denoising.

5.2. Thresholding

We propose to first focus on thresholding the primary part – which is the most intuitive. Thresholding will be done in the following way:

- Estimation of noise level from the first scale (by assumption it is the sole scale containing only noise).

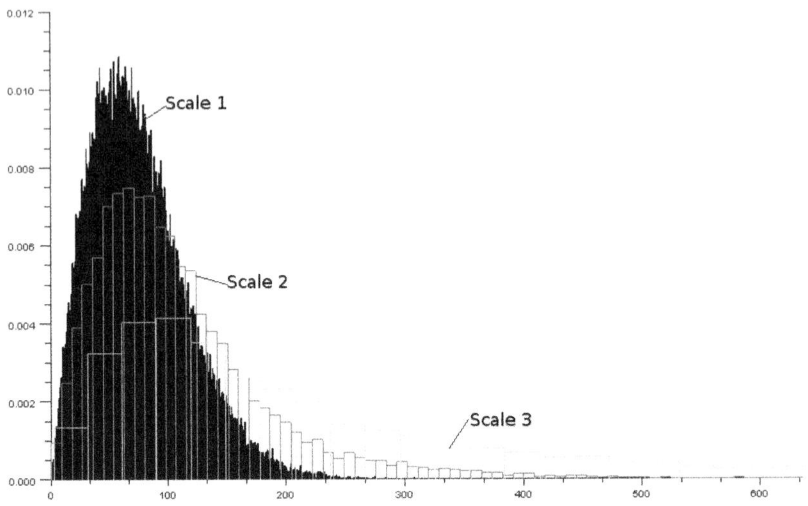

FIGURE 10. Experimental distribution of the Riesz-part modulus on the first three scales.

- Soft thresholding of the different scales with the threshold equal to three times the estimated noise level.

Using the first scale to estimate the threshold is classical (see [7]). By assuming that the first scale contains mainly noise, one can use the experimental median of the coefficients to estimate the standard deviation of the Gaussian noise distribution well. So we have the estimate $\hat{\sigma} = \text{median}(|W_1|)/0.6745$ with W_1 being the coefficients of the first scale.

Recall that the soft thresholding is defined as follows:

$$\text{thresholding}(z, g) = \text{sign}(z)\big\lfloor |z| - g \big\rfloor_+ , \tag{5.3}$$

where z is the coefficient, g the threshold and $\lfloor . \rfloor_+$ is $\max(., 0)$.

As already discussed in Section 4.4, about the synthesis part of the colour monogenic filterbank, there are several ways to reconstruct the denoised image. Within our work we must ensure that the Riesz part *controls* the denoising process. Otherwise, this would be equivalent to marginal wavelet denoising without any use of the monogenic analysis. That means, it would consider the colorimetric information (primary part) without taking into account the geometric information (Riesz part). But how to combine both the primary and the Riesz decomposition in a unified reconstruction is an open issue. In this work we propose to use the Riesz part jointly with the primary part in the *thresholding process* rather than in the *filterbank synthesis*. So the filterbank synthesis will be done for the thresholded primary part only. Finally, the denoising scheme actually fully uses the colour monogenic analysis.

Thresholding from the modulus of the Riesz coefficients is not straightforward because of the Rayleigh distribution for which classical schemes do not hold. A similar issue is dealt with in [6] with the modulus of Gabor wavelets, where setting the following threshold is suggested:

$$E(Z) + k * \sqrt{V(Z)}, \tag{5.4}$$

where $E(Z)$ and $V(Z)$ are the moments of the Rayleigh distribution already given above, and estimated from the modulus of the first scale Riesz part. We propose to use $k = 3$ which gave satisfactory experimental results. In the context of the histograms plotted in Figure 10, this corresponds to a threshold of 193. We can see that it actually discards most of the noise from the first scale, while preserving higher coefficients on the following scales.

The combination of both thresholding processes is done through a simple logical AND operator, *i.e.*, at a given position, we keep the primary coefficient only if both the primary part and the Riesz part lie above their respective thresholds. This step may be easily improved through a wider study about how to combine colorimetric information with geometric information for this kind of colour monogenic analysis.

5.3. Experimental Results

We finally show different results of colour image restoration in Figures 11, 12, 13 and 14. Note that our test images are strongly altered so as to easily reveal the particularities of different denoising methods. With less substantial levels of noise, the visual comparison is too difficult. We compare our approach to the two classical techniques that have become the reference in wavelet based denoising [2]:

- Soft thresholding of the decimated orthogonal wavelet transform (filter Daubechies 4),
- Hard thresholding of the undecimated wavelet transform (Daubechies 4).

Thresholds are established by the estimation of the noise variance as described above, whereas the low frequency subband is not modified.

We can see that the colour monogenic wavelet transform performance is intermediate between those of decimated wavelets and of undecimated wavelets. The latter has a large redundancy (L scales on an image of N pixels will give $(3L+1)N$ coefficients) while preserving orthogonality properties between interlaced decompositions, which makes it one of the most efficient denoising tools at the moment. On the other hand, decimated wavelets are known to produce unwanted oscillations – the so-called pseudo-Gibbs phenomenon – as the price to pay for their very fast implementation.

We observe that the colour monogenic approach allows us to preserve the main colour information of a colour image with limited redundancy. However, these experimental results hint that this is not fully satisfactory, since unwanted oscillation artifacts appear due to the thresholding process. Contrary to classical wavelets, the artifacts are smoother and more rounded, which may be thought of

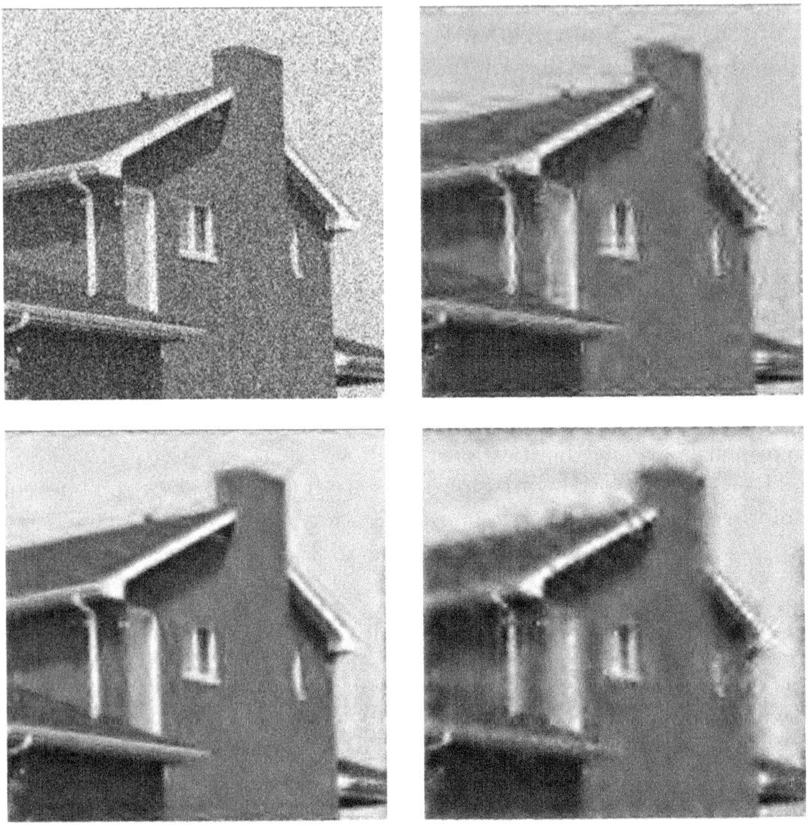

FIGURE 11. Restoration of image `house`. Upper row: Noisy image and decimated wavelet based denoising. Lower row: Undecimated wavelet based denoising and proposed approach.

as visually less annoying. The 'artifact shape' is due to the reconstruction functions, which is independent of the monogenic analysis. All the information is well selected and preserved. First we can see that image discontinuities are retained well. Textures are substantially smoothed – see, *e.g.*, the `mandrill` image shown in Figure 14 – which is usual whatever the denoising method. Note that no false colour is introduced, contrary to the decimated wavelet method. See for example the `parrot` image shown in Figure 13, where the decimated wavelet method introduces many coloured artifacts: green artifacts around the black area of the red parrot's beak, yellow artifacts on the right of the beak, green artifacts on the large yellow area on the right side of the parrot.

Given the results, it is clear that one of the possible future directions of this work is about the numerical scheme used to perform the monogenic wavelet

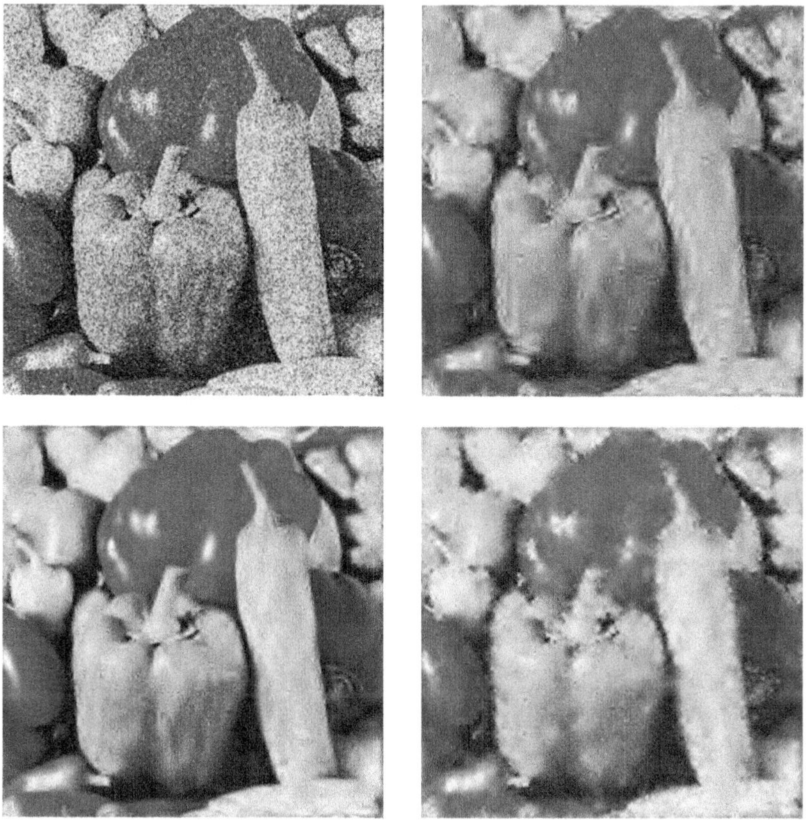

FIGURE 12. Restoration of image `peppers`. Upper row: Noisy image and decimated wavelet based denoising. Lower row: Undecimated wavelet based denoising and proposed approach.

analysis, since the current version is very sensitive to coefficient modification. The shape of the basis functions would be explored as well.

We think that the main way to improve this work is to focus on the physical interpretation when designing the key monogenic colour concept. According to us, this transform suffers from a lack of unity around its different components, as well as a poor link between them and the visual features. Although the generalization is not strictly marginal, it has a marginal style since it reduces to apply the Riesz transform to the intensity of the image. We are currently working on a new definition of colour monogenic wavelets, where the physical interpretation is taken more into account, and the local geometry is studied more deeply from a vector differential geometry viewpoint.

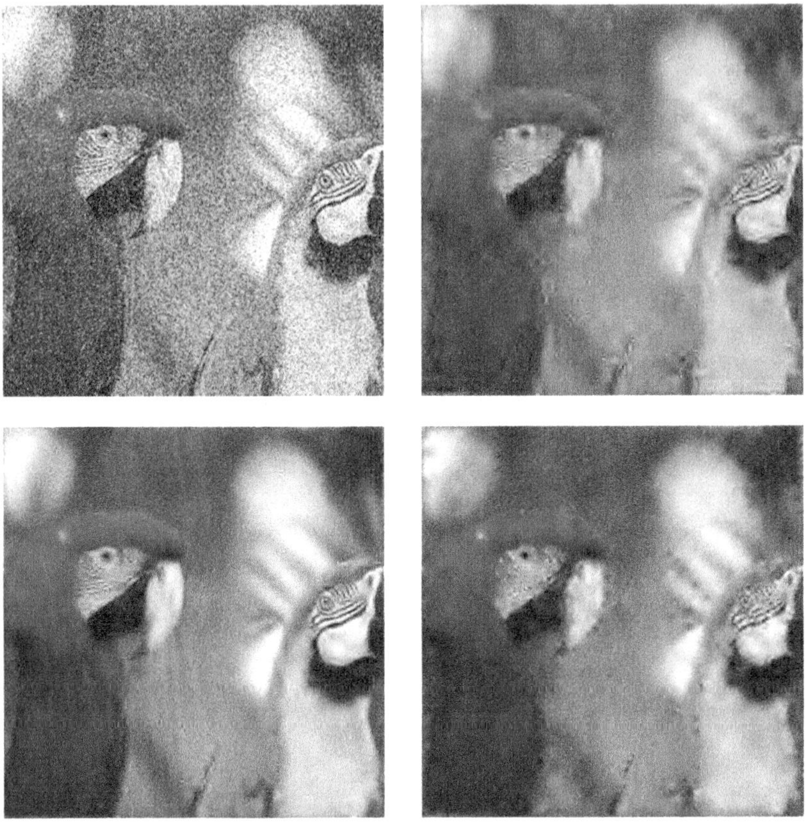

FIGURE 13. Restoration of image `parrot`. Upper row: Noisy image and decimated wavelet based denoising. Lower row: Undecimated wavelet based denoising and proposed approach.

6. Conclusion

We have defined a colour extension of the recent monogenic wavelet transform proposed in [9]. This extension is non-marginal since we have taken care to consider a vector signal at the very beginning of the fundamental construction and it leads to a definition fundamentally different from the marginal approach. The use of non-separable wavelets together with the monogenic framework permits a good orientation analysis, well separated from the colour information. This colour transform can be a great colour image analysis tool thanks to its good separation of information into various data. A statistical modelling of the coefficients for thresholding as well as a full denoising scheme is given and compared to state-of-the art wavelet based denoising methods.

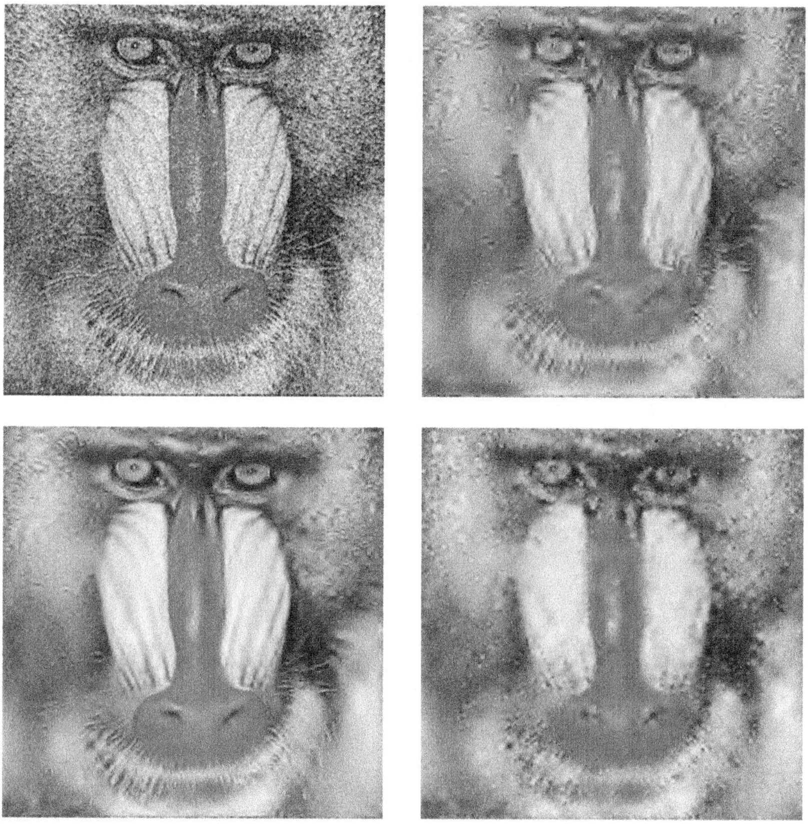

FIGURE 14. Restoration of image `mandrill`. Upper row: Noisy image and decimated wavelet based denoising. Lower row: Undecimated wavelet based denoising and proposed approach.

Although it is not marginal the colour generalization has a marginal style, since it reduces to applying the Riesz transform to the intensity of the image. So the geometric analysis is done without considering the colour information and it would be much more attractive to have a complete representation of the colour monogenic signal in terms of magnitude and phase(s) with both colour and geometric interpretation. The numerical scheme used is fragile with respect to the visual impact of modifying wavelet coefficients. Our future work includes a new definition of monogenic filterbanks by focusing on reconstruction.

References

[1] W.L. Chan, H.H. Choi, and R.G. Baraniuk. Coherent multiscale image processing using dual-tree quaternion wavelets. *IEEE Transactions on Image Processing*, 17(7):1069–1082, July 2008.

[2] I. Daubechies. *Ten Lectures on Wavelets*. Society for Industrial and Applied Mathematics, 1992.

[3] G. Demarcq, L. Mascarilla, and P. Courtellemont. The color monogenic signal: A new framework for color image processing. application to color optical flow. In *16th IEEE International Conference on Image Processing (ICIP)*, pages 481–484, 2009.

[4] D.L. Donoho. De-noising by soft-thresholding. *IEEE Transactions on Information Theory*, 41(3):613–627, 1995.

[5] M. Felsberg. *Low-Level Image Processing with the Structure Multivector*. PhD thesis, Christian-Albrechts-Universität, Institut für Informatik und Praktische Mathematik, Kiel, 2002.

[6] P. Kovesi. Image features from phase congruency. *VIDERE: Journal of Computer Vision Research*, 1(3):2–26, 1999.

[7] S. Mallat. *A Wavelet Tour of Signal Processing*. Academic Press, third edition, 2008. First edition published 1998.

[8] I.W. Selesnick, R.G. Baraniuk, and N.G. Kingsbury. The dual-tree complex wavelet transform. *IEEE Signal Processing Magazine*, 22(6):123–151, Nov. 2005.

[9] M. Unser, D. Sage, and D. Van De Ville. Multiresolution monogenic signal analysis using the Riesz-Laplace wavelet transform. *IEEE Transactions on Image Processing*, 18(11):2402–2418, Nov. 2009.

[10] M. Unser and D. Van De Ville. The pairing of a wavelet basis with a mildly redundant analysis via subband regression. *IEEE Transactions on Image Processing*, 17(11):2040–2052, Nov. 2008.

Raphaël Soulard and Philippe Carré
Xlim-SIC laboratory
University of Poitiers, France
e-mail: `raphael.soulard@univ-poitiers.fr`
`carre@sic.univ-poitiers.fr`

Quaternion and Clifford–Fourier Transforms and Wavelets
Trends in Mathematics, 269–284

13 Seeing the Invisible and Maxwell's Equations

Swanhild Bernstein

Abstract. In this chapter we study inverse scattering for Dirac operators with scalar, vector and quaternionic potentials. For that we consider factorizations of the Helmholtz equation and related fundamental solutions; the standard Green's function and Faddeev's Green function. This chapter is motivated by optical coherence tomography.

Mathematics Subject Classification (2010). Primary 30G35; secondary 45B05.

Keywords. Optical coherence tomography, Dirac operator, inverse scattering, Faddeev's Green function.

1. Preliminaries

Let $\mathbb{H}$ be the algebra of real quaternions and $\mathbb{B}$ the complex quaternions or bi-quaternions. The vectors $\boldsymbol{e}_1$, $\boldsymbol{e}_2$, $\boldsymbol{e}_3$ are the generating vectors with

$$\boldsymbol{e}_j\boldsymbol{e}_k + \boldsymbol{e}_k\boldsymbol{e}_j = -2\delta_{jk}$$

and $\boldsymbol{e}_0$ the unit element. An arbitrary element $a \in \mathbb{H}$ is given by

$$a = \sum_{j=0}^{3} a_j\boldsymbol{e}_j, \quad a_j \in \mathbb{R}$$

and an arbitrary element $b \in \mathbb{B}$ by

$$b = \sum_{j=0}^{3} b_j\boldsymbol{e}_j, \quad b_j \in \mathbb{C}.$$

We denote by $\mathrm{S}(q)$ the scalar part q_0, by $\mathbf{V}(q) = \mathbf{q} = \sum_{j=1}^{3} q_j\boldsymbol{e}_j$ the vector part and define the conjugated quaternion by

$$\overline{q} = \mathrm{S}(q) - \mathbf{V}(q)$$

for a quaternion $q \in \mathbb{H}$ or a biquaternion $q \in \mathbb{B}$. The algebra of quaternions is free of zero divisors, *i.e.*, if $a_1 a_2 = 0$ for $a_1, a_2 \in \mathbb{H}$ then $a_1 = 0$ or $a_2 = 0$. This is not true for biquaternions, for example

$$(1 + i\boldsymbol{e}_3)(1 - i\boldsymbol{e}_3) = 1 - i^2\boldsymbol{e}_3^2 = 1 - 1 = 0 \quad \text{and}$$
$$(\boldsymbol{e}_1 + i\boldsymbol{e}_3)(\boldsymbol{e}_1 + i\boldsymbol{e}_3) = \boldsymbol{e}_1^2 + i(\boldsymbol{e}_1\boldsymbol{e}_3 + \boldsymbol{e}_3\boldsymbol{e}_1) + i^2\boldsymbol{e}_3^2 = -1 + 1 = 0.$$

A zero divisor $a \in \mathbb{B}$ is an element $a \neq 0$ such that there exists a $b \in \mathbb{B}$ with $b \neq 0$ and $ab = 0$.

Let $G \in \mathbb{R}^3$ be a domain and $G^c = \{x \in \mathbb{R}^3 : x \notin \overline{G}\} = \mathbb{R}^3 \backslash \overline{G}$. A (bi)quaternion-valued function u belongs to $C(G)$, $L^p(G)$, $H^1(G)$ *etc.* when each real respectively complex-valued component u_j belongs to that function space. For $s \in \mathbb{R}$, $L^{2,s}$ denotes the set of (scalar-valued) functions u such that

$$\|u\|_s = \left\|(1 + |x|^2)^{s/2} u\right\|_{L^2(\mathbb{R}^3)} < \infty.$$

By H^α, $\alpha \geq 0$, we denote the Sobolev space of (scalar-valued) functions u such that

$$\|u\|_{H^\alpha} = \left\|(1 + |\xi|^{\alpha/2})\hat{u}\right\|_{L^2(\mathbb{R}^3)} < \infty,$$

and the weighted Sobolev spaces $H^{\alpha,s}$, $s \in \mathbb{R}$

$$\left\|(1 + |x|^{s/2}\, u)\right\|_{H^\alpha} < \infty.$$

Furthermore, we identify a vector $\mathbf{k} \in \mathbb{C}^3$ with the bi-quaternion

$$k_1\boldsymbol{e}_1 + k_2\boldsymbol{e}_2 + k_3\boldsymbol{e}_3.$$

We will denote the vector and the element in $\mathbb{B}$ or $\mathbb{H}$ with $\mathbf{k}$. By $\cdot$ we denote the inner product in $\mathbb{R}^3$

$$\mathbf{k} \cdot \mathbf{k} = \sum_{j=1}^{3} k_j^2,$$

whereas

$$\mathbf{k}\,\mathbf{k} = \mathbf{k}^2 = -\sum_{j=1}^{3} k_j^2$$

is the product of the quaternion $\mathbf{k}$ with itself. By D we denote the Dirac operator

$$D = \sum_{j=1}^{3} \boldsymbol{e}_j \frac{\partial}{\partial x_j} = \sum_{j=1}^{3} \boldsymbol{e}_j \partial_j.$$

In particular, we have for a vector field $\mathbf{u} = (u_1, u_2, u_3) \sim u_1\boldsymbol{e}_1 + u_2\boldsymbol{e}_2 + u_3\boldsymbol{e}_3$:

$$D\mathbf{u} = \begin{pmatrix} -\operatorname{div}\mathbf{u} \\ \operatorname{curl}\mathbf{u} \end{pmatrix},$$

where

$$\operatorname{div}\mathbf{u} = \nabla \cdot \mathbf{u} = \sum_{j=1}^{3} \partial_j u_j \quad \text{and} \quad \operatorname{curl}\mathbf{u} = \nabla \times \mathbf{u} = \begin{pmatrix} \partial_2 u_3 - \partial_3 u_2 \\ \partial_3 u_1 - \partial_1 u_3 \\ \partial_1 u_2 - \partial_2 u_1 \end{pmatrix}.$$

Because the multiplication in the algebra of quaternions is not commutative we introduce two different multiplication operators with vectors $\mathbf{k} \in \mathbb{H}$ or $\mathbf{k} \in \mathbb{B}$:

$${}^{\mathbf{k}}Mq = \mathbf{k}\,q \quad \text{but} \quad M^{\mathbf{k}}q = q\,\mathbf{k}, \;\; q \in \mathbb{B}.$$

2. Motivation

2.1. Optical Coherence Tomography

This chapter is motivated by optical coherence tomography (OCT). Methods in tomography usually use diffraction of beams to reconstruct an image. OCT is different because it uses the interference of light waves. Therefore the mathematical model is given by scattering of waves, see for example [7] and [8].

In this section we present the mathematical treatment of single OCT following [7]. Unscattered photons like x-rays and γ-rays have been used to obtain tomographic ray projections for a long time. The mathematical problem of reconstructing a function from its straight ray projections has already been presented by Radon in 1917 [31]. Its solution, the Fourier slice theorem, shows that some of the three-dimensional Fourier data of the object can be obtained from two-dimensional Fourier transforms of its projections.

Optical tomography techniques, in particular, OCT, deviate in several respects from the better known coherence tomography (CT) concept.

- Diffraction optical tomography (DOT) uses highly diffracted and scattered radiation, straight ray propagation can only be assumed for a fraction of the photons.
- OCT images are synthesized from a series of adjacent interferometric depth-scans performed by a straight propagating low-coherence probing beam. This leads to an advantageous decoupling of transversal resolution from depth resolution.
- OCT uses backscattering, *i.e.*, light propagates twice through the same object.

Let us consider a weakly inhomogeneous sample illuminated by the Rayleigh length around the beam waist where we can assume plane-wave illumination with incident waves:

$$V^{(i)}(\mathbf{r}, \mathbf{k}, t) = A^{(i)} e^{i\mathbf{k}^{(i)} \cdot \mathbf{r} - i\omega t},$$

$\mathbf{k}^{(i)}$ is the wave vector of the illumination wave, $|\mathbf{k}^{(i)}| = k = 2\pi/\lambda$ the wave number. Using the outgoing free-space Green's function

$$F_k(\mathbf{r}, \mathbf{r}') = \frac{e^{ik(\mathbf{r}-\mathbf{r}')}}{4\pi\,|\mathbf{r}-\mathbf{r}'|}$$

of the Helmholtz operator, the first-order Born approximation yields the scattered wave as an approximate solution of the Helmholtz equation

$$V_s(\mathbf{r},\mathbf{k}^{(s)},t) = V^{(i)}(\mathbf{r},\mathbf{k}^{(i)},t) + \int_{V(')} V^{(i)}(\mathbf{r},\mathbf{k}^{(i)},t)\cdot F_s(\mathbf{r}',k)\cdot F_k(\mathbf{r},\mathbf{r}')\,d\mathbf{r}'.$$

$\mathbf{k}^{(s)}$ is the wave vector of the scattered wave, $|\mathbf{k}^{(s)}| = k$. This integral is extended over wavelets originating from the illuminated sample volume $V(\mathbf{r}^{(\prime)})$. The relative amplitudes of these wavelets are determined by the scattering potential of the sample

$$F_s(\mathbf{r},k) = k^2[m^2(\mathbf{r},k) - 1],$$

where m is the complex refractive index distribution of the sample structure:

$$m(\mathbf{r}) = n(\mathbf{r})[1 + i\kappa(\mathbf{r})],$$

with $n(\mathbf{r})$ being the phase refractive index, and $\kappa(\mathbf{r})$ the attenuation index. In OCT, backscattered light originating from the coherently illuminated sample volume is detected at a distance d much larger than the linear dimensions of that volume. Therefore

$$V_s(\mathbf{r},\mathbf{k},t) = \frac{A^{(i)}}{4\pi\, d} e^{i\mathbf{k}^{(s)}\cdot\mathbf{r}-i\omega t} \int_{V(')} F_s(\mathbf{r}')e^{-\mathbf{k}\cdot\mathbf{r}'}\,d\mathbf{r}' = A_s(\mathbf{r},\mathbf{k}^{(s)},t)e^{i\mathbf{k}^{(s)}\cdot\mathbf{r}-i\omega t},$$

where the amplitude $A^{(i)}$ of the illuminating wave has been assumed constant within the coherent probe volume.

3. Helmholtz Equation and Maxwell's Equation

The propagation of waves can be described by Maxwell's equations which constitute a first-order differential system. That system can be reduced to second-order differential equations. Due to their importance, Maxwell's equations under different boundary conditions, radiation conditions and other properties are widely studied.

In this chapter we want to discuss the scattering problem for Maxwell's equations in terms of the Dirac operator. The description of Maxwell's equations with the Dirac operator and Clifford algebras as well as quaternions has been studied in several papers ([12, 16, 15, 23, 3, 11, 10, 27]). Usually, time harmonic equations in homogeneous media were investigated. Inhomogeneous media and in particular chiral media are considered in [15] and [19].

The Helmholtz equation and its application in acoustics and electromagnetics have been widely investigated (see for example [30, 32, 25]). Also some kind of Dirac equation and scattering has been considered [13].

We will study some scattering problems in connection with the Dirac-type operator $D + M^{i\mathbf{k}}$, where $\mathbf{k}$ is a vector, described with the aid of quaternions.

This problem is also motivated by so-called Lax pairs and the Abiowitz, Kaup, Newell and Segur (AKNS) method (see [1]). In very simple terms this method means that a nonlinear partial differential equation can be written as a pair of linear equations where one operator describes the spectral problem and the other operator is the operator governing the associated time evolution. Most known methods use the complex $\bar{\partial}$-operator and can be reduced to a scalar or matrix Riemann–Hilbert problem [9]. But it can also be done with a Dirac operator. The factorization of the Helmholtz equation is also an important tool, because we have to consider the operator $D + M^{i\mathbf{k}}$ and the scattering data.

4. Maxwell's Equations

The well-known Maxwell's equations are usually used to describe electromagnetic phenomena. In optics the interaction of light with a medium is characterized by these equations. Maxwell's equations in optics can be found in [28]. Under the assumption of an isotropic material that obeys Ohm's law for electric conduction and can act in a para- or diamagnetic manner, Maxwell's equationsare as follows:

$$\operatorname{curl}\mathbf{E} = -\frac{\partial \mathbf{B}}{\partial t}, \qquad \operatorname{div}\mathbf{D} = \rho,$$
$$\operatorname{curl}\mathbf{H} = \sigma\mathbf{E} + \varepsilon\frac{\partial \mathbf{E}}{\partial t}, \qquad \operatorname{div}\mathbf{B} = 0,$$

where $\mathbf{E}$ is the electric field, $\mathbf{H}$ the magnetic field, $\mathbf{D}$ the electric induction, $\mathbf{B}$ the magnetic induction, ρ the charge density not due to polarization of the medium, μ the permeability, σ describes the electric conductivity and ε the permittivity of the medium. The permittivity, permeability and electric conductivity are parameters that are related to material properties of the medium and do not change with time. In optics $\mu \approx 1$ can be used. If we consider a perfectly transparent insulator then its electric conductivity and its external charge density can be taken to be zero.

Maxwell's equations are considered together with so-called constitutive relations, which describe the relation between induction vectors and field vectors:

$$\mathbf{D} = \varepsilon\mathbf{E} \quad \text{and} \quad \mathbf{B} = \mu\mathbf{H}.$$

Interference occurs when radiation follows more than one path from its source to the point of detection. The simplest example of interference is that between plane waves. We restrict our consideration to plane waves, which means that we can consider time-harmonic Maxwell's equations, *i.e.*,

$$\mathbf{E}(x,t) = \operatorname{Re}\left(\mathbf{E}(x)e^{-i\omega t}\right), \quad \mathbf{H}(x,t) = \operatorname{Re}\left(\mathbf{H}(x)e^{-i\omega t}\right).$$

If we take everything into account we obtain the system

$$\operatorname{curl}\mathbf{E} = i\omega\mu\mathbf{H}, \qquad \operatorname{div}(\varepsilon\mathbf{E}) = 0$$
$$\operatorname{curl}\mathbf{H} = -i\omega\varepsilon\mathbf{E}, \qquad \operatorname{div}(\mu\mathbf{H}) = 0.$$

5. Maxwell's Equations Written as Dirac Equations

5.1. Quaternionic Formulation

Let ε_0 be the free space permittivity and μ_0 be the free space permeability, then $\frac{1}{\sqrt{\mu_0\varepsilon_0}} = c$, the speed of light, and $k = \omega\sqrt{\mu\varepsilon}$ is the wave number. We further introduce the complex refractive index N by

$$N^2 = \mu_r\varepsilon_r = \left(\frac{\mu}{\mu_0}\right)\left(\frac{\varepsilon}{\varepsilon_0}\right),$$

where μ_r and ε_r are called relative permittivity and permeability.

$$N = n + i\kappa,$$

where the real part n is the conventional refractive index, whereas κ is called the extinction coefficient, which describes the attenuation of the electric field in the medium. Maxwell's equations, medium properties in optics, and their mathematical description are studied in [28].

Remark 5.1. Force-free magnetic fields are an important special solution of nonlinear equations of magnetohydrodynamics. They are characterized by

$$\operatorname{div}\mathbf{B} = 0 \quad \text{and} \quad \operatorname{curl}\mathbf{B} + \alpha(x)\mathbf{B} = 0,$$

where $\alpha(x)$ is a scalar-valued function. This system is equivalent to

$$D_\alpha B = 0. \tag{5.1}$$

This relation between force-free fields and (5.1) can be found in [17].

Remark 5.2. In case of static electric and magnetic fields in an inhomogeneous medium we obtain the decoupled system

$$D_\varepsilon\mathbf{E} = -\frac{\rho}{\sqrt{\varepsilon}}, \quad D_\mu\mathbf{H} = \sqrt{\mu}\mathbf{j}.$$

A very elegant quaternionic reformulation of the Maxwell's equations (see [14]) in inhomogeneous media can now be obtained by the substitution

$$\mathcal{E} := \sqrt{\varepsilon}\mathbf{E} \quad \text{and} \quad \mathcal{H} := \sqrt{\mu}\mathbf{H},$$

which leads to the system

$$\begin{cases} D_{\boldsymbol{\varepsilon}}\mathcal{E} & = -ik\mathcal{H} - \frac{\rho}{\sqrt{\varepsilon}}, \\ D_\mu\mathcal{H} & = ik\mathcal{E} + \sqrt{\mu}j, \end{cases}$$

where $\boldsymbol{\varepsilon} = \frac{\operatorname{grad}\sqrt{\varepsilon}}{\sqrt{\varepsilon}}$ ($\mu = \frac{\operatorname{grad}\sqrt{\mu}}{\sqrt{\mu}}$) and $D_\varepsilon = D + \varepsilon$ ($D_\mu = D + \mu$). This system can also be written using matrices:

$$\begin{pmatrix} D_{\boldsymbol{\varepsilon}} & -ik \\ ik & D_\mu \end{pmatrix}\begin{pmatrix} \mathcal{E} \\ \mathcal{H} \end{pmatrix} = \left[\begin{pmatrix} D & -ik \\ ik & D \end{pmatrix} + \begin{pmatrix} \boldsymbol{\varepsilon} & 0 \\ 0 & \mu \end{pmatrix}\right]\begin{pmatrix} \mathcal{E} \\ \mathcal{H} \end{pmatrix} = \begin{pmatrix} -\frac{\rho}{\sqrt{\varepsilon}} \\ \sqrt{\mu}\mathbf{j} \end{pmatrix}.$$

6. Factorization of the Wave Equation

Using vector and scalar potentials, Maxwell's equations can be rewritten as second-order differential equations, that, in turn can be factorized with the help of Dirac operators:

$$\frac{1}{c^2}\frac{\partial^2}{\partial t^2} - \Delta = \left(\frac{1}{c}\frac{\partial}{\partial t} + iD\right)\left(\frac{1}{c}\frac{\partial}{\partial t} + iD\right),$$

where $i \in \mathbb{C}$ is the complex unit. If we just take the time-harmonic case, *i.e.*, the time dependence is given by the expontial $e^{i\omega t}$, differentiation by t is simply given by multiplication with $i\omega$ and we can consider the operator

$$-\frac{\omega^2}{c^2} - \Delta = \left(D + \frac{\omega}{c}\right)\left(D - \frac{\omega}{c}\right).$$

These factorizations show the deep connection between the Helmholtz equation and Maxwell's equations. But there is yet another factorization. Let $\mathbf{k}$ be a vector, then

$$(D + M^{\mathbf{k}})(D - M^{\mathbf{k}}) = (D - M^{\mathbf{k}})(D + M^{\mathbf{k}}) = -\Delta - \mathbf{k}^2 = -\Delta + |\mathbf{k}|^2,$$

and a factorization of the Helmholtz equation arises from

$$(D + M^{i\mathbf{k}})(D - M^{i\mathbf{k}}) = -\Delta - i^2\mathbf{k}^2 = -\Delta - |\mathbf{k}|^2.$$

7. Scattering

We will need some properties of the fundamental solution of the Helmholtz equation.

Proposition 7.1 ([5]). *Let*

$$F_k(x-y) := -\frac{e^{ik|x-y|}}{4\pi|x-y|}, \quad x, y \in \mathbb{R}^3, \ x \neq y.$$

$F_k(\cdot, y)$ solves the Helmholtz equation $\Delta u + k^2 u = 0$ in $\mathbb{R}^3 \backslash \{y\}$, and $y \in G$ for every bounded subset $G \subset \mathbb{R}^3$.

$$F_k(x-y) = -\frac{e^{ik|x|}}{4\pi|x|}e^{-ik\hat{x}\cdot y} + \mathbf{O}(|x|^{-2}),$$

$$D_x F_k(x-y) = (D_x F_k(x))e^{-ik\hat{x}\cdot y} + \mathbf{O}(|x|^{-2}),$$

uniform in $\hat{x} = \frac{x}{|x|} \in S^2$ and $y \in G$ for every bounded subset $G \subset \mathbb{R}^3$.

Let $k = \text{const}$ and $\Im(k) \geq 0$. Then application of $-D_{-k}$ to the fundamental solution of the Helmholtz operator gives

$$-D_{-k}F_k(x) = C_k(x) = \left(k + \frac{x}{|x|^2} - ik\frac{x}{|x|}\right)\frac{e^{ik|x|}}{4\pi}.$$

This fundamental solution was obtained in [20].

Proposition 7.2 ([4]). *Let G be a bounded Lipschitz domain with boundary ∂G and outward pointing unit normal $\mathbf{n}$. Let $\mathbf{u} \in H^1(G)$. If $\mathbf{u} \in \ker D_k(G)$, $\Im(k) \geq 0$, then*

$$\mathbf{u}(x) = \mathcal{C}_k[\mathbf{u}](x) = -\int_{\partial G} C_k(x-y)\mathbf{n}(y)\mathbf{u}(y)dy.$$

If $\mathbf{u} \in \ker D_{\mathbf{k}} = \ker(D + M^{\mathbf{k}})$, then

$$\mathbf{u}(x) = \mathcal{C}_{\mathbf{k}}[\mathbf{u}](x) = -\int_{\partial G} (-D_x F_k)(x-y)\mathbf{n}(y)\mathbf{u}(y) + \mathbf{n}(y)\mathbf{u}(y)F_k(x-y)\mathbf{k}\,dy,$$

$$= -\int_{\partial G} \left(\frac{x-y}{|x-y|^2} - ik\frac{(x-y)}{|x-y|} \right) \frac{e^{ik|x-y|}}{4\pi\,|x-y|}\mathbf{n}(y)\mathbf{u}(y)$$
$$- \frac{e^{ik|x-y|}}{4\pi\,|x-y|}\mathbf{n}(y)\mathbf{u}(y)\mathbf{k}\,dy,$$

where $k = \sqrt{\mathbf{k}^2} \in \mathbb{C}$ is chosen such that $\Im(k) \geq 0$.

Proposition 7.3 (Radiation condition [18]). *Let $f \in H^1_{\text{loc}}(G^c)$, $f \in \ker D_k(G^c)$ and f satisfy the radiation condition*

$$\left(k - \frac{x}{|x|^2} + ik\frac{x}{|x|} \right) f(x) = \mathbf{o}(|x|^{-1}) \quad \textit{as } |x| \to \infty,$$

then

$$f(x) = -\mathcal{C}_k[f](x) \quad \forall x \in G^c.$$

If f satisfies the radiation condition

$$kf(x) + \frac{ix}{|x|}f(x)\mathbf{k} = \mathbf{o}(|x|^{-1}) \quad \textit{as } |x| \to \infty,$$

where $k = \sqrt{\mathbf{k}^2}$ and $\Im(k) \geq 0$, then

$$f(x) = -\mathcal{C}_{\mathbf{k}}[f](x), \quad \textit{for all } x \in G^c.$$

If $\mathbf{k}$ is a zero divisor we suppose additionally $f(x) = \mathbf{o}(|x|^{-1})$.

Remark 7.4. It looks as if the condition for a scalar k can be written without the term $x/|x|^2$ which apparently gives a faster decay. But this is not true because $(1 + ix/|x|)$ is a zero divisor. See also [18].

Remark 7.5. This proposition is true for a quaternion-valued function, but we will restrict ourselves to vector fields $\mathbf{u}$, *i.e.*, quaternion-valued functions with zero scalar part.

8. The Scattering Problem

In this section we want to consider the scattering problem for the operator $D+M^{\mathbf{k}}$. The case $D + k$ can be treated similarly.

Statement of the problem: Let $\mathbf{m}(x)$ be a quaternion-valued potential with compact support $= G$. Let $\mathbf{k}$ be a vector which will be identified with $k_1e_1 + k_2e_2 + k_3e_3 \in \mathbb{H}$ and $k = \sqrt{\mathbf{k}^2}$ with $\Im(k) \geq 0$.

Further, let $\mathbf{u}^i(x)$ be a solution of

$$D\mathbf{u}^i(x) + \mathbf{u}^i(x)\mathbf{k} = 0 \quad \text{in } \mathbb{R}^3. \tag{8.1}$$

The scattering problem then consists in determining a scattering solution $\mathbf{u}^s(x)$ such that

$$D\mathbf{u} + \mathbf{u}\mathbf{k} = \mathbf{u}\mathbf{m}(x)\mathbf{k} \quad \text{in } \mathbb{R}^3, \tag{8.2}$$

$$\mathbf{u} = \mathbf{u}^i + \mathbf{u}^s, \tag{8.3}$$

and $\mathbf{u}^s$ fullfils the radiation condition

$$k\mathbf{u}^s(x) + \frac{ix}{|x|}\mathbf{u}^s(x)\mathbf{k} = \mathbf{o}(|x|^{-1}), \quad \text{as } |x| \to \infty. \tag{8.4}$$

If $\mathbf{k}$ is a zero divisor we additionally assume $\mathbf{u}^s(x) = \mathbf{o}(1)$.

Remark 8.1. The unknown vector field $\mathbf{m}$ with compact support could also be quaternion-valued, *i.e.*, it could have a non-zero scalar part. We write $\mathbf{m}$ to indicate that we consider a scalar- or vector- or quaternion-valued function and not only a scalar-valued function.

Theorem 8.2. *If* $\mathbf{u}$ *is a solution of the scattering problem* (8.1)–(8.4) *then* $\mathbf{u}\,|_G$ *solves the Lippmann–Schwinger integral equation*

$$\mathbf{u}(x) = \mathbf{u}^i(x) - \int_G (-D_x F_k(x-y)\mathbf{u}(y))\mathbf{m}(y)\mathbf{k} + \mathbf{u}(y)\mathbf{m}(y)\mathbf{k}F_k(x-y)\mathbf{k}\,dy. \tag{8.5}$$

Conversely, if $\mathbf{u}$ *is a solution of the Lippmann–Schwinger equation then* $\mathbf{u}$ *is a solution of the scattering problem.*

Proof. Let $\mathbf{u}$ be a solution of the scattering problem and $\mathbf{v}$ the integral on the right-hand side of (8.5). Because

$$D\mathbf{v} - \mathbf{v}\mathbf{k} = \mathbf{u}\mathbf{m}(x)\mathbf{k} = D\mathbf{u} + \mathbf{u}(x)\mathbf{k},$$

$\mathbf{w} = \mathbf{u} - \mathbf{v}$ satisfies

$$D\mathbf{w} + \mathbf{w}\mathbf{k} = 0 \quad \text{in } \mathbb{R}^3.$$

Furthermore,

$$\begin{aligned} \mathbf{w}(x) &= (\mathbf{u}^i(x) - \mathbf{u}(x)) \\ &\quad - \int_G (-D_x F_k(x-y))\mathbf{u}(y)\mathbf{m}(y)\mathbf{k} + \mathbf{u}(y)\mathbf{m}(y)\mathbf{k}F_k(x-y)\mathbf{k}\,dy, \end{aligned}$$

$x \in \mathbb{R}^3$, and it satisfies the radiation condition which will be seen as follows: $\mathbf{u}^i(x) - \mathbf{u}(x) = \mathbf{u}^s(x)$, which satisfies the radiation condition by assumption. Let us now consider the integral. First we look at the kernels

$$\begin{aligned} F_k(x-y) &= F_k(x)e^{-ik\hat{x}\cdot y} + \mathbf{O}(|x|^{-2}) \quad \text{and} \\ (D_xF_k)(x-y) &= (D_xF_k)(x)e^{-ik\hat{x}\cdot y} + \mathbf{O}(|x|^{-2}). \end{aligned}$$

We conclude that

$$\begin{aligned}&(-D_xF_k(x-y)\mathbf{u}(y))\mathbf{m}(y)\mathbf{k}+\mathbf{u}(y)\mathbf{m}(y)\mathbf{k}F_k(x-y)\mathbf{k}\\&\quad=\left((-D_xF_k(x)\mathbf{u}(y))\mathbf{m}(y)\mathbf{k}+k^2\mathbf{u}(y)\mathbf{m}(y)F_k(x-y)\right)e^{-ik\hat{x}\cdot y}+\mathbf{O}(|x|^{-2}).\end{aligned}$$

Therefore it is enough to consider

$$(-D_xF_k(x)\mathbf{u}(y))\mathbf{m}(y)\mathbf{k}+k^2\mathbf{u}(y)\mathbf{m}(y)F_k(x-y).$$

We obtain

$$\begin{aligned}&k\left((-D_xF_k(x)\mathbf{u}(y))\mathbf{m}(y)\mathbf{k}+k^2\mathbf{u}(y)\mathbf{m}(y)F_k(x-y)\right)\\&\quad+\frac{ix}{|x|}\left((-D_xF_k(x)\mathbf{u}(y))\mathbf{m}(y)\mathbf{k}+k^2\mathbf{u}(y)\mathbf{m}(y)F_k(x-y)\right)\mathbf{k}\\&\quad+k\left((-D_xF_k(x)\mathbf{u}(y))\mathbf{m}(y)\mathbf{k}+k^2\mathbf{u}(y)\mathbf{m}(y)F_k(x-y)\right)\\&\quad+\frac{ix}{|x|}\left((-D_xF_k(x)\mathbf{u}(y))\mathbf{m}(y)\mathbf{k}^2+\frac{ix}{|x|}k^2\mathbf{u}(y)\mathbf{m}(y)F_k(x-y)\mathbf{k}\right)\\&=\left(k(-D_xF_k)(x)+\frac{ik^2x}{|x|}F_k(x)\right)\mathbf{u}(y)\mathbf{m}(y)\mathbf{k}\\&\quad+k^2\left(\frac{ix}{|x|}(-D_xF_k)(x)+kF_k(x)\right)\mathbf{u}(y)\mathbf{m}(y)\\&=-\left(\frac{kx}{|x|^3}-ik^2\frac{x}{|x|^2}+ik^2\frac{x}{|x|^2}\right)\frac{e^{ik|x|}}{4\pi}\mathbf{u}(y)\mathbf{m}(y)\mathbf{k}\\&\quad-k^2\left(\frac{ix^2}{|x|^4}+k\frac{x^2}{|x|^3}+k\frac{1}{|x|}\right)\frac{e^{ik|x|}}{4\pi}\mathbf{u}(y)\mathbf{m}(y)\\&=-\frac{kxe^{ik|x|}}{4\pi|x|^3}\mathbf{u}(y)\mathbf{m}(y)\mathbf{k}-k^2\frac{ix^2e^{ik|x|}}{4\pi|x|^4}\mathbf{u}(y)\mathbf{m}(y)\\&=\mathbf{O}(|x|^{-2}).\end{aligned}$$

□

Remark 8.3. We can replace the region of integration by any domain G such that the support of $\mathbf{m}$ is contained in $\overline{G}$

Using the unique continuation principle (8.5) we will prove that the homogeneous equation has only the trivial solution and by the Fredholm theory the existence of a solution for functions in $C(G)$ or $L^p(G)$, $1<p<\infty$. Then the solution of the Lippmann–Schwinger equation can be computed and we obtain the solution $\mathbf{u}$ from

$$\mathbf{u}(x)=\mathbf{u}^i(x)-\int_G(-D_xF_k(x-y))\mathbf{u}(y)\mathbf{m}(y)\mathbf{k}+\mathbf{u}(y)\mathbf{m}(y)\mathbf{k}F_k(x-y)\mathbf{k}\,dy.$$

Lemma 8.4. *We have*

$$|F_k(x-y)|\le\frac{c}{|x-y|}\quad\text{and}\quad|DF_k(x-y)|\le\frac{c_1}{|x-y|^2}+\frac{c_2}{|x-y|}.$$

Because the kernels of the Lippmann–Schwinger equation are weakly singular, the integral operators are compact as mappings in spaces of continuous functions as well as L^p, $1 < p < \infty$, but only for *bounded domains.*

To get information about the solution of the Lippmann–Schwinger equation we would like to apply the Fredholm theory for compact operators. So far we know that the integral operator is a compact operator. Now we need to prove that the homogeneous equation has only the trivial solution. For that we need the unique continuation principle. The proof of this principle goes back to [24]. For a proof we refer to [5, Lemma 8.5], which also uses ideas from [29] and [21].

Lemma 8.5 (Unique continuation principle, [5]). *Let G be a domain in $\mathbb{R}^3$ and let $u_1, \ldots, u_p \in C^2(G)$, be real-valued functions satisfying*

$$|\Delta u_p| \leq c \sum_{q=1}^{P} \{|u_q| + |\mathrm{grad} u_q|\} \quad \text{in } G,$$

for $p = 1, \ldots, P$ and some constant c. Assume that u_p vanishes in a neighborhood of some $x_0 \in G$ for $p = 1, \ldots, P$. Then u_p is identically zero in G for $p = 1, \ldots, P$.

To be able to apply the unique continuation principle to

$$(D + M^{\mathbf{k}})\mathbf{u} - \mathbf{m}(x)\mathbf{u}\mathbf{k} = 0$$

we apply the operator D again to obtain

$$D((D + M^{\mathbf{k}})\mathbf{u} - \mathbf{m}(x)\mathbf{u}\mathbf{k}) = -\Delta\mathbf{u} + (D\mathbf{u})\mathbf{k} - D(\mathbf{m}(x)\mathbf{u}\mathbf{k}).$$

and we can use the equation

$$\Delta\mathbf{u} = (D\mathbf{u})\mathbf{k} - D(\mathbf{m}(x)\mathbf{u}\mathbf{k}).$$

The last equation equation can be rewritten as the following system

$$\begin{aligned} \Delta\mathbf{u} &= -(\mathrm{div}\,\mathbf{u})\mathbf{k} + (\mathrm{curl}\,\mathbf{u}) \times \mathbf{k} + (\mathrm{div}\,\mathbf{m})(\mathbf{u} \times \mathbf{k}) \\ &\quad -(\mathrm{curl}\,\mathbf{m})(\mathbf{u} \cdot \mathbf{k}) - 2\left(\sum_{j=1}^{3} m_j \partial_j \mathbf{u}\right)\mathbf{k}. \end{aligned}$$

If there are constants C_1, $C_2 > 0$ such that

$$\max_{x \in \overline{G}} |m_j(x)| \leq C_1 \quad \text{and} \quad \max_{x \in \overline{G}} |\mathrm{grad}\, m_j(x)| \leq C_2,$$

then we can apply the unique continuation principle to conclude that the homogeneous Lippmann–Schwinger equation over a bounded domain has only the trivial solution.

Theorem 8.6. *Let $\mathbf{k} \in \mathbb{C}^3 \backslash \{0\}$. There exists a unique solution to the inverse scattering problem (8.1)–(8.4) and the solution $\mathbf{u}$ depends continuously, with respect to the maximum norm, on the incident field $\mathbf{u}^i$.*

Proof. Due to the Fredholm theory it is enough to prove that the homogeneous equation has only the trivial solution. If that is proven, the integral equation is a bounded invertible operator in $C(\bar{B}_R)$. From this it follows that $\mathbf{u}$ depends continuously on the incident field $\mathbf{u}^i$ with respect to the maximum norm. Let $B_R :=$

$\{x \in \mathbb{R}^3 : |x| \leq R\}$. Due to Proposition 7.2 the function $\mathbf{u}$ can be represented as an integral over ∂B_R, and due to the properties of $F_k(x-y)$ we have $\mathbf{u}(x) = \mathbf{o}(1)$ as $|x| \to \infty$. Now,

$$\begin{aligned}
\mathbf{u}(x) = \mathcal{C}_{\mathbf{k}}[f](x) &= \int_{\partial B_R} (-D_x F_k)(x-y)\mathbf{n}(y)\mathbf{u}(y) + \mathbf{n}(y)\mathbf{u}(y)F_k(x-y)\mathbf{k}\, dy \\
&\sim \int_{\partial B_R} (D_y F_k)(x-y)\frac{y}{|y|}\mathbf{u}(y) + \frac{y}{|y|}\mathbf{u}(y)F_k(x-y)\mathbf{k}\, dy \\
&\sim \int_{\partial B_R} F_k(y)\left\{\left(\frac{1}{|y|} - ik\right)\mathbf{u}(y) + \frac{y}{|y|}\mathbf{u}(y)\mathbf{k}\right\} dy \\
&\sim \int_{\partial B_R} F_k(y)\left\{\frac{1}{|y|}\mathbf{u}(y) - i\left(kf(y) + i\frac{y}{|y|}\mathbf{u}(y)\mathbf{k}\right)\right\} dy \\
&\sim \int_{\partial B_R} F_k(y)\left\{\frac{1}{|y|}\mathbf{u}(y) + \mathbf{o}(|y|^{-1})\right\} dy \to 0 \quad \text{as } y \to \infty,
\end{aligned}$$

because $\mathbf{u}$ fulfils the radiation condition and $\mathbf{u}(y) = \mathbf{o}(1)$. We have proven that $\mathbf{u} = 0$ for $|x| \geq R$. The unique continuation principle (Lemma 8.5) implies that $\mathbf{u} = 0$ in $\mathbb{R}^3$. □

Up to now we have used the usual Green's function for the Helmholtz equation $F_k(x)$ and the Green's functions $C_k(x)$ and $C_{\mathbf{k}}(x)$ for the Dirac operators $D + k$ and $D + M^{\mathbf{k}}$. Another type of Green's functions are the exponentially growing Green's functions. These Green's functions were introduced by Faddeev [6] and later on used by Nachman and Ablowitz [25], Beals and Coifman [2], Sylvester and Uhlmann [32], Päivärinta [30] and Isozaki [13] in inverse scattering. Some inverse scattering problems for the Dirac operator are also discussed in [26] and [22].

The main idea is to consider plane waves

$$e^{i\mathbf{k}\cdot x}, \ \mathbf{k} \in \mathbb{C}^3, \ x \in \mathbb{R}^3, \quad \mathbf{k}\cdot\mathbf{k} = k^2.$$

Therefore we analyze how the operators change when $\mathbf{u}$ is replaced by $e^{i\mathbf{k}\cdot x}\mathbf{v}$. We obtain

$$\begin{aligned}
(\Delta + k^2)\mathbf{u} = (\Delta + k^2)(e^{i\mathbf{k}\cdot x}\mathbf{v}) &= e^{i\mathbf{k}\cdot x}(\Delta + 2i\mathbf{k}\cdot\nabla)\mathbf{v}, \\
(D - i\mathbf{k})\mathbf{u} = (D - i\mathbf{k})(e^{i\mathbf{k}\cdot x}\mathbf{v}) &= e^{i\mathbf{k}\cdot x}\left(i\mathbf{k}\mathbf{v} + D\mathbf{v} - i\mathbf{k}\mathbf{v}\right) = e^{i\mathbf{k}\cdot x}D\mathbf{v} \\
(D + M^{i\mathbf{k}})\mathbf{u} = (D + M^{i\mathbf{k}})(e^{i\mathbf{k}\cdot x}\mathbf{v}) &= e^{i\mathbf{k}\cdot x}\left(i\mathbf{k}\mathbf{v} + D\mathbf{v} + \mathbf{v}(i\mathbf{k})\right).
\end{aligned}$$

In the last equation it does matter whether $\mathbf{v}$ is truly quaternion-valued or just a vector. In the latter case we see that $\mathbf{k}\mathbf{v} + \mathbf{v}\mathbf{k} = -\mathbf{k}\cdot\mathbf{v} + \mathbf{k}\times\mathbf{v} - \mathbf{v}\cdot\mathbf{k} + \mathbf{v}\times\mathbf{k} = -2\mathbf{k}\cdot\mathbf{v}$. Hence

$$(D + M^{i\mathbf{k}})\mathbf{u} = (D + M^{i\mathbf{k}})(e^{i\mathbf{k}\cdot x}\mathbf{v}) = e^{i\mathbf{k}\cdot x}(D\mathbf{v} - 2i\mathbf{k}\cdot\mathbf{v}).$$

We will relate these new operators to the operator

$$-\Delta - 2i\boldsymbol{\gamma}\cdot\nabla - \lambda^2,$$

and Faddeev's Green function.

8.1. Faddeev's Green Function

The idea of Faddeev to obtain a nice Green function starts with decomposing the vector $\mathbf{k} = \boldsymbol{\eta} + t\boldsymbol{\gamma}$, where $\boldsymbol{\gamma} \in S^2$ is an arbitrary direction and $\boldsymbol{\eta} \cdot \boldsymbol{\gamma} = 0$. If we apply $\Delta + k^2$ to $e^{it\boldsymbol{\gamma}\cdot x}\mathbf{w}(x)$ we obtain

$$(\Delta + k^2)(e^{it\boldsymbol{\gamma}\cdot x}\mathbf{w}(x)) = e^{it\boldsymbol{\gamma}\cdot x}(-t^2 + k^2 + 2it\boldsymbol{\gamma} \cdot \nabla + \Delta)\mathbf{w}(x).$$

Let $\lambda^2 = k^2 - t^2$, take the Fourier transform of the differential operator $-\Delta - 2it\boldsymbol{\gamma} \cdot \nabla - \lambda^2$. We then get Faddeev's Green operator defined as

$$(g_{\boldsymbol{\gamma}}(\lambda, z)f)(x) = \frac{1}{(2\pi)^3} \int \frac{e^{ix\cdot\xi}}{\xi^2 + 2z\boldsymbol{\gamma} \cdot \xi - \lambda^2} \hat{f}(\xi)\, d\xi,$$

where $\boldsymbol{\gamma} \in S^2, \lambda \geq 0$, and $z \in \mathbb{C}_+ = \{z \in \mathbb{C} : \Im(z) > 0\}$. If $\Im(z) \neq 0$, $(\xi^2 + 2z\boldsymbol{\gamma} \cdot \xi - \lambda^2)^{-1} \in L^1_{\text{loc}}(\mathbb{R}^3)$. Therefore the integral is absolutely convergent for $f \in \mathcal{S}$. For $t \in \mathbb{R}$, $g_{\boldsymbol{\gamma}}(\lambda, t)$ is defined as the boundary value $g_{\boldsymbol{\gamma}}(\lambda, t + i0)$.

Proposition 8.7 ([13]). *Let $s > \frac{1}{2}$. Then*

1. *$g_{\boldsymbol{\gamma}}(\lambda, z)$ is continuous with respect to $\lambda \geq 0, \boldsymbol{\gamma} \in S^2, z \in \overline{\mathbb{C}_+}$ except for $(\lambda, z) = (0, 0)$.*
2. *$g_{\boldsymbol{\gamma}}(\lambda, z)$ is analytic in $z \in \mathbb{C}_+$.*
3. *For any $\delta_0 > 0$, there exists a constant $C > 0$ such that*

$$\|g_{\boldsymbol{\gamma}}(\lambda, z)\|_{(L^{2,s}, H^{\alpha,s})} \leq \frac{C}{(\lambda + |z|)^{1-\alpha}},$$

with $\lambda + |z| \geq \delta_0$, and $0 \leq \alpha \leq 2$.

We would like to have a similar Faddeev's Green function for the operators $D + i\mathbf{k} = D + {}^{i\mathbf{k}}M$ and $D + M^{i\mathbf{k}}$. There are differences between both cases due to the fact that

$$\begin{aligned}(D + i\mathbf{k})(D + i\mathbf{k})\mathbf{u} &= -\Delta\mathbf{u} + i\mathbf{k}D\mathbf{u} + iD(\mathbf{k}\mathbf{u}) - \mathbf{k}^2\mathbf{u} \\ &= -\Delta\mathbf{u} + i\mathbf{k}D\mathbf{u} - i\mathbf{k}D\mathbf{u} - 2i\sum_{j=1}^{3} k_j\partial_j\mathbf{u} + \mathbf{k}\cdot\mathbf{k}\mathbf{u} \\ &= (-\Delta - 2i\mathbf{k}\cdot\nabla + k^2)\mathbf{u}.\end{aligned}$$

With $i\mathbf{k} = iz\boldsymbol{\gamma}$ we have

$$(D + iz\boldsymbol{\gamma})(D + iz\boldsymbol{\gamma})\mathbf{u} = (-\Delta - 2iz\boldsymbol{\gamma} \cdot \nabla + z^2)\mathbf{u},$$

and hence

$$\begin{aligned}&(D + iz\boldsymbol{\gamma} - k)(D + iz\boldsymbol{\gamma} + k)\mathbf{u} \\ &\quad = (D + iz\boldsymbol{\gamma})(D + iz\boldsymbol{\gamma})\mathbf{u} + (D + iz\boldsymbol{\gamma})(k\mathbf{u}) + k(D + iz\boldsymbol{\gamma})\mathbf{u} - k^2\mathbf{u} \\ &\quad = (-\Delta - 2iz\boldsymbol{\gamma} \cdot \nabla - k^2 + z^2)\mathbf{u} \\ &\quad = (-\Delta - 2iz\boldsymbol{\gamma} \cdot \nabla - \lambda^2)\mathbf{u}.\end{aligned}$$

This is the way that Isozaki got a Faddeev's Green function. We obtain

$$G_{\boldsymbol{\gamma}}(\lambda, z) = (D + iz\boldsymbol{\gamma} - k)g(\lambda, z).$$

The multiplication from the other side leads to a different Faddeev's Green function. In this case we have

$$(D +^{iz\boldsymbol{\gamma}} M + M^{i(z\boldsymbol{\gamma}+\boldsymbol{\eta})})(D +^{iz\boldsymbol{\gamma}} M - M^{i(z\boldsymbol{\gamma}+\boldsymbol{\eta})})\mathbf{u} = (-\Delta - 2iz\boldsymbol{\gamma}\cdot\nabla - \lambda^2)\mathbf{u},$$

and thus with $i\mathbf{k} = iz\boldsymbol{\gamma} + i\boldsymbol{\eta}$ we obtain

$$G_{\boldsymbol{\gamma}}(\lambda, z) = (D +^{iz\boldsymbol{\gamma}} M - M^{i\mathbf{k}})g(\lambda, z).$$

This easily shows that for both operators $G_{\boldsymbol{\gamma}}(\lambda, z)$ the following is true.

Theorem 8.8. *Let $s > \frac{1}{2}$. Then*

1. $g_{\boldsymbol{\gamma}}(\lambda, z)$ *is continuous with respect to* $\lambda \geq 0, \boldsymbol{\gamma} \in S^2, z \in \overline{\mathbb{C}_+}$ *except for* $(\lambda, z) = (0, 0)$.
2. $g_{\boldsymbol{\gamma}}(\lambda, z)$ *is analytic in* $z \in \mathbb{C}_+$.
3. *For any* $\delta_0 > 0$, *and* $0 \leq \alpha \leq 1$, *there exists a constant* $C > 0$ *such that*
$$\|g_{\boldsymbol{\gamma}}(\lambda, z)\|_{(L^{2,s}, H^{\alpha,s})} \leq C(\lambda + |z|)^{\alpha},$$
with $\lambda + |z| \geq \delta_0$.

We conclude that we can also use Faddeev's Green function to solve the inverse scattering problem and we can assume the solution $\mathbf{u}$ to have the structure $\mathbf{u} = e^{i\mathbf{k}\cdot x}\mathbf{v}$, which is the appropriate setting for optical coherence tomography.

Acknowledgment

I would like to thank Dr. B. Heise and the Christian Doppler Laboratory for Microscopic and Spectroscopic Material Characterization, Johannes Kepler University Linz, Austria, for making me realize the importance of Maxwell's equations and inverse scattering for OCT and for the opportunity to learn the physics behind the mathematical formulae.

References

[1] M.J. Ablowitz and A.P. Clarkson. *Solitons, Nonlinear Evolution Equations and Inverse Scattering*, volume 149 of *London Mathematical Society Lecture Note Series*. Cambridge University Press, 1991.

[2] R. Beals and R.R. Coifman. Multidimensional inverse scattering and nonlinear partial differential equations. In F. Trèves, editor, *Pseudodifferential Operators and Applications*, volume 43 of *Proceedings of Symposia in Pure Mathematics*, pages 45–70. American Mathematical Society, 1984.

[3] S. Bernstein. Factorization of the Schrödinger operator. In W. Sprößig, editor, *Proceedings of the Symposium Analytical and Numerical Methods in Quaternionic and Clifford analysis*, pages 1–6, 1996.

[4] S. Bernstein. Lippmann–Schwinger's integral equation for quaternionic Dirac operators. unpublished, available at `http://euklid.bauing.uni-weimar.de/ikm2003/papers/46/M_46.pdf`, 2003.

[5] D. Colton and R. Kress. *Inverse Acoustic and Electromagnetic Scattering Theory*, volume 93 of *Applied Mathematical Sciences*. Springer-Verlag, Berlin, Heidelberg, New York, 1992.

[6] L.D. Faddeev. Increasing solutions of the Schrödinger equation. *Doklady Akademii Nauk SSSR*, 165:514–517, 1965.

[7] A.F. Fercher. Optical coherence tomography. *Journal of Biomedical Optics*, 1(2):153–173, 1996.

[8] A.F. Fercher, W. Drexler, C.K. Hitzenberger, and T. Lasser. Optical coherence tomography – principles and applications. *Reports on Progress in Physics*, 66:239–303, 2003.

[9] A.S. Fokas. A unified transform method for solving linear and certain nonlinear PDEs. *Proceedings of the Royal Society A, Mathematical, Physical & Engineering Sciences*, 453:1411–1443, 1997.

[10] K. Gürlebeck and W. Sprößig. *Quaternionic and Clifford Calculus for Physicists and Engineers*. Wiley, Aug. 1997.

[11] K. Gürlebeck. Hypercomplex factorization of the Helmholtz equation. *Zeitschrift für Analysis und ihre Anwendungen*, 5(2):125–131, 1986.

[12] K. Imaeda. A new formulation of classical electrodynamics. *Nuovo Cimento*, 32(1):138–162, 1976.

[13] H. Isozaki. Inverse scattering theory for Dirac operators. *Annales de l'I.H.P., section A*, 66(2):237–270, 1997.

[14] V.V. Kravchenko. On a new approach for solving Dirac equations with some potentials and Maxwell's system in inhomogeneous media. In *Operator theory: Advances and Applications*, volume 121 of *Operator Theory: Advances and Applications*, pages 278–306. Birkhäuser Verlag, 2001.

[15] V.V. Kravchenko. Quaternionic Reformulation of Maxwell's Equations for Inhomogeneous Media and New Solutions. *Zeitschrift für Analysis und ihre Anwendungen*, 21(1):21–26, 2002.

[16] V.V. Kravchenko. *Applied Quaternionic Analysis*, volume 28 of *Research and Exposition in Mathematics*. Heldermann Verlag, 2003.

[17] V.V. Kravchenko. On force-free magnetic fields: Quaternionic approach. *Mathematical Methods in the Applied Sciences*, 28:379–386, 2005.

[18] V.V. Kravchenko and R.P. Castillo. An analogue of the Sommerfeld radiation condition for the Dirac operator. *Mathematical Methods in the Applied Sciences*, 25:1383–1394, 2002.

[19] V.V. Kravchenko and M.P. Ramirez. New exact solutions of the massive Dirac equation with electric or scalar potential. *Mathematical Methods in the Applied Sciences*, 23:769–776, 2000.

[20] V.V. Kravchenko and M.V. Shapiro. On a generalized system of Cauchy-Riemann equations with a quaternionic parameter. *Russian Academy of Sciences, Doklady.*, 47:315–319, 1993.

[21] R. Leis. *Initial Boundary Value Problems in Mathematical Physics*. John Wiley, New York, 1986.

[22] X. Li. On the inverse problem for the Dirac operator. *Inverse Problems*, 23:919–932, 2007.

[23] A. McIntosh and M. Mitrea. Clifford algebras and Maxwell's equations in Lipschitz domains. *Mathematical Methods in the Applied Sciences*, 22(18):1599–1999, 1999.

[24] C. Müller. On the behavior of solutions of the differential equation $\delta u = f(x, u)$ in a neighborhood of a point. *Communications on Pure and Applied Mathematics*, 7:505–515, 1954.

[25] A.I. Nachman and M.J. Ablowitz. A multidimensional inverse scattering method. *Studies in Applied Mathematics*, 71:243–250, 1984.

[26] G. Nakamura and T. Tsuchida. Uniqueness for an inverse boundary value problem for Dirac operators. *Communications in Partial Differential Equations*, 25:7-8:557–577, 2000.

[27] E.I. Obolashvili. *Partial Differential Equations in Clifford Analysis*, volume 96 of *Pitman Monographs and Surveys in Pure and Applied Mathematics.* Harlow: Addison Wesley Longman Ltd., 1998.

[28] K.-E. Peiponen, E.M. Vartiainen, and T. Asakura. *Dispersion, Complex Analysis and Optical Spectroscopy.* Springer Tracts in Modern Physics. Springer Verlag, Berlin, Heidelberg, 1999.

[29] M.H. Protter. Unique continuation principle for elliptic equations. *Transactions of the American Mathematical Society*, 95:81–90, 1960.

[30] L. Päivärinta. Analytic methods for inverse scattering theory. In Y.K.K. Bingham and E. Somersalo, editors, *New Analytic and Geometric Methods in Inverse Problems*, pages 165–185. Springer, Berlin, Heidelberg, New York, 2003.

[31] J. Radon. Über die Bestimmung von Funktionen durch ihre Integralwerte längs gewisser Mannigfaltigkeiten. *Berichte über die Verhandlungen der Sächsischen Akademie der Wissenschaften* (*Reports on the proceedings of the Saxony Academy of Science*), 69:262–277, 1917.

[32] J. Sylvester and G. Uhlmann. A global uniqueness theorem for an inverse boundary value problem. *Annals of Mathematics*, 125(1):153–169, Jan. 1987.

Swanhild Bernstein
TU Bergakademie Freiberg
Institute of Applied Analysis
Prüferstr. 9
D-09596 Freiberg, Germany
e-mail: swanhild.bernstein@math.tu-freiberg.de

Quaternion and Clifford–Fourier Transforms and Wavelets
Trends in Mathematics, 285–298

14 A Generalized Windowed Fourier Transform in Real Clifford Algebra $C\ell_{0,n}$

Mawardi Bahri

Abstract. The Clifford–Fourier transform in $C\ell_{0,n}$ (CFT) can be regarded as a generalization of the two-dimensional quaternionic Fourier transform (QFT), which was first introduced from the mathematical aspect by Brackx. In this chapter, we propose the Clifford windowed Fourier transform using the kernel of the CFT. Some important properties of the transform are investigated.

Mathematics Subject Classification (2010). Primary 15A66; secondary 42B10.

Keywords. Multivector-valued function, Clifford algebra, Clifford–Fourier transform.

1. Introduction

Recently researchers have paid much attention to the generalization of the classical windowed Fourier transform (WFT) using the quaternion algebra and Clifford algebra. The first attempt to extend the WFT to the quaternion algebra was by Bülow and Sommer [6, 7]. They introduced a special case of the quaternionic windowed Fourier transform (QWFT) known as quaternionic Gabor filters. They applied these filters to obtain a local two-dimensional quaternionic phase. Their generalization was obtained using the inverse (two-sided) quaternion Fourier kernel. A further extension of the quaternionic Gabor filter to quaternionic Gabor wavelets was introduced by Bayro-Corrochano [3] and Xi *et al.*. [20]. Hahn [12] constructed a Fourier-Wigner distribution of 2D quaternionic signals, which is in fact closely related to the QWFT.

The WFT has been also studied in the quaternion algebra framework. The generalization uses the kernel of the (right-sided) quaternion Fourier transform [18, 13] which was introduced in [2] and the kernel of the (two-sided) quaternion Fourier transform recently proposed by Fu *et al.* in [9]. In [15], we applied the $C\ell_{n,0}$ Clifford windowed Fourier transform to linear time-varying systems. In this chapter, we continue the generalization of the quaternionic windowed Fourier transform

This work was partially supported by Bantuan Seminar Luar Negeri oleh DP2M DIKTI 2011, Indonesia.

to the real Clifford algebra $C\ell_{0,n}$ called the $C\ell_{0,n}$ Clifford windowed Fourier transform (CWFT) which differs appreciably[1] from the $C\ell_{n,0}$ Clifford windowed Fourier transform (see [17, 16]).

This chapter is organized as follows. A brief review of real Clifford algebra is given in § 2. § 3 introduces the Clifford–Fourier transform in $C\ell_{0,n}$ (CFT) and derives its important properties. § 4 discusses the basic ideas for constructing the Clifford windowed Fourier transform in $C\ell_{0,n}$ (CWFT) using the kernel of the Clifford–Fourier transform in $C\ell_{0,n}$. We show that some properties of the two-dimensional quaternionic windowed Fourier transform (see [2]) are not valid in the CWFT such as the shift property and the Heisenberg uncertainty principle.

2. Preliminaries

We will be working with real Clifford algebras. Let $\{e_1, e_2, e_3, \ldots, e_n\}$ be an orthonormal vector basis of the real n-dimensional Euclidean vector space $\mathbb{R}^n$. The real Clifford algebra over $\mathbb{R}^n$ denoted by $C\ell_{0,n}$ then has the graded 2^n-dimensional basis

$$\{1, e_1, e_2, \ldots, e_n, e_{12}, e_{31}, e_{23}, \ldots, i_n = e_1 e_2 \cdots e_n\}. \tag{2.1}$$

Obviously, for $n = 2 \pmod 4$ the pseudoscalar $i_n = e_1 e_2 \cdots e_n$ anti-commutes with each basis of the Clifford algebra while $i_n^2 = -1$. The associative geometric multiplication of the basis vectors is governed by the rules:

$$\begin{aligned} e_k e_l &= -e_l e_k \qquad \text{for} \quad k \neq l, \qquad 1 \leq k, l \leq n, \\ e_k^2 &= 1 \qquad \text{for} \qquad 1 \leq k \leq n. \end{aligned} \tag{2.2}$$

An element of a Clifford algebra is called a *multivector* and has the following form

$$f = \sum_A e_A f_A, \tag{2.3}$$

where $f_A \in \mathbb{R}$, $e_A = e_{\alpha_1 \alpha_2 \cdots \alpha_k} = e_{\alpha_1} e_{\alpha_2} \cdots e_{\alpha_k}$, and $1 \leq \alpha_1 \leq \alpha_2 \leq \cdots \leq \alpha_k \leq n$ with $\alpha_j \in \{1, 2, \ldots n\}$. For convenience, we introduce $\langle f \rangle_k = \sum_{|A|=k} f_A e_A$ to denote the *k-vector part* of f $(k = 0, 1, 2, \ldots, n)$, then

$$f = \sum_{k=0}^{k=n} \langle f \rangle_k = \langle f \rangle + \langle f \rangle_1 + \langle f \rangle_2 + \cdots + \langle f \rangle_n, \tag{2.4}$$

where $\langle \ldots \rangle_0 = \langle \ldots \rangle$.

The Clifford conjugate $\overline{f}$ of a multivector f is defined as the anti-automorphism for which

$$\overline{f} = \sum_A \overline{e_A} f_A, \quad \overline{e_A} = (-1)^{k(k+1)/2}, \quad \overline{e_k} = -e_k, k = 1, 2, 3, \ldots, n. \tag{2.5}$$

[1]The CWFT presented here is constructed using the kernel of the $C\ell_{0,n}$ Clifford–Fourier transform whose properties are quite different from the kernel of the $C\ell_{n,0}$ Clifford–Fourier transform (see [1, 14]).

and hence

$$\overline{fg} = \overline{g}\,\overline{f} \quad \text{for arbitrary } f, g \in C\ell_{0,n}. \tag{2.6}$$

The scalar product of multivectors f and g and its associated norm is defined by, respectively,

$$\langle f\, g\rangle = f * \overline{g} = \sum_A f_A \overline{g}_B, \quad \text{and} \quad |f|^2 = \sum_A f_A^2. \tag{2.7}$$

The product of two Clifford multivectors $\boldsymbol{x} = \sum_{i=1}^n x_i \boldsymbol{e}_i$ and $\boldsymbol{y} = \sum_{j=1}^n y_j \boldsymbol{e}_j$ splits into a scalar part and a 2-vector or so-called bivector part

$$\boldsymbol{xy} = \boldsymbol{x} \cdot \boldsymbol{y} + \boldsymbol{x} \wedge \boldsymbol{y} \tag{2.8}$$

where

$$\boldsymbol{x} \cdot \boldsymbol{y} = \sum_{i=1}^n x_i y_i, \quad \text{and} \quad \boldsymbol{x} \wedge \boldsymbol{y} = \sum_{i=1}^n \sum_{j=i+1}^n \boldsymbol{e}_i \boldsymbol{e}_j (x_i y_j - x_j y_i). \tag{2.9}$$

In the following we give the definition of a $C\ell_{0,n}$-valued function.

Definition 2.1. Let $\Omega \subset \mathbb{R}^n$ be an open connected set. Functions f defined in Ω with values in $C\ell_{0,n}$ can be expressed as:

$$f : \Omega \longrightarrow C\ell_{0,n}.$$

They are of the form:

$$f(\boldsymbol{x}) = \sum_A \boldsymbol{e}_A f_A(\boldsymbol{x}), \tag{2.10}$$

where f_A are real-valued functions in Ω.

More specifically, we let $L^p(\Omega; C\ell_{0,n}), 1 \le p < \infty$ and $L^\infty(\Omega; C\ell_{0,n})$ denote the usual Lebesgue space of integrable or essentially bound $C\ell_{0,n}$-valued function on Ω. Notice that $L^p(\Omega; C\ell_{0,n})$ is a $C\ell_{0,n}$-bimodule. Moreover, it can be proved to be a Banach module. The norms on the space are denoted $\|\cdot\|_{L^p(\mathbb{R}^n;C\ell_{0,n})}$ and $\|\cdot\|_{L^\infty p(\mathbb{R}^n;C\ell_{0,n})}$, respectively. The set of C^k-functions in Ω with values in $C\ell_{0,n}$ is denoted by

$$C^k(\Omega; C\ell_{0,n}) = \{f | f : \Omega \longrightarrow C\ell_{0,n}, \quad f(\boldsymbol{x}) = \sum_A \boldsymbol{e}_A f_A(\boldsymbol{x})\} \tag{2.11}$$

Notice that if $f_A \in C^k(\Omega)$ then we say $f \in C^k(\Omega; C\ell_{0,n})$.

Let us consider $L^2(\mathbb{R}^n; C\ell_{0,n})$ as a left module. For two multivector functions $f, g \in L^2(\mathbb{R}^n; C\ell_{0,n})$, an inner product is defined by

$$(f, g)_{L^2(\mathbb{R}^n;C\ell_{0,n})} = \int_{\mathbb{R}^n} f(\boldsymbol{x})\overline{g(\boldsymbol{x})}\, d^n\boldsymbol{x} = \sum_{A,B} \boldsymbol{e}_A \overline{\boldsymbol{e}_B} \int_{\mathbb{R}^n} f_A(\boldsymbol{x}) g_B(\boldsymbol{x})\, d^n\boldsymbol{x} \tag{2.12}$$

In particular, if $f = g$, then the scalar part of the above inner product gives the L^2-norm

$$\|f\|^2_{L^2(\mathbb{R}^n;C\ell_{0,n})} = \int_{\mathbb{R}^n} \sum_A f_A^2(\boldsymbol{x})\, d^n\boldsymbol{x} \tag{2.13}$$

3. Clifford–Fourier Transform (CFT)

In the following, we introduce the Clifford–Fourier transform (CFT) (see [8, 5, 19]). We can regard the CFT as an alternative representation of the classical tensorial Fourier transform, *i.e.*, we apply a one-dimensional Fourier transform n times (each time with a different imaginary unit).

3.1. Definition of CFT

Definition 3.1. The CFT of a multivector function $f \in L^1(\mathbb{R}^n; C\ell_{0,n})$ is the function $\mathcal{F}\{f\}\colon \mathbb{R}^n \to C\ell_{0,n}$ given by

$$\mathcal{F}\{f\}(\boldsymbol{\omega}) = \int_{\mathbb{R}^n} f(\boldsymbol{x}) \prod_{k=1}^{n} e^{-\boldsymbol{e}_k \omega_k x_k} d^n \boldsymbol{x}, \tag{3.1}$$

with $\boldsymbol{\omega}, \boldsymbol{x} \in \mathbb{R}^n$.

Note that

$$d^n \boldsymbol{x} = \frac{d\boldsymbol{x}_1 \wedge d\boldsymbol{x}_2 \wedge \cdots \wedge d\boldsymbol{x}_n}{i_n} \tag{3.2}$$

and is scalar valued ($d\boldsymbol{x}_k = dx_k \boldsymbol{e}_k$, $k = 1, 2, 3, \ldots, n$, no summation). Notice also that the Clifford–Fourier kernel $\prod_{k=1}^{n} e^{-\boldsymbol{e}_k \omega_k x_k}$ in general does not commute with elements of $C\ell_{0,n}$. Furthermore, the product has to be performed in a fixed order.

The existence of the inverse CFT is given by the following theorem. For more detail and for proofs see [5, 7].

Theorem 3.2. *Suppose that $f \in L^1(\mathbb{R}^n; C\ell_{0,n})$ and $\mathcal{F}\{f\} \in L^1(\mathbb{R}^n; C\ell_{0,n})$. Then the CFT is invertible and its inverse is calculated by*

$$f(\boldsymbol{x}) = \frac{1}{(2\pi)^n} \int_{\mathbb{R}^n} \mathcal{F}\{f\}(\boldsymbol{\omega}) \prod_{k=0}^{n-1} e^{\boldsymbol{e}_{n-k} \omega_{n-k} x_{n-k}} \, d^n \boldsymbol{\omega}. \tag{3.3}$$

The CFT is a generalization of the quaternionic Fourier transform (QFT), so most of the properties of the QFT such as the shift property, convolution, and Plancherel's theorem, have their corresponding CFT generalizations. Observe that this Fourier transform can be extended to the whole of $L^2(\mathbb{R}^n; C\ell_{0,n})$ in the usual way by considering it in the weak or distributional sense.

The following subsection investigates some important properties of the CFT, which will be necessary to establish the Clifford windowed Fourier transform in $C\ell_{0,n}$ (CWFT).

3.2. Main Properties of CFT

We first establish a *scalar Plancherel theorem*. It states that for every $f \in L^2(\mathbb{R}^n; C\ell_{0,n})$

$$\|f\|^2_{L^2(\mathbb{R}^n; C\ell_{0,n})} = \frac{1}{(2\pi)^n} \|\mathcal{F}\{f\}\|^2_{L^2(\mathbb{R}^n; C\ell_{0,n})}. \tag{3.4}$$

This shows that the total signal energy computed in the spatial domain is equal to the total signal energy computed in the Clifford domain. The Plancherel theorem allows the energy of a Clifford-valued signal to be considered in either the spatial domain or the Clifford domain and the change of domains for convenience of computation.

Let us now formulate the Clifford–Parseval theorem, which is needed to prove the orthogonality relation of the CWFT.

Theorem 3.3 (CFT Parseval). *The inner product* (2.12) *of two Clifford functions* $f, g \in L^2(\mathbb{R}^n; C\ell_{0,n})$ *and their CFTs are related by*

$$(f, g)_{L^2(\mathbb{R}^n;C\ell_{0,n})} = \frac{1}{(2\pi)^n}(\mathcal{F}\{f\}, \mathcal{F}\{g\})_{L^2(\mathbb{R}^n;C\ell_{0,n})}. \tag{3.5}$$

Proof. We have

$$\begin{aligned}
&(f, g)_{L^2(\mathbb{R}^n;C\ell_{0,n})} \\
&= \int_{\mathbb{R}^n} f(\boldsymbol{x})\overline{g(\boldsymbol{x})}\, d^n\boldsymbol{x} \\
&\overset{(3.3)}{=} \frac{1}{(2\pi)^n}\int_{\mathbb{R}^n}\left[\int_{\mathbb{R}^n}\mathcal{F}\{f\}(\boldsymbol{\omega})\prod_{k=0}^{n-1} e^{e_{n-k}\omega_{n-k}x_{n-k}}\, d^n\boldsymbol{\omega}\right]\overline{g(\boldsymbol{x})}d^n\boldsymbol{x} \\
&\overset{(2.6)}{=} \frac{1}{(2\pi)^n}\int_{\mathbb{R}^n}\mathcal{F}\{f\}(\boldsymbol{\omega})\overline{\left[\int_{\mathbb{R}^n} g(\boldsymbol{x})\prod_{k=1}^{n} e^{-e_k\omega_k x_k}\, d^n\boldsymbol{x}\right]}\, d^n\boldsymbol{\omega} \\
&= \frac{1}{(2\pi)^n}\int_{\mathbb{R}^n}\mathcal{F}\{f\}(\boldsymbol{\omega})\overline{\mathcal{F}\{g\}(\boldsymbol{\omega})}\, d^n\boldsymbol{\omega} \\
&= \frac{1}{(2\pi)^n}(\mathcal{F}\{f\}, \mathcal{F}\{g\})_{L^2(\mathbb{R}^n;C\ell_{0,n})}.
\end{aligned}$$

This proves the theorem. □

Notice that equation (3.5) is multivector valued. In particular, with $f = g$, we get the multivector version of the Plancherel theorem, *i.e.*,

$$(f, f)_{L^2(\mathbb{R}^n;C\ell_{0,n})} = \frac{1}{(2\pi)^n}(\mathcal{F}\{f\}, \mathcal{F}\{f\})_{L^2(\mathbb{R}^n;C\ell_{0,n})}. \tag{3.6}$$

Due to the non-commutativity of the Clifford exponential product factors we have a left linearity property for general linear combinations with Clifford constants and a scaling property.

Theorem 3.4 (Left linearity property). *The CFT of two functions in the Clifford module* $f, g \in L^1(\mathbb{R}^n; C\ell_{0,n})$ *is a left linear operator*[2], *i.e.*,

$$\mathcal{F}\{\alpha f + \beta g\}(\boldsymbol{\omega}) = \alpha\mathcal{F}\{f\}(\boldsymbol{\omega}) + \beta\mathcal{F}\{g\}(\boldsymbol{\omega}), \tag{3.7}$$

where α *and* $\beta \in C\ell_{0,n}$ *are Clifford constants.*

[2] The CFT is also right linear for real constants $\mu, \lambda \in \mathbb{R}$.

Theorem 3.5 (Scaling property). *Suppose $f \in L^1(\mathbb{R}^n; C\ell_{0,n})$. Let a be a positive real constant. Then the CFT of the function $f_a(\boldsymbol{x}) = f(a\boldsymbol{x})$ is*

$$\mathcal{F}\{f_a\}(\boldsymbol{\omega}) = \frac{1}{a^n}\mathcal{F}\{f\}\left(\frac{\boldsymbol{\omega}}{a}\right). \tag{3.8}$$

Remark 3.6. The usual form of the shift and modulation properties of the complex FT does not hold for the CFT because of the non-commutativity of the Clifford–Fourier kernel

$$\prod_{k=1}^{n} e^{\boldsymbol{e}_k \omega_k x_k} \prod_{k-0}^{n-1} e^{\boldsymbol{e}_{n-k}\omega_{n-k}x_{n-k}} \neq \prod_{k=0}^{n-1} e^{\boldsymbol{e}_{n-k}\omega_{n-k}x_{n-k}} \prod_{k=1}^{n} e^{\boldsymbol{e}_k \omega_k x_k}. \tag{3.9}$$

The following properties are extensions of the QFT, which are very useful in solving partial differential equations in Clifford algebra. First, let us give an explicit proof of the derivative properties stated in Table 1.

TABLE 1. Properties of the CFT of Clifford functions $f, g \in L^2(\mathbb{R}^n; C\ell_{0,n})$.
The constants are $\alpha, \beta \in C\ell_{0,n}$, $a \in \mathbb{R} \setminus \{0\}$, and $n \in \mathbb{N}$.

Property	**Clifford Function**	**Clifford–Fourier Transform**
Left linearity	$\alpha f(\boldsymbol{x})+\beta g(\boldsymbol{x})$	$\alpha\mathcal{F}\{f\}(\boldsymbol{\omega})+ \beta\mathcal{F}\{g\}(\boldsymbol{\omega})$
Scaling	$f(a\boldsymbol{x})$	$\frac{1}{a^n}\mathcal{F}\{f\}(\frac{\boldsymbol{\omega}}{a})$
Partial derivative	$\frac{\partial^n}{\partial x_1^n} f(\boldsymbol{x})\, \boldsymbol{e}_1^{-n}$	$\omega_1^n \mathcal{F}\{f\}(\boldsymbol{\omega})$
	$\frac{\partial^n}{\partial x_1^n} f(\boldsymbol{x})$	$(\boldsymbol{e}_1\omega_1)^n \mathcal{F}\{f\}(\boldsymbol{\omega})$, $f = f_0 + \boldsymbol{e}_1 f_1 + i_n f_{123\cdots n}$ if $n = 3 \pmod 4$
	$\frac{\partial^m f}{\partial x_r^m}$	$(\omega_r \boldsymbol{e}_r)^m \mathcal{F}\{f\}(\boldsymbol{\omega})$, $r = 2, \ldots, n-1, m = 2s, s \in \mathbb{N}$
	$\frac{\partial^m f}{\partial x_n^m}$	$\mathcal{F}\{f\}(\boldsymbol{\omega})(\boldsymbol{e}_n\omega_n)^m$
Power	$f(\boldsymbol{x})(x_1\boldsymbol{e}_1)^m$	$(-1)^m \frac{\partial^m}{\partial\omega_1^m}\mathcal{F}\{f\}(\boldsymbol{\omega})$
	$f(\boldsymbol{x})(x_r\boldsymbol{e}_r)^m$	$\frac{\partial^m}{\partial\omega_r^m}\mathcal{F}\{f\}(\boldsymbol{\omega})$, $r = 2, \ldots, n-1, m = 2s, s \in \mathbb{N}$
	$f(\boldsymbol{x})x_n^m$	$\frac{\partial^m}{\partial\omega_n^m}\mathcal{F}\{f\}(\boldsymbol{\omega})\boldsymbol{e}_n^m$
Plancherel	$(f, g)_{L^2(\mathbb{R}^n;C\ell_{0,n})} =$	$\frac{1}{(2\pi)^n}(\mathcal{F}\{f\}, \mathcal{F}\{f\})_{L^2(\mathbb{R}^n;C\ell_{0,n})}$
Scalar Parseval	$\frac{1}{(2\pi)^n}\ \lVert f\rVert^2_{L^2(\mathbb{R}^n;C\ell_{0,n})} =$	$\frac{1}{(2\pi)^n}\ \lVert\mathcal{F}\{f\}\rVert^2_{L^2(\mathbb{R}^n;C\ell_{0,n})}$

Theorem 3.7. *Suppose $f \in L^1(\mathbb{R}^n; C\ell_{0,n})$. Then the CFT of the nth partial derivative of $x_1 f \in L^1(\mathbb{R}^n; C\ell_{0,n})$ with respect to the variable x_1 is given by*

$$\mathcal{F}\left\{\frac{\partial^n f}{\partial x_1^n}\, \boldsymbol{e}_1^{-n}\right\}(\boldsymbol{\omega}) = \omega_1^n \mathcal{F}\{f\}(\boldsymbol{\omega}), \quad \forall n \in \mathbb{N} \tag{3.10}$$

and if $x_r f \in L^1(\mathbb{R}^n; C\ell_{0,n})$, then for $r = 2, 3, \ldots, \mathbb{N}-1$ we have

$$\mathcal{F}\left\{\frac{\partial^m f}{\partial x_r^m}\right\}(\boldsymbol{\omega}) = \int_{\mathbb{R}^n} f(\boldsymbol{x})\, e^{-\boldsymbol{e}_1\omega_1 x_1} e^{-\boldsymbol{e}_2\omega_2 x_2} \cdots (\omega_r\, \boldsymbol{e}_r)^m \times \prod_{k=r}^{n} e^{-\boldsymbol{e}_k\omega_k x_k}\, d^n\boldsymbol{x}, \quad m = 2s+1, s \in \mathbb{N}, \tag{3.11}$$

and

$$\mathcal{F}\left\{\frac{\partial^m f}{\partial x_r^m}\right\}(\boldsymbol{\omega}) = (\omega_r \boldsymbol{e}_r)^m \mathcal{F}\{f\}(\boldsymbol{\omega}), \quad m = 2s, s \in \mathbb{N}. \tag{3.12}$$

Proof. We only prove (3.10) of Theorem 3.7, the others being similar. In this proof, we first prove the theorem for $n = 1$. Applying integration by parts and using the fact that f tends to zero for $x_1 \to \infty$ we immediately obtain

$$\begin{aligned}
&\mathcal{F}\left\{\frac{\partial}{\partial x_1} f\, \boldsymbol{e}_1^{-1}\right\}(\boldsymbol{\omega}) \\
&= \int_{\mathbb{R}^n} \left(\frac{\partial}{\partial x_1} f(\boldsymbol{x})\, \boldsymbol{e}_1^{-1}\right) \prod_{k=1}^{n} e^{-\boldsymbol{e}_k\omega_k x_k}\, d^n\boldsymbol{x} \\
&= \int_{\mathbb{R}^{n-1}} \left[\int_{\mathbb{R}} \left(\frac{\partial}{\partial x_1} f(\boldsymbol{x})\, \boldsymbol{e}_1^{-1}\right) e^{-\boldsymbol{e}_1\omega_1 x_1}\, dx_1\right] \prod_{k=2}^{n} e^{-\boldsymbol{e}_k\omega_k x_k}\, d^{n-1}\boldsymbol{x} \\
&= \int_{\mathbb{R}^{n-1}} \Bigg[f(\boldsymbol{x})\, \boldsymbol{e}_1^{-1} e^{-\boldsymbol{e}_1\omega_1 x_1} \Big|_{x_1=-\infty}^{x_1=\infty} \\
&\quad - \int_{\mathbb{R}} f(\boldsymbol{x})\, \boldsymbol{e}_1^{-1} \frac{\partial}{\partial x_1} e^{-\boldsymbol{e}_1\omega_1 x_1}\, dx_1 \Bigg] \prod_{k=2}^{n} e^{-\boldsymbol{e}_k\omega_k x_k}\, d^{n-1}\boldsymbol{x} \\
&= \int_{\mathbb{R}^n} f(\boldsymbol{x})\, \omega_1 \prod_{k=1}^{n} e^{-\boldsymbol{e}_k\omega_k x_k}\, d^n\boldsymbol{x} \\
&= \omega_1 \mathcal{F}\{f\}(\boldsymbol{\omega}).
\end{aligned} \tag{3.13}$$

In a similar way we get

$$\mathcal{F}\left\{\frac{\partial}{\partial x_1} f\right\}(\boldsymbol{\omega}) = \omega_1 \mathcal{F}\{f \boldsymbol{e}_1\}(\boldsymbol{\omega}). \tag{3.14}$$

For $n=2$ we obtain

$$\begin{aligned}
&\mathcal{F}\left\{\frac{\partial^2}{\partial x_1^2} f\, \boldsymbol{e}_1^{-2}\right\}(\boldsymbol{\omega})\\
&= \int_{\mathbb{R}^n} \left(\frac{\partial}{\partial x_1}(\frac{\partial}{\partial x_1} f(\boldsymbol{x}))\, \boldsymbol{e}_1^{-2}\right) \prod_{k=1}^{n} e^{-\boldsymbol{e}_k \omega_k x_k}\, d^n\boldsymbol{x}\\
&= \int_{\mathbb{R}^{n-1}} \left[\int_{\mathbb{R}} \left(\frac{\partial}{\partial x_1}(\frac{\partial}{\partial x_1} f(\boldsymbol{x}))\, \boldsymbol{e}_1^{-2}\right) e^{-\boldsymbol{e}_1\omega_1 x_1}\, dx_1\right] \prod_{k=2}^{n} e^{-\boldsymbol{e}_k \omega_k x_k}\, d^{n-1}\boldsymbol{x}\\
&= \int_{\mathbb{R}^{n-1}} \left[(\frac{\partial}{\partial x_1} f(\boldsymbol{x}))\, \boldsymbol{e}_1^{-2} e^{-\boldsymbol{e}_1\omega_1 x_1}\big|_{x_1=-\infty}^{x_1=\infty}\right.\\
&\quad \left. -\omega_1 \int_{\mathbb{R}} \frac{\partial}{\partial x_1} f(\boldsymbol{x})\, \boldsymbol{e}_1^{-1} e^{-\boldsymbol{e}_1\omega_1 x_1}\, dx_1\right] \prod_{k=2}^{n} e^{-\boldsymbol{e}_k \omega_k x_k}\, d^{n-1}\boldsymbol{x}\\
&= \omega_1^2 \int_{\mathbb{R}^n} f(\boldsymbol{x}) \prod_{k=1}^{n} e^{-\boldsymbol{e}_k \omega_k x_k}\, d^n\boldsymbol{x}\\
&= \omega_1^2 \mathcal{F}\{f\}(\boldsymbol{\omega}). \qquad (3.15)
\end{aligned}$$

By repeating this process $n-2$ additional times we finish the proof of Theorem 3.7. □

Remark 3.8. Observe that when we assume that $f = f_0 + \boldsymbol{e}_1 f_1 + i_n f_{123\cdots n}$, $n = 3 \pmod 4$, then equation (3.14) takes the form

$$\mathcal{F}\left\{\frac{\partial^n f}{\partial x_1^n}\right\}(\boldsymbol{\omega}) = (\boldsymbol{e}_1\omega_1)^n \mathcal{F}\{f\}(\boldsymbol{\omega}), \quad n \in \mathbb{N}. \qquad (3.16)$$

Theorem 3.9. *Let $x_n f \in L^1(\mathbb{R}^n; C\ell_{0,n})$. Then the CFT of the mth partial derivative of a Clifford-valued function $f \in L^1(\mathbb{R}^n; C\ell_{0,n})$ with respect to the variable x_n is given by*

$$\mathcal{F}\left\{\frac{\partial^m f}{\partial x_n^m}\right\}(\boldsymbol{\omega}) = \mathcal{F}\{f\}(\boldsymbol{\omega})(\boldsymbol{e}_n\omega_n)^m, \quad m \in \mathbb{N}. \qquad (3.17)$$

Proof. For $m=1$ direct calculation gives

$$\begin{aligned}
\frac{\partial f(\boldsymbol{x})}{\partial x_n} &= \frac{\partial}{\partial x_n}\frac{1}{(2\pi)^n} \int_{\mathbb{R}^n} \mathcal{F}\{f\}(\boldsymbol{\omega}) \prod_{k=0}^{n-1} e^{\boldsymbol{e}_{n-k}\omega_{n-k} x_{n-k}}\, d^n\boldsymbol{\omega}\\
&= \frac{1}{(2\pi)^n} \int_{\mathbb{R}^n} \mathcal{F}\{f\}(\boldsymbol{\omega}) \left(\frac{\partial}{\partial x_n} e^{\boldsymbol{e}_n\omega_n x_n}\right) \prod_{k=1}^{n-1} e^{\boldsymbol{e}_{n-k}\omega_{n-k} x_{n-k}}\, d^n\boldsymbol{\omega}\\
&= \frac{1}{(2\pi)^n} \int_{\mathbb{R}^n} [\mathcal{F}\{f\}(\boldsymbol{\omega})\, \boldsymbol{e}_n\omega_n] \prod_{k=0}^{n-1} e^{\boldsymbol{e}_{n-k}\omega_{n-k} x_{n-k}}\, d^n\boldsymbol{\omega}\\
&= \mathcal{F}^{-1}\left[\mathcal{F}\{f\}(\boldsymbol{\omega})\, \boldsymbol{e}_n\omega_n\right]. \qquad (3.18)
\end{aligned}$$

We therefore get

$$\mathcal{F}\left\{\frac{\partial f}{\partial x_n}\right\}(\boldsymbol{\omega}) = \mathcal{F}\{f\}(\boldsymbol{\omega})\boldsymbol{e}_n\omega_n. \tag{3.19}$$

By successive differentiation with respect to the variable x_n and by induction we easily obtain

$$\mathcal{F}\left\{\frac{\partial^m f}{\partial x_n^m}\right\}(\boldsymbol{\omega}) = \mathcal{F}\{f\}(\boldsymbol{\omega})(\boldsymbol{e}_n\omega_n)^m, \forall m \in \mathbb{N}. \tag{3.20}$$

This ends the proof of (3.17). □

Next we derive the *power properties* of the CFT stated in Table 1.

Theorem 3.10. *If we assume that $x_1 f \in L^1(\mathbb{R}^n; C\ell_{0,n})$. Then the CFT of the nth partial derivative of $f \in L^1(\mathbb{R}^n; C\ell_{0,n})$ with respect to the variable x_1 is given by*

$$\mathcal{F}\{f(\boldsymbol{x})(x_1\boldsymbol{e}_1)^n\}(\boldsymbol{\omega}) = (-1)^n \frac{\partial^n}{\partial \omega_1^n}\mathcal{F}\{f\}(\boldsymbol{\omega}), \quad \forall n \in \mathbb{N}. \tag{3.21}$$

If $x_r^m f \in L^1(\mathbb{R}^n; C\ell_{0,n})$, then for $r = 2, 3, \ldots, n-1$ we have

$$\begin{aligned}\frac{\partial^m}{\partial \omega_r^m}\mathcal{F}\{f\}(\boldsymbol{\omega}) = &\int_{\mathbb{R}^n} f(\boldsymbol{x})\, e^{-\boldsymbol{e}_1\omega_1 x_1} e^{-\boldsymbol{e}_2\omega_2 x_2} \cdots (-\boldsymbol{e}_r x_r)^m \\ &\times \prod_{k=r}^{n} e^{-\boldsymbol{e}_k\omega_k x_k}\, d^n\boldsymbol{x}, m = 2s+1, s \in \mathbb{N},\end{aligned} \tag{3.22}$$

and for $m = 2s, s \in \mathbb{N}$

$$\mathcal{F}\{(\boldsymbol{e}_r x_r)^m f\}(\boldsymbol{\omega}) = \frac{\partial^m}{\partial \omega_r^m}\mathcal{F}\{f\}(\boldsymbol{\omega}). \tag{3.23}$$

If $x_n f \in L^1(\mathbb{R}^n; C\ell_{0,n})$, then

$$\mathcal{F}\{x_n^m f\}(\boldsymbol{\omega}) = \frac{\partial^m}{\partial \omega_n^m}\mathcal{F}\{f\}(\boldsymbol{\omega})\, \boldsymbol{e}_n^m, \quad m = 1, 2, 3, \ldots, n. \tag{3.24}$$

Proof. We only prove (3.21) of Theorem 3.10. It is not difficult to check that

$$f(\boldsymbol{x})(x_1\boldsymbol{e}_1)^n \prod_{k=1}^{n} e^{-\boldsymbol{e}_k\omega_k x_k} = (-1)^n \frac{\partial^n}{\partial \omega_1^n} f(\boldsymbol{x}) \prod_{k=1}^{n} e^{-\boldsymbol{e}_k\omega_k x_k}. \tag{3.25}$$

We immediately obtain

$$\int_{\mathbb{R}^n} f(\boldsymbol{x})(x_1\boldsymbol{e}_1)^n \prod_{k=1}^{n} e^{-\boldsymbol{e}_k\omega_k x_k}\, d^n\boldsymbol{x} = (-1)^n \frac{\partial^n}{\partial \omega_1^n}\int_{\mathbb{R}^n} f(\boldsymbol{x}) \prod_{k=1}^{n} e^{-\boldsymbol{e}_k\omega_k x_k}\, d^n\boldsymbol{x}, \tag{3.26}$$

which gives the desired result. □

4. Clifford Windowed Fourier Transform (CWFT)

In this section, we introduce the Clifford windowed Fourier transform as a generalization of the two-dimensional quaternionic Fourier transform to higher dimensions. For this let us define the Clifford–Gabor filter, which is a special case of the Clifford atom operator [9].

4.1. Two-dimensional Clifford–Gabor Filters

The two-dimensional Clifford–Gabor filter[3] is the extension of the complex Gabor filter to the two-dimensional Clifford algebra. It takes the form

$$\begin{aligned} G_c(\boldsymbol{x}, \sigma_1, \sigma_2) &= e^{e_2 u_0 x_1} e^{e_1 v_0 x_2} g(\boldsymbol{x}, \boldsymbol{\sigma}) \\ &= e^{e_2 u_0 x_1} e^{e_1 v_0 x_2} e^{-\left[(x_1/\sigma_1)^2 + (x_2/\sigma_2)^2\right]/2}. \end{aligned} \tag{4.1}$$

Equation (4.1) is often called the*quaternionic Gabor filter.* Bülow and Sommer [6, 7] have applied it to get the local quaternionic phase of a two-dimensional real signal. From this, we get the following facts:

- It is generated using the kernel of the $Cl(0,2)$ CFT.
- If the Gaussian function $g(\boldsymbol{x}, \boldsymbol{\sigma})$ is replaced by the Clifford window function $\phi(\boldsymbol{x} - \boldsymbol{b})$, then it becomes the Clifford atom operator, *i.e.*,

$$\phi_{\boldsymbol{\omega}, \boldsymbol{b}}(\boldsymbol{x}) = e^{e_2 v_0 x_2} e^{e_1 u_0 x_1} \phi(\boldsymbol{x} - \boldsymbol{b}), \quad \boldsymbol{x}, \boldsymbol{b} \in \mathbb{R}^2. \tag{4.2}$$

- Since the modulation property does not hold for the CFT, (4.2) can not be expressed in terms of the CFT.

4.2. Definition of CWFT

Definition 4.1. The CWFT of a multivector function $f \in L^2(\mathbb{R}^n; C\ell_{0,n})$ with respect to the non-zero Clifford window function $\phi \in L^2(\mathbb{R}^n; C\ell_{0,n})$ such that $|\boldsymbol{x}|^{1/2} \phi(\boldsymbol{x}) \in L^2(\mathbb{R}^n; C\ell_{0,n})$ is given by

$$\begin{aligned} G_\phi f(\boldsymbol{\omega}, \boldsymbol{b}) &= (f, \phi_{\boldsymbol{\omega}, \boldsymbol{b}})_{L^2(\mathbb{R}^n; C\ell_{0,n})} \\ &= \int_{\mathbb{R}^n} f(\boldsymbol{x}) \, \overline{\phi_{\boldsymbol{\omega}, \boldsymbol{b}}(\boldsymbol{x})} \, d^n\boldsymbol{x} \\ &= \int_{\mathbb{R}^n} f(\boldsymbol{x}) \, \overline{\phi(\boldsymbol{x} - \boldsymbol{b})} \prod_{k=1}^{n} e^{-e_k \omega_k x_k} \, d^n\boldsymbol{x}. \end{aligned} \tag{4.3}$$

We then call

$$\phi_{\boldsymbol{\omega}, \boldsymbol{b}}(\boldsymbol{x}) = \prod_{k=0}^{n-1} e^{e_{n-k} \omega_{n-k} x_{n-k}} \phi(\boldsymbol{x} - \boldsymbol{b}), \tag{4.4}$$

the *atom operator* as the kernel of the CWFT in (4.3). Notice that for $n = 2$ the CWFT above is identical to the two-dimensional quaternionic windowed Fourier

[3]Here we start with the Clifford–Gabor filter to obtain the Clifford atom operator, which is needed to construct the Clifford windowed Fourier transform.

transform (see [2]) and for $n = 1$ is the classical windowed Fourier transform (see [10, 11]).

Example. Consider the Clifford Gaussian window $\phi \in L^2(\mathbb{R}^2; C\ell_{0,2})$ given by:

$$\phi(\boldsymbol{x}) = (2 + \boldsymbol{e}_1 + \boldsymbol{e}_2 - \boldsymbol{e}_{12})e^{-(x_1^2+x_2^2)}. \tag{4.5}$$

Thus we obtain the Clifford window daughter functions (4.4) of the form

$$\begin{aligned}\phi_{\boldsymbol{\omega},\boldsymbol{b}}(\boldsymbol{x}) &= \{e^{\boldsymbol{e}_2\omega_2 x_2} e^{\boldsymbol{e}_1\omega_1 x_1}(2 + \boldsymbol{e}_1 + \boldsymbol{e}_2 - \boldsymbol{e}_{12})\}\, e^{-((x_1-b_1)^2+(x_2-b_2)^2)} \\ &= \{(2e^{\boldsymbol{e}_2\omega_2 x_2} e^{\boldsymbol{e}_1\omega_1 x_1} + \boldsymbol{e}_1 e^{-\boldsymbol{e}_2\omega_2 x_2} e^{\boldsymbol{e}_1\omega_1 x_1} + \boldsymbol{e}_2 e^{\boldsymbol{e}_2\omega_2 x_2} e^{-\boldsymbol{e}_1\omega_1 x_1} \\ &\quad - \boldsymbol{e}_{12} e^{-\boldsymbol{e}_2\omega_2 x_2} e^{-\boldsymbol{e}_1\omega_1 x_1})\} e^{-((x_1-b_1)^2+(x_2-b_2)^2)}.\end{aligned} \tag{4.6}$$

We first notice that, for fixed $\boldsymbol{b}$,

$$G_\phi f(\boldsymbol{\omega}, \boldsymbol{b}) = \mathcal{F}\{f\,\overline{\phi}(\cdot - \boldsymbol{b})\}(\boldsymbol{\omega}) = \mathcal{F}\{f\,T_{\boldsymbol{b}}\overline{\phi}\}(\boldsymbol{\omega}), \tag{4.7}$$

where $T_{\boldsymbol{b}}$ is the translation operator defined by $T_{\boldsymbol{b}} f = f(\boldsymbol{x} - \boldsymbol{b})$. It thus means that the CWFT can be regarded as the CFT of the product of a multivector-valued function f and a shifted and Clifford reversion of the Clifford atom operator (4.4).

4.3. Properties of the CWFT

The following proposition describes the elementary properties of the CWFT. Its proof can be easily obtained.

Proposition 4.2. *Let $\phi \in L^2(\mathbb{R}^n; C\ell_{0,n})$ be a Clifford window function.*

Left linearity:

$$[G_\phi(\lambda f + \mu g)](\boldsymbol{\omega}, \boldsymbol{b}) = \lambda G_\phi f(\boldsymbol{\omega}, \boldsymbol{b}) + \mu G_\phi g(\boldsymbol{\omega}, \boldsymbol{b}), \tag{4.8}$$

for arbitrary Clifford constants $\lambda, \mu \in C\ell_{0,n}$.

Parity:

$$G_{P\phi}(Pf)(\boldsymbol{\omega}, \boldsymbol{b}) = G_\phi f(\boldsymbol{\omega}, -\boldsymbol{b}), \tag{4.9}$$

where P is the parity operator defined by $Pf(x) = f(-x)$.

Theorem 4.3 (Orthogonality relation). *Let ϕ, ψ be Clifford window functions and $f, g \in L^2(\mathbb{R}^n; C\ell_{0,n})$ be arbitrary. Then we have*

$$\begin{aligned}\int_{\mathbb{R}^n}\int_{\mathbb{R}^n} G_\phi f(\boldsymbol{\omega}, \boldsymbol{b})\, \overline{G_\psi g(\boldsymbol{\omega}, \boldsymbol{b})}\, d^n\boldsymbol{\omega}\, d^n\boldsymbol{b} \\ = (2\pi)^n (f(\overline{\phi}, \overline{\psi})_{L^2(\mathbb{R}^n; C\ell_{0,n})}, g)_{L^2(\mathbb{R}^n; C\ell_{0,n})}.\end{aligned} \tag{4.10}$$

Proof. Applying (4.7) we have

$$\begin{aligned}\int_{\mathbb{R}^n} G_\phi f(\boldsymbol{\omega}, \boldsymbol{b})\, \overline{G_\psi g(\boldsymbol{\omega}, \boldsymbol{b})}\, d^n\boldsymbol{\omega} &= \int_{\mathbb{R}^n} \mathcal{F}\{f T_{\boldsymbol{b}}\overline{\phi}\} \overline{\mathcal{F}\{g T_{\boldsymbol{b}}\overline{\psi}\}}\, d^n\boldsymbol{\omega} \\ &= \left(\mathcal{F}\{f T_{\boldsymbol{b}}\overline{\phi}\}, \mathcal{F}\{g T_{\boldsymbol{b}}\overline{\psi}\}\right)_{L^2(\mathbb{R}^n; C\ell_{0,n})}.\end{aligned} \tag{4.11}$$

We assume that Clifford windows $\phi, g \in L^1(\mathbb{R}^n; C\ell_{0,n}) \bigcap L^\infty(\mathbb{R}^n; C\ell_{0,n})$ so that $f T_{\boldsymbol{b}}\overline{\phi}, g T_{\boldsymbol{b}}\overline{\psi} \in L^2(\mathbb{R}^n; C\ell_{0,n})$. We know [6, 13] that Parseval's theorem is valid for the CFT. So, applying it to the right-hand side of (4.11) we easily get (compare to Gröchenig [10])

$$\begin{aligned}\int_{\mathbb{R}^n} G_\phi f(\boldsymbol{\omega}, \boldsymbol{b}) \overline{G_\psi g(\boldsymbol{\omega}, \boldsymbol{b})}\, d^n\boldsymbol{\omega} &= (\mathcal{F}\{fT_{\boldsymbol{b}}\overline{\phi}\}, \mathcal{F}\{gT_{\boldsymbol{b}}\overline{\psi}\})_{L^2(\mathbb{R}^n;C\ell_{0,n})}\\ &= (2\pi)^n \left(fT_{\boldsymbol{b}}\overline{\phi}, gT_{\boldsymbol{b}}\overline{\psi}\right)_{L^2(\mathbb{R}^n;C\ell_{0,n})}\\ &= (2\pi)^n \int_{\mathbb{R}^n} f(\boldsymbol{x})\overline{\phi(\boldsymbol{x}-\boldsymbol{b})}\psi(\boldsymbol{x}-\boldsymbol{b})\overline{g(\boldsymbol{x})}\, d^n\boldsymbol{x}. \end{aligned} \tag{4.12}$$

Observe that $f\overline{\phi}$ and $\psi\overline{g}$ are in $L^1(\mathbb{R}^n; C\ell_{0,n})$. Then integrating (4.12) with respect to $d^n\boldsymbol{b}$ we immediately get

$$\begin{aligned}&\int_{\mathbb{R}^n}\int_{\mathbb{R}^n} G_\phi f(\boldsymbol{\omega}, \boldsymbol{b}) \overline{G_\psi g(\boldsymbol{\omega}, \boldsymbol{b})}\, d^n\boldsymbol{\omega}\, d^n\boldsymbol{b}\\ &\quad= (2\pi)^n \int_{\mathbb{R}^n}\int_{\mathbb{R}^n} f(\boldsymbol{x})\overline{\phi(\boldsymbol{x}-\boldsymbol{b})}\psi(\boldsymbol{x}-\boldsymbol{b})\overline{g(\boldsymbol{x})}\, d^n\boldsymbol{x}\, d^n\boldsymbol{b}\\ &\quad= (2\pi)^n \int_{\mathbb{R}^n}\int_{\mathbb{R}^n} f(\boldsymbol{x})\overline{\phi(\boldsymbol{x}-\boldsymbol{b})}\psi(\boldsymbol{x}-\boldsymbol{b})\, d^n\boldsymbol{b}\,\overline{g(\boldsymbol{x})}\, d^n\boldsymbol{x}\\ &\quad= (2\pi)^n \int_{\mathbb{R}^n} f(\boldsymbol{x})\int_{\mathbb{R}^n} \overline{\phi(\boldsymbol{x}-\boldsymbol{b})}\psi(\boldsymbol{x}-\boldsymbol{b})\, d^n\boldsymbol{b}\,\overline{g(\boldsymbol{x})}\, d^n\boldsymbol{x}, \end{aligned} \tag{4.13}$$

where from the second to the third line of (4.13) we applied Fubini's theorem to interchange the order of integration. Using a standard density argument we can now extend the result to the $L^2(\mathbb{R}^n; C\ell_{0,n})$-case. This proves the theorem. □

From the above theorem, we obtain the following consequences.

(i) If $\phi = \psi$ and $C_\phi = (\overline{\phi}, \overline{\phi})_{L^2(\mathbb{R}^n;C\ell_{0,n})}$ is a multivector constant, then

$$\begin{aligned}\int_{\mathbb{R}^n}\int_{\mathbb{R}^n} G_\phi f(\boldsymbol{\omega}, \boldsymbol{b}) \overline{G_\phi g(\boldsymbol{\omega}, \boldsymbol{b})}\, d^n\boldsymbol{b}\, d^n\boldsymbol{\omega} &= (2\pi)^n \langle C_\phi\rangle (f, g)_{L^2(\mathbb{R}^n;C\ell_{0,n})}\\ &\quad+ (2\pi)^n (f\langle C_\phi\rangle_1, g)_{L^2(\mathbb{R}^n;C\ell_{0,n})} + \cdots\\ &\quad+ (2\pi)^n (f\langle C_\phi\rangle_n, g)_{L^2(\mathbb{R}^n;C\ell_{0,n})}. \end{aligned} \tag{4.14}$$

(ii) If $f = g$ is a paravector, then (4.14) reduces to

$$\begin{aligned}\int_{\mathbb{R}^n}\int_{\mathbb{R}^n} G_\phi f(\boldsymbol{\omega}, \boldsymbol{b}) \overline{G_\phi f(\boldsymbol{\omega}, \boldsymbol{b})}\, d^n\boldsymbol{b}\, d^n\boldsymbol{\omega} &= (2\pi)^n \langle C_\phi\rangle \left\| f\right\|^2_{L^2(\mathbb{R}^n;C\ell_{0,n})}\\ &\quad+ (2\pi)^n (f\langle C_\phi\rangle_1, f)_{L^2(\mathbb{R}^n;C\ell_{0,n})} + \cdots\\ &\quad+ (2\pi)^n (f\langle C_\phi\rangle_n, f)_{L^2(\mathbb{R}^n;C\ell_{0,n})}. \end{aligned} \tag{4.15}$$

Theorem 4.4 (Reconstruction formula). *Let $\phi, \psi \in L^2(\mathbb{R}^n; C\ell_{0,n})$ be two Clifford window functions with $(\overline{\phi}, \overline{\psi})_{L^2(\mathbb{R}^n;C\ell_{0,n})} \neq 0$. Then every n-D Clifford signal $f \in L^2(\mathbb{R}^n; C\ell_{0,n})$ can be fully reconstructed by*

$$f(\boldsymbol{x}) = \frac{1}{(2\pi)^n} \int_{\mathbb{R}^n}\int_{\mathbb{R}^n} G_\phi f(\boldsymbol{\omega}, \boldsymbol{b})\psi_{\boldsymbol{\omega},\boldsymbol{b}}(\boldsymbol{x})\, (\overline{\phi}, \overline{\psi})^{-1}_{L^2(\mathbb{R}^n;C\ell_{0,n}})d^n\boldsymbol{b}\, d^n\boldsymbol{\omega}. \tag{4.16}$$

Proof. By direct calculation, we obtain for every $g \in L^2(\mathbb{R}^n; C\ell_{0,n})$

$$\begin{aligned}
&\int_{\mathbb{R}^n}\int_{\mathbb{R}^n} G_\phi f(\boldsymbol{\omega},\boldsymbol{b})\,\overline{G_\psi g(\boldsymbol{\omega},\boldsymbol{b})}\,d^n\boldsymbol{\omega}\,d^n\boldsymbol{b}\\
&= \int_{\mathbb{R}^n}\int_{\mathbb{R}^n}\int_{\mathbb{R}^n} G_\phi f(\boldsymbol{\omega},\boldsymbol{b})\,\psi_{\boldsymbol{\omega},\boldsymbol{b}}(\boldsymbol{x})\overline{g(\boldsymbol{x})}\,d^n\boldsymbol{\omega}\,d^n\boldsymbol{b}\,d^n\boldsymbol{x}\\
&= \left(\int_{\mathbb{R}^n}\int_{\mathbb{R}^n} G_\phi f(\boldsymbol{\omega},\boldsymbol{b})\psi_{\boldsymbol{\omega},\boldsymbol{b}}\,d^n\boldsymbol{\omega}\,d^n\boldsymbol{b}, g\right)_{L^2(\mathbb{R}^n;C\ell_{0,n})}. \qquad (4.17)
\end{aligned}$$

Applying equation (4.10) of Theorem 4.3 to the left-hand side of (4.17) gives for every $g \in L^2(\mathbb{R}^n; C\ell_{0,n})$

$$\begin{aligned}
&(2\pi)^n (f(\overline{\phi},\overline{\psi})_{L^2(\mathbb{R}^n;C\ell_{0,n})}, g)_{L^2(\mathbb{R}^n;C\ell_{0,n})}\\
&= \left(\int_{\mathbb{R}^n}\int_{\mathbb{R}^n} G_\phi f(\boldsymbol{\omega},\boldsymbol{b})\psi_{\boldsymbol{\omega},\boldsymbol{b}}\,d^n\boldsymbol{\omega}\,d^n\boldsymbol{b}, g\right)_{L^2(\mathbb{R}^n;C\ell_{0,n})}. \qquad (4.18)
\end{aligned}$$

Because the inner product identity (4.18) holds for every $g \in L^2(\mathbb{R}^n; C\ell_{0,n})$ we conclude that

$$(2\pi)^n f(\overline{\phi},\overline{\psi})_{L^2(\mathbb{R}^n;C\ell_{0,n})} = \int_{\mathbb{R}^n}\int_{\mathbb{R}^n} G_\phi f(\boldsymbol{\omega},\boldsymbol{b})\psi_{\boldsymbol{\omega},\boldsymbol{b}}\,d^n\boldsymbol{\omega}\,d^n\boldsymbol{b}. \qquad (4.19)$$

If it is assumed that the inner product $(\overline{\phi},\overline{\psi})_{L^2(\mathbb{R}^n;C\ell_{0,n})}$ is invertible. Then multiplying both sides of (4.19) from the right side by $(\overline{\phi},\overline{\psi})^{-1}_{L^2(\mathbb{R}^n;C\ell_{0,n})}$ we immediately obtain

$$(2\pi)^n f = \int_{\mathbb{R}^n}\int_{\mathbb{R}^n} G_\phi f(\boldsymbol{\omega},\boldsymbol{b})\psi_{\boldsymbol{\omega},\boldsymbol{b}}\,(\overline{\phi},\overline{\psi})^{-1}_{L^2(\mathbb{R}^n;C\ell_{0,n})}\,d^n\boldsymbol{\omega}\,d^n\boldsymbol{b}, \qquad (4.20)$$

which was to be proved. □

Acknowledgment

The author would like to thank the reviewer whose deep and extensive comments greatly contributed to improve this chapter. He thanks Ass. Prof. Eckhard Hitzer for his helpful guidance. He also wants to thank ICCA9 organizer Professor Klaus Gürlebeck.

References

[1] M. Bahri. Generalized Fourier transform in real clifford algebra $cl(0, n)$. *Far East Journal of Mathematical Sciences*, 48(1):11–24, Jan. 2011.

[2] M. Bahri, E. Hitzer, R. Ashino, and R. Vaillancourt. Windowed Fourier transform of two-dimensional quaternionic signals. *Applied Mathematics and Computation*, 216(8):2366–2379, June 2010.

[3] E. Bayro-Corrochano. The theory and use of the quaternion wavelet transform. *Journal of Mathematical Imaging and Vision*, 24(1):19–36, 2006.

[4] E. Bayro-Corrochano and G. Scheuermann, editors. *Applied Geometric Algebras in Computer Science and Engineering*. Springer, London, 2010.

[5] F. Brackx, R. Delanghe, and F. Sommen. *Clifford Analysis*, volume 76. Pitman, Boston, 1982.

[6] T. Bülow. *Hypercomplex Spectral Signal Representations for the Processing and Analysis of Images.* PhD thesis, University of Kiel, Germany, Institut für Informatik und Praktische Mathematik, Aug. 1999.

[7] T. Bülow, M. Felsberg, and G. Sommer. Non-commutative hypercomplex Fourier transforms of multidimensional signals. In G. Sommer, editor, *Geometric computing with Clifford Algebras: Theoretical Foundations and Applications in Computer Vision and Robotics*, pages 187–207, Berlin, 2001. Springer.

[8] T. Bülow and G. Sommer. Hypercomplex signals – a novel extension of the analytic signal to the multidimensional case. *IEEE Transactions on Signal Processing*, 49(11):2844–2852, Nov. 2001.

[9] Y. Fu, U. Kähler, and P. Cerejeiras. The Balian–Low theorem for the windowed quaternionic Fourier transform. *Advances in Applied Clifford Algebras*, page 16, 2012. published online 3 February 2012.

[10] K. Gröchenig. *Foundations of Time-Frequency Analysis.* Applied and Numerical Harmonic Analysis. Birkhäuser, Boston, 2001.

[11] K. Gröchenig and G. Zimmermann. Hardy's theorem and the short-time Fourier transform of Schwartz functions. *Journal of the London Mathematical Society*, 2:205–214, 2001.

[12] S.L. Hahn. Wigner distributions and ambiguity functions of 2-D quaternionic and monogenic signals. *IEEE Transactions on Signal Processing*, 53:3111–3128, 2005.

[13] E. Hitzer. Quaternion Fourier transform on quaternion fields and generalizations. *Advances in Applied Clifford Algebras*, 17(3):497–517, May 2007.

[14] E.M.S. Hitzer and B. Mawardi. Clifford Fourier transform on multivector fields and uncertainty principles for dimensions $n = 2(\bmod 4)$ and $n = 3(\bmod 4)$. *Advances in Applied Clifford Algebras*, 18(3-4):715–736, 2008.

[15] B. Mawardi. Clifford windowed Fourier transform applied to linear time-varying systems. *Applied Mathematical Sciences*, 6:2857–2864, 2012.

[16] B. Mawardi, S. Adji, and J. Zhao. Real Clifford windowed Fourier transform. *Acta Mathematica Sinica*, 27:505–518, 2011.

[17] B. Mawardi, E. Hitzer, and S. Adji. Two-dimensional Clifford windowed Fourier transform. In Bayro-Corrochano and Scheuermann [4], pages 93–106.

[18] B. Mawardi, E. Hitzer, A. Hayashi, and R. Ashino. An uncertainty principle for quaternion Fourier transform. *Computers and Mathematics with Applications*, 56(9):2411–2417, 2008.

[19] F. Sommen. A product and an exponential function in hypercomplex function theory. *Applicable Analysis*, 12:13–26, 1981.

[20] Y. Xi, X. Yang, L. Song, L. Traversoni, and W. Lu. QWT: Retrospective and new application. In Bayro-Corrochano and Scheuermann [4], pages 249–273.

Mawardi Bahri
Department of Mathematics
Universitas Hasanuddin
Tamalanrea Makassar 90245, Indonesia
e-mail: `mawardibahri@gmail.com`

Quaternion and Clifford–Fourier Transforms and Wavelets
Trends in Mathematics, 299–319

15 The Balian–Low Theorem for the Windowed Clifford–Fourier Transform

Yingxiong Fu, Uwe Kähler and Paula Cerejeiras

Abstract. In this chapter, we provide the definition of the Clifford–Zak transform associated with the discrete version of the kernel of a windowed Clifford–Fourier transform. We proceed with deriving several important properties of such a transform. Finally, we establish the Balian–Low theorem for a Clifford frame under certain natural assumptions on the window function.

Mathematics Subject Classification (2010). 15A66; 30G35.

Keywords. Clifford–Zak transform, Clifford frame, Balian–Low theorem, Clifford–Fourier transform, windowed Clifford–Fourier transform.

1. Introduction

The last decade has seen a growing interest in generalizations of the Fourier transform, motivated by applications to higher-dimensional signal processing. First steps in that direction where already made in Brackx *et al.* [10], where a Fourier transform for multivector-valued distributions in $C\ell_{0,n}$ with compact support was presented. Also worth mention is the alternative definition of the Fourier transform of Sommen [27], based on a generalization of the exponential function to $\mathbb{R}^n \times \mathbb{R}^{n+1}$. A quaternionic Fourier transform was given in Bülow *et al.* [12] in the context of quaternionic-valued two-dimensional signals (the so-called hypercomplex signals). Shortly after, motivated by spectral analysis of colour images Sangwine *et al.* proposed in [26] a quaternionic Fourier transform in which the imaginary unit i was replaced by a unit quaternion. Almost in parallel, Felsberg defined in [17] his Clifford–Fourier transform (CFT) for the low-dimensional Clifford algebras $C\ell_{2,0}$ and $C\ell_{3,0}$, using the pseudoscalar i_n as imaginary unit. Following this approach, several authors have extended this transform to a three-dimensional setting and successfully detected vector-valued patterns in the frequency domain (*cf.* [15, 16, 24]). However, a major problem did remain: as the classical Fourier transform, the Clifford–Fourier transform is ineffective for representing and computing local information of signals. In fact, the harmonic analysis version of Heisenberg's uncertainty principle states that it is impossible to localize simultaneously a function and its Fourier transform.

One way to overcome this difficulty is by means of Clifford–Gabor filters. They were initially proposed in [11], and later on also in [8], which extended the applications of the complex Gabor filters. In general, they correspond to modulations of Gaussians. A good account on this subject can be found in [9].

A more general approach to this problem is by means of the windowed Fourier transform (WFT), also called continuous Gabor transform or short-time Fourier transform. Given $f \in L^2(\mathbb{R})$ and a fixed non-zero window function $g \in L^2(\mathbb{R})$, one defines the WFT $V_g\{f\} \in L^2(\mathbb{R}^2)$ as

$$V_g\{f\}(t,\omega) = \int_{\mathbb{R}} f(x)\overline{g(x-t)}e^{-2\pi i x\omega}dx. \tag{1.1}$$

Mawardi *et al.* extended the theory of WFT to the Clifford case [1–4, 23]. In [3] the definition of a windowed Clifford–Fourier transform (WCFT) was established and several important properties were obtained such as shift, modulation, reconstruction formulae, *etc.* Furthermore, a Heisenberg type uncertainty principle for the WCFT was derived.

Practical applications require a discrete version of this continuous transform. One can establish a discrete form of the WFT V_g linked to a given countable set $\Lambda \subset \mathbb{R}^2 = \mathbb{R} \times \mathbb{R}$ as the transformation that assigns to every function $f \in L^2(\mathbb{R})$ the number sequence

$$f \mapsto \{\, \langle f, g_{m,n}\rangle : (m,n) \in \Lambda\}, \tag{1.2}$$

where $\langle\cdot,\cdot\rangle$ denotes the ordinary inner product in $L^2(\mathbb{R})$ and the functions $g_{m,n}$ are shifts and modulations of the window g given by

$$g_{m,n}(x) = e^{2\pi i\, mx} g(x-n) \tag{1.3}$$

for all $(m,n) \in \Lambda \subset \mathbb{R}^2$. Such a collection $\{g_{m,n} : (m,n) \in \Lambda\}$ is called Gabor system, or Weyl-Heisenberg system, generated by g and Λ. To recover the original function f from the number sequence $\{\langle f, g_{m,n}\rangle : (m,n) \in \Lambda\}$ it is necessary that the system forms an orthonormal basis or at least a frame.

Gabor systems are related to the classical uncertainty principle by the Balian–Low theorem (a stronger version of the said principle), as it expresses the fact that time-frequency concentration and non-redundancy are incompatible properties of a Gabor system if such a system is a frame for $L^2(\mathbb{R})$ [7, 13, 14, 18, 21]. Specifically, if the window g is such that the Gabor system $\{g_{m,n} : m,n \in \mathbb{Z}\}$ constitutes an exact frame for $L^2(\mathbb{R})$, *i.e.*, if there exist constants $0 < B \leq C < \infty$ such that

$$B\|f\|^2 \leq \sum_{m,n\in\mathbb{Z}} |\langle f, g_{m,n}\rangle|^2 \leq C\,\|f\|^2, \quad \forall f \in L^2(\mathbb{R}), \tag{1.4}$$

and the system ceases to be a frame when any of its elements is removed then, it holds

$$\left(\int_{-\infty}^{+\infty} t^2|g(t)|^2 dt\right)\left(\int_{-\infty}^{+\infty} \xi^2|\hat{g}(\xi)^2 d\xi\right) = \infty. \tag{1.5}$$

In other words, the window function g maximizes the uncertainty principle in some sense. This result has been extended to higher dimensions and to a more general set of time-frequency shifts in the standard coordinate system [6, 19]. It was also proved to be valid in a multi-window setting [25, 28] and in the case of superframes [5].

Regarding the important question of discretizing the WCFT it is natural to ask if the Balian–Low theorem holds for Gabor systems generated by certain Clifford-valued window functions g and countable sets $\Lambda \subset \mathbb{R}^{2n}$. Our goal in this chapter is to obtain it for Gabor systems which form a Clifford frame arising from the discrete version of the kernel of the WCFT. To this end it is necessary to introduce a Clifford–Zak transform and study some of its properties.

This chapter is organized as follows: Section 2 is devoted to the review of the necessary results on Clifford algebra and the definitions of both the CFT and the WCFT. In Section 3 we provide the definition of a Clifford frame associated with the discrete version of the kernel of the WCFT, establish the definition of a Clifford–Zak transform and derive some properties of it, which will play a key role in the proof of the Balian–Low theorem. In Section 4 we demonstrate the Balian–Low theorem for Gabor systems which form a Clifford frame. Some conclusions are drawn in Section 5.

2. Preliminaries

Let $C\ell_{n,0}$ be the 2^n-dimensional universal real Clifford algebra over $\mathbb{R}^n$ constructed from the basis $\{\boldsymbol{e}_1, \boldsymbol{e}_2, \ldots, \boldsymbol{e}_n\}$ under the usual relations

$$\boldsymbol{e}_k\boldsymbol{e}_l + \boldsymbol{e}_l\boldsymbol{e}_k = 2\delta_{kl}, \ 1 \le k, l \le n, \tag{2.1}$$

where δ_{kl} is the Kronecker delta function. An element $f \in C\ell_{n,0}$ can be represented as $f = \sum_A f_A \boldsymbol{e}_A$, $f_A \in \mathbb{R}$, where $\boldsymbol{e}_A = \boldsymbol{e}_{j_1 j_2 \cdots j_k} = \boldsymbol{e}_{j_1}\boldsymbol{e}_{j_2}\cdots\boldsymbol{e}_{j_k}$, $A = \{j_1, j_2, \ldots j_k\}$ with $1 \le j_1 \le j_2 \le \cdots \le j_k \le n$, and $\boldsymbol{e}_0 = \boldsymbol{e}_\emptyset = 1$ is the identity element of $C\ell_{n,0}$. The elements of the algebra $C\ell_{n,0}$ for which $|A| = k$ are called k-vectors. We denote the space of all k-vectors by

$$C\ell_{n,0}^k := \operatorname{span}_{\mathbb{R}}\{\boldsymbol{e}_A : |A| = k\}. \tag{2.2}$$

It is clear that the spaces $\mathbb{R}$ and $\mathbb{R}^n$ can be identified with $C\ell_{n,0}^0$ and $C\ell_{n,0}^1$, respectively.

Of interest for this work is the (unit oriented) pseudoscalar element $i_n = \boldsymbol{e}_1\boldsymbol{e}_2\cdots\boldsymbol{e}_n$. Observe that $i_n^2 = -1$ for $n = 2, 3 (\text{mod } 4)$. For the sake of simplicity, if not otherwise stated, n is always assumed to be $n = 2, 3 (\text{mod } 4)$ for the remaining of this chapter.

We define the anti-automorphism reversion $\tilde{}: C\ell_{n,0} \to C\ell_{n,0}$ by its action on the basis elements $\widetilde{e_A} = (-1)^{\frac{k(k-1)}{2}} e_A$, for $|A| = k$, and its reversion property $\widetilde{fg} = \tilde{g}\tilde{f}$ for every $f, g \in C\ell_{n,0}$. In particular, we remark that $\widetilde{i_n} = -i_n$.

In what follows, we will require two types of scalar products. First, we introduce the (real-valued) scalar product of $f, g \in C\ell_{n,0}$ as the scalar part of their geometric product

$$f \cdot g := [f\,\tilde{g}]_0 = \sum_A f_A g_A. \tag{2.3}$$

As usual, when we set $f = g$ we obtain the square of the modulus (or magnitude) of the multivector $f \in C\ell_{n,0}$,

$$|f|^2 = \left[f\tilde{f}\right]_0 = \sum_A f_A^2. \tag{2.4}$$

Also, an additional useful property of the scalar part $[\ ]_0$ is the cyclic symmetric product

$$[pqr]_0 = [qrp]_0, \quad \forall p, q, r \in C\ell_{n,0}. \tag{2.5}$$

Second, we require an inner product in the function space under consideration. We denote by $L^p(\mathbb{R}^n; C\ell_{n,0})$ the left module of all Clifford-valued functions $f : \mathbb{R}^n \to C\ell_{n,0}$ with finite norm

$$\|f\|_p = \begin{cases} \left(\int_{\mathbb{R}^n} |f(\mathbf{x})|^p d^n\mathbf{x}\right)^{\frac{1}{p}}, & 1 \leq p < \infty \\ \operatorname{ess\,sup}_{\mathbf{x}\in\mathbb{R}^n} |f(\mathbf{x})|, & p = \infty \end{cases}, \tag{2.6}$$

where $d^n\mathbf{x} = dx_1 dx_2 \cdots dx_n$ represents the usual Lebesgue measure in $\mathbb{R}^n$. In the particular case of $p = 2$, we shall denote this norm by $\|f\|$.

Given two functions $f, g \in L^2(\mathbb{R}^n; C\ell_{n,0})$, we define a Clifford-valued bilinear form

$$f, g \to (f, g) := \int_{\mathbb{R}^n} f(\mathbf{x})\widetilde{g(\mathbf{x})} d^n\mathbf{x}, \tag{2.7}$$

from which we construct the scalar inner product

$$\langle f, g\rangle := [(f, g)]_0 = \int_{\mathbb{R}^n} \left[f(\mathbf{x})\widetilde{g(\mathbf{x})}\right]_0 d^n\mathbf{x}. \tag{2.8}$$

We remark that (2.8) satisfies the (Clifford) Cauchy–Schwarz inequality

$$|\langle f, g\rangle| \leq \|f\|\,\|g\|, \quad \forall f, g \in L^2(\mathbb{R}^n; C\ell_{n,0}). \tag{2.9}$$

In the following, we recall the CFT, originally introduced by M. Felsberg (see [17]).

Definition 2.1. Let $f \in L^1(\mathbb{R}^n; C\ell_{n,0})$. The CFT of f at the point $\omega \in \mathbb{R}^n$ is defined as the $C\ell_{n,0}$-valued (Lebesgue) integral

$$\mathcal{F}\{f\}(\omega) = \int_{\mathbb{R}^n} f(\mathbf{x}) e^{-2\pi i_n \omega\cdot\mathbf{x}} d^n\mathbf{x}. \tag{2.10}$$

The function $\omega \to \mathcal{F}\{f\}(\omega)$ s called the CFT of f.

Lemma 2.2 (Parseval's equality for CFT). *If $f \in L^1(\mathbb{R}^n; C\ell_{n,0}) \cap L^2(\mathbb{R}^n; C\ell_{n,0})$, then*

$$\|f\| = \|\mathcal{F}\{f\}\|. \tag{2.11}$$

Lemma 2.2 asserts that the CFT is a bounded linear operator on $L^1(\mathbb{R}^n; C\ell_{n,0}) \cap L^2(\mathbb{R}^n; C\ell_{n,0})$. Hence, standard density arguments allow us to extend the CFT in an unique way to the whole of $L^2(\mathbb{R}^n; C\ell_{n,0})$. In what follows we always consider the properties of the CFT as an operator from $L^2(\mathbb{R}^n; C\ell_{n,0})$ into $L^2(\mathbb{R}^n; C\ell_{n,0})$.

Definition 2.3. Let $g \in L^2(\mathbb{R}^n; C\ell_{n,0})$ be a non-zero window function such that $|\mathbf{x}|^{1/2} g(\mathbf{x})$ is in $L^2(\mathbb{R}^n; C\ell_{n,0})$. Then, the WCFT of $f \in L^2(\mathbb{R}^n; C\ell_{n,0})$ with respect to g is defined by

$$\begin{aligned} Q_g f(\omega, \mathbf{b}) :&= \int_{\mathbb{R}^n} f(\mathbf{x}) \widetilde{g(\mathbf{x} - \mathbf{b})} e^{-2\pi i_n \omega \cdot \mathbf{x}} d^n \mathbf{x} \\ &= (f, g_{\omega, \mathbf{b}}), \end{aligned} \tag{2.12}$$

where $g_{\omega,\mathbf{b}}(\mathbf{x}) := e^{2\pi i_n \omega \cdot \mathbf{x}} g(\mathbf{x} - \mathbf{b})$ denotes the kernel of the WCFT.

3. Clifford Frame and Clifford–Zak Transform

The Balian–Low theorem is regarded as a strong version of the uncertainty principle for the Gabor system associated with the discrete version of the kernel of the classical WFT. To establish the Balian–Low theorem for a Gabor system in the Clifford algebra module $L^2(\mathbb{R}^n; C\ell_{n,0})$ we consider the following discrete version of the kernel of the WCFT

$$g_{\mathbf{m},\mathbf{n}}(\mathbf{x}) := e^{2\pi i_n \mathbf{m} \cdot \mathbf{x}} g(\mathbf{x} - \mathbf{n}), \quad \mathbf{x} \in \mathbb{R}^n, \quad \mathbf{m}, \mathbf{n} \in \mathbb{Z}^n. \tag{3.1}$$

A frame for a vector space equipped with an inner product allows each element in the space to be written as a linear combination of the elements in the frame. In general frame elements are neither orthogonal to each other nor linearly independent. We now introduce the definition and properties of a Clifford frame in $L^2(\mathbb{R}^n; C\ell_{n,0})$ as follows.

Definition 3.1. $\{g_{\mathbf{m},\mathbf{n}} : \mathbf{m}, \mathbf{n} \in \mathbb{Z}^n\}$ is a Clifford frame for $L^2(\mathbb{R}^n; C\ell_{n,0})$ if there exist real constants $0 < B \leq C < \infty$ such that

$$\begin{aligned} B \|f\|^2_{L^2(\mathbb{R}^n; C\ell_{n,0})} \leq \sum_{\mathbf{m},\mathbf{n} \in \mathbb{Z}^n} |\langle f, g_{\mathbf{m},\mathbf{n}} \rangle|^2 \leq C \|f\|^2_{L^2(\mathbb{R}^n; C\ell_{n,0})}, \\ \forall f \in L^2(\mathbb{R}^n; C\ell_{n,0}), \end{aligned} \tag{3.2}$$

where $g_{\mathbf{m},\mathbf{n}}$ is defined by (3.1) and the scalar inner product $\langle \cdot, \cdot \rangle$ is defined by (2.8).

Any two constants B, C satisfying condition (3.1) are called frame bounds. If $B = C$, then $\{g_{\mathbf{m},\mathbf{n}} : \mathbf{m}, \mathbf{n} \in \mathbb{Z}^n\}$ is called a tight frame.

To understand frames and reconstruction methods better, we study some important associated operators.

Definition 3.2. For any subset $\{g_{\mathbf{m},\mathbf{n}} : \mathbf{m},\mathbf{n} \in \mathbb{Z}^n\} \subseteq L^2(\mathbb{R}^n; C\ell_{n,0})$, the coefficient operator F is defined by

$$Ff = \{\langle f, g_{\mathbf{m},\mathbf{n}}\rangle : \mathbf{m},\mathbf{n} \in \mathbb{Z}^n\}. \tag{3.3}$$

The reconstruction operator R for a sequence $c = (c_{\mathbf{m},\mathbf{n}})_{\mathbf{m},\mathbf{n}\in\mathbb{Z}^n}$ is given by

$$Rc = \sum_{\mathbf{m},\mathbf{n}\in\mathbb{Z}^n} c_{\mathbf{m},\mathbf{n}} g_{\mathbf{m},\mathbf{n}} \in L^2(\mathbb{R}^n; C\ell_{n,0}). \tag{3.4}$$

Finally, the frame operator S in $L^2(\mathbb{R}^n; C\ell_{n,0})$ is defined by

$$Sf = \sum_{\mathbf{m},\mathbf{n}\in\mathbb{Z}^n} \langle f, g_{\mathbf{m},\mathbf{n}}\rangle\, g_{\mathbf{m},\mathbf{n}}. \tag{3.5}$$

Based on the classic frame theory [18], under the assumption that $\{g_{\mathbf{m},\mathbf{n}} : \mathbf{m},\mathbf{n} \in \mathbb{Z}^n\}$ is a frame defined by (3.1) for $L^2(\mathbb{R}^n; C\ell_{n,0})$, we know that the frame operator S maps $L^2(\mathbb{R}^n; C\ell_{n,0})$ to $L^2(\mathbb{R}^n; C\ell_{n,0})$ and it is a self-adjoint, positive and invertible operator satisfying

$$BI \le S \le CI, \quad C^{-1}I \le S^{-1} \le B^{-1}I \tag{3.6}$$

with I being the identity operator. Moreover, observe that

$$\langle Sf, f\rangle = \sum_{\mathbf{m},\mathbf{n}\in\mathbb{Z}^n} |\langle f, g_{\mathbf{m},\mathbf{n}}\rangle|^2 \tag{3.7}$$

and

$$\begin{aligned}\sum_{\mathbf{m},\mathbf{n}\in\mathbb{Z}^n} \left|\langle f, S^{-1}g_{\mathbf{m},\mathbf{n}}\rangle\right|^2 &= \sum_{\mathbf{m},\mathbf{n}\in\mathbb{Z}^n} \left|\langle S^{-1}f, g_{\mathbf{m},\mathbf{n}}\rangle\right|^2 \\ &= \left\langle S(S^{-1}f), S^{-1}f\right\rangle = \left\langle S^{-1}f, f\right\rangle.\end{aligned} \tag{3.8}$$

Therefore, we have

$$C^{-1}\|f\|^2 \le \left\langle S^{-1}f, f\right\rangle = \sum_{\mathbf{m},\mathbf{n}\in\mathbb{Z}^n} \left|\langle f, S^{-1}g_{\mathbf{m},\mathbf{n}}\rangle\right|^2 \le B^{-1}\|f\|^2. \tag{3.9}$$

Thus the collection $\{S^{-1}g_{\mathbf{m},\mathbf{n}} : \mathbf{m},\mathbf{n} \in \mathbb{Z}^n\}$ is a so-called dual frame with frame bounds C^{-1} and B^{-1}. Using the factorizations $I = S^{-1}S = SS^{-1}$, we obtain the series expansions

$$\begin{aligned}f = S(S^{-1}f) &= \sum_{\mathbf{m},\mathbf{n}\in\mathbb{Z}^n} \left\langle S^{-1}f, g_{\mathbf{m},\mathbf{n}}\right\rangle g_{\mathbf{m},\mathbf{n}} \\ &= \sum_{\mathbf{m},\mathbf{n}\in\mathbb{Z}^n} \left\langle f, S^{-1}g_{\mathbf{m},\mathbf{n}}\right\rangle g_{\mathbf{m},\mathbf{n}}\end{aligned} \tag{3.10}$$

and

$$f = S^{-1}Sf = \sum_{\mathbf{m},\mathbf{n}\in\mathbb{Z}^n} \langle f, g_{\mathbf{m},\mathbf{n}}\rangle\, S^{-1}g_{\mathbf{m},\mathbf{n}}. \tag{3.11}$$

Furthermore, it is well known that the classic Zak transform is a very useful tool to analyze Gabor systems $\{e^{2\pi i mt}g(t-n) : m, n \in \mathbb{Z}\}$ in $L^2(\mathbb{R})$. The classic Zak transform, $Z : L^2(\mathbb{R}) \to L^2([0,1)^2)$, is defined by

$$f = f(x), \quad x \in \mathbb{R} \mapsto Zf = Zf(t,\omega) = \sum_{k\in\mathbb{Z}} e^{2\pi i\, k\omega} f(t-k), \quad (t,\omega) \in \mathbb{R}^2. \tag{3.12}$$

Moreover, the Zak transform is a unitary transformation from $L^2(\mathbb{R})$ to $L^2([0,1)^2)$. Interest in this transform has been revived in recent years due to its relationship to different types of coherent states, one of which is the affine coherent state, commonly known as wavelet [14, 20].

To analyse Clifford–Gabor systems $\{e^{2\pi i_n \mathbf{m}\cdot\mathbf{x}} g(\mathbf{x}-\mathbf{n}) : \mathbf{m}, \mathbf{n} \in \mathbb{Z}^n\}$ in $L^2(\mathbb{R}^n; C\ell_{n,0})$, we need to establish a definition of the Clifford–Zak transform and to show some of its properties.

First, we define the Clifford–Zak transform pointwisely. The Clifford–Zak transform of a Clifford-valued function $f \in L^2(\mathbb{R}^n; C\ell_{n,0})$, with $\mathbf{t} \mapsto f(\mathbf{t})$, at the point $(\mathbf{x}, \omega) \in \mathbb{R}^n \times \mathbb{R}^n$, is given as

$$Z_c f(\mathbf{x},\omega) := \sum_{\mathbf{k}\in\mathbb{Z}^n} e^{2\pi i_n \mathbf{k}\cdot\omega} f(\mathbf{x}-\mathbf{k}). \tag{3.13}$$

As in the one-dimensional case, periodicity properties allow us to consider a smaller domain for the variables $(\mathbf{x}, \omega)$ in this transform. In fact, the Clifford–Zak transform Z_c satisfies the following relations

$$Z_c f(\mathbf{x}, \omega + \mathbf{n}) = Z_c f(\mathbf{x}, \omega), \qquad \mathbf{n} \in \mathbb{Z}^n, \tag{3.14}$$

$$Z_c f(\mathbf{x} + \mathbf{n}, \omega) = e^{2\pi i_n \mathbf{n}\cdot\omega} Z_c f(\mathbf{x}, \omega), \qquad \mathbf{n} \in \mathbb{Z}^n. \tag{3.15}$$

Thus, $Z_c f$ is uniquely determined by its values on $[0,1)^n \times [0,1)^n := Q^{2n} \subset \mathbb{R}^n \times \mathbb{R}^n$. Henceforward, we consider $Z_c f$ as a function on Q^{2n}, where its extension to $\mathbb{R}^n \times \mathbb{R}^n$ is trivially obtained by the quasiperiodic properties (3.14) and (3.15) of the transform.

Finally, we prove that the series defining $Z_c f$ converges in $L^2(Q^{2n}; C\ell_{n,0})$. This will be achieved by showing that Z_c is a unitary map from $L^2(\mathbb{R}^n; C\ell_{n,0})$ onto $L^2(Q^{2n}; C\ell_{n,0})$.

Theorem 3.3. *The Clifford–Zak transform Z_c is a unitary map of $L^2(\mathbb{R}^n; C\ell_{n,0})$ onto $L^2(Q^{2n}; C\ell_{n,0})$.*

Proof. Let $f, g \in L^2(\mathbb{R}^n; C\ell_{n,0})$. In order to show that $Z_c f$ is a unitary mapping we consider the auxiliary functions in $L^2(Q^{2n}; C\ell_{n,0})$

$$(\mathbf{x}, \omega) \mapsto F_{\mathbf{k}}(\mathbf{x}, \omega) := e^{2\pi i_n \mathbf{k}\cdot\omega} f(\mathbf{x}-\mathbf{k}), \tag{3.16}$$

and

$$(\mathbf{x}, \omega) \mapsto G_{\mathbf{k}}(\mathbf{x}, \omega) := e^{2\pi i_n \mathbf{k}\cdot\omega} g(\mathbf{x}-\mathbf{k}), \tag{3.17}$$

each obtained from the original $f, g \in L^2(\mathbb{R}^n; C\ell_{n,0})$ by a specific modulation, and a translation, dependent on the parameter $\mathbf{k} \in \mathbb{Z}^n$. Note that we have

$$\begin{aligned} Z_c f(\mathbf{x}, \omega) &= \sum_{\mathbf{k}\in\mathbb{Z}^n} F_{\mathbf{k}}(\mathbf{x}, \omega) \\ Z_c g(\mathbf{x}, \omega) &= \sum_{\mathbf{k}\in\mathbb{Z}^n} G_{\mathbf{k}}(\mathbf{x}, \omega), \end{aligned} \tag{3.18}$$

for all $(\mathbf{x}, \omega) \in [0,1)^n \times [0,1)^n$. It is easy to see that $F_{\mathbf{k}}, G_{\mathbf{k}} \in L^2(Q^{2n}; C\ell_{n,0})$, for all $\mathbf{k} \in \mathbb{Z}^n$.

We have

$$\begin{aligned} &\langle F_{\mathbf{k}}, G_{\mathbf{m}} \rangle_{L^2(Q^{2n};C\ell_{n,0})} \\ &= \left[\int_{Q^{2n}} e^{2\pi i_n \mathbf{k}\cdot\omega} f(\mathbf{x}-\mathbf{k}) \widetilde{g(\mathbf{x}-\mathbf{m}) e^{-2\pi i_n \mathbf{m}\cdot\omega}} d^n\mathbf{x} d^n\omega \right]_0 \\ &= \left[\int_{[0,1)^n\times[0,1)^n} f(\mathbf{x}-\mathbf{k}) \widetilde{g(\mathbf{x}-\mathbf{m})} e^{-2\pi i_n (\mathbf{m}-\mathbf{k})\cdot\omega} d^n\mathbf{x} d^n\omega \right]_0 \\ &= \left[\left(\int_{[0,1)^n} f(\mathbf{x}-\mathbf{k}) \widetilde{g(\mathbf{x}-\mathbf{m})} d^n\mathbf{x} \right) \left(\int_{[0,1)^n} e^{-2\pi i_n (\mathbf{m}-\mathbf{k})\cdot\omega} d^n\omega \right) \right]_0, \end{aligned} \tag{3.19}$$

due to the cyclic property (2.5) and the relation $\widetilde{i_n} = -i_n$. Hence

$$\langle F_{\mathbf{k}}, G_{\mathbf{m}} \rangle_{L^2(Q^{2n};C\ell_{n,0})} = \begin{cases} \langle f(\cdot-\mathbf{k}), g(\cdot-\mathbf{k}) \rangle_{L^2([0,1)^n;C\ell_{n,0})}, & \mathbf{m} = \mathbf{k} \\ 0, & \mathbf{m} \neq \mathbf{k}. \end{cases} \tag{3.20}$$

Hence,

$$\begin{aligned} \langle Z_c f, Z_c g \rangle_{L^2(Q^{2n};C\ell_{n,0})} &= \sum_{\mathbf{k},\mathbf{m}\in\mathbb{Z}^n} \langle F_{\mathbf{k}}, G_{\mathbf{m}} \rangle_{L^2(Q^{2n};C\ell_{n,0})} \\ &= \sum_{\mathbf{k}\in\mathbb{Z}^n} \langle F_{\mathbf{k}}, G_{\mathbf{k}} \rangle_{L^2(Q^{2n};C\ell_{n,0})} \\ &= \sum_{\mathbf{k}\in\mathbb{Z}^n} \langle f(\cdot-\mathbf{k}), g(\cdot-\mathbf{k}) \rangle_{L^2([0,1)^n;C\ell_{n,0})} \\ &= \sum_{\mathbf{k}\in\mathbb{Z}^n} \left[\int_{[0,1)^n} f(\mathbf{x}-\mathbf{k}) \widetilde{g(\mathbf{x}-\mathbf{k})} d^n\mathbf{x} \right]_0 \\ &= \int_{\mathbb{R}^n} f(\mathbf{y}) \widetilde{g(\mathbf{y})} d^n\mathbf{y} = \langle f, g \rangle_{L^2(\mathbb{R}^n;C\ell_{n,0})}, \end{aligned} \tag{3.21}$$

therefore, completing the proof of the unitary property of the Clifford–Zak transform. □

Consequently, based on the above theorem, we get the following corollary.

Corollary 3.4. *In particular, it holds*

$$\|Z_c f\|^2_{L^2(Q^{2n};C\ell_{n,0})} = \|f\|^2_{L^2(\mathbb{R}^n;C\ell_{n,0})}, \tag{3.22}$$

where $\|Z_c f\|^2_{L^2(Q^{2n};C\ell_{n,0})} = \int_{Q^{2n}} |Z_c f(\mathbf{x},\omega)|^2 d^n\mathbf{x} d^n\omega$.

The unitary nature of the Clifford–Zak transform allows us to translate conditions on Clifford frames for $L^2(\mathbb{R}^n;C\ell_{n,0})$ into those for $L^2(Q^{2n};C\ell_{n,0})$, where things are frequently easier to deal with.

Let us now define the space $\mathcal{Z}$ as the set of all $F:\mathbb{R}^n \to C\ell_{n,0}$ such that

$$\begin{aligned} F(\mathbf{x}+\mathbf{n},\omega) &= e^{2\pi i_n \mathbf{n}\cdot\omega} F(\mathbf{x},\omega), \\ F(\mathbf{x},\omega+\mathbf{n}) &= F(\mathbf{x},\omega), \\ \|F\|^2_{L^2(Q^{2n};C\ell_{n,0})} &= \int_{Q^{2n}} |F(\mathbf{x},\omega)|^2 d^n\mathbf{x} d^n\omega < \infty. \end{aligned} \tag{3.23}$$

In consequence, as $\mathcal{Z}$ is a subset of $L^2(Q^{2n};C\ell_{n,0})$ the Clifford–Zak transform Z_c is a unitary mapping between $L^2(\mathbb{R}^n;C\ell_{n,0})$ and $\mathcal{Z}$.

The following theorem provides some inversion formulas.

Theorem 3.5. *If* $f \in L^2(\mathbb{R}^n;C\ell_{n,0}) \cap L^1(\mathbb{R}^n;C\ell_{n,0})$, *then the following relations,*

$$f(\mathbf{x}) = \int_{[0,1)^n} Z_c f(\mathbf{x},\omega) d^n\omega, \qquad \mathbf{x} \in \mathbb{R}^n, \tag{3.24}$$

and

$$\mathcal{F}\{\widetilde{f}\}(-\omega) = \int_{[0,1)^n} \widetilde{Z_c f(\mathbf{x},\omega)} e^{2\pi i_n \omega\cdot\mathbf{x}} d^n\mathbf{x}, \qquad \omega \in \mathbb{R}^n, \tag{3.25}$$

hold true, where $\mathcal{F}$ *denotes the CFT operator given by* (2.10).

Before proceeding with the proof, we remark that the Zak transform can be extended in both arguments to the whole of $\mathbb{R}^n$ by relations (3.14) and (3.15). In consequence, the above identities state that both the signal f, and the CFT $\mathcal{F}\{\widetilde{f}\}$, can be reconstructed on the whole of $\mathbb{R}^n$ *via* the Clifford–Zak transform. Standard density arguments allow us to extend this result in an unique way to the whole of $L^2(\mathbb{R}^n;C\ell_{n,0})$.

Proof. By definition, there exist unique $\mathbf{y} \in [0,1)^n$ and $\mathbf{n} \in \mathbb{Z}^n$ such that $\mathbf{x} = \mathbf{y}+\mathbf{n}$, and

$$\begin{aligned} \int_{[0,1)^n} Z_c f(\mathbf{x},\omega) d^n\omega &= \int_{[0,1)^n} Z_c f(\mathbf{y}+\mathbf{n},\omega) d^n\omega \\ &= \int_{[0,1)^n} e^{2\pi i_n \mathbf{n}\cdot\omega} Z_c f(\mathbf{y},\omega) d^n\omega \\ &= \int_{[0,1)^n} \sum_{\mathbf{k}\in\mathbb{Z}^n} e^{2\pi i_n (\mathbf{k}+\mathbf{n})\cdot\omega} f(\mathbf{y}-\mathbf{k}) d^n\omega \end{aligned}$$

$$= \int_{[0,1)^n} \sum_{\mathbf{m}\in\mathbb{Z}^n} e^{2\pi i_n \mathbf{m}\cdot\omega} f(\mathbf{y}+\mathbf{n}-\mathbf{m}) d^n\omega$$

$$= \int_{[0,1)^n} \sum_{\mathbf{m}\in\mathbb{Z}^n} e^{2\pi i_n \mathbf{m}\cdot\omega} f(\mathbf{x}-\mathbf{m}) d^n\omega$$

$$= \int_{[0,1)^n} f(\mathbf{x}) d^n\omega + \int_{[0,1)^n} \sum_{\mathbf{m}\neq 0} e^{2\pi i_n \mathbf{m}\cdot\omega} f(\mathbf{x}-\mathbf{m}) d^n\omega$$

$$= f(\mathbf{x}) + \int_{[0,1)^n} \sum_{\mathbf{m}\neq 0} e^{2\pi i_n \mathbf{m}\cdot\omega} f(\mathbf{x}-\mathbf{m}) d^n\omega. \qquad (3.26)$$

To calculate the remaining integral, we use Fubini's Theorem to validate the interchange between integration and summation so that

$$\int_{[0,1)^n} \sum_{\mathbf{m}\neq 0} e^{2\pi i_n \mathbf{m}\cdot\omega} f(\mathbf{x}-\mathbf{m}) d^n\omega$$
$$= \sum_{\mathbf{m}\neq 0} \left(\int_{[0,1)^n} e^{2\pi i_n \mathbf{m}\cdot\omega} d^n\omega \right) f(\mathbf{x}-\mathbf{m}) = 0. \qquad (3.27)$$

This completes the proof of the first identity. For the second, a direct calculation leads to

$$\int_{[0,1)^n} Z_c\widetilde{f(\mathbf{x},\omega)} e^{2\pi i_n \omega\cdot\mathbf{x}} d^n\mathbf{x} = \int_{[0,1)^n} \sum_{\mathbf{k}\subset\mathbb{Z}^n} \widetilde{f(\mathbf{x}-\mathbf{k})} e^{-2\pi i_n \mathbf{k}\cdot\omega} e^{2\pi i_n \omega\cdot\mathbf{x}} d^n\mathbf{x}$$

$$= \sum_{\mathbf{k}\in\mathbb{Z}^n} \int_{[0,1)^n} \widetilde{f(\mathbf{x}-\mathbf{k})} e^{2\pi i_n (\mathbf{x}-\mathbf{k})\cdot\omega} d^n\mathbf{x} \qquad (3.28)$$

$$= \int_{\mathbb{R}^n} \widetilde{f(\mathbf{y})} e^{2\pi i_n \mathbf{y}\cdot\omega} d^n\mathbf{y} = \mathcal{F}\{\widetilde{f}\}(-\omega). \qquad \square$$

In particular, we obtain the following reconstruction formula: for any $F \in \mathcal{Z}$, we have

$$\left(Z_c^{-1}F\right)(\mathbf{x}) = \int_{[0,1)^n} F(\mathbf{x},\omega) d^n\omega, \quad \mathbf{x}\in\mathbb{R}^n. \qquad (3.29)$$

Lemma 3.6. *If* $g_{\mathbf{m},\mathbf{n}}$ *is defined by* (3.1), *then*

$$Z_c g_{\mathbf{m},\mathbf{n}}(\mathbf{x},\omega) = e^{-2\pi i_n \mathbf{n}\cdot\omega} e^{2\pi i_n \mathbf{m}\cdot\mathbf{x}} Z_c g(\mathbf{x},\omega), \quad (\mathbf{x},\omega)\in Q^{2n}. \qquad (3.30)$$

Proof. From the definitions of the Clifford–Zak transform and of $g_{\mathbf{m},\mathbf{n}}$ we obtain

$$Z_c g_{\mathbf{m},\mathbf{n}}(\mathbf{x},\omega) = \sum_{\mathbf{k}\in\mathbb{Z}^n} e^{2\pi i_n \mathbf{k}\cdot\omega} e^{2\pi i_n \mathbf{m}\cdot(\mathbf{x}-\mathbf{k})} g(\mathbf{x}-\mathbf{k}-\mathbf{n})$$
$$= \sum_{\mathbf{k}\in\mathbb{Z}^n} e^{2\pi i_n \mathbf{k}\cdot\omega} e^{2\pi i_n \mathbf{m}\cdot\mathbf{x}} g(\mathbf{x}-\mathbf{k}-\mathbf{n})$$
$$= e^{2\pi i_n \mathbf{m}\cdot\mathbf{x}} \sum_{\mathbf{k}\in\mathbb{Z}^n} e^{2\pi i_n \mathbf{k}\cdot\omega} g(\mathbf{x}-\mathbf{k}-\mathbf{n})$$

$$= e^{2\pi i_n \mathbf{m}\cdot\mathbf{x}} \sum_{\mathbf{k}\in\mathbb{Z}^n} e^{2\pi i_n(\mathbf{k}-\mathbf{n})\cdot\omega} g(\mathbf{x}-\mathbf{k})$$

$$= e^{-2\pi i_n \mathbf{n}\cdot\omega} e^{2\pi i_n \mathbf{m}\cdot\mathbf{x}} \sum_{\mathbf{k}\in\mathbb{Z}^n} e^{2\pi i_n \mathbf{k}\cdot\omega} g(\mathbf{x}-\mathbf{k}) \tag{3.31}$$

$$= e^{-2\pi i_n \mathbf{n}\cdot\omega} e^{2\pi i_n \mathbf{m}\cdot\mathbf{x}} Z_c g(\mathbf{x},\omega).$$

□

Theorem 3.7. *If $f, g \in L^2(\mathbb{R}^n; \mathbb{R} \oplus i_n\mathbb{R})$ and $g_{\mathbf{m},\mathbf{n}}$ is defined as in (3.1) then we have*

$$\sum_{\mathbf{m},\mathbf{n}\in\mathbb{Z}^n} \left|\langle f, g_{\mathbf{m},\mathbf{n}}\rangle_{L^2(\mathbb{R}^n;C\ell_{n,0})}\right|^2 = \left\|Z_c f \widetilde{Z_c g}\right\|^2_{L^2(Q^{2n};C\ell_{n,0})}. \tag{3.32}$$

Proof. We remark that $f \in L^2(\mathbb{R}^n; \mathbb{R} \oplus i_n\mathbb{R}) \subset L^2(\mathbb{R}^n; C\ell_{n,0})$ so that we have $\|f\|_{L^2(\mathbb{R}^n;\mathbb{R}\oplus i_n\mathbb{R})} = \|f\|_{L^2(\mathbb{R}^n;C\ell_{n,0})}$. Then, based on Theorem 3.3 and Lemma 3.6, we obtain that

$$\begin{aligned}
&\sum_{\mathbf{m},\mathbf{n}\in\mathbb{Z}^n} \left|\langle f, g_{\mathbf{m},\mathbf{n}}\rangle_{L^2(\mathbb{R}^n;C\ell_{n,0})}\right|^2 \\
&= \sum_{\mathbf{m},\mathbf{n}\in\mathbb{Z}^n} \left|\langle Z_c f, Z_c g_{\mathbf{m},\mathbf{n}}\rangle_{L^2(Q^{2n};C\ell_{n,0})}\right|^2 \\
&= \sum_{\mathbf{m},\mathbf{n}\in\mathbb{Z}^n} \left|\left[\int_{Q^{2n}} Z_c f \widetilde{Z_c g_{\mathbf{m},\mathbf{n}}} d^n\mathbf{x} d^n\omega\right]_0\right|^2 \\
&= \sum_{\mathbf{m},\mathbf{n}\in\mathbb{Z}^n} \left|\left[\int_{Q^{2n}} Z_c f \widetilde{Z_c g} e^{2\pi i_n \mathbf{n}\cdot\omega} e^{-2\pi i_n \mathbf{m}\cdot\mathbf{x}} d^n\mathbf{x} d^n\omega\right]_0\right|^2 \qquad (3.33)\\
&= \sum_{\mathbf{m},\mathbf{n}\in\mathbb{Z}^n} \left|\left\langle Z_c f \widetilde{Z_c g}, E_{\mathbf{m},\mathbf{n}}\right\rangle_{L^2(Q^{2n};C\ell_{n,0})}\right|^2 \\
&= \left\|Z_c f \widetilde{Z_c g}\right\|^2_{L^2(Q^{2n};C\ell_{n,0})},
\end{aligned}$$

where the set of all $E_{\mathbf{m},\mathbf{n}} := e^{-2\pi i_n \mathbf{n}\cdot\omega} e^{2\pi i_n \mathbf{m}\cdot\mathbf{x}}$ constitutes an orthonormal basis for $L^2(Q^{2n}; \mathbb{R} \oplus i_n\mathbb{R})$. □

4. Balian–Low Theorem for WCFT

Before proceeding with the Balian–Low theorem for a Clifford–Gabor frame, let us recall some basic facts on the modulus of a Clifford number. In general, the modulus of arbitrary Clifford numbers is not multiplicative. In fact, for any two elements $a, b \in C\ell_{n,0}$ we have $|ab| \le 2^{\frac{n}{2}} |a|\,|b|$. However, in some special cases, the multiplicative property does hold. An easy calculation leading to such a case is described in the following lemma.

Lemma 4.1. *Let $b \in C\ell_{n,0}$ be such that $b\tilde{b} = |b|^2$. Then*

$$|ab| = |a|\,|b|, \quad \forall a \in C\ell_{n,0}. \tag{4.1}$$

In this section, and since we need the Clifford–Zak transform Z_c of the window function g to satisfy $|fZ_cg| = |f|\,|Z_cg|$ for any multivector $f \in C\ell_{n,0}$, it is necessary to impose some restrictions on the non-zero window function g. In what follows we will require

$$g \in L^2(\mathbb{R}^n; \mathbb{R} \oplus i_n\mathbb{R}). \tag{4.2}$$

Some simple examples of possible real-valued window functions are the two-dimensional Gaussian function and the two-dimensional first-order B-spline [23], which can be generalized to n dimensions and to cases of linear combinations of real-valued window functions and pseudoscalars.

Lemma 4.2. *Suppose that the non-zero window function g satisfies* (4.2). *Then we have*

$$Z_cg\widetilde{Z_cg} = |Z_cg|^2, \tag{4.3}$$

and for any multivector $a \in C\ell_{n,0}$

$$|aZ_cg| = |a|\,|Z_cg|. \tag{4.4}$$

Proof. Since $g \in L^2(\mathbb{R}^n; \mathbb{R} \oplus i_n\mathbb{R})$, an easy computation shows that there exist $G, H \in L^2(Q^{2n}; \mathbb{R})$ such that

$$Z_cg = G + i_nH. \tag{4.5}$$

Thus, based on the fact that $\widetilde{i_n} = -i_n$ and $\widetilde{G} = G, \widetilde{H} = H$ for $G, H \in L^2(Q^{2n}; \mathbb{R})$, it follows that

$$\begin{aligned} Z_cg\widetilde{Z_cg} = (G + i_nH)(\widetilde{G} + \widetilde{H}\widetilde{i_n}) &= (G + i_nH)(G - Hi_n) \\ &= GG - GHi_n + i_nHG - i_nHHi_n \\ &= G^2 - HGi_n + i_nHG - i_nH^2i_n. \end{aligned} \tag{4.6}$$

Note that the pseudoscalar i_n commutes with the scalar elements of the algebra. Thus,

$$HGi_n = i_nHG, \quad G^2 = |G|^2, \quad H^2 = |H|^2. \tag{4.7}$$

Substituting (4.7) into (4.6) leads to

$$Z_cg\,\widetilde{Z_cg} = |G|^2 + |H|^2 = |Z_cg|^2. \tag{4.8}$$

Moreover, by Lemma 4.1 we see that for any $a \in C\ell_{n,0}$

$$|aZ_cg| = |a|\,|Z_cg|, \tag{4.9}$$

which completes the proof. □

Based on the properties of the Clifford–Zak transform discussed in the previous section, we are going to study the time-frequency localization property of a Clifford–Gabor system $\{g_{\mathbf{m},\mathbf{n}} : \mathbf{m}, \mathbf{n} \in \mathbb{Z}^n\}$ for $L^2(\mathbb{R}^n; \mathbb{R} \oplus i_n\mathbb{R})$, which is the content of the following theorem. This theorem will enable us later on to derive the Clifford version of the Balian–Low theorem.

Theorem 4.3. *Suppose that* $\{g_{\mathbf{m},\mathbf{n}}(\mathbf{x}) : \mathbf{m},\mathbf{n} \in \mathbb{Z}^n\}$ *constitutes a frame for* $L^2(\mathbb{R}^n; \mathbb{R} \oplus i_n\mathbb{R})$ *with a non-zero window function* g *satisfying* (4.2). *Then we have*

$$\triangle x_k \triangle \omega_k = \infty, \quad k = 1, 2, \dots, n, \tag{4.10}$$

where

$$\triangle x_k = \int_{\mathbb{R}^n} x_k^2 \left|g(\mathbf{x})\right|^2 d^n\mathbf{x}, \quad \triangle \omega_k = \int_{\mathbb{R}^n} \omega_k^2 \left|\mathcal{F}g(\omega)\right|^2 d^n\omega \tag{4.11}$$

and the CFT of g, $\mathcal{F}g(\omega)$, *is defined by* (2.10).

We will divide the proof by demonstrating a sequence of lemmas. Denote by D_k and M_k the following partial derivative and multiplication operators

$$D_k f(\mathbf{x}) := \partial_{x_k} f(\mathbf{x}) \frac{1}{2\pi i_n}, \qquad M_k f(\mathbf{x}) := x_k f(\mathbf{x}), \quad k = 1, 2, \dots, n, \tag{4.12}$$

where $\partial_{x_k} := \frac{\partial}{\partial x_k}$. One can easily see that the product of these operators depends on their order. In fact, they satisfy the commutation relation

$$[M_k, D_k] f(\mathbf{x}) := (M_k D_k - D_k M_k) f(\mathbf{x}) = -f(\mathbf{x}) \frac{1}{2\pi i_n}. \tag{4.13}$$

This is traditionally expressed by saying that the time and frequency variables are canonically conjugate. In the first place, we review the following lemma which was shown in [22, 24].

Lemma 4.4 (CFT partial derivative). *The CFT of* $\partial_{x_k} f(\mathbf{x}) \in L^2(\mathbb{R}^n; C\ell_{n,0})$ *is given by*

$$\mathcal{F}\{\partial_{x_k} f(\mathbf{x})\}(\omega) = 2\pi \omega_k \mathcal{F}\{f\}(\omega) i_n, \tag{4.14}$$

that is,

$$\mathcal{F}\{D_k f(\mathbf{x})\}(\omega) = M_k \mathcal{F}\{f\}(\omega), \quad k = 1, 2, \dots, n. \tag{4.15}$$

Lemma 4.5. *Let* $g, \mu \in L^2(\mathbb{R}^n; \mathbb{R} \oplus i_n\mathbb{R})$. *If we have*

$$D_k g, D_k \mu, M_k g, M_k \mu \in L^2(\mathbb{R}^n; \mathbb{R} \oplus i_n\mathbb{R}), \quad k = 1, 2, \dots, n, \tag{4.16}$$

then the following relation holds

$$\langle M_k g, D_k \mu \rangle - \langle D_k g, M_k \mu \rangle = \pm \frac{1}{2\pi i_n} \langle g, \mu \rangle, \tag{4.17}$$

where the scalar inner product $\langle \cdot, \cdot \rangle$ *is given by* (2.8).

Proof. Let us choose $\varphi_j, \psi_j \in \mathcal{S}(\mathbb{R}^n; \mathbb{R} \oplus i_n\mathbb{R})$, where $\mathcal{S}(\mathbb{R}^n; \mathbb{R} \oplus i_n\mathbb{R})$ denotes the Schwartz class. We recall that $\mathcal{S}(\mathbb{R}^n; \mathbb{R} \oplus i_n\mathbb{R})$ is a dense subspace in $L^2(\mathbb{R}^n; \mathbb{R} \oplus i_n\mathbb{R})$ and it is defined as the set of all smooth functions from $\mathbb{R}^n$ to $\mathbb{R} \oplus i_n\mathbb{R}$ such that all of its partial derivatives are rapidly decreasing. Therefore, we have that if $\varphi_j \to g, \psi_j \to \mu$, then $D_k\varphi_j \to D_k g, M_k\varphi_j \to M_k g$ and $D_k\psi_j \to D_k\mu, M_k\psi_j \to M_k\mu, k = 1, 2, \dots, n$. Moreover, the convergence is in the L^2-sense. Since it is easy to check that for fixed k the operators D_k and M_k are self-adjoint, we obtain

$$\langle M_k\varphi_j, D_k\psi_j \rangle - \langle D_k\varphi_j, M_k\psi_j \rangle = \langle D_k M_k\varphi_j, \psi_j \rangle - \langle M_k D_k\varphi_j, \psi_j \rangle$$

$$= - \langle [M_k, D_k] \varphi_j, \psi_j \rangle = \left\langle \varphi_j \frac{1}{2\pi i_n}, \psi_j \right\rangle = \pm \frac{1}{2\pi i_n} \langle \varphi_j, \psi_j \rangle , \tag{4.18}$$

where we use the fact that the pseudoscalar i_n commutes with the scalar elements of the algebra. Moreover, since the scalar inner product is continuous, the desired result holds in the limit. □

The utility of the Clifford–Zak transform Z_c for constructing Gabor bases stems from the following result.

Lemma 4.6. *Suppose that the non-zero window function g satisfies (4.2). Then we have*

1. $\{g_{\mathbf{m},\mathbf{n}} : \mathbf{m},\mathbf{n} \in \mathbb{Z}^n\}$ *is a frame for* $L^2(\mathbb{R}^n; \mathbb{R} \oplus i_n\mathbb{R})$ *if and only if*
$$0 < B \leq |Z_c g(\mathbf{x}, \omega)|^2 \leq C < \infty \quad a.e. \quad (\mathbf{x}, \omega) \in Q^{2n}. \tag{4.19}$$
2. $\{g_{\mathbf{m},\mathbf{n}} : \mathbf{m},\mathbf{n} \in \mathbb{Z}^n\}$ *is an orthonormal basis for* $L^2(\mathbb{R}^n; \mathbb{R} \oplus i_n\mathbb{R})$ *if and only if* $|Z_c g(\mathbf{x}, \omega)|^2 = 1$ *for almost all* $(\mathbf{x}, \omega) \in Q^{2n}$.

Proof. Let us first prove statement 1. Assume that $\{g_{\mathbf{m},\mathbf{n}} : \mathbf{m},\mathbf{n} \in \mathbb{Z}^n\}$ is a frame for $L^2(\mathbb{R}^n; \mathbb{R} \oplus i_n\mathbb{R})$. According to the definition, there exist $0 < B \leq C < \infty$ such that for all $f \in L^2(\mathbb{R}^n; \mathbb{R} \oplus i_n\mathbb{R})$, it holds

$$B \|f\|^2_{L^2(\mathbb{R}^n;\mathbb{R}\oplus i_n\mathbb{R})} \leq \sum_{\mathbf{m},\mathbf{n}\in\mathbb{Z}^n} |\langle f, g_{\mathbf{m},\mathbf{n}} \rangle|^2 \leq C \|f\|^2_{L^2(\mathbb{R}^n;\mathbb{R}\oplus i_n\mathbb{R})} . \tag{4.20}$$

Then $F = Z_c f \in L^2(Q^{2n}; \mathbb{R} \oplus i_n\mathbb{R})$ and by Corollary 3.4 and Theorem 3.7 we get

$$B \|F\|^2_{L^2(Q^{2n};\mathbb{R}\oplus i_n\mathbb{R})} \leq \left\| F \widetilde{Z_c g} \right\|^2_{L^2(Q^{2n};\mathbb{R}\oplus i_n\mathbb{R})} \leq C \|F\|^2_{L^2(Q^{2n};\mathbb{R}\oplus i_n\mathbb{R})} , \tag{4.21}$$

which implies that $B \leq |Z_c g|^2 \leq C$ *a.e.* due to Lemma 4.2. Conversely, if $B \leq |Z_c g|^2 \leq C$ *a.e.*, then by Theorem 3.7 and Corollary 3.4, we conclude that for all $f \in L^2(\mathbb{R}^n; \mathbb{R} \oplus i_n\mathbb{R})$

$$\begin{aligned} B \|f\|^2_{L^2(\mathbb{R}^n;\mathbb{R}\oplus i_n\mathbb{R})} &\leq \sum_{\mathbf{m},\mathbf{n}\in\mathbb{Z}^n} |\langle f, g_{\mathbf{m},\mathbf{n}} \rangle|^2 = \left\| Z_c f \widetilde{Z_c g} \right\|^2_{L^2(Q^{2n};\mathbb{R}\oplus i_n\mathbb{R})} \\ &\leq C \|f\|^2_{L^2(\mathbb{R}^n;\mathbb{R}\oplus i_n\mathbb{R})} . \end{aligned} \tag{4.22}$$

Now, let us take a look at statement 2. If $\{g_{\mathbf{m},\mathbf{n}} : \mathbf{m},\mathbf{n} \in \mathbb{Z}^n\}$ is an orthonormal basis for $L^2(\mathbb{R}^n; \mathbb{R} \oplus i_n\mathbb{R})$ then, for all $f \in L^2(\mathbb{R}^n; \mathbb{R} \oplus i_n\mathbb{R})$, it holds

$$\begin{aligned} \sum_{\mathbf{m},\mathbf{n}\in\mathbb{Z}^n} |\langle f, g_{\mathbf{m},\mathbf{n}} \rangle|^2 &= \|f\|^2_{L^2(\mathbb{R}^n;\mathbb{R}\oplus i_n\mathbb{R})} = \|f\|^2_{L^2(\mathbb{R}^n;C\ell_{n,0})} \\ &= \|Z_c f\|^2_{L^2(Q^{2n};C\ell_{n,0})} . \end{aligned} \tag{4.23}$$

Thus, by Theorem 3.7 we get

$$\left\| Z_c f \widetilde{Z_c g} \right\|^2_{L^2(Q^{2n};C\ell_{n,0})} = \|Z_c f\|^2_{L^2(Q^{2n};C\ell_{n,0})} , \tag{4.24}$$

which implies that $|Z_c g(\mathbf{x},\omega)|^2 = 1$ for almost all $(\mathbf{x},\omega) \in Q^{2n}$ due to Lemma 4.2. Conversely, if $|Z_c g(\mathbf{x},\omega)|^2 = 1$ for almost all $(\mathbf{x},\omega) \in Q^{2n}$, then by Theorem 3.7, $\{g_{\mathbf{m},\mathbf{n}} : \mathbf{m},\mathbf{n} \in \mathbb{Z}^n\}$ is a tight frame for $L^2(\mathbb{R}^n;\mathbb{R}\oplus i_n\mathbb{R})$. Moreover, Corollary 3.4 yields $\|g\|^2_{L^2(\mathbb{R}^n;C\ell_{n,0})} = \|Z_c g\|^2_{L^2(Q^{2n};C\ell_{n,0})} = 1$, which leads to $\|g_{\mathbf{m},\mathbf{n}}\|^2_{L^2(\mathbb{R}^n;C\ell_{n,0})} = 1$. Consequently, $\{g_{\mathbf{m},\mathbf{n}} : \mathbf{m},\mathbf{n} \in \mathbb{Z}^n\}$ is an orthonormal basis for $L^2(\mathbb{R}^n;\mathbb{R}\oplus i_n\mathbb{R})$. □

Combining Corollary 3.4, Theorem 3.7 and (3.7), we can assert that

$$\langle Z_c Sf, Z_c f\rangle = \left\| Z_c f \widetilde{Z_c g} \right\|^2_{L^2(Q^{2n};C\ell_{n,0})}, \tag{4.25}$$

where S is the frame operator defined by (3.2). Thus, we get that $Z_c S Z_c^{-1}$ corresponds to a multiplication by $|Z_c g|^2$ on the space $\mathcal{Z}$ defined by (3.23).

The following lemma shows that $\partial_{x_1} Z_c g \in L^2(Q^{2n};\mathbb{R}\oplus i_n\mathbb{R})$ implies that $\partial_{x_1} |Z_c g| \in L^2(Q^{2n};\mathbb{R})$ under a certain condition. In a similar way, we can get the corresponding results for $\partial_{x_2} Z_c g, \partial_{\omega_1} Z_c g$ and $\partial_{\omega_2} Z_c g$ in $L^2(Q^{2n};\mathbb{R}\oplus i_n\mathbb{R})$.

Lemma 4.7. *Under the hypotheses of Theorem 4.3 we see that if* $\partial_{x_1} Z_c g \in L^2(Q^{2n};\mathbb{R}\oplus i_n\mathbb{R})$ *then* $\partial_{x_1} |Z_c g| \in L^2(Q^{2n};\mathbb{R}) \subset L^2(Q^{2n};\mathbb{R}\oplus i_n\mathbb{R})$.

Proof. Since $g \in L^2(\mathbb{R}^n;\mathbb{R}\oplus i_n\mathbb{R})$, an easy computation shows that there exist $G, H \in L^2(Q^{2n};\mathbb{R}\oplus i_n\mathbb{R})$ such that

$$Z_c g = G + i_n H, \quad G, H \in \mathbb{R}. \tag{4.26}$$

Thus $\partial_{x_1} Z_c g = \partial_{x_1} G + \partial_{x_1} H i_n$ and

$$\begin{aligned} \|\partial_{x_1} Z_c g\|^2 &= \int_{Q^{2n}} |\partial_{x_1} Z_c g|^2 \, d^n\mathbf{x} d^n\omega \\ &= \int_{Q^{2n}} \left((\partial_{x_1} G)^2 + (\partial_{x_1} H)^2\right) d^n\mathbf{x} d^n\omega < \infty, \end{aligned} \tag{4.27}$$

which means that

$$\|\partial_{x_1} G\| < \infty, \quad \|\partial_{x_1} H\| < \infty. \tag{4.28}$$

Moreover, since $\{g_{\mathbf{m},\mathbf{n}} : \mathbf{m},\mathbf{n} \in \mathbb{Z}^n\}$ is a frame for $L^2(\mathbb{R}^n;\mathbb{R}\oplus i_n\mathbb{R})$ by Lemma 4.6 we have

$$0 < B \le |Z_c g(\mathbf{x},\omega)|^2 \le C < \infty, \quad a.e.\ (\mathbf{x},\omega) \in Q^{2n}, \tag{4.29}$$

which tells us that

$$G^2 \le C < \infty, \quad H^2 \le C < \infty. \tag{4.30}$$

A simple calculation leads to

$$\partial_{x_1} |Z_c g| = |Z_c g|^{-1} (G \partial_{x_1} G + H \partial_{x_1} H). \tag{4.31}$$

Now, based on (4.28), (4.30), and the Minkowski inequality in $L^2(Q^{2n};\mathbb{R})$, it follows that

$$\|\partial_{x_1} |Z_c g|\|^2 = \int_{Q^{2n}} |Z_c g|^{-2} |(G \partial_{x_1} G + H \partial_{x_1} H)|^2 \, d^n\mathbf{x} d^n\omega$$

$$\leq B^{-1} \int_{Q^{2n}} |(G\partial_{x_1} G + H\partial_{x_1} H)|^2 \, d^n\mathbf{x} d^n\omega$$
$$\leq B^{-1}(\|G\partial_{x_1} G\| + \|H\partial_{x_1} H\|)^2 \tag{4.32}$$
$$\leq B^{-1}C(\|\partial_{x_1} G\| + \|\partial_{x_1} H\|)^2 < \infty. \qquad \square$$

Now, let us consider the dual frame $\breve{g}_{\mathbf{m},\mathbf{n}} := S^{-1}g_{\mathbf{m},\mathbf{n}}$ given by (3.9). Since $Z_cSZ_c^{-1}$ corresponds to a multiplication by $| Z_cg |^2$ on the space $\mathcal{Z}$, by Lemma 4.6 it follows that

$$Z_c\breve{g}_{\mathbf{m},\mathbf{n}} = Z_cS^{-1}Z_c^{-1}Z_cg_{\mathbf{m},\mathbf{n}} = |Z_cg|^{-2} Z_cg_{\mathbf{m},\mathbf{n}} \tag{4.33}$$

or

$$\begin{aligned} Z_c\breve{g}_{\mathbf{m},\mathbf{n}} &= |Z_cg|^{-2} e^{-2\pi i_n \mathbf{n}\cdot\omega} e^{2\pi i_n \mathbf{m}\cdot\mathbf{x}} Z_cg \\ &= e^{-2\pi i_n \mathbf{n}\cdot\omega} e^{2\pi i_n \mathbf{m}\cdot\mathbf{x}} Z_c\breve{g}, \end{aligned} \tag{4.34}$$

which belongs to the space $\mathcal{Z}$ with $Z_c\breve{g} = |Z_cg|^{-2} Z_cg$. In particular, (4.34) implies that

$$\breve{g}_{\mathbf{m},\mathbf{n}}(\mathbf{x}) := e^{2\pi i_n \mathbf{m}\cdot\mathbf{x}} \breve{g}(\mathbf{x} - \mathbf{n}), \tag{4.35}$$

which proves the following lemma associated with Lemma 4.2.

Lemma 4.8. *Under the hypotheses of Theorem* 4.3 *the dual window function* $\breve{g}$ *satisfies* $Z_c\breve{g} = |Z_cg|^{-2} Z_cg$ *and* $Z_c\breve{g}\,\widetilde{Z_cg} = 1$.

Analogous to the Minkowski inequality in $L^2(\mathbb{R}^n;\mathbb{R})$, by the definition of the norm (2.6) and the Clifford–Cauchy–Schwarz inequality (2.9), the following Clifford–Minkowski inequality holds true in $L^2(Q^{2n}; C\ell_{n,0})$. This will be necessary in the proof of Theorem 4.3.

Lemma 4.9. *For* $\varphi, \psi \in L^2(Q^{2n}; C\ell_{n,0})$, *we have* $\varphi + \psi \in L^2(Q^{2n}; C\ell_{n,0})$ *and*

$$\|\varphi + \psi\| \leq \|\varphi\| + \|\psi\|. \tag{4.36}$$

Now, we are in position to demonstrate Theorem 4.3.

Proof. We prove the theorem by contradiction. Suppose that $\{g_{\mathbf{m},\mathbf{n}} : \mathbf{m}, \mathbf{n} \in \mathbb{Z}^n\}$ is a frame for $L^2(\mathbb{R}^n; \mathbb{R} \oplus i_n\mathbb{R})$ with a non-zero window function g satisfying (4.2), and, furthermore, suppose that

$$\triangle x_k = \int_{\mathbb{R}^n} x_k^2 \, |g(\mathbf{x})|^2 \, d^n\mathbf{x} < \infty, \tag{4.37}$$

and

$$\triangle\omega_k = \int_{\mathbb{R}^n} \omega_k^2 \, |\mathcal{F}g(\omega)|^2 \, d^n\omega < \infty, \quad k = 1, 2, \ldots, n. \tag{4.38}$$

Now, based on Lemmas 2.2 and 4.4, we get $M_kg, D_kg \in L^2(\mathbb{R}^n; \mathbb{R} \oplus i_n\mathbb{R})$, where the multiplication and partial derivative operators M_k, D_k are defined by (4.12).

One has that

$$[Z_c(M_kg)]\,(\mathbf{x}, \omega) = x_kZ_cg - \frac{1}{2\pi i_n}\partial_{\omega_k} Z_cg, \tag{4.39}$$

which means that $M_k g \in L^2(\mathbb{R}^n; \mathbb{R}\oplus i_n\mathbb{R})$ if and only if $\partial_{\omega_k} Z_c g \in L^2(Q^{2n}; \mathbb{R}\oplus i_n\mathbb{R})$ by Theorem 3.4 and Lemma 4.9. In a similar way, we find that $D_k g \in L^2(\mathbb{R}^n; \mathbb{R} \oplus i_n\mathbb{R})$ if and only if $\partial_{x_k} Z_c g \in L^2(Q^{2n}; \mathbb{R} \oplus i_n\mathbb{R})$. By Lemma 4.8 we know that the dual window function $\breve{g}$ satisfies $Z_c\breve{g} = |Z_c g|^{-2} Z_c g$. Consequently, we see that

$$\begin{aligned}\partial_{x_k}(Z_c\breve{g}) &= \partial_{x_k}(|Z_c g|^{-2})Z_c g + |Z_c g|^{-2}\partial_{x_k} Z_c g \\ &= -2\,|Z_c g|^{-3} Z_c g\,\partial_{x_k}|Z_c g| + |Z_c g|^{-2}\partial_{x_k} Z_c g \qquad (4.40)\end{aligned}$$

and

$$\partial_{\omega_k}(Z_c\breve{g}) = -2\,|Z_c g|^{-3} Z_c g\,\partial_{\omega_k}|Z_c g| + |Z_c g|^{-2}\partial_{\omega_k} Z_c g \qquad (4.41)$$

are in $L^2(Q^{2n}; \mathbb{R} \oplus i_n\mathbb{R})$ by Lemmas 4.9, 4.6, and 4.7. Hence, we conclude that $M_k\breve{g}, D_k\breve{g} \in L^2(\mathbb{R}^n; \mathbb{R} \oplus i_n\mathbb{R})$. For the functions g and $\breve{g}$, we shall next prove the fact

$$\langle M_k g, D_k\breve{g}\rangle = \langle D_k g, M_k\breve{g}\rangle\,, \qquad (4.42)$$

from which we derive the contradiction. In fact, in the first place, by Corollary 3.4 and Lemma 4.8 we find that

$$\begin{aligned}\langle \breve{g}, g_{\mathbf{m},\mathbf{n}}\rangle &= \langle Z_c\breve{g}, Z_c g_{\mathbf{m},\mathbf{n}}\rangle = \left[\int_{Q^{2n}} Z_c\breve{g}\widetilde{Z_c g} e^{2\pi i_n \mathbf{n}\cdot\omega} e^{-2\pi i_n \mathbf{m}\cdot\mathbf{x}} d^n\mathbf{x} d^n\omega\right]_0 \\ &= \left[\int_{Q^{2n}} e^{2\pi i_n \mathbf{n}\cdot\omega} e^{-2\pi i_n \mathbf{m}\cdot\mathbf{x}} d^n\mathbf{x} d^n\omega\right]_0 = \delta_{\mathbf{m},0}\delta_{\mathbf{n},0}, \qquad (4.43)\end{aligned}$$

where $\delta_{\mathbf{m},\mathbf{k}}$ denotes the Kronecker delta function. Similarly, we can check that

$$\langle g, \breve{g}_{\mathbf{m},\mathbf{n}}\rangle = \delta_{\mathbf{m},0}\delta_{\mathbf{n},0}. \qquad (4.44)$$

Now, since $M_k g, D_k\breve{g} \in L^2(\mathbb{R}^n; \mathbb{R} \oplus i_n\mathbb{R})$ and $\{g_{\mathbf{m},\mathbf{n}}\}, \{\breve{g}_{\mathbf{m},\mathbf{n}}\}$ constitute dual frames for $L^2(\mathbb{R}^n; \mathbb{R} \oplus i_n\mathbb{R})$, we obtain

$$\langle M_k g, D_k\breve{g}\rangle = \sum_{\mathbf{m},\mathbf{n}\in\mathbb{Z}^n} \langle M_k g, \breve{g}_{\mathbf{m},\mathbf{n}}\rangle \langle g_{\mathbf{m},\mathbf{n}}, D_k\breve{g}\rangle\,. \qquad (4.45)$$

Based on (4.35), (4.44), and (2.5), a simple calculation leads to

$$\begin{aligned}\langle g_{-\mathbf{m},-\mathbf{n}}, M_k\breve{g}\rangle &= \left[\int_{\mathbb{R}^n} e^{-2\pi i_n \mathbf{m}\cdot\mathbf{x}} g(\mathbf{x}+\mathbf{n})\widetilde{x_k\breve{g}(\mathbf{x})} d^n\mathbf{x}\right]_0 \\ &= \left[\int_{\mathbb{R}^n} e^{-2\pi i_n \mathbf{m}\cdot\mathbf{x}} g(\mathbf{x}+\mathbf{n}) x_k\widetilde{\breve{g}(\mathbf{x})} d^n\mathbf{x}\right]_0 \\ &= \left[\int_{\mathbb{R}^n} x_k g(\mathbf{x}+\mathbf{n})\widetilde{\breve{g}(\mathbf{x})} e^{-2\pi i_n \mathbf{m}\cdot\mathbf{x}} d^n\mathbf{x}\right]_0 \\ &= \left[\int_{\mathbb{R}^n} (x_k - n_k) g(\mathbf{x})\widetilde{\breve{g}(\mathbf{x}-\mathbf{n})} e^{-2\pi i_n \mathbf{m}\cdot(\mathbf{x}-\mathbf{n})} d^n\mathbf{x}\right]_0 \\ &= \langle M_k g, \breve{g}_{\mathbf{m},\mathbf{n}}\rangle - n_k\langle g, \breve{g}_{\mathbf{m},\mathbf{n}}\rangle = \langle M_k g, \breve{g}_{\mathbf{m},\mathbf{n}}\rangle\,. \qquad (4.46)\end{aligned}$$

On the other hand, by (4.43) and (2.5) we obtain

$$\langle D_k g, \breve{g}_{-\mathbf{m},-\mathbf{n}}\rangle = \left[\int_{\mathbb{R}^n} \partial_{x_k} g(\mathbf{x}) \frac{1}{2\pi i_n} \widetilde{\breve{g}(\mathbf{x}+\mathbf{n})} e^{2\pi i_n \mathbf{m}\cdot\mathbf{x}} d^n\mathbf{x}\right]_0$$

$$
\begin{aligned}
&= 0 - \left[\int_{\mathbb{R}^n} g(\mathbf{x}) \frac{1}{2\pi i_n} \partial_{x_k} \left(\widetilde{\breve{g}(\mathbf{x}+\mathbf{n})} e^{2\pi i_n \mathbf{m}\cdot\mathbf{x}} \right) d^n\mathbf{x} \right]_0 \\
&= - \left[\int_{\mathbb{R}^n} g(\mathbf{x}) \frac{1}{2\pi i_n} \begin{pmatrix} \partial_{x_k} \widetilde{\breve{g}(\mathbf{x}+\mathbf{n})} e^{2\pi i_n \mathbf{m}\cdot\mathbf{x}} \\ + \\ \widetilde{\breve{g}(\mathbf{x}+\mathbf{n})} m_k 2\pi i_n e^{2\pi i_n \mathbf{m}\cdot\mathbf{x}} \end{pmatrix} d^n\mathbf{x} \right]_0 \\
&= \left[\int_{\mathbb{R}^n} e^{2\pi i_n \mathbf{m}\cdot\mathbf{x}} g(\mathbf{x}-\mathbf{n}) \widetilde{D_k \breve{g}(\mathbf{x})} d^n\mathbf{x} \right]_0 \\
&\quad - m_k \left[\int_{\mathbb{R}^n} g(\mathbf{x}) \frac{1}{2\pi i_n} \widetilde{\breve{g}(\mathbf{x}+\mathbf{n})} 2\pi i_n e^{2\pi i_n \mathbf{m}\cdot\mathbf{x}} d^n\mathbf{x} \right]_0 \\
&= \langle g_{\mathbf{m},\mathbf{n}}, D_k \breve{g} \rangle \pm m_k \langle g_{\mathbf{m},\mathbf{n}}, \breve{g} \rangle = \langle g_{\mathbf{m},\mathbf{n}}, D_k \breve{g} \rangle .
\end{aligned} \tag{4.47}
$$

Consequently, substituting (4.46) and (4.47) into (4.45), we get

$$
\begin{aligned}
\langle M_k g, D_k \breve{g} \rangle &= \sum_{\mathbf{m},\mathbf{n}\in\mathbb{Z}^n} \langle g_{-\mathbf{m},-\mathbf{n}}, M_k \breve{g} \rangle \langle D_k g, \breve{g}_{-\mathbf{m},-\mathbf{n}} \rangle \\
&= \sum_{\mathbf{m},\mathbf{n}\in\mathbb{Z}^n} \langle D_k g, \breve{g}_{-\mathbf{m},-\mathbf{n}} \rangle \langle g_{-\mathbf{m},-\mathbf{n}}, M_k \breve{g} \rangle = \langle D_k g, M_k \breve{g} \rangle ,
\end{aligned} \tag{4.48}
$$

which means that

$$\langle g, \breve{g} \rangle = 0 \tag{4.49}$$

by Lemma 4.5. However, by (4.44), taking $\mathbf{m} = \mathbf{n} = 0$, we have

$$\langle g, \breve{g} \rangle = 1, \tag{4.50}$$

which leads to the contradiction. Thus, we get

$$\triangle x_k \triangle \omega_k = \infty, \quad k = 1, 2, \ldots, n. \tag{4.51}$$

□

We are now in a position to extend Theorem 4.3 to the case of an arbitrary Clifford frame.

Theorem 4.10 (Balian–Low theorem). *Suppose that $\{g_{\mathbf{m},\mathbf{n}}(\mathbf{x}) : \mathbf{m}, \mathbf{n} \in \mathbb{Z}^n\}$ constitutes a frame for $L^2(\mathbb{R}^n; C\ell_{n,0})$ associated to a non-zero window function g in $L^2(\mathbb{R}^n; C\ell_{n,0})$. Then we have*

$$\triangle x_k \triangle \omega_k = \infty, \quad k = 1, 2, \ldots, n, \tag{4.52}$$

where

$$\triangle x_k = \int_{\mathbb{R}^n} x_k^2 \, |g(\mathbf{x})|^2 \, d^n\mathbf{x}, \quad \triangle \omega_k = \int_{\mathbb{R}^n} \omega_k^2 \, |\mathcal{F}g(\omega)|^2 \, d^n\omega \tag{4.53}$$

and the CFT of g, $\mathcal{F}g(\omega)$, is defined by (2.10).

Proof. If $\{g_{\mathbf{m},\mathbf{n}} : \mathbf{m}, \mathbf{n} \in \mathbb{Z}^n\}$ constitutes a frame for $L^2(\mathbb{R}^n; C\ell_{n,0})$ then the system $\{h_{\mathbf{m},\mathbf{n}} = [g_{\mathbf{m},\mathbf{n}}]_0 + i_n [g_{\mathbf{m},\mathbf{n}}]_n : \mathbf{m}, \mathbf{n} \in \mathbb{Z}^n\}$ is a frame for $L^2(\mathbb{R}^n; \mathbb{R} \oplus i_n \mathbb{R})$. By Theorem 4.3 the result follows. □

5. Conclusions

The classical Balian–Low theorem is a strong form of the uncertainty principle for Gabor systems which can be obtained by discretization of the kernel of the WFT. The WCFT is a generalization of the WFT in the framework of Clifford analysis. In this chapter, associated with the discretization of the kernel of the WCFT, we established a new kind of Gabor system

$$g_{\mathbf{m},\mathbf{n}}(\mathbf{x}) := e^{2\pi i_n \mathbf{m}\cdot\mathbf{x}} g(\mathbf{x}-\mathbf{n}), \quad \mathbf{m},\mathbf{n} \in \mathbb{Z}^n, \tag{5.1}$$

which constitutes a Clifford frame satisfying the frame condition

$$B\|f\|^2 \leq \sum_{\mathbf{m},\mathbf{n}\in\mathbb{Z}^n} |\langle f, g_{\mathbf{m},\mathbf{n}}\rangle|^2 \leq C\|f\|^2, \qquad B, C > 0, \quad \forall f \in L^2(\mathbb{R}^n; C\ell_{n,0}). \tag{5.2}$$

To analyse Gabor systems $\{g_{\mathbf{m},\mathbf{n}}(\mathbf{x}) : \mathbf{m},\mathbf{n} \in \mathbb{Z}^n\}$ in $L^2(\mathbb{R}^n, C\ell_{n,0})$, we established the definition of the Clifford–Zak transform for $f \in L^2(\mathbb{R}^n; C\ell_{n,0})$ by

$$Z_c f(\mathbf{x},\omega) := \sum_{\mathbf{k}\in\mathbb{Z}^n} e^{2\pi i_n \mathbf{k}\cdot\omega} f(\mathbf{x}-\mathbf{k}), \quad (\mathbf{x},\omega) \in Q^{2n}, \tag{5.3}$$

and showed some properties of it. Furthermore, we proved the corresponding Balian–Low theorem for such Gabor systems which form a Clifford frame.

Acknowledgment

The first author is the recipient of a postdoctoral grant from *Fundação para a Ciência e a Tecnologia*, ref. SFRH/BPD/46250/2008. This work was supported by *FEDER* funds through *COMPETE* – Operational Programme Factors of Competitiveness ('Programa Operacional Factores de Competitividade') and by Portuguese funds through the *Center for Research and Development in Mathematics and Applications* (University of Aveiro) and the Portuguese Foundation for Science and Technology ('FCT – Fundação para a Ciência e a Tecnologia'), within project PEst-C/MAT/UI4106/2011 with COMPETE number FCOMP-01-0124-FEDER-022690. The work was also supported by the Foundation of Hubei Educational Committee (No. Q20091004) and the NSFC (No. 11026056).

References

[1] M. Bahri. Generalized Fourier transform in Clifford algebra $Cl(0,3)$. *Far East Journal of Mathematical Science*, 44(2):143–154, Sept. 2010.

[2] M. Bahri. Generalized Fourier transform in real clifford algebra $cl(0,n)$. *Far East Journal of Mathematical Sciences*, 48(1):11–24, Jan. 2011.

[3] M. Bahri, S. Adji, and J.M. Zhao. Real Clifford windowed Fourier transform. *Acta Mathematica Sinica*, 27(3):505–518, 2011.

[4] M. Bahri, E. Hitzer, R. Ashino, and R. Vaillancourt. Windowed Fourier transform of two-dimensional quaternionic signals. *Applied Mathematics and Computation*, 216(8):2366–2379, June 2010.

[5] R. Balan. Extensions of no-go theorems to many signal systems. In A. Aldroubi and E.B. Lin, editors, *Wavelets, Multiwavelets, and Their Applications*, volume 216 of *Contemporary Mathematics*, pages 3–14. American Mathematical Society, 1998.

[6] I.J. Benedetto, W. Czaja, and A.Y. Maltsev. The Balian–Low theorem for the symplectic form on $\mathbb{R}^{2d}$. *Journal of Mathematical Physics*, 44(4):1735–1750, 2003.

[7] J.J. Benedetto, C. Heil, and D.F. Walnut. Differentiation and the Balian–Low theorem. *Journal of Fourier Analysis and Applications*, 1(4):355–402, 1994.

[8] F. Brackx, N. De Schepper, and F. Sommen. The two-dimensional Clifford–Fourier transform. *Journal of Mathematical Imaging and Vision*, 26(1):5–18, 2006.

[9] F. Brackx, N. De Schepper, and F. Sommen. The Fourier transform in Clifford analysis. *Advances in Imaging and Electron Physics*, 156:55–201, 2009.

[10] F. Brackx, R. Delanghe, and F. Sommen. *Clifford Analysis*, volume 76. Pitman, Boston, 1982.

[11] T. Bülow, M. Felsberg, and G. Sommer. Non-commutative hypercomplex Fourier transforms of multidimensional signals. In G. Sommer, editor, *Geometric computing with Clifford Algebras: Theoretical Foundations and Applications in Computer Vision and Robotics*, pages 187–207, Berlin, 2001. Springer.

[12] T. Bülow and G. Sommer. Quaternionic Gabor filters for local structure classification. In *Proceedings Fourteenth International Conference on Pattern Recognition*, volume 1, pages 808–810, 16–20 Aug 1998.

[13] W. Czaja and A.M. Powell. Recent developments in the Balian–Low theorem. In C. Heil, editor, *Harmonic Analysis and Applications*, pages 79–100. Birkhäuser, Boston, 2006.

[14] I. Daubechies. *Ten Lectures on Wavelets*. Society for Industrial and Applied Mathematics, 1992.

[15] J. Ebling and G. Scheuermann. Clifford Fourier transform on vector fields. *IEEE Transactions on Visualization and Computer Graphics*, 11(4):469–479, JulyAug. 2005.

[16] J. Ebling and J. Scheuermann. Clifford convolution and pattern matching on vector fields. In *Proceedings IEEE Visualization*, volume 3, pages 193–200, Los Alamitos, CA, 2003. IEEE Computer Society.

[17] M. Felsberg. *Low-Level Image Processing with the Structure Multivector*. PhD thesis, Christian-Albrechts-Universität, Institut für Informatik und Praktische Mathematik, Kiel, 2002.

[18] K. Gröchenig. *Foundations of Time-Frequency Analysis*. Applied and Numerical Harmonic Analysis. Birkhäuser, Boston, 2001.

[19] K. Gröchenig, D. Han, and G. Kutyniok. The Balian–Low theorem for the symplectic lattice in higher dimensions. *Applied and Computational Harmonic Analysis*, 13:169–176, 2002.

[20] C. Heil and D.F. Walnut. Continuous and discrete wavelet transforms. *SIAM Review*, 31:628–666, 1989.

[21] S. Held, M. Storath, P. Massopust, and B. Forster. Steerable wavelet frames based on the Riesz transform. *IEEE Transactions on Image Processing*, 19(3):653–667, 2010.

[22] E.M.S. Hitzer and B. Mawardi. Clifford Fourier transform on multivector fields and uncertainty principles for dimensions $n = 2(\bmod 4)$ and $n = 3(\bmod 4)$. *Advances in Applied Clifford Algebras*, 18(3-4):715–736, 2008.

[23] B. Mawardi, E. Hitzer, and S. Adji. Two-dimensional Clifford windowed Fourier transform. In E. Bayro-Corrochano and G. Scheuermann, editors, *Applied Geometric Algebras in Computer Science and Engineering*, pages 93–106. Springer, London, 2010.

[24] B. Mawardi and E.M.S. Hitzer. Clifford Fourier transformation and uncertainty principle for the Clifford algebra $C\ell_{3,0}$. *Advances in Applied Clifford Algebras*, 16(1):41–61, 2006.

[25] M. Porat, Y.Y. Zeevi, and M. Zibulski. Multi-window Gabor schemes in signal and image representations. In H.G. Feichtinger and T. Strohmer, editors, *Gabor Analysis and Algorithms*, pages 381–408. Birkhäuser, 1998.

[26] S.J. Sangwine and T.A. Ell. Hypercomplex Fourier transforms of color images. In *IEEE International Conference on Image Processing* (*ICIP* 2001), volume I, pages 137–140, Thessaloniki, Greece, 7–10 Oct. 2001. IEEE.

[27] F. Sommen. Hypercomplex Fourier and Laplace transforms I. *Illinois Journal of Mathematics*, 26(2):332–352, 1982.

[28] Y.Y. Zeevi and M. Zibulski. Analysis of multiwindow Gabor-type schemes by frame methods. *Applied and Computational Harmonic Analysis*, 4:188–221, 1997.

Yingxiong Fu
Hubei Key Laboratory of Applied Mathematics
Hubei University

and

Faculty of Mathematics and Computer Science
Hubei University, Hubei, China

and

Department of Mathematics
University of Aveiro, Portugal
e-mail: `fyx@ua.pt`

Uwe Kähler and Paula Cerejeiras
Department of Mathematics
University of Aveiro, Portugal
e-mail: `ukaehler@ua.pt`
`pceres@ua.pt`

Quaternion and Clifford–Fourier Transforms and Wavelets
Trends in Mathematics, 321–332

16 Sparse Representation of Signals in Hardy Space

Shuang Li and Tao Qian

Abstract. Mathematically, signals can be seen as functions in certain spaces. And processing is more efficient in a sparse representation where few coefficients reveal the information. Such representations are constructed by decomposing signals into elementary waveforms. A set of all elementary waveforms is called a dictionary. In this chapter, we introduce a new kind of sparse representation of signals in Hardy space $H^2(\mathbb{D})$ *via* the compressed sensing (CS) technique with the dictionary

$$\mathscr{D} = \{e_a : e_a(z) = \frac{\sqrt{1-|a|^2}}{1-\bar{a}z}, a \in \mathbb{D}\}.$$

where $\mathbb{D}$ denotes the unit disk. In addition, we give examples exhibiting the algorithm.

Mathematics Subject Classification (2010). 30H05, 42A50, 42A38.

Keywords. Hardy space, compressed sensing, analytic signals, reproducing kernels, sparse representation, redundant dictionary, l_1 minimization.

1. Introduction

A basis gives unique representations for signals in some certain space. However, it does not always give sparse expressions. One of the problems of approximation theory is to approximate functions with elements from a large candidate set called a dictionary. Let H be a Hilbert space. Using terminology introduced by Mallat and Zhang [18], a dictionary is defined as a family of parameterized vectors $\mathscr{D} = \{g_\gamma\}_{\gamma\in\Gamma}$ in H such that $\|g_\gamma\| = 1$ and $\overline{\text{span}(g_\gamma)} = H$. The g_γ are usually called *atoms*. For the discrete-time situation, the approximation problem can be written as

$$s = \mathscr{D}x \tag{1.1}$$

where s is the discrete signal, matrix $\mathscr{D}$ represents the dictionary with atoms as columns and x is the vector of coefficients. Notice that we adopt the vector inner product instead of the Hilbert inner product. In general, $\mathscr{D}$ has more columns than rows because of redundancy. A natural question is: can we find the best M-term approximation in a redundant dictionary for a given signal? That is an optimization problem:

$$\min \|s - \mathscr{D}x\|_{l^2} \quad \text{subject to} \quad \|x\|_{l^0} \leq M \tag{1.2}$$

where $\|x\|_{l^0}$ is the number of nonzero coefficients of x. Unfortunately, finding an optimal M-term approximation in redundant dictionaries is computationally intractable because it is NP-hard [8, 9, 17]. Thus, it is necessary to rely on good but not optimal approximations with computational algorithms. Until now, three main strategies have been investigated, they are *matching pursuit*, *basis pursuit* and *compressed sensing*.

1.1. Matching Pursuit and Basis Pursuit

Matching pursuit (MP) introduced by Mallat and Zhang computes signal approximations from a redundant dictionary by iteratively selecting one vector at a time. It is an example of a greedy algorithm. For a detailed description, please refer to [17, 18]. A recent development is the *Adaptive Fourier Decomposition* (AFD), which is a variation of the greedy algorithm in the particular context of Hardy spaces [20]. An intrinsic feature of the algorithm is that when stopped after a few steps, it yields an approximation using only a few atoms. If the dictionary is orthogonal, the method works perfectly. If the dictionary is not orthogonal, the situation is less clear [6]. The MP algorithm often yields locally optimal solutions depending on initial values. In contrast, *basis pursuit* (BP) performs a more global search. It finds signal representations by solving the following problem

$$\min \|x\|_{l^1} \quad \text{subject to} \quad s = \mathscr{D}x. \tag{1.3}$$

Given s and $\mathscr{D}$, we find x with minimal l_1 norm. Notice that (1.3) is a convex optimization program which is not NP-hard. Actually, the use of l_1 minimization to promote sparsity has a long history, dating back at least to the work of Beurling [1] on Fourier transform extrapolation from partial observations. Basis pursuit is an optimization principle, not an algorithm. Empirical evidence suggests that BP is more powerful than MP [6]. And the stability of BP has been proved in the presence of noise for sufficiently sparse representations [11]. BP is closely connected with convex programming. The interior-point method and the homotopy method can be applied to BP in nearly linear time [6, 13].

1.2. Compressed Sensing

If the original signal is sparse in some sense, compressed sensing (CS) gives an excellent recovery of the signal. CS is a new concept in signal processing. The ideas have their origins in certain abstract results by Kashin [7, 15] but were brought into the forefront by the work of Candes, Romberg and Tao [2–5] and Donoho [10]. The core idea behind CS is that a signal or image, unknown but supposed to be compressible by a known transform, can be subjected to fewer measurements than the nominal number of pixels, and yet be accurately reconstructed [12]. Basically, CS relies on random projection and BP. Suppose we have

$$y = \Phi x, \tag{1.4}$$

where x is a finite vector, Φ is observation matrix and y is the vector of available measurements. Then the BP solution x^* of

$$\min \|x\|_{l^1} \quad \text{subject to} \quad y = \Phi x \tag{1.5}$$

recovers x exactly provided that x is sufficiently sparse and the matrix obeys the Restricted Isometry Property (RIP) [3,4]. However the RIP of a fixed matrix is very hard to check, thus in practice we use random matrices instead. A Gaussian matrix $\Phi \in R^{m\times N}$ whose entries $\Phi_{i,j}$ are independent and follow a normal distribution with expectation 0 and variance $1/m$ is often adopted because we have [21] that:

Theorem 1.1. *Let $\Phi \in \mathbb{R}^{m\times N}$ be a Gaussian random matrix. Let $\varepsilon, \delta \in (0,1)$ and assume*

$$m \geq C\delta^{-2}\left(s\log(N/s) + \log\varepsilon^{-1}\right)$$

for a universal constant $C > 0$. Then with probability at least $1-\varepsilon$ the restricted isometry constant of Φ satisfies $\delta_s \leq \delta$.

The theorem tells us that we can raise the probability by increasing the number of rows of the matrix Φ. We will not discuss details of CS, for more about this technique, please see [2–5, 10, 12, 13, 21].

1.3. Hardy Space $H^2(\mathbb{D})$

Hardy space is an important class of spaces connected to analytic signals. Control theory and rational approximation theory are also bound with the research in this field. In this chapter, we discuss signal decomposition in $H^2(\mathbb{D})$. $H^2(\mathbb{D})$ contains analytic signals with finite energy. Denote $\mathbb{D} = \{z \in \mathbb{C}\colon |z| < 1\}$, and let $\mathrm{Hol}(\mathbb{D})$ be the space of analytic functions on $\mathbb{D}$. For $p > 0$, Hardy space $H^p(\mathbb{D})$ is defined as follows:

$$H^p(\mathbb{D}) = \left\{ f \in \mathrm{Hol}(\mathbb{D})\colon \|f\|^p_{H^p} = \sup_{0\leq r<1} \int_0^{2\pi} |f(re^{ix})|^p\, dx/2\pi < \infty \right\}, \tag{1.6}$$

and

$$H^\infty(\mathbb{D}) = \left\{ f \in \mathrm{Hol}(\mathbb{D})\colon \|f\|_{H^\infty} = \sup_{z\in\mathbb{D}} |f(z)| < \infty \right\}. \tag{1.7}$$

Indeed, $H^2(\mathbb{D})$ is a Hilbert space consisting of all functions $f(z) = \sum_{n=0}^{\infty} a_n z^n$ analytic in the unit disc $\mathbb{D}$ such that $\|f\|^2 = \sum_{n=0}^{\infty} |a_n|^2 < \infty$. It has reproducing kernels $k_a(z) = \frac{1}{1-\bar{a}}z$, $a \in \mathbb{D}$. Besides the Fourier basis $\{1, z, z^2, \ldots, z^n, \ldots\}$, $H^2(\mathbb{D})$

has another orthonormal basis $\{g_n\}_{n=1}^{\infty}$ named after Takenaka and Malmquist (TM):

$$g_1(z) = \frac{\sqrt{1-|z_1|^2}}{1-\overline{z_1}z} \tag{1.8}$$

and

$$g_n(z) = \frac{\sqrt{1-|z_n|^2}}{1-\overline{z_n}z} \prod_{k=1}^{n-1} \frac{z-z_k}{1-\overline{z_k}z} \tag{1.9}$$

for $n \geq 2$, where $\{z_n\}_{n=1}^{\infty}$ must satisfy

$$\sum_{n=1}^{\infty} 1-|z_n| = \infty. \tag{1.10}$$

It is clear that the TM basis can be obtained by applying the Gram–Schmidt procedure to the reproducing kernels $k_{z_n} = 1/(1-\overline{z_n}z)$ under condition (1.10). TM systems have long been associated with fruitful results in applied mathematics such as control theory, signal processing and system identification. Qian *et al.* recently proposed an adaptive Fourier decomposition algorithm (AFD) that results in a TM system (not necessarily a basis) with selected poles according to the given signal [20]. However AFD is valid only for one-dimensional signals which naturally raises the question: what happens in higher dimensions? Inspired by the aforementioned works, the adaptive decomposition of any function of two or three variables is obtained in [19] by means of quaternionic analysis. Unfortunately, the result can not be directly generalized into $(m+1)$-dimensional space in the Clifford setting because a Clifford number is not invertible in general.

It is clear that the normalized reproducing kernel

$$\mathscr{D} = \{e_a \colon e_a(z) = \frac{\sqrt{1-|a|^2}}{1-\bar{a}z}, \quad a \in \mathbb{D}\} \tag{1.11}$$

forms a dictionary of $H^2(\mathbb{D})$, and this redundant dictionary does give a sparse representation of signals in Hardy space. In [19] Qian *et al.* construct a dictionary for decomposition of quaternionic-valued signals of finite energy. It is a generalization of (1.11). This chapter will focus on the method based on CS to decompose signals in $H^2(\mathbb{D})$. Later we will generalize it into quaternionic Hardy space.

The chapter is concisely arranged as follows. We discuss the main results in Section 2. In Section 3, examples are given to illustrate our algorithm.

2. Main Results

The singular value decomposition (SVD) is an effective tool for the analysis of linear operators. It is also useful in this chapter. We recall that:

Theorem 2.1 (Singular Value Decomposition [16]). *Let A denote an arbitrary matrix with elements in $\mathbb{C}^{m\times n}$ and let $\{s_i\}_{i=1}^r$ be the nonzero singular values of A. Then A can be represented in the form*

$$A = UDV^*, \tag{2.1}$$

where $U \in \mathbb{C}^{m\times m}, V \in \mathbb{C}^{n\times n}$ are unitary and the $m \times n$ matrix D has elements s_i in the i,i position $(1 \le i \le r)$ and zeros elsewhere. The s_i are singular values of A, $s_i = \sqrt{\lambda_i(A^*A)}$. In fact, the diagonal matrix D can be written as

$$\begin{pmatrix} \Sigma_r & 0 \\ 0 & 0 \end{pmatrix} \tag{2.2}$$

where $\Sigma_r = \operatorname{diag}(s_1, s_2, \dots, s_r)$ with $s_1 \ge s_2 \ge \cdots, s_r$.

In the model (1.1), (1.11), we select N points $\{a_k\}_{k=1}^N$ in $\mathbb{D}$, and sample M points for each e_{a_k} equally spaced in $[0, 2\pi]$. So we get a matrix $\widetilde{\mathscr{D}} \in \mathbb{C}^{M\times N}$. For a signal in $H^2(\mathbb{D})$, we also sample M points on its boundary value equally spaced to form a vector $s \in \mathbb{C}^M$. Thus, the representation problem is

$$s = \widetilde{\mathscr{D}}x. \tag{2.3}$$

Suppose that $\widetilde{\mathscr{D}} = UDV^*$, then $U^*s = DV^*x$. Note that D is a diagonal matrix and $\|U^*s\|_2 = \|s\|_2, \|V^*x\|_2 = \|x\|_2$. If the singular values of $\widetilde{\mathscr{D}}$ decay fast, we can expect a relatively sparse representation x with a small energy error. The most important thing is that the assertion should be established in the sense of $N \to \infty$ because one must explain what the situation would be when the number of atoms is large.

First, we give two lemmas, they are also the simple cases of our theorem.

Lemma 2.2. *Suppose N points $\{a_k\}_{k=0}^{N-1}$ are selected equally spaced on the circle of radius r. Let $\lambda_1 \ge \lambda_2 \ge \cdots \ge \lambda_N$ be eigenvalues of the matrix $\widetilde{\mathscr{D}}^*\widetilde{\mathscr{D}}$. Then we have*

$$\lim_{N\to\infty} \frac{1}{N} \sum_{j=1}^m \lambda_j \ge 1 - r^{2m}. \tag{2.4}$$

Proof. (Sketch of proof.) Let

$$\mathscr{D} = \begin{pmatrix} e_{a_0} & e_{a_1} & \cdots & e_{a_{N-1}} \end{pmatrix} \tag{2.5}$$

then the Hermitian matrix H can be written as

$$H = \mathscr{D}^*\mathscr{D} = \begin{pmatrix} \overline{e_{a_0}} \\ \overline{e_{a_1}} \\ \vdots \\ \overline{e_{a_{N-1}}} \end{pmatrix} \begin{pmatrix} e_{a_0} & e_{a_1} & \cdots & e_{a_{N-1}} \end{pmatrix}. \tag{2.6}$$

with elements $\overline{H}_{ij} = \langle e_{a_i}, e_{a_j} \rangle = \langle e_{a_0}, e_{a_{j-i}} \rangle = b_{j-i}$. Notice that

$$b_k = \langle e_{a_0}, e_{a_k} \rangle = \frac{1-r^2}{1-r^2e^{ik\theta}} = \frac{1-r^2}{1-r^2e^{i\theta_k}}, \qquad (k = 0, 1, \dots, N-1) \tag{2.7}$$

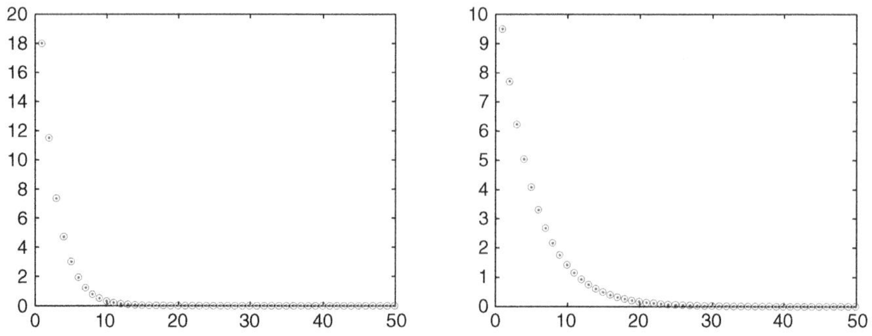

(a) $N=50$, $r=0.8$ (b) $N=50$, $r=0.9$

FIGURE 1. Selecting N points on the circle of radius r (2.9) gives that $\lambda_j \approx (1-r^2)r^{2(j-1)}N$ when N is large. The small red circles are the corresponding estimation whereas the blue points represent the real eigenvalues given by numerical calculation. They fit amazingly well even though N is not very large.

where $\theta_k = k\theta$ is the argument of a_k. Recall Ky Fan's maximum principle: let A be any Hermitian operator, then for $k = 1, 2, \ldots, n$, we have

$$\sum_{j=1}^{k} \lambda_j(A) = \max \sum_{j=1}^{k} \langle Ax_j, x_j \rangle \tag{2.8}$$

where the eigenvalues $\lambda_1(A) \geq \lambda_2(A) \geq \cdots \geq \lambda_n(A)$, and the maximum are taken over all orthonormal k-tuples $\{x_1, \ldots, x_k\}$.

Sampling N points equally spaced on the orthonormal functions $\{e^{-ijt}\}_{j=0}^{m-1}$, we prove that

$$\lim_{N\to\infty} \frac{\lambda_{j+1}}{N} = (1-r^2)r^{2j} \tag{2.9}$$

where $\lambda_1 \geq \lambda_2 \geq \cdots \geq \lambda_m$. Hence

$$\lim_{N\to\infty} \frac{1}{N} \sum_{k=1}^{m} \lambda_k \geq \sum_{k=1}^{m} (1-r^2)r^{2(k-1)} = 1 - r^{2m}. \tag{2.10}$$

$\square$

Remark 2.3. Note that $\mathrm{trace}(H) = N$, thus (2.9) shows that the eigenvalues decay as a geometric series with common ratio r^2 as $N \to \infty$. See Figure 1.

Lemma 2.4. *Suppose N points $\{a_k\}_{k=1}^N$ are selected equally spaced on the segment $[0, 1)$, and λ_1 is the largest eigenvalue of the matrix $\widetilde{\mathscr{D}}^* \widetilde{\mathscr{D}}$. Then we have*

$$\lim_{N\to\infty} \frac{\lambda_1}{N} \geq \int_0^1 \int_0^1 \frac{\sqrt{1-s^2}\sqrt{1-r^2}}{1-sr}\, \mathrm{d}r\, \mathrm{d}s \approx 0.815784. \tag{2.11}$$

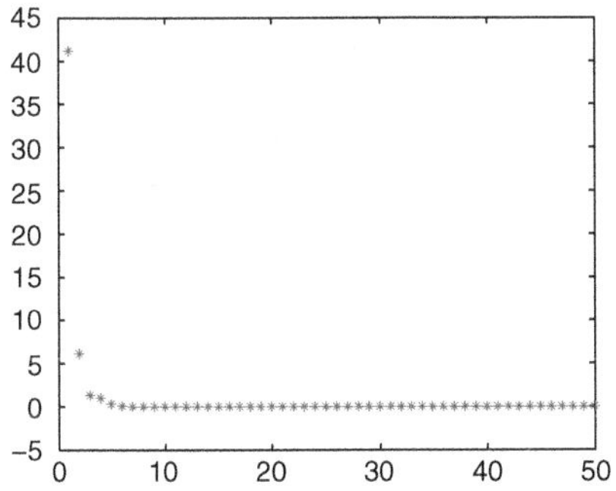

(a) $N = 50$. The maximum eigenvalue is 41.1816.

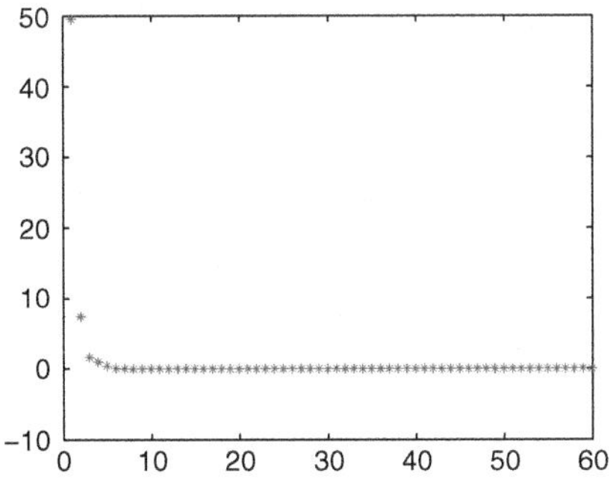

(b) $N = 60$. The maximum eigenvalue is 49.5133.

FIGURE 2. Eigenvalue distribution

Sketch of Proof. The proof is similar to the previous lemma. In this situation, the entries of the Hermitian matrix H in (2.6) are

$$H_{ij} = \frac{\sqrt{1-r_i^2}\sqrt{1-r_j^2}}{1-r_ir_j}, \tag{2.12}$$

where r_i is the ith point on $[0,1)$. Let $x = \left(\frac{1}{\sqrt{N}}, \frac{1}{\sqrt{N}}, \dots, \frac{1}{\sqrt{N}}\right) \in \mathbb{C}^N$ with $\|x\|_2 = 1$. We use Ky Fan's principle to estimate the maximum eigenvalue. We find that $\langle Hx, x\rangle$ is a double Riemann sum of

$$\int_0^1\int_0^1 \frac{\sqrt{1-s^2}\sqrt{1-r^2}}{1-sr}\,\mathrm{d}r\,\mathrm{d}s. \tag{2.13}$$

Hence,

$$\frac{\lambda_1(H)}{N} \geq \frac{\langle Hx, x\rangle}{N} \approx 0.815784. \tag{2.14}$$

□

Remark 2.5. Select N points on the interval $[0,1)$. The maximum eigenvalues are 41.1816 and 49.5133 respectively. $\lambda_1(H)/N$ satisfy (2.11). See Figure 2.

Generally, we select $\{a_k\}$ in the whole disc as follows. Divide $[0,2\pi]$ and $[0,1)$ into N_1 and N_2 parts respectively. Hence, the number of a_ks is $\mathcal{O}(N_1N_2)$. We obtain the main theorem as follows.

Theorem 2.6. *Let $\lambda_1 \geq \lambda_2 \geq \lambda_3 \geq \cdots$ be eigenvalues of $H = \widetilde{\mathscr{D}}^*\widetilde{\mathscr{D}}$. Then we have*

$$\lim_{\substack{N_1\to\infty\\ N_2\to\infty}} \frac{\sum_{k=1}^{n}\lambda_k}{N_1N_2} \geq 1 - \frac{1}{2n+1}. \tag{2.15}$$

Sketch of Proof. H is actually a block matrix in this situation

$$H = \begin{pmatrix} B_1^*B_1 & B_1^*B_2 & B_1^*B_3 & \dots & B_1^*B_{N_2} \\ B_2^*B_1 & B_2^*B_2 & B_2^*B_3 & \dots & B_2^*B_{N_2} \\ B_3^*B_1 & B_3^*B_2 & B_3^*B_3 & \dots & B_3^*B_{N_2} \\ \vdots & \vdots & \vdots & \ddots & \vdots \\ B_{N_2}^*B_1 & B_{N_2}^*B_2 & B_{N_2}^*B_3 & \dots & B_{N_2}^*B_{N_2} \end{pmatrix} \tag{2.16}$$

with each block $B_i^*B_j \in \mathbb{C}^{N_1\times N_1}$. Let

$$f_n(r) = r^n\sqrt{1-r^2} \tag{2.17}$$

and

$$g_n(r) = \frac{f_n(r)}{\|f_n(r)\|_{L^2(0,1)}} \tag{2.18}$$

where $r \in (0,1)$, $n \geq 0$. Denote $e_n(\theta) = e^{-in\theta}$, $n \geq 0$.

Sample N_2 points on $g_n(r)$ and get the vector $G_n \in \mathbb{C}^{N_2}$. Sample N_1 points on e_n to get $E_n \in \mathbb{C}^{N_1}$. Consider

$$\frac{\langle H(G_n \bigotimes E_n), G_n \bigotimes E_n\rangle}{N_1N_2} \tag{2.19}$$

where the tensor product $G_n \bigotimes E_n$ is a vector in $\mathbb{C}^{N_1N_2}$. We prove that

$$\lim_{\substack{N_1\to\infty \\ N_2\to\infty}} \frac{\langle H(G_n \bigotimes E_n), G_n \bigotimes E_n\rangle}{N_1N_2} = \frac{1}{2n+1} - \frac{1}{2n+3}, \quad n \geq 0. \tag{2.20}$$

Hence, Ky Fan's maximum principle gives

$$\lim_{\substack{N_1\to\infty \\ N_2\to\infty}} \frac{\sum_{k=1}^{n} \lambda_k}{N_1N_2} \geq \sum_{k=1}^{n} \left(\frac{1}{2k-1} - \frac{1}{2k+1}\right) = 1 - \frac{1}{2n+1}. \tag{2.21}$$

□

Remark 2.7. Note that trace$(H) = N_1N_2$, the theorem shows that the eigenvalues decay rapidly. Figure 3 shows the numerical calculation when $N_1 = N_2 = 40$. We have that $\lambda_1/(N_1N_2) = 0.6544178 \approx 2/(1\times 3)$, $\lambda_2/(N_1N_2) = 0.1312327 \approx 2/(3\times 5)$, $\lambda_3/(N_1N_2) = 0.057843 \approx 2/(5\times 7)$, $\lambda_4/(N_1N_2) = 0.03319499 \approx 2/(7\times 9)$, and so on. The several largest eigenvalues contribute a large part of the sum of all eigenvalues. Since singular value $s_i = \sqrt{\lambda_i(H)}$, we assert that s_i tends to zero rapidly.

3. Numerical Examples

We give two numerical examples. In our examples, $\mathscr{D} \in \mathbb{C}^{900\times 3000}$. This means the dictionary has 3000 atoms and each atom vector is of size 900. We embed $\mathscr{D}$ into

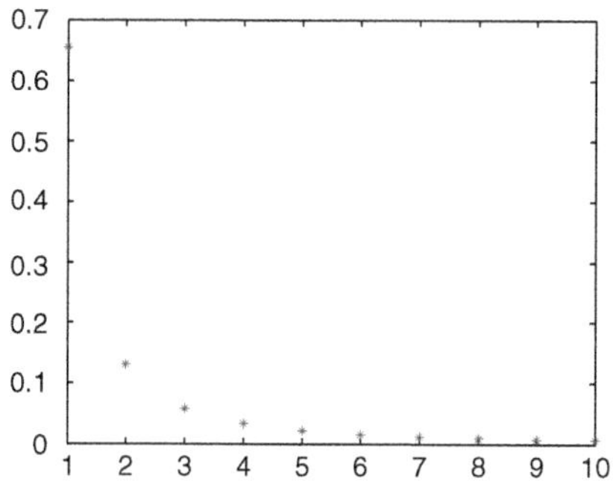

FIGURE 3. Decay of eigenvalues (see Remark 2.7).

$\mathbb{R}^{1800\times 6000}$ for programming convenience by

$$\widehat{\mathscr{D}} = \begin{pmatrix} \Re(\mathscr{D}) & -\Im(\mathscr{D}) \\ \Re(\mathscr{D}) & \Re(\mathscr{D}) \end{pmatrix}. \tag{3.1}$$

Therefore,

$$s = \mathscr{D}x \Longleftrightarrow \widehat{s} = \begin{pmatrix} \Re(s) \\ \Im(s) \end{pmatrix} = \widehat{\mathscr{D}} \begin{pmatrix} \Re(x) \\ \Im(x) \end{pmatrix} = \widehat{\mathscr{D}}\,\widehat{x}. \tag{3.2}$$

where $\Re(s)$ and $\Im(s)$ are the real and imaginary parts respectively.

Consider the minimization problem

$$\min \|\widehat{x}\|_1 \quad \text{subject to} \quad y = \Phi\widehat{\mathscr{D}}\,\widehat{x} \tag{3.3}$$

where $\Phi \in \mathbb{R}^{n\times N}$ is a Gaussian random matrix satisfying $\Phi_{ij} \sim \mathcal{N}(0, 1/n)$. We recover the signal by $\widehat{\mathscr{D}}x^*$ when the solution x^* of (3.3) is obtained by convex programming. As we mentioned above, s belongs to the vector space of $900 \times 2 = 1800$ dimensions with a 900-dimensional real part and 900-dimensional imaginary part. $\widehat{\mathscr{D}}$ has $3000 \times 2 = 6000$ columns and x is a 6000-element coefficient vector.

3.1. Example 1

$$s(z) = \frac{0.247z^4 + 0.0355z^3}{(1 - 0.9048z)(1 - 0.3679z)} \tag{3.4}$$

Set the Gaussian random matrix Φ with 160 rows and its entries $\Phi_{ij} \sim \mathcal{N}(0, 1/160)$. Runtime is 16.437 seconds and relative error is 3.7188×10^{-4}. See Figures 4 and 5.

3.2. Example 2

$$s(z) = z^{20} + z^{10} + z^5. \tag{3.5}$$

Choose the Gaussian random matrix Φ with 160 rows and entries $\Phi_{i,j} \sim \mathcal{N}(0,1/160)$. Runtime is 34.891 seconds and relative error is 1.7×10^{-3}. See Figures 6 and 7.

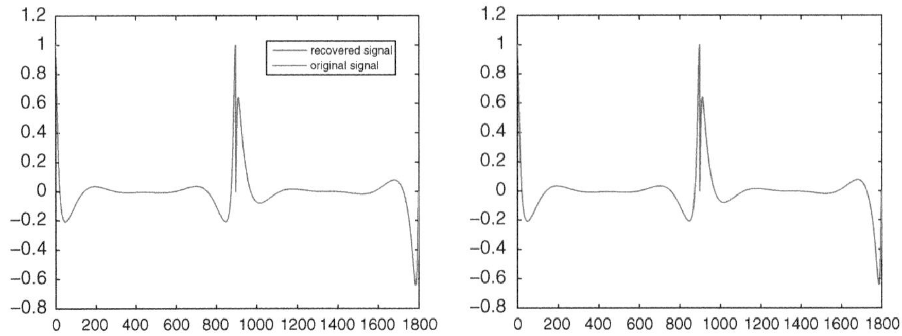

(a) Recovered and original signals. (b) Recovered signal $\widehat{\mathscr{D}}x^*$.

FIGURE 4. Example 1. The recovered signal almost coincides with the original signal. See Section 3.1.

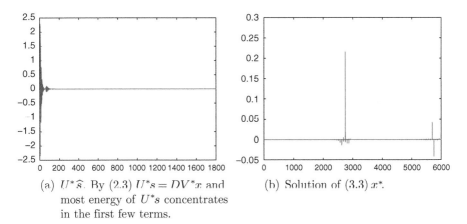

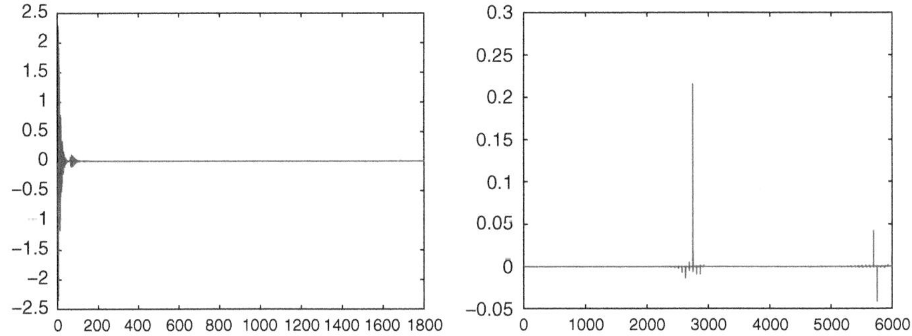

(a) $U^*\widehat{s}$. By (2.3) $U^*s = DV^*x$ and most energy of U^*s concentrates in the first few terms. (b) Solution of (3.3) x^*.

FIGURE 5. Example 1. See Section 3.1.

References

[1] A. Beurling. Sur les intégrales de Fourier absolument convergentes et leur application à une transformation functionelle. In *Proceedings Scandinavian Mathematical Congress*, Helsinki, Finland, 1938.

[2] E. Candes, J. Romberg, and T. Tao. Robust uncertainty principles: Exact signal reconstruction from highly incomplete frequency information. *IEEE Transactions on Information Theory*, 52(2):489–509, 2006.

[3] E. Candes, J. Romberg, and T. Tao. Stable signal recovery from incomplete and inaccurate measurements. *Communications on Pure Applied Mathematics*, 59(8):1207–1223, 2006.

[4] E. Candes and T. Tao. Decoding by linear programming. *IEEE Transactions on Information Theory*, 51(12):4203–4215, 2005.

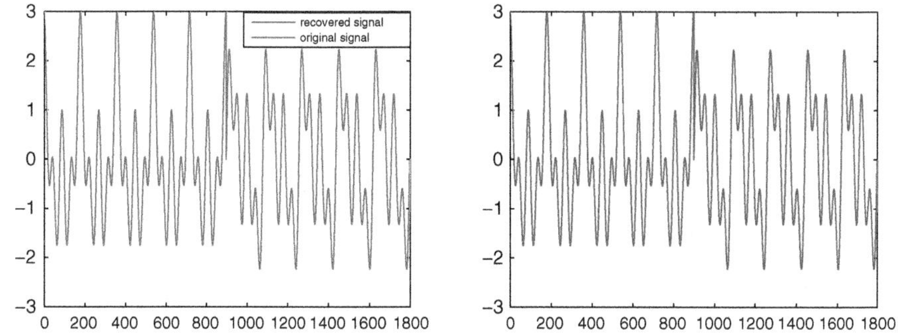

(a) Recovered and original signals. (b) Recovered signal $\widehat{\mathscr{D}}x^*$.

FIGURE 6. Example 2. The recovered signal almost coincides with the original signal. See Section 3.2.

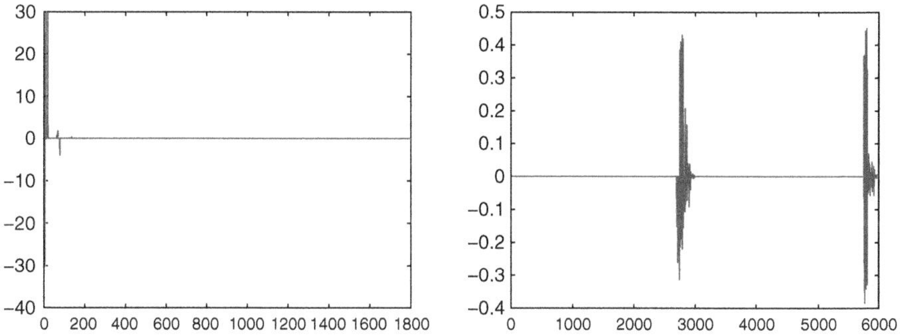

(a) $U^*\widehat{s}$. Most of the energy of U^*s concentrates in the first few terms. (b) Solution of (3.3) x^*.

FIGURE 7. Example 2. See Section 3.2.

[5] E. Candes and T. Tao. Near optimal signal recovery from random projections: Universal encoding stategies? *IEEE Transactions on Information Theory*, 52(12):5406–5425, 2006.

[6] S. Chen, D. Donoho, and M. Saunders. Atomic decomposition by basis pursuit. *SIAM Journal on Scientific Computing*, 20(1):33–61, 1999.

[7] A. Cohen, W. Dahmen, and R. DeVore. Compressed sensing and best k-term approximation. *Journal of the American Mathematical Society*, 22:211–231, 2009.

[8] G. Davis. *Adaptive Nonlinear Approximations.* PhD thesis, New York University, Courant Institute, 1994.

[9] G. Davis and S. Mallat. Adaptive greedy approximations. *Constructive Approximation*, 13(1):57–98, 1997.

[10] D. Donoho. Compressed sensing. *IEEE Transactions on Information Theory*, 52(4): 1289–1306, 2006.

[11] D. Donoho and M. Elad. On the stability of the basis pursuit in the presence of noise. *Signal Processing*, 86(3):511–532, 2006.

[12] D. Donoho and Y. Tsaig. Extensions of compressed sensing. *Signal Processing*, 86(3):533–548, 2006.

[13] M. Fornasier. Numerical methods for sparse recovery. In *Theoretical Foundations and Numerical Methods for Sparse Recovery* [14], pages 93–200.

[14] M. Fornasier, editor. *Theoretical Foundations and Numerical Methods for Sparse Recovery*, volume 9 of *Radon Series on Computational and Applied Mathematics*. De Gruyter, Germany, 2010.

[15] B. Kashin. The widths of certain finite dimensional sets and classes of smooth functions. *Izvestia*, 41:334–351, 1977.

[16] P. Lancaster and M. Tismenetsky. *The Theory of Matrices with Applications*. Academic Press, second edition, 1985.

[17] S. Mallat. *A Wavelet Tour of Signal Processing*. Academic Press, third edition, 2008. First edition published 1998.

[18] S. Mallat and Z. Zhang. Matching pursuit with time-frequency dictionaries. *IEEE Transactions on Signal Processing*, 41(12):3397–3415, Dec. 1993.

[19] T. Qian, W. Sprößig, and J. Wang. Adaptive Fourier decomposition of functions in quaternionic Hardy spaces. *Mathematical Methods in the Applied Sciences*, 35(1):43–64, 2012.

[20] T. Qian and Y.-B. Wang. Adaptive Fourier series – a variation of greedy algorithm. *Advances in Computational Mathematics*, 34(3):279–293, 2011.

[21] H. Rauhut. Compressive sensing and structured random matrices. In Fornasier [14], pages 1–92.

Shuang Li and Tao Qian
Department of Mathematics
University of Macau, Macau
e-mail: `ya97418@umac.mo`
`fsttq@umac.mo`

Quaternion and Clifford–Fourier Transforms and Wavelets
Trends in Mathematics, 333–338

Index

CPSIA information can be obtained at www.ICGtesting.com
Printed in the USA
LVOW02s1620290915

456189LV00002B/4/P

9 783034 807777